PSAM 12
Probabilistic Safety Assessment and Management
22–27 June 2014 • Sheraton Waikiki, Honolulu, Hawaii, USA

Probabilistic Safety Assessment and Management (PSAM) 12 Conference

Volume 7 - Wednesday AM

PSAM 12

Probabilistic Safety Assessment and Management

22 - 27 June, 2014

Sheraton Waikiki, Honolulu, Hawaii USA

CONFERENCE PROCEEDINGS

Volume 7

Wednesday AM

Foreword

It is was our honor to welcome you to Honolulu, Hawaii, for the twelfth rendition of the Probabilistic Safety Assessment and Management (PSAM) Conference. The planning for PSAM Honolulu began back in 2007 (before PSAM 9 in Hong Kong), when we looked at several locations around the United States, included Arizona, California, Boston, and even considered locations in Oceania. Based upon the feedback both during and after the conference, PSAM 12 proved to be a great success.

We would like to thank all of the volunteers, those that served before, during, and after the Conference. Members of the Technical Program Committee, the Organizing Committee, the session chairs, and the presenters have our gratitude for making PSAM 12 the most memorable PSAM yet.

This publication represents the technical proceedings for the Conference. Due to the large number of published papers (a total of 391), we have subdivided the technical content (papers) into five volumes, one for each day of the conference.

On behalf of the International Association for Probabilistic Safety Assessment and Management Board of Directors, we hope that this publication will provide a valuable technical resource in addition to a reminder of the memorable stay in the Hawaiian Islands.

Dr. Curtis Smith
Technical Program Chairs

Dr. Todd Paulos
General Chair

Sponsors

Sponsors

PSAM 12 - Probabilistic Safety Assessment and Management
JUNE 22-27, 2014

Technical Program Committee

Technical Program Chair: Curtis Smith, INL USA
Assistant Technical Program Chairs: Steve Epstein, Lloyd's Register Japan
Vinh Dang, PSI Switzerland
Ted Steinberg, QUT Australia

We would like to thank the members of the PSAM 12 Technical Program Committee. These individuals helped to make PSAM 12 a success by reviewing abstracts, technical papers, organizing sessions, and providing technical leadership for the conference.

Technical Committee Members:

- Roland Akselsson
- S. Massoud (Mike) Azizi
- Tito Bonano
- Ronald Boring
- Roger Boyer
- Mario Brito
- Kaushik Chatterjee
- Vinh Dang
- Claver Diallo
- Nsimah Ekanem
- Steve Epstein
- Fernando Ferrante
- Federico Gabriele
- Ray Gallucci
- S. Tina Ghosh
- David Grabaskas
- Katrina Groth
- Seth Guikema
- Steve Hess
- Christopher J. Jablonowski
- Moosung Jae
- Jeffrey Joe

- Vyacheslav S. Kharchenko
- James Knudsen
- Zoltan Kovacs
- Ping Li
- Harry Liao
- Francois van Loggerenberg
- Jerome Lonchampt
- Soliman A. Mahmoud
- Diego Mandelli
- Donoval Mathias
- Zahra Mohaghegh
- Thor Myklebust
- Cen Nan
- Mohammad Pourgolmohammad
- Marina Roewekamp
- Clayton Smith
- Shawn St. Germain
- Ted Steinberg
- Kurt Vedros
- Smain Yalaoui
- Robert Youngblood
- Enrico Zio

Organizing Committee

General Chair: Dr. Todd Paulos
General Vice Chair: Prof. Stephen Hora, USC
Technical Program Chair: Curtis Smith, INL USA
Webmaster, Registration, Support for Papers/Abstracts Submission and Review: Hanna Shapira, TICS

Table of Content

Page	Paper	
10	20	**Quantifying Risk in Commercial Aviation with Fault Trees and Event Sequence Diagrams** Robin L. Dillon-Merrill (a), Vicki Bier (b), Sherry S. Borener, Mindy J. Robinson (c), Kandi K. Mitchell (d), Poornima Balakrishna (e), Amanda Hepler (f), Aleta Best (c) _{a) Georgetown University, Washington, DC, United States, b) University of Wisconsin-Madison, Madison, WI, United States, c) Federal Aviation Administration, Washington DC, United States, d) Crown Consulting, Inc., Arlington, VA, e) Saab Sensis Corporation, Washington, DC, f) Innovative Decisions Inc., Vienna, VA}
18	209	**Reliability-Based Design Optimization of Space Tether Considering Hybrid Uncertainty** Liping He, Jian Xiao, Tao Zhao (a), Yi Chen (b), Shuchun Duan (a) _{a) School of Mechanics, Electronic, and Industrial Engineering, University of Electronic Science and Technology of China, Chengdu, China, b) School of Engineering and Built Environment, Glasgow Caledonian University, Glasgow, UK}
31	280	**Using Subset Simulation to Quantify Stakeholder Contribution to Runway Overrun** Ludwig Drees, Chong Wang, and Florian Holzapfel _{Institute of Flight System Dynamics, Technische Universität München, Garching, Germany}
42	476	**International Space Station End-of-Life Probabilistic Risk Assessment** Gary Duncan _{ARES Technical Services, Houston, TX, USA}
49	533	**MCSS Based Numerical Simulation for Reliability Evaluation of Repairable System in NPP** Daochuan Ge (a,b), Ruoxing Zhang, Qiang Chou (b), Yanhua Yang (a) _{a) School of Nuclear Science and Engineering, Shanghai Jiao Tong University, Shanghai, China, b) Software Development Center, State Nuclear Power Technology Corporation, Beijing, China}
59	410	**Design for Reliability of Complex System with Limited Failure Data; Case Study of a Horizontal Drilling Equipment** Morteza Soleimani (a), Mohammad Pourgol-Mohammad (b) _{a) Tabriz University, Tabriz, Iran, b) Sahand University of Technology, Tabriz, Iran}
67	514	**Analyzing Simulation-Based PRA Data Through Clustering: a BWR Station Blackout Case Study** Dan Maljovec, Shusen Liu, BeiWang, Valerio Pascucci (a), Peer-Timo Bremer (b), Diego Mandelli, and Curtis Smith (c) _{a) SCI Institute, University of Utah, Salt Lake City, USA, b) Lawrence Livermore National Laboratory, Livermore, USA, c) Idaho National Laboratory, Idaho Falls, USA}
79	456	**Quantification of MCS with BDD, Accuracy and Inclusion of Success in the Calculation – the RiskSpectrum MCS BDD Algorithm** Wei Wang, Ola Bäckström (a), and Pavel Krcal (a,b) _{a) Lloyd's Register Consulting, Stockholm, Sweden, b) Uppsala University, Uppsala, Sweden}
91	573	**Developing a New Fire PRA Framework by Integrating Probabilistic Risk Assessment with a Fire Simulation Module** Tatsuya Sakurahara, Seyed A. Reihani, Zahra Mohaghegh, Mark Brandyberry (a), Ernie Kee (b), David Johnson (c), Shawn Rodgers (d), and Mary Anne Billings (d) _{a) The University of Illinois at Urbana-Champaign, Urbana, IL, USA, b) YK.risk, LLC, Bay City, TX, USA, c) ABS Consulting Inc., Irvine, CA, USA, d) South Texas Project Nuclear Operating Company, Wadsworth, TX, USA}
104	347	**Lessons Learned from the US HRA Empirical Study** Huafei Liao (a), John Forester (a,b), Vinh N. Dang (c), Andreas Bye (d), Erasmia Lois, Y. James Chang (e) _{a) Sandia National Laboratories, Albuquerque, NM, USA, b) Idaho National Laboratory, Idaho Falls, ID, USA, c) Paul Scherrer Institute, Villigen PSI, Switzerland, d) OECD Halden Reactor Project, Institute for Energy Technology, IFE, Halden, Norway, e) U.S. Nuclear Regulatory Commission, Washington, DC, USA}
114	360	**Extracting Human Reliability Information from Data Collected at Different Simulators: A Feasibility Test on Real Data** Salvatore Massaiu _{OECD Halden Reactor Project, Halden, Norway}
126	380	**Simplified Human Reliability Analysis Process for Emergency Mitigation Equipment (EME) Deployment** Don E. MacLeod, Gareth W. Parry, Barry D. Sloane (a), Paul Lawrence (b), Eliseo M. Chan (c), and Alexander V. Trifanov (d) _{a) ERIN Engineering and Research, Inc., Walnut Creek, USA, b) Ontario Power Generation, Inc., Pickering, Canada, c) Bruce Power, Toronto, Canada, d) Kinectrics, Inc., Pickering, Canada}
137	126	**Study on Operator Reliability of Digital Control System in Nuclear Power Plants Based on Boolean Network** Yanhua Zou, Li Zhang (a,b,c), Licao Dai, Pengcheng Li (c) _{a) Institute of Human Factors Engineering and Safety Management, Hunan Institute of Technology, Hengyang, China, b) School of Nuclear Science and Technology, University of South China, Hengyang, China, c) Human Factor Institute, University of South China, Hengyang, China}

PSAM12 - Probabilistic Safety Assessment and Management

Table of Content

Page	Paper	
146	392	**Toward Modelling of Human Performance of Infrastructure Systems** Cen Nan (a,c) and Wolfgang Kröger (b) *a) Reliability and Risk Engineering Group (RRE), ETH Zürich, Switzerland, b) ETH Risk Center, ETH Zürich, Switzerland, c) Land Using Engineering Group (LUE), ETH Zürich, Switzerland*
158	26	**A Bayesian Network Model for Accidental Oil Outflow in Double Hull Oil Product Tanker Collisions** Floris Goerlandt and Jakub Montewka *Aalto University, Department of Applied Mechanics, Marine Technology, Research Group on Maritime Risk and Safety, P.O. Box 15300, FI-00076 AALTO, Finland*
169	37	**Ship Grounding Damage Estimation Using Statistical Models** Otto-Ville Sormunen *Aalto University, Department of Applied Mechanics, Marine Technology, Research Group on Maritime Risk and Safety, Espoo, Finland*
180	61	**Effects of the Background and Experience on the Experts' Judgments through Knowledge Extraction from Accident Reports** Noora Hyttinen (a), Arsham Mazaheri (b), and Pentti Kujala (c) *a) Aalto University, Department of Applied Mathematics, School of Science, Espoo, Finland, b) Aalto University, Department of Applied Mechanics, School of Engineering, Espoo, Finland Kotka Maritime Research Center (Merikotka), Kotka, Finland, c) Aalto University, Department of Applied Mechanics, School of Engineering, Espoo, Finland*
192	327	**A Study for Adapting a Human Reliability Analysis Technique to Marine Accidents** Kenji Yoshimura (a), Takahiro Takemoto (b), Shin Murata (c), and Nobuo Mitomo (d) *a) National Maritime Research Institute, Mitaka, Japan, b) Tokyo University of Marine Science and Technology, Tokyo, Japan, c) National Institute for Sea Training, Yokohama, Japan, d) Nihon University, Funabashi, Japan*
199	364	**Quantifying the Effect of Noise, Vibration and Motion on Human Performance in Ship Collision and Grounding Risk Assessment** Jakub Montewka, Floris Goerlandt (a), Gemma Innes-Jones, Douglas Owen (b), Yasmine Hifi (c), Markus Porthin (d) *a) Aalto University, Department of Applied Mechanics, Marine Technology, Research Group on Maritime Risk and Safety, Espoo, Finland, b) – Lloyd's Register, EMEA, Bristol, UK, c) – Brookes Bell R&D, Glasgow, UK, d) - VTT Technical Research Centre of Finland, Espoo, Finland*
212	10	**Further Development of the GRS Common Cause Failure Quantification Method** Jan Stiller, Albert Kreuser, Claus Verstegen *Gesellschaft für Anlagen- und Reaktorsicherheit mbH (GRS), Cologne, Germany*
224	75	**Plant-Specific Uncertainty Analysis for a Severe Accident Pressure Load Leading to a Late Containment Failure** S.Y.Park and K.I.Ahn *aKorea Atomic Energy Research Institute, Daejeon, KOREA*
233	87	**Comparison of Uncertainty and Sensitivity Analyses Methods Under Different Noise Levels** David Esh and Christopher Grossman *US Nuclear Regulatory Commission, Washington, DC, USA*
244	115	**Understanding Relative Risk: An Analysis of Uncertainty and Time at Risk** A. El-Shanawany (a,b) *a) Imperial College London, London, United Kingdom, b) Corporate Risk Associates, London, United Kingdom*
256	257	**Understanding the Long-term Behavior of Sealing Systems and Neutron Shielding Material for Extended Dry Cask Storage** Dietmar Wolff, Matthias Jaunich, Ulrich Probst, and Sven Nagelschmidt *Federal Institute for Materials Research and Testing (BAM), Berlin, Germany*
265	259	**Gap Analysis Examples from Periodical Reviews of Transport Package Design Safety Reports of SNF/HLW Dual Purpose Casks** Steffen Komann, Frank Wille, Bernhard Droste *Federal Institute for Materials Research and Testing (BAM), Berlin, Germany*
274	417	**The Evolution of Safety Related Parameters and their Influence on Long-Term Dry Cask Storage** Klemens Hummelsheim (a), Jörn Stewering (b), Sven Keßen and Florian Rowold (a) *a) Gesellschaft für Anlagen und Reaktorsicherheit (GRS) mbH, Garching, Germany, b) Gesellschaft für Anlagen und Reaktorsicherheit (GRS) mbH, Cologne, Germany*
284	542	**Aging Management of Dual-Purpose Casks on the Example of CASTOR® KNK** Iris Graffunder (a), Ralf Schneider-Eickhoff and Rainer Nöring (b) *a) EWN Energiewerke Nord GmbH, Lubmin, Germany, b) GNS Gesellschaft für Nuklear-Service mbH, Essen, Germany*

Table of Content

Page	Paper	
296	513	**Overview of New Tools to Perform Safety Analysis: BWR Station Black Out Test Case** D. Mandelli, C. Smith (a), T. Riley (c), J. Nielsen, J. Schroeder, C. Rabiti, A. Alfonsi, J. Cogliati, R. Kinoshita (a), V. Pascucci, B. Wang, D. Maljovec (b) *a) Idaho National Laboratory, Idaho Falls (ID), USA, b) University of Utah, Salt Lake City (UT), USA, c) Oregon State University, Corvallis (OR), USA*
308	80	**Simulation Methods to Assess Long-Term Hurricane Impacts to U.S. Power Systems** Andrea Staid, Seth D. Guikema (a), Roshanak Nateghi (a,b), Steven M. Quiring (c), and Michael Z. Gao (a) *a) Johns Hopkins University, Baltimore, MD USA, b) Resources for the Future, Washington, DC USA, c) Texas A&M University, College Station, TX USA*
318	238	**Towards Reliability Evaluation of AFDX Avionic Communication Systems With Rare-Event Simulation** Armin Zimmermann, Sven Jäger (a), and Fabien Geyer (b) *a) Software and Systems Engineering, Ilmenau University of Technology; Ilmenau, Germany, b) Airbus Group Innovations, Dept. TX4CP; Munich, Germany*
330	486	**Extension of DMCI to Heterogeneous Infrastructures: Model and Pilot Application** Paolo Trucco, Massimiliano De Ambroggi, Pablo Fernandez Campos (a), Ivano Azzini, and Georgios Giannopoulos (b) *a) Department of Management, Economics and Industrial Engineering, Politecnico di Milano, Milan, Italy, b) European Commission - DG Joint Research Centre (JRC), Ispra, Italy*
341	532	**A Longitudinal Analysis of the Drivers of Power Outages During Hurricanes: A Case Study with Hurricane Isaac** Gina Tonn, Seth Guikema (a), Celso Ferreira (b), and Steven Quiring (c) *a) Johns Hopkins University, Baltimore, MD, US, b) George Mason University, Fairfax, VA, US, c) Texas A&M University, College Station, TX*

Quantifying Risk in Commercial Aviation with Fault Trees and Event Sequence Diagrams

Robin L. Dillon-Merrill[a*], Vicki Bier[b], Sherry S. Borener[c], Mindy J. Robinson[c], Kandi K. Mitchell[d], Poornima Balakrishna[e], Amanda Hepler[f], Aleta Best[c]

[a] *presenting author*, Georgetown University, Washington, DC, United States
[b] University of Wisconsin-Madison, Madison, WI, United States
[c] Federal Aviation Administration, Washington DC, United States
[d] Crown Consulting, Inc., Arlington, VA
[e] Saab Sensis Corporation, Washington, DC
[f] Innovative Decisions Inc., Vienna, VA

Abstract: The mission of the Federal Aviation Administration (FAA) is to provide the safest and most efficient aerospace system in the world. As the FAA plans and develops the Next Generation (NextGen) Air Transportation System, quantitative risk assessments can help evaluate the impacts of new technologies and changed procedures. The FAA needs to ensure that NextGen changes that could potentially increase capacity or efficiency also maintain or improve safety. A systematic quantitative view of risk of the air transportation system provides the opportunity to fully understand how possible improvements can impact the overall safety of the system. This FAA modeling effort, led by the System Safety Management Transformation program (Sherry Borener, Program Manager), is called the Integrated Safety Assessment Model (ISAM). Within ISAM, event sequence diagrams (ESDs) describe the sequence of events that a flight must encounter for an accident scenario to occur, and a fault tree is developed for each of the pivotal events in the ESDs. The risks identified by the fault trees are linked to identifiable hazards with the goal of managing the hazards and improving system safety. This paper describes the process being used to develop the event sequence diagram-fault tree model, including lessons learned from applying probabilistic risk analysis modeling in the commercial aviation context.

Keywords: PRA, Commercial Aviation, Event Sequence Diagrams, Fault Trees.

1. INTRODUCTION

The System Safety Management Transformation program (SSMT) is an integral part of meeting FAA Aviation Safety (AVS) responsibilities for the implementation of NextGen. As the FAA plans and develops new operational improvements (OIs), quantitative risk assessments can help evaluate the impacts of these OIs, and ensure that changes maintain or improve safety while delivering capacity or efficiency benefits. For example, in an effort to reduce runway overruns caused by excessive tailwinds, one option is to switch the landing direction on runways more often to be more sensitive to shifting wind directions. While potentially decreasing one risk (i.e., runway overruns caused by excessive tailwinds), this increased switching could result in more confusion and more errors by air traffic controllers and flight crews which could thus result in more runway incursions. A systematic quantitative view of risk of the air transportation system provides the opportunity to fully understand how possible changes can impact the overall safety of the system.

---

[*] rld9@georgetown.edu

The overall modeling effort to create this systematic quantitative view of risk of the air transportation system is called the Integrated Safety Assessment Model (ISAM). The goal of ISAM is to produce a baseline risk for the National Aerospace System (NAS) using data collected across the NAS and through subject matter expert (SME) input. ISAM allows users to evaluate air traffic, airport and air vehicle systems and operators' individual and integrated impacts [1].

In January 2013, the SSMT program established a Fault Tree Working Group (FTWG) to develop event sequence diagrams and fault trees in support of ISAM. The FTWG began by reviewing event sequence diagrams and fault trees previously developed by the National Aerospace Laboratory of the Netherlands (NLR) as part of a project named the Causal Model for Air Transport Safety (CATS) [2] and the European Organization for the Safety of Air Navigation's (EUROCONTROL) Integrated Risk Picture (IRP) [3]. The FTWG modified and expanded the earlier European effort in particular to capture US versus European system differences. Additionally, the European effort was developed based on actual accident data, so the FTWG expanded the models to capture accident events that could occur but may not yet have occurred.

2. EVENT SEQUENCE DIAGRAMS

Table 1 lists the 35 ESDs developed for ISAM by the FTWG. Note that there are several numbers missing in the list. NLR originally developed 37 accident scenarios of which 33 were ultimately incorporated in the Causal Model for Air Transport Safety (CATS) [2]. The FTWG adopted 30 as applicable to the US system, and then added five scenarios (ESD 39-43) not addressed by NLR. In order to maintain correspondence of the ESDs to CATS scenarios, the original numbering was maintained on relevant ESDs and new ones were added to the end of the list.

Each ESD begins with the initiating event described in Table 1. Following the initiating event, the ESD includes the pivotal events and the possible outcomes, where the pivotal events are those events in the scenario that may occur or not occur depending on action (e.g., does rejected take-off occur, does the flight crew maintain control of aircraft, etc.). The possible scenario outcomes include successful outcomes (e.g., aircraft continues flight), failure outcomes (e.g., runway overrun, runway veer-off, aircraft collides with ground, etc.), and partially successful outcomes (e.g., aircraft stops on the runway).

Figure 1 shows as an example ESD US06 – *aircraft takes off with contaminated flight surface*. A more descriptive definition for each node in each ESD is captured in a "data dictionary" that documents the ESDs/fault tree models. For example, in the data dictionary for ESD06, an aircraft takes off with a contaminated flight surface if: aircraft wing, horizontal stabilizer, tail and/or flight control surfaces (i.e. ailerons, elevator, trim, rudder) are contaminated with frost, ice, slush or snow, as the aircraft commences take-off. An event in which the contaminated wing results in engine problems due to ice/snow ingestion is considered in the scope of this ESD. Occurrences in which ice, snow or slush from the runway or landing gear enters the engine(s) and causes problems are excluded from this initiating event but are included in ESD US09.

Each node in the ESD has only two possible outcomes: occurs (or yes) and does not occur (or no). For the initiating event, there is not a "no" branch because the ESD is only applicable if the initiating event occurs. The probability of the scenario going in either direction at a particular node is determined by the corresponding fault tree developed for that node.

Table 1: Accident Event Sequence Diagrams

ESD	Initiating Event	Flight Phase	Number of Fault Tree Nodes
US 01	Aircraft system failure during take-off	take-off	425
US 02	ATC event during take-off	take-off	162
US 03	Directional control by flight crew inappropriate during take-off	take-off	134
US 04	Aircraft directional control related system failure during take-off	take-off	165
US 05	Incorrect configuration during take-off	take-off	132
US 06	Aircraft takes off with contaminated flight surface	take-off	57
US 08	Aircraft encounters wind shear after rotation	take-off	108
US 09	Single engine failure during take off	take-off	124
US 10	Pitch control problem during take-off	take-off	176
US 11	Fire on-board aircraft	in flight	223
US 12	Flight crew member spatially disoriented	in flight	52
US 13	Flight control system failure	in flight	152
US 14	Flight crew member incapacitation	in flight	20
US 15	Ice accretion on aircraft in flight	in flight	30
US 16	Airspeed, altitude or attitude display failure	in flight	128
US 17	Aircraft encounters adverse weather	in flight	67
US 18	Single engine failure in flight	in flight	57
US 19	Unstable approach	approach & landing	200
US 21	Aircraft weight and balance outside limits during approach	approach & landing	234
US 23	Aircraft encounters wind shear during approach or landing	approach & landing	199
US 25	Aircraft handling by flight crew inappropriate during flare	approach & landing	176
US 26	Aircraft handling by flight crew inappropriate during landing roll	approach & landing	59
US 27	Aircraft directional control related systems failure during landing roll	approach & landing	157
US 31	Aircraft are positioned on collision course in flight	in flight	114
US 32	Runway incursion involving a conflict	take-off/landing	82
US 33	Cracks in aircraft pressure boundary	in flight	149
US 35	Conflict with terrain or obstacle imminent	in flight	61
US 36	Conflict on taxiway or apron	take-off/landing	145
US 37	Wake vortex encounter	in flight	53
US 38	Loss of control due to poor airmanship	in flight	29
US 39	Runway incursion - incorrect presence of single aircraft for take-off	take-off	177
US 40	ATC event during landing	approach & landing	202
US 41	Taking off from a taxiway	take-off	177
US 42	Landing on a taxiway	approach & landing	63
US 43	Landing on the wrong runway	approach & landing	63

Figure 1: ESD US06 – Aircraft takes off with contaminated flight surface

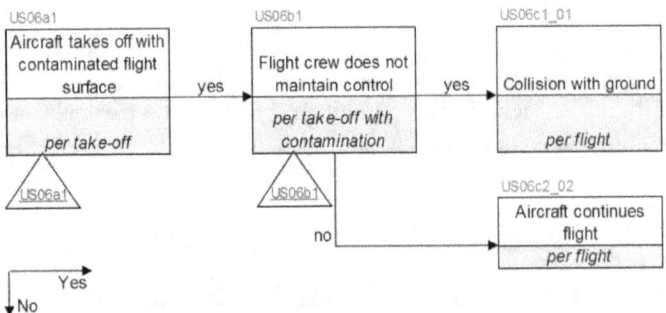

3. FAULT TREE DIAGRAMS

The triangles in Figure 1 denote the presence of a fault tree for that event node. Figure 2 shows part of the fault tree for the initiating event – *aircraft takes off with contaminated flight surface*. Figure 2 shows both types of gates used in the fault trees: OR gates and AND gates. The AND gate is the rounded red shape with the dot and means that both lower level events need to be true for the higher level event to be true. The OR gate is the more pointed yellow shape with the plus and means that the higher level event will be true if any of the lower level events are true. In order to increase the readability of the figure, not all nodes are shown in Figure 2. Figure 2 and all fault tree figures in this paper are drawn with Syncopation Software's Decision Programming Language Fault Tree package (DPL-f). In order for an aircraft to take off with a contaminated flight surface, ice needs to be present and either the flight crew or the ground crew need to perform an incorrect action (i.e., someone needs to fail to notice the icing on the flight surface since if all correct procedures are performed, ice present on a flight surface should be detected and treated before take-off). As shown in Figure 2, three reasons are identified for the failure to identify and correct the contaminated flight surface prior to take-off: *incorrect flight crew actions, incorrect ground crew actions,* or *communications failure* including technical difficulties such as equipment failure, or miscommunications between flight crew and ground crew.

Figure 2: US06a1 – Fault tree for initiating event: aircraft takes off with contaminated flight surface (Not all nodes shown)

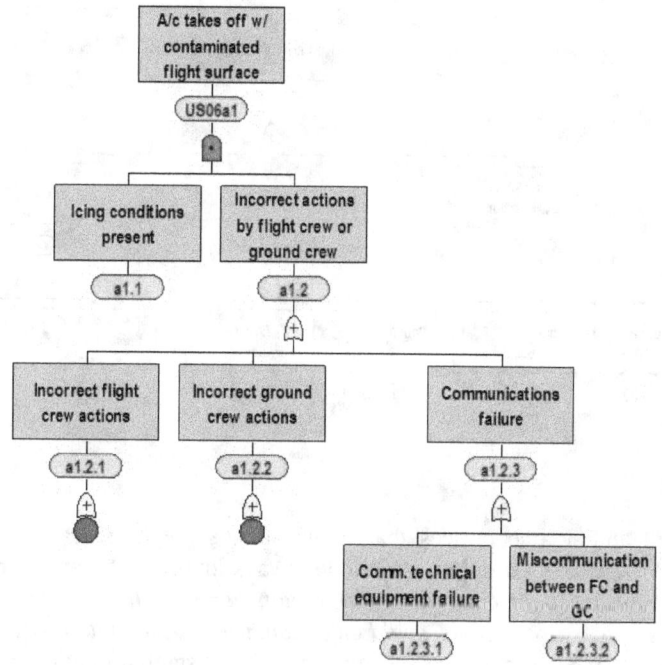

The units for each node in the ESD appear in Figure 1. At this stage, the fault tree structures do not incorporate dependency among nodes or among ESDs beyond the conditional relationships defined by the units, so having clearly identified units are important for the quantification task. The initiating event in US06a1 is per take-off. Then, the second node, the pivotal event US06b1 – *flight crew does not maintain control*, is conditional on the initiating event occurring, i.e., the aircraft taking off with a contaminated flight surface. This conditional relationship is captured in the units for the pivotal event – per take-off with contamination. Within the fault tree, the nodes are interpreted as conditional with respect to the top event. In the data dictionary that accompanies the model (not shown here), units are described for each node. In the US06a1 tree shown in Figure 2, icing conditions present are in units of per take-off. All units in the incorrect actions by flight crew or ground crew sub-tree are per take-off in icing conditions. In the future, careful modeling is needed to capture critical dependencies. For example, some of the events that might cause the need to abort a take-off could also affect the aircraft's ability to stop after an aborted take-off (such as a landing gear system failure), and this failure would increase the likelihood of a runway overrun at a later pivotal event in the ESD.

Figure 3 shows part of the fault tree for the pivotal event in US06b1 - *flight crew does not maintain control*. On take-off with a contaminated flight surface, a significant concern is the aircraft not getting enough lift and stalling. Other ways besides a stall where the flight crew does not maintain control are modeled in the b1.2 sub-tree (not shown here). Focusing on the b1.1 sub-tree as an example, the two ways that a stall is not avoided are: *incorrect flight crew actions* or the *situation exceeded the capability of the flight crew to correct* (i.e., correct actions by flight crew are performed, but the stall is unavoidable).

Figure 3: US06b1 – Fault tree for pivotal event: flight crew does not maintain controls (Not all nodes shown)

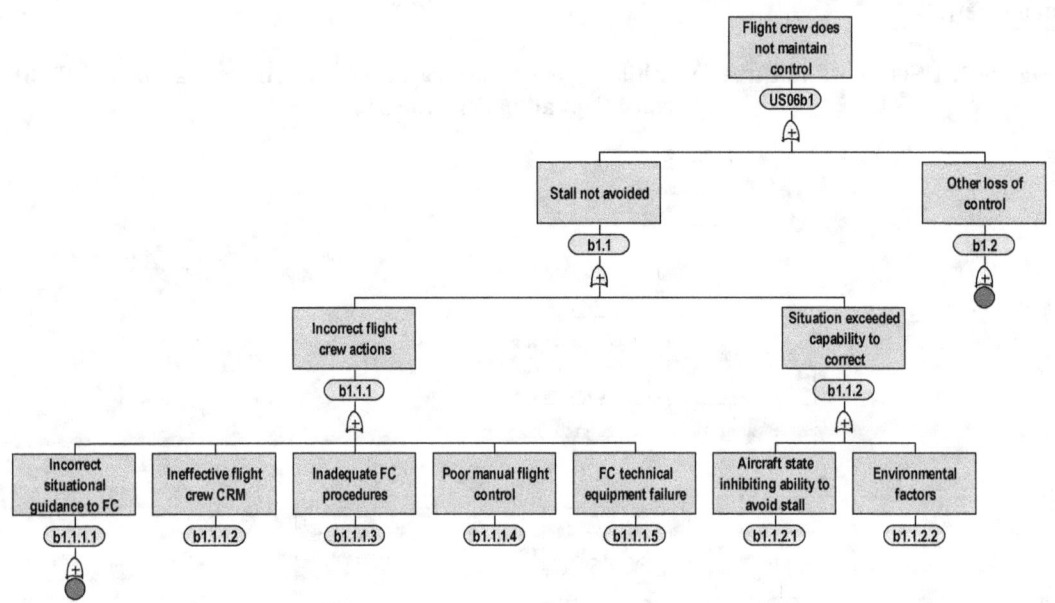

Reviewing Figures 2 and 3, one notices that human errors are a critical component in aviation accidents. In all 35 ESD's, at least one pivotal node always involves human intervention: *flight crew does not maintain control, flight crew does not regain control, flight crew does not resolve the conflict*, or *flight crew does not execute avoidance maneuver successfully*. While humans generally play an important role in most critical systems, their role is substantial in aviation accidents. Figure 4 provides the sub-tree for node a1.2.1 - *incorrect flight crew actions*. Similar sub-trees for incorrect flight crew actions appear 167 times across the 35 ESDs. The five identified contributing factors that could cause incorrect flight crew actions are: 1) *incorrect situational guidance* (i.e., receives inadequate information, leading to incorrect action), 2) *ineffective flight crew Crew Resource Management (CRM)* (i.e., follows inadequate procedures, leading to an incorrect action), 3) *inadequate flight crew procedures* (i.e., inadequate procedural guidelines, leading to an incorrect action), 4) *flight crew technical equipment failure* (i.e., experiences technical equipment failure, leading to an incorrect action), and 5) *environmental/other factors* (i.e., experiences environmental or other factors not otherwise accounted for in the fault tree lead to incorrect or insufficient flight crew instructions).

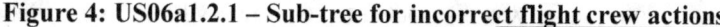

Figure 4: US06a1.2.1 – Sub-tree for incorrect flight crew actions

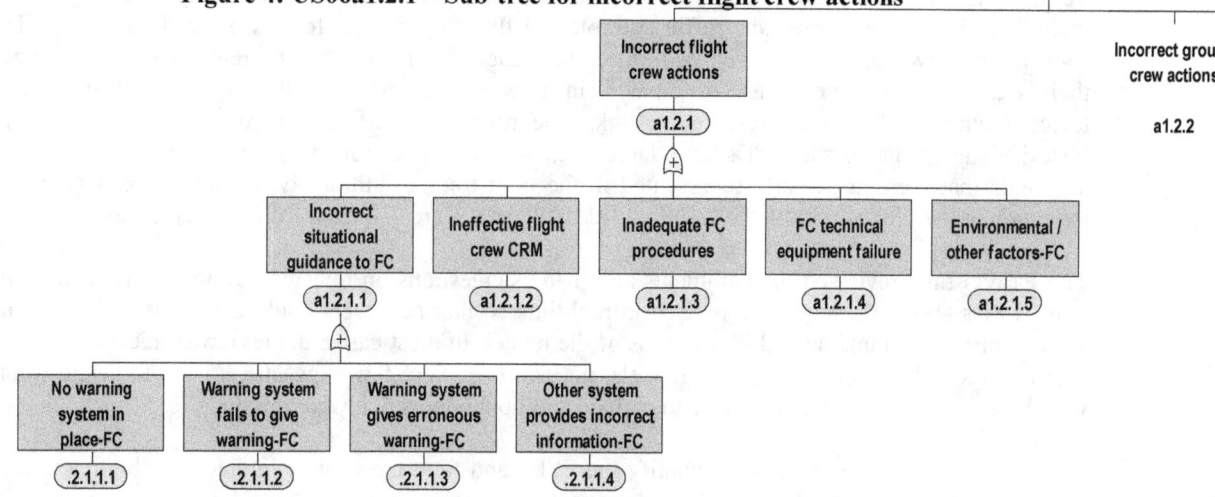

4. HAZARDS ASSESSMENT AND ISAM INTEGRATION

Implementing changes to reduce risks identified in the fault trees requires an additional step – linking the failure modes to identifiable hazards. As part of the ISAM effort, one single hazard database for hazards relevant to aviation safety was created from multiple previously existing efforts. This process resulted in a database of about 500 hazards. The hazards in this resultant list are organized into one of four categories: technical, environmental, organizational, and human factors. For the purposes of ISAM integration and validation, these groups were mapped to events in the ESDs and fault trees, such as Runway Incursion. During ISAM workshops, subject matter experts evaluated the validity of the hazard groupings and assignments, and determined which groups would be impacted by NextGen. In addition, they also had the opportunity to add new hazards or comment on definitions for existing hazards. This hazard assessment and assignment process is continuing as existing or new hazard databases are updated or identified. Thoroughly understanding the roles of hazards in the system provides the best opportunity to intervene and further improve the safety of the system. Figure 5 provides an example mapping of relevant hazards to the base event – *miscommunication between flight crew and air traffic control*.

Figure 5: Example mapping of hazards to base event in the fault tree model

5. MODEL VALIDATION AND QUANTIFICATION

At the completion of the initial version of the ESD/fault trees model, the Microsoft Excel formatted workbook containing a separate worksheet for each accident scenario (including ESD and all fault trees, along with node definitions) was distributed to relevant SMEs from a variety of backgrounds including airport operations, air traffic control, and flight operations for review and validation. To start, ten review meetings occurred with different participants reviewing different scenarios based on their expertise. Some experts participated in person, but most of the reviews occurred via teleconference with a web-presentation link. The reviews specifically focused on mapping past historical accidents to the ESDs and fault trees to identify factors that contributed to historical accidents that were not clearly represented in the fault trees. Additionally, comments regarding the structure and wordings of both the ESDs and all fault trees were collected from the reviewers.

The FTWG then reviewed the comments and made suggestions on revisions to both ESDs and fault trees. ESDs and fault trees were revised in real-time so that reviewers could comment on how their suggestions were implemented in revisions of the model. In most cases, the reviewers identified more detail that should be included in the models, not less (i.e., very few suggestions were to delete what was there, and most suggestions were to include additional items that were missing).

The next step in this project is to quantify the ESDs and fault trees by assigning probabilities to the nodes based on historical data. One source of data being used is the FAA's Aviation Safety Information Analysis and Sharing (ASIAS) system. ASIAS has 42 member airlines sharing data integrating both data from the Aviation Safety Reporting System (ASRS) (i.e., voluntary reports) and Flight Operational Quality Assurance (FOQA) data (i.e., recorded data) with other data sources such as the National Transportation Safety Board (NTSB) investigations [4]. The quantification task is on-going, and it is expected that many additional lessons will be learned from the quantification step.

6. LESSONS LEARNED

In total, the 35 ESDs and corresponding fault trees have 4,552 nodes. The final column in Table 1 shows the number of nodes in the fault trees per ESD. From reviewing this list, it is apparent that some accident scenarios are more complex than others.

Some of the lessons learned identified by the FTWG that would be helpful to others undertaking such a task are [5]:

- The task is best performed with a team of contributors. Breaking off the team to work on different ESDs/fault trees may prove ineffective and produce inconsistent models especially if types of events are similar across accident scenarios (e.g., *incorrect flight crew actions*).
- Review of the European CATS and IRP models included higher-level "luck" or "providence" nodes. These nodes likely represent situations when ATC and flight crews do everything right (as detailed by the sub-trees), but the aircraft still find themselves on a collision course or vice versa when ATC and flight crews do everything wrong but still do not collide. Initially, luck was not included in the 35 ESDs, but in a later iteration, in several ESDs a node named *Avoidance Essential* was added to reflect that subsequent to some failure events occurring, action by the flight crew or vehicle driver would still have been necessary to avoid a collision (i.e., good luck alone does not resolve the conflict for example).
- Actions by ATC to resolve conflicts are still dependent on flight crew actions in many cases. For example, ATC's instructions alone cannot bring about a successful outcome. The flight crew has to successfully respond to the ATC instructions, and that needs to be clearly modelled in the fault trees.

Several significant challenges remain in completing this quantitative risk model. These challenges include:
- Quantification of the probabilities of the 4,552 nodes from available data, recognizing that some data may not exist in any searchable database. This could be a driver for the development of new data sets.

- Developing a common taxonomy. Existing safety reporting databases currently utilize ad-hoc language to describe events with little standardization. This problem makes the quantification task even more challenging.
- Understanding the role of luck (as described above).
- Appropriately capturing dependencies among nodes and among ESDs (e.g., in ESD US01 – *aircraft system failure during takeoff*: if system failure is associated with landing gear system, this will affect a later pivot event: *sufficient braking not accomplished*)
- Learning from near-misses. Quantifying the nodes relying on available data from events that have happened is hard enough, but since fault trees are binary (happened or did not happen), how can the risk model capture events that almost happened but did not happen, differently from those that never almost happened.
- Keeping it simple. As the team applied recommended changes to the fault tree structures, the structures became increasingly complex. Additionally, some ESDs were altered to reflect this complexity. While increasing complexity, these changes were necessary to ensure the model represents all possible events in a particular scenario; so simplicity versus completeness will always be a significant trade-off, and commonly more complex ESDs can result in simpler fault trees, another trade-off.

7. CONCLUSION

The FTWG developed a set of 35 accident ESDs and corresponding fault trees. Development and validation of the trees included input from more than 20 SMEs from a wide range of aviation and industry backgrounds. A series of workgroup sessions resulted in the initial development of the trees and a data dictionary. The data dictionary serves as a change tracking tool, hazard mapping tool, and integration mechanism for the web-based ISAM implementation. A single-source hazard database was developed and hazards are currently mapped to initiating-event fault trees for a priority set of ESDs and integrated in ISAM. Those hazard connections were validated by additional SMEs during multiple workshops. Validation of the fault trees occurred indirectly during the ISAM workshops and directly through a series of accident scenario mapping sessions. The validation process has revealed gaps in the development process. Many of those gaps are already fixed, but some questions remain for future work.

Acknowledgements

ISAM is funded by the System Safety Management Transformation program (Sherry Borener, Program Manager). This paper reflects the opinions of the authors only and does not reflect the opinion of the Federal Aviation Administration or any other organization affiliated with the authors.

References

[1] S. Borener, S. Trajkov, P. Balakrishna, "Design and Development of an Integrated Safety Assessment Model for NextGen", American Society for Engineering Management, Proceedings of the 33rd International Annual Conference, Virginia Beach, (2012).
[2] B. Ale, L.J. Bellamy, R. Cooke, M. Duyvis, D. Kurowicka, P.H. Lin, O. Morales, A. Roelen, J. Spouge, *Causal Model for Air Transport Safety, Final Report*, (2009).
[3] Eurocontrol, D2.4.3-02, *SESAR Top-down Systematic Risk Assessment Version 1.01*, http://www.episode3.aero/public-documents, (accessed June 24, 2013).
[4] ASIAS (2013), FAA AVIATION SAFETY INFORMATION ANALYSIS AND SHARING (ASIAS) SYSTEM, http://www.asias.faa.gov/pls/apex/f?p=100:1:, (accessed February 26, 2014).
[5] M. Robinson, S. Borener, K. Mitchell, P. Balakrishna, R. Dillon-Merrill, A. Hepler, A. Best, "Development of an Event Sequence Diagram-Fault Tree Model for Integrated Safety Assessment in U.S. Commercial Aviation," Federal Aviation Administration, White Paper, Washington DC, (2013).

Reliability-Based Design Optimization of Space Tether Considering Hybrid Uncertainty

Liping He[a], Jian Xiao[a], Tao Zhao[a], Yi Chen[b], Shuchun Duan[a]

[a] School of Mechanics, Electronic, and Industrial Engineering, University of Electronic Science and Technology of China, Chengdu, China

[b] School of Engineering and Built Environment, Glasgow Caledonian University, Glasgow, UK

Email: lipinghe@uestc.edu.cn

ABSTRACT: Space tether is widely used in the field of global space and its reliability problem increasingly become one of research hotspots in the space field. In order to address such problems as low model versatility, high computational complexity and heavy computation workload, referring to the deployable motorized momentum exchange tether (DMMET) as the engineering background, this paper aims to investigate and study a methodology framework which can deal with hybrid uncertainty factors in design optimization After introducing the structure and application characteristics of DMMET, this paper firstly analyses the performance characteristics and uncertain factors of the deployable mechanism, which is used to establish the neural network surrogate model of the deployable mechanism's strength and can enable scalability of deployable mechanism. Specifically, it mainly includes the load uncertainty, the uncertainties of system parameters and calculation model. In this paper, we mainly discuss the uncertainty from system parameters, including the various bar size of the developable mechanism (length, width and thickness) and material performance parameters (elastic modulus, density, strength, etc.). Finally, we obtain a more satisfactory optimization results by reliability-based design optimization of a specific developable mechanisms. Numerical examples verify that this method is feasible and has higher solution accuracy, which can offer reference to the DMMET engineering design.

Keywords: Space tether, parameters uncertainty, design optimization, finite element analysis

1. Introduction

The Deployable Motorized Momentum Exchange Tether (DMMET), on the basis of (MMET) The Motorized Momentum Exchange Tether, is a new kind of tethered system. Compared with the traditional flexible space tether, this system is a scalable parallelogram hinge mechanism which is composed of a series of connecting rod and the pin shaft. By controlling the angle of the hinge on the rod, it can implement the function of stretching and contraction of the organization. The schematic is shown in figure 1[1].

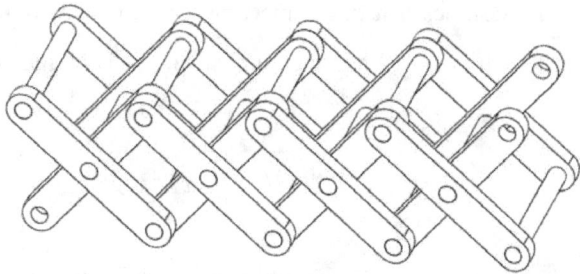

Figure 1. DMMET' scalable parallelogram hinge mechanism

Because of many uncertainty factors in complex space environment, there are many different faults or accident when the space mission is in the execution. Especially some mechanical or structural system of the key tasks, in the event of accident, can cause unbearable serious consequences, not only cause huge economic losses, even threaten human own safety and human society in production order. Therefore, it is necessary to research all kinds of space system reliability. Space tether as a system is widely used in the field of global space, which its reliability problem increasingly become one of research hotspots in the field of global space. DMMET as the research object, this chapter study the reliability modeling and optimization problems of space tether under the condition of all kinds of uncertainty [2][3].

Reliability based design optimization (RBDO) can obtain good results under aleatory uncertainty obviously, and it has become increasingly mature. While under epistemic uncertainty, such as lack of knowledge, fuzzy information or insufficient data, possibility based design optimization (PBDO) has presented its own advantages, and it is still in the process of development. With both aleatory and epistemic uncertainty, there is no a general methodology of reliability design optimization. This paper mainly emphasizes the method of reliability optimization design in such cases, by integrating the probability theory and possibility theory[4]-[8].

Guided by the analytical process as "uncertainty processing - model simplified - variable selection the key parameters - design optimization and model validation", we can obtain the simplified organization strength finite element analysis model.

2. Structural Description of Space Tether

Developable institutions are the main stress components, connecting two spacecraft. Design needs to consider the stress characteristics in the process of institutions' scalability. At the same time it also needs to satisfy such demands of structure, reliability, mechanical strength, scalability, as well volume and weight of the structure. Considering the volume is always proportional to the weight, here we only discuss the influence of volume. Assume a unit of DMMET with 4 connecting rods and 3 pins. Assumes that the connecting length is L, width is B, thickness is H; Pin shaft diameter is B/2, length is L_{xz}. When the number of unit is N, the volume of the body V can be expressed as

$$V = N\left(4BHL + \frac{3}{16}\pi B^2 L_{xz}\right) \quad (1)$$

Figure 2 and 3 respectively show the scalable organizations' two states of the limit of stretching and contraction.

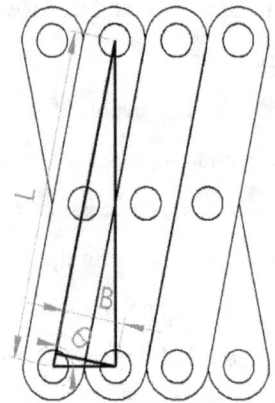

Figure 2. The limit of contraction

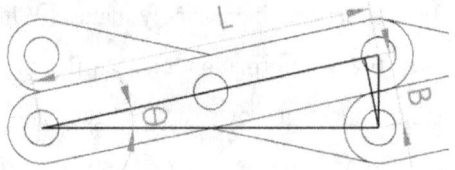

Figure 3. The limit of stretching

From the two figures, we can get some relation:

$$l_{min} = NL\cos\theta_{max} \quad (2)$$

$$l_{max} = NL\cos\theta_{min} \quad (3)$$

$$L\cos\theta_{min}\sin\theta_{min} = \frac{L}{2}\sin 2\theta_{min} = B \quad (4)$$

$$\begin{cases} \theta_{max} + \theta_{min} = \dfrac{\pi}{2} \\ \theta_{max} = \dfrac{1}{2}\arcsin\dfrac{2B}{L}, \dfrac{\pi}{4} < \theta_{max} < \dfrac{\pi}{2} \end{cases} \quad (5)$$

$$i = \frac{l_{min}}{l_{max}} = \frac{NL\cos\theta_{max}}{NL\cos\theta_{min}} = \cot\theta_{max} \propto \frac{B}{L}, 0 < \frac{B}{L} < \frac{1}{2} \tag{6}$$

In the above equation, the scalability function index i is smaller, the scalability function is better. In addition, the index has nothing to do with the number of telescopic unit, which just has the proportional relationship with the connecting rod's aspect ratio. So in the process of design, it only needs to meet the requirements of aspect ratio [9][10].

The strength of the mechanical structure is an important index to measure the reliability of the institution. Failure caused by insufficient strength is one of the main failure mode of the telescopic institutions. In general, the strength of the mechanical structure are closely related to the shape, size, material and other load of the structure, also there are many uncertainty factors. According to engineering experience, the strength of the developable mechanism may be related to institutions include the telescopic unit number N, the length L, width B, thickness of the connecting rod H, the length of the pin shaft Lxz and radius R, the elastic modulus of materials used F, mechanism of load and state of the expansion and other factors.

Build the strength of the telescopic mechanism model, we first need to understand the working situation of the organization. Usually, there are contractions and rotary motion in the work process of DMMET. Ignoring the space gravity gradient and velocity fluctuation's influence, DMMET is mainly caused by the end of the tether payload tension or stress; In the process of the rotary motion, the institutions is mainly caused by pulling force and stress from shaft, as shown in figure 4.

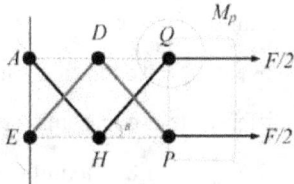

Figure 4. The stress diagram of expandable telescopic mechanism

3. Finite Element Model of Space Tether

After obtaining the stress of the telescopic mechanism, we use the finite element method to calculate the body stress value of each state. For finite element modeling is convenient, physical model of telescopic mechanism can be appropriately simplified, so the mechanism of the connecting rod and the pin are considered to be a slender bar. In ANSYS software, simulation connecting rod with 3 nodes BEAM189 beam element, the unit length is connecting rod length L, unit section is rectangular, cross section size of the width of the connecting rod thickness is B and H; 2 nodes BEAM188 beam element was used to simulate pin, and unit length is the pin length L_{xz}, unit cross section is circular, the radius of the section is pin shaft radius R. Finite element model of telescopic mechanism is as shown in figure 7, the boundary conditions of the model is that the left four nodes of the model degrees of freedom is 0. Finite element model of telescopic mechanism as shown in figure 2 to 6, the boundary conditions of the model for the model of the left four nodes to degrees of freedom is 0, and the right side four nodes bear load f from x direction.

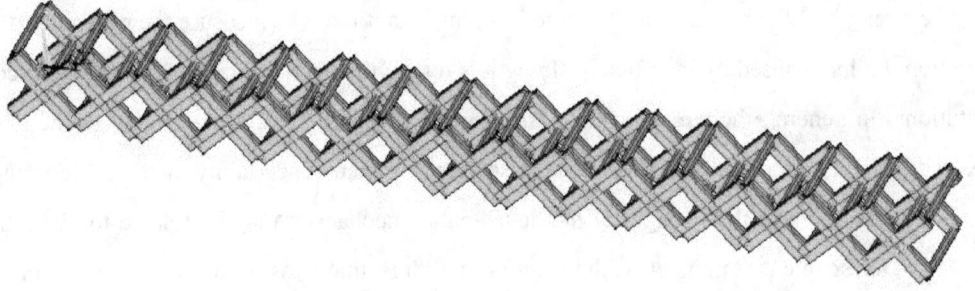

Figure 5. The finite element model of space tethered developable mechanism

We firstly research on the influence of mechanism stress distribution from telescopic unit number. In the case of other conditions unchanged, change the number of telescopic unit corresponding finite element analysis. When the telescopic unit numbers are 5, 10, 20 and 50 respectively, the stress distribution is shown in figure (a), (b), (c) and (d). Observing the stress distribution cloud, the maximum stress is located in the last one expansion unit, and other telescopic unit stress distribution is more even, and the stress is far less than the maximum stress. The stress of pin shaft is also small, almost no stress.

As showed in table 1, in the case of other conditions unchanged, we can gain the changes of maximum stress when it contain different telescopic unit number of institution. We can see from the table, with the increase of telescopic unit, maximum stress σ_{max} has remained steady at around 260.2 MPa.

Accordingly, σ_{max} is always in the last expansion unit, and has nothing to do with the telescopic unit number; Stress is mainly concentrated on the connecting rod. Pin shaft is only connected connecting rod, so we only need to analyze a telescopic unit number less representative institutions.

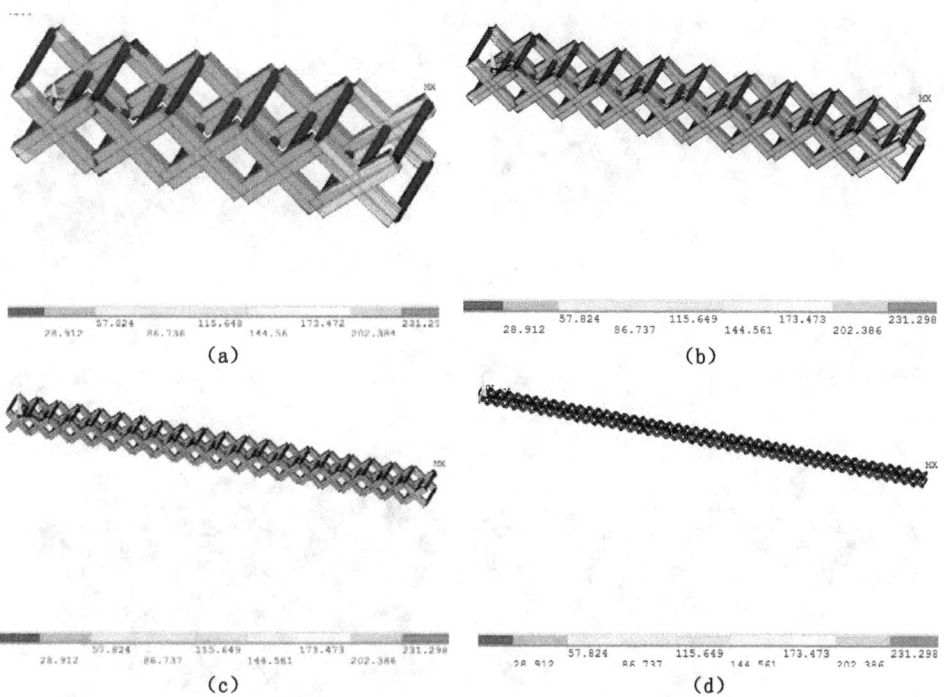

Figure 6. The stress distribution nephogram of different telescopic unit number

Table 1. Different telescopic unit maximum stress of the number of institutions

N	4	5	10	20	30	40	50	100
σ_{max}	260.225	260.208	260.21	260.21	260.21	260.21	260.21	260.21

Next, in the intensity of the developable mechanism model, we research the influence on the stress distribution when telescopic state changes. There are three of the following: Ultimate elongation, ultimate shrinkage and middle state. Because the scaling state is only associated with the angle of connecting rod. Here will change the angle of connecting rod only without considering other factors, and the corresponding finite element analysis will carry on, then discuss the relationship between the state and the stress distribution. Figure 7 (a)~ figure 7 (d) is the stress distribution nephogram when the connecting rod angle is respectively $25°, 40°, 55°, 70°$.

We know from figure 8, when change the situation of the telescopic unit number and change the angle of connecting rod, maximum stress of mechanism is still in the last telescopic unit's connecting rod, and other telescopic unit stress is far less than the maximum stress. The difference is that with the change of the connecting rod angle, maximum stress will change accordingly.

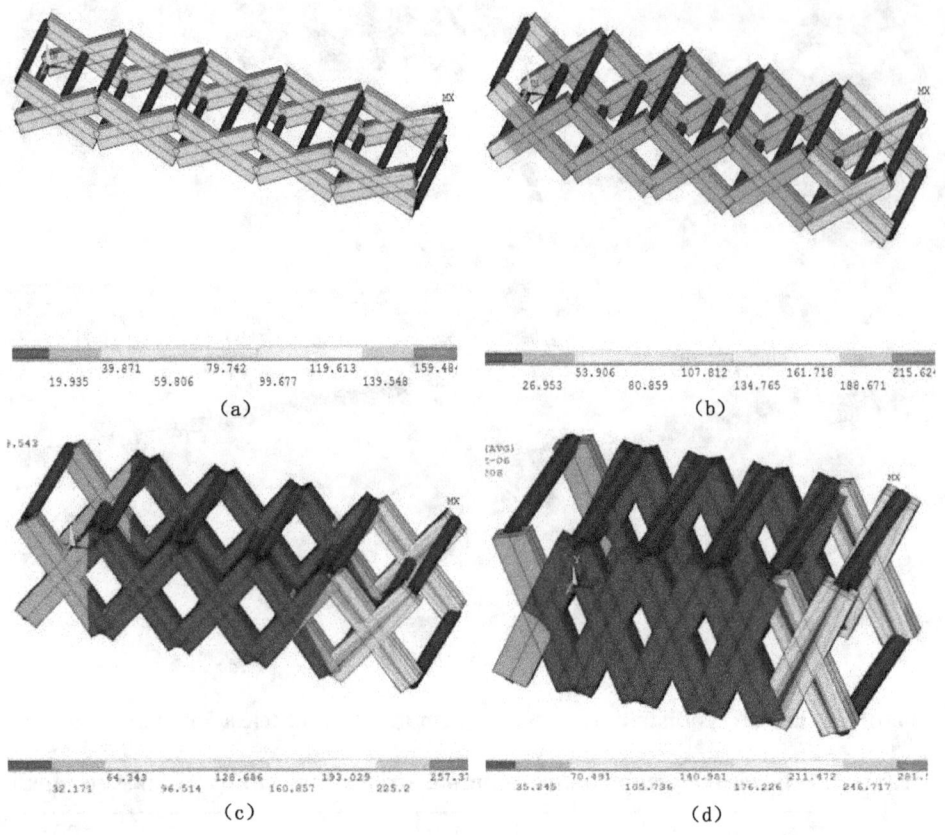

Figure 7. stress nephogram of different connecting rod angle

Table 2 lists the different connecting rod angle corresponding to the maximum stress. From the table 2, with the increase of Angle, maximum stress also increase accordingly. Accordingly when institutions are in a state of extreme contraction, the maximum stress is the highest number in all states.

Table 2. Maximum stress under different connecting rod obliquity of institutions

θ	10	15	20	25	30	35	40	45
σ_{max}	104.275	130.355	155.467	179.419	202.033	223.137	242.577	260.208
θ	50	55	60	65	70	75	80	
σ_{max}	275.901	289.543	301.035	310.288	317.208	321.655	323.361	

From figure 8, σ_{max} increased with the increase of θ: when $\theta < 60°$, σ_{max} approximate linearly increases with θ, the curve slope is bigger, it shows that the growth rate of σ_{max} is faster, namely σ_{max} is sensitive about θ; As θ continues to increase, curve slope gradually decreases, and the growth rate of σ_{max} gradually decreases, and when $75° < \theta < 80°$, the curve slope is almost parallel to the abscissa axis, and σ_{max} increased by less than $2 MPa$ at this time.

This shows σ_{max} sensitivity is small at this time, when θ is in this range, influence can be ignored, which the uncertainty has done on the largest stress σ_{max} (the corresponding connecting rod width to length is 0.1~0.25).

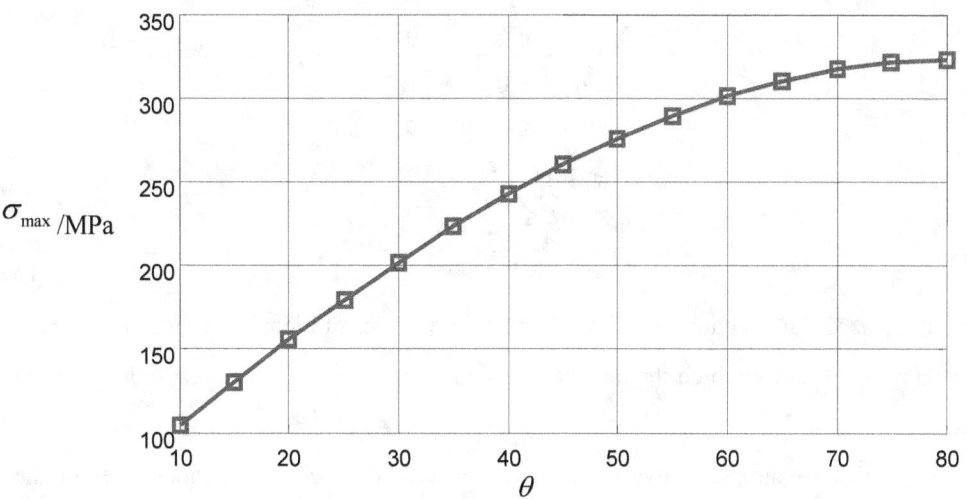

Figure 8. relation curve of θ - σ_{max}

Based on the finite element analysis model about the developable mechanism, and combined with the previous discussion, we can determine that σ_{max} is basicly connected with the connecting rod length, width, thickness and material of elastic modulus, and these five uncertainty factors connect with σ_{max} by a implicit function. So wo can build a neural network approximation model:

$$\sigma_{max} = \hat{\sigma}(B, H, L, E, F) + \delta \tag{9}$$

In the equation, $\hat{\sigma}(L, B, H, E, F)$ is the the maximum stress of the neural network model, δ is the error due to the uncertainty of the model.

4. Design Optimization of Space Tether

Based on the above discussion and analysis, when designing developable institutions, under the condition of meeting the minimizing volume of the institution, we also need to consider breadth length ratio of connecting rod and maximum extension distance, etc. Therefore, the reliability of the proposed developable mechanism design optimization model are as follows:

$$\begin{aligned}
\text{find} \quad & \mathbf{x} = [B, H, L, N] \\
\min \quad & f(\mathbf{x}) = V = N\left(4BHL + \frac{3}{16}\pi B^2 L_{xz}\right) \\
\text{s.t.} \quad & P\left(\hat{\sigma}(B, H, L, F, E,) + \delta - [\sigma] \geq 0\right) \leq \alpha \\
& l_{\max} - NL\cos\theta_{\min} \leq 0 \\
& B - i_{B/L} L \leq 0 \\
& B_{\min} \leq B \leq B_{\max} \\
& H_{\min} \leq H \leq H_{\max} \\
& L_{\min} \leq L \leq L_{\max} \\
& N_{\min} \leq N \leq N_{\max}
\end{aligned} \qquad (10)$$

In the equation, α is the reliability index of the failure of institutional strength, $l_{\max}$ is the maximum extension distance which the institutions need to meet, $i_{B/L}$ is the breadth length ratio for connecting rod.

Next we work on example analysis. Known as uncertainty variables, various sizes of the connecting rod in the tolerance range is normal distribution. We assume that $B \in [4, 40]\,mm$, $H \in [3, 30]\,mm$, standard deviation is $0.05\,mm$; $L \in [60, 140]\,mm$, standard deviation is $0.05\,mm$; $i_{B/L} \leq 0.2$ and the maximum extension distance of not less than $10m$; axis pin $L_{xz} = 50$, which is a determined value; Material elastic modulus E is normal distribution, mean is $70\,GPa$, standard deviation is $1.5\,GPa$, poisson's ratio is 0.33; material allowable stress obeys normal distribution, the average is , standard deviation is ; Maximum load obey an isosceles triangle possibility distribution, median possibility is $5000N$, distribution of interal range between $[4000, 6000]\,N$. Using the foregoing method, design and optimize the developable mechanism to improve the overall performance of the mechanism.

After obtain the neural network model for meet the accuracy requirement of maximum stress, the next step of work is according to the type (10) model to do the developable mechanism reliability design optimization.

Type (10) in the reliability of the deployable structure design optimization model including four design variables and a reliability constraint, the telescopic unit number is deterministic design variables, the remaining three design variables is for the uncertainty. The reliability of the model constraints contains seven uncertain parameters, including load F for a cognitive uncertainty parameters, which obey the probability distribution; The rest of the six parameters are random uncertainty, which obey the normal distribution. In this mixed uncertainty can be used to solve the optimization problem under the condition of the third chapter presents methods to solve. Considering the expandable momentum exchange tethered in space work the objective requirement need high reliability, and failure probability α of a set developable mechanism is not greater than 0.001, and the optimization model for the initial value is

$$x = [B, H, L, N] = [10, 5, 100, 101]$$

The above model after six deterministic optimization and reliability analysis of the results obtained were as table 3 and table 4.

Table 3. Deterministic optimization results of the developable mechanism optimization model

k	B	H	L	N	V
0	10	5	100	101	2317445
1	9.4148	2.0000	60.00	168.81	1203555
2	8.4455	3.1758	60.00	168.38	1437523
3	8.5874	3.0906	60.00	168.44	1438685
4	8.5688	3.1020	60.00	168.43	1438692
5	8.5713	3.1005	60.00	168.43	1438693
6	8.5713	3.1005	60.00	168.43	1438693

In the table 3, compared to the initial value of developable mechanism connecting rod size such as length, width and thickness, which were decreased, but the mechanism of telescopic unit number increased from 101 to about 168, the volume of the body from 2 to 1, reduced a total of 1, decline in percentage of 1. This shows that the optimized developable institutions gain obvious effects in reducing the volume and weight.

Table 4. MPP data results of the reliability analysis in developable mechanism

k	B	H	L	F	E	δ	$[\sigma]$	$\sigma-[\sigma]$
1	9.2115	1.9016	60.3032	5999	6.997372e10	2.4214	560.314	214.39
2	8.2375	3.0939	60.3578	5999	7.000160e10	3.1836	557.266	0.9227
3	8.3817	3.0071	60.3536	5999	7.000113e10	3.1846	557.262	6.520e-3
4	8.3629	3.0187	60.3541	5999	7.000138e10	3.1852	557.260	1.136e-4
5	8.3653	3.0172	60.3541	5999	7.000138e10	3.1852	557.260	4.402e-7
6	8.3653	3.0172	60.3541	5999	7.000138e10	3.1852	557.260	2.479e-7

In the process of optimization, organization of the design variables are regarded as a continuous design variables. Obviously institutions telescopic unit number must be a positive integer, at the same time, considering the size of the connecting rod to conform to the requirements of the standardized process precision, so it can will be rounded processing optimization results.

After the roundness the size of the developable mechanism and the results of performance parameters are as shown in table 5. Visibly in table 6, the performance indexes such as the maximum extension distance and breadth length ratio of developable mechanism after the roundness meet the design requirements.

Table 5. Optimization results of developable mechanism and performance parameters after roundness

$B(mm)$	$H(mm)$	$L(mm)$	N	$V(m^3)$	$i_{B/L}$	$\theta_{max}(°)$	$l_{max}(m)$
8.6	3.1	60	169	1.45	0.143	81.67	10.03

Figure 9 and 10, respectively is the optimization result of institutions in the nominal value (round) and MPP stress distribution nephogram. By figure 9 and 10 we can see that the maximum stress value of the expandable body is located at the end of the telescopic unit on the connecting rod, thus verify institutional strength simplified finite element model is feasible.

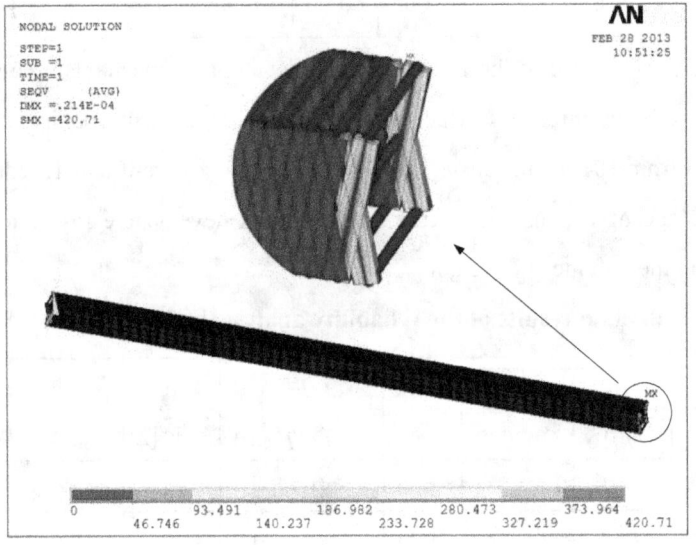

Figure 9. the stress distribution of nominal value of organization optimization results

In order to study the effectiveness of the application examples of optimization calculation by neural network model, respectively, we obtain the developable mechanism maximum stress value through the finite element model and the neural network model, as shown in table 6. Table 6 used neural network model instead of the finite element model for the result error is smaller, the percentage error is less than 2%, which is acceptable in engineering, it verify the validity adopted by the the model and method.

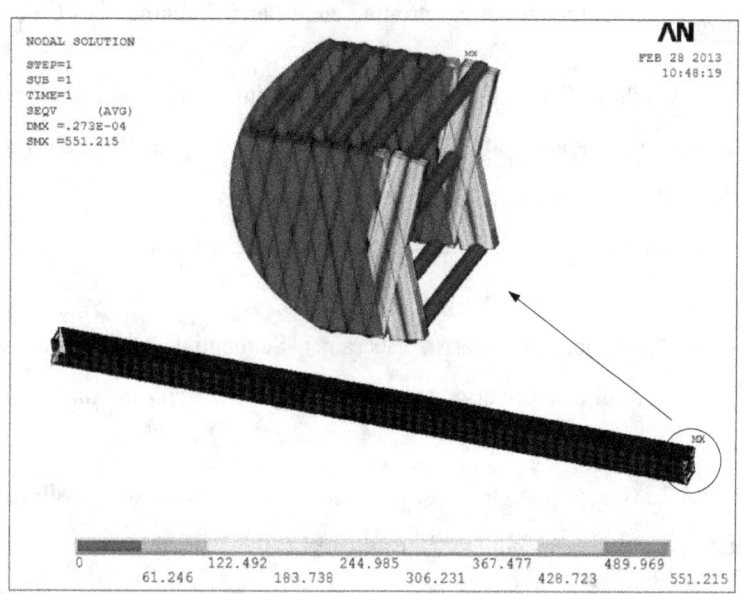

Figure 10. structure optimization results MPP point stress distribution

Table 6. Maximum stress value results in error of expandable structure optimization

	$\sigma_{NN,\max}(MPa)$	$\sigma_{FE,\max}(Mpa)$	error (MPa)	Percentage error
Nominal value	413.122	420.710	7.589	1.81%
MPP value	557.260	551.215	6.045	1.01%

5. Conclusion

Through the understanding of the structure of developable tethered system, analysis the uncertainty factors in the design of space tether. Under the condition of the minimizing volume of the institution, considering institutional scalability (The smaller the scalability index i is, the scalability of the structure is better, and the index has nothing to do with the number of telescopic unit), the stress based on the intensity (The biggest stress value of telescopic mechanism is always located at the last expansion units) and other aspects, establish the reliability design optimization model of developable mechanism, and then we got the design optimization results for developable mechanism (table 6). Through comparing the finite element model with the neural network model, we found that the percentage error is not more than 2%, which is acceptable in engineering, verifing the validity of the model.

6. Acknowledgments

The authors would like to acknowledge the partial supports provided by the National Natural

Science Foundation of China under the contract number 51275077 and 51105061, and the Oversea Academic Training Fund, University of Electronic Science and Technology of China.

References

[1]. Chen, Yi, and Matthew Cartmell. "Hybrid fuzzy sliding mode control for motorised space tether spin-up when coupled with axial and torsional oscillation." *Astrophysics and Space Science* 326.1 (2010): 105-118.

[2]. Cartmell, M. P., and D. J. McKenzie. "A review of space tether research."*Progress in Aerospace Sciences* 44.1 (2008): 1-21.

[3]. Du, Xiaoping, Jia Guo, and Harish Beeram. "Sequential optimization and reliability assessment for multidisciplinary systems design." *Structural and Multidisciplinary Optimization* 35.2 (2008): 117-130.

[4]. Zhou, Jun, and Zissimos P. Mourelatos. "A sequential algorithm for possibility-based design optimization." *Journal of Mechanical Design* 130.1 (2008): 011001.

[5]. Lee, Ikjin, K. K. Choi, and David Gorsich. "System reliability-based design optimization using the MPP-based dimension reduction method." *Structural and Multidisciplinary Optimization* 41.6 (2010): 823-839.

[6]. Youn, Byeng D., Kyung K. Choi, and Liu Du. "Enriched Performance Measure Approach for Reliability-Based Design Optimization." *AIAA journal*43.4 (2005): 874-884.

[7]. Du Xiaoping, Wei Chen. "A most probable point-based method for efficient uncertainty analysis." *Journal of Design and Manufacturing Automation* 4.1 (2001): 47-66.

[8]. Li, Ying, et al. "Sequential optimisation and reliability assessment for multidisciplinary design optimisation under hybrid uncertainty of randomness and fuzziness." *Journal of Engineering Design* 24.5 (2013): 363-382.

[9]. Yi Chen, Rui Huang, Xianlin Ren, Liping He, Ye He. "History of the tether concept and tether missions: a review". *ISRN Astronomy and Astrophysics,* 2013, 502973: 1-7.

[10]. Li-Ping He, H-Z Huang, Yu Pang, Y Li, Y Liu. "Importance identification for fault trees based on possibilistic information measurements". *Journal of Intelligent and Fuzzy Systems,* 2013, 25 (4): 1013-1026.

USING SUBSET SIMULATION TO QUANTIFY STAKEHOLDER CONTRIBUTION TO RUNWAY OVERRUN

Ludwig Drees[a*], Chong Wang[a], and Florian Holzapfel[a]

[a] Institute of Flight System Dynamics, Technische Universität München, Garching, Germany

Abstract: This paper studies the use of sensitivities to quantify the extent to which individual airline departments (stakeholders) contribute to the incident of runway overrun. For that purpose, we present a model of the incident runway overrun. The incident model is based on the dynamics of aircraft and describes the functional relationship between contributing factors leading to the incident. The incident model also takes operational dependencies into account. Model input are the probability distributions of the contributing factors, which are obtained by fitting distributions to data of a fictive airline. By propagating the probability distributions through our incident model, we are able to make statistical valid statements of the occurrence probability of the incident itself. Therefore, we use the subset simulation method. By estimating the design point using the samples of the subset simulation we obtain the sensitivities by applying the First-Order Reliability method. The sensitivities are used to quantify stakeholder contribution to the incident runway overrun by allocating the various stakeholders to the contributing factors.

Keywords: Runway Excursion, Incident Model, Subset Simulation, Sensitivity, First-Order Reliablity Method

1. INTRODUCTION

Certain incidents (e.g. runway excursion) occur rarely if at all for a single airline. Yet, the probability of such an incident is also not equal to zero, making it difficult to calculate the incident probability based purely on historical rates. If the airline increases its sample size with data from worldwide statistics (e.g. annual safety reports), a second problem arises, namely when data is collected across multiple airlines, it is currently impossible to correct for the effects of airline-specific safety cultures, flight procedures, types of aircraft, routes, training, etc.

Our hypothesis is as follows: even if the incident itself cannot be observed within the flight operation of the single airline, it is possible to measure the contributing factors leading to the incident. By modeling the contributing factors with probability distributions and knowing the functional relationship between them, we are able to calculate the probability that the actual incident will occur. We use the subset simulation method to compute the incident probability.

However, the individual contribution of each factor to the incident has not yet been quantified. In addition, it is desirable for safety management systems to know the contribution of stakeholders (e.g. departments of an airline such as training, or flight operation) to the safety level of an airline. To overcome this fact, this paper studies the use of sensitivities obtained from the subset simulation method to quantify such contributions. By tagging the contributing factors with the stakeholders, we are also able to assess the contribution of stakeholders to the risk of experiencing an incident that is analyzed.

The remainder of this paper is as follows: First, we describe in Sec. 2 the incident model including the physical and operational dependencies. Furthermore, we will describe the modeling process of the contributing factors that are the model input. In Sec. 3, we will present the method of subset simulation. Section 4 describes the estimation of the sensitivities and Sec. 5 concludes the paper.

---

[*] Email address: ludwig.drees@tum.de

2. RUNWAY OVERRUN EXAMPLE

2.1. General concept

Runway excursion is one of the most frequently occurring incidents worldwide [1]. Therefore, many studies have focused on determining the typical contributing factors leading to runway overruns and analyzing the dependencies of the contributing factors [2, 3]. Examples of typical contributing factors are: high-speed deviations from the target approach speed, high tailwind component, landings on a short runway, long landings (touching down late), or a wet runway. All of these contributing factors can be measured based on the operational flight data of an airline. The physical relationship between these factors is known and equal for each airline. The statistical variance of each contributing factor, however, heavily depends on the flight operation of an airline. Figure 1 shows some contributing factors of the incident type runway overrun.

Figure 1: General Concept

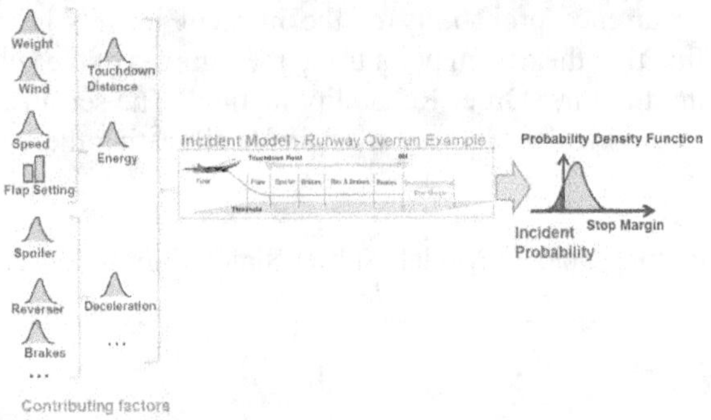

As indicated on the left of Fig. 1, many of the contributing factors vary during flight operation (e.g. wind) or even exceed limits imposed by the airline. We input the measured distributions into the incident model that contains the functional relationships between the contributing factors. This allows us to quantify an estimate for the incident probability, even if an airline has not observed such an incident in their flight operation.

First, we present an incident model of the runway overrun that is based on the dynamics of an aircraft (physical approach) and also includes operational dependencies. Then, we identify for each contributing factor a corresponding probability distribution that describes the contributing factor based on flight operational data. For that purpose, we present measures, which allow us to determine the goodness of fit. Third, we input the distributions of the contributing factors to the incident model in order to quantify the occurrence probability of an overrun. However, when using classical Monte-Carlo techniques, a large sample size is required for small probabilities since the sample size is inversely proportional to the failure probability that is to be obtained. To overcome this fact, we use the subset simulation method [4].

2.2. Runway Overrun Model

In this section, we present the runway-overrun model. The first step is to define a metric that describes the closeness of a single flight to ending in a specific incident. We call such metrics incident metrics. Put it another way, the incident metric describes a safety margin of a single flight with respect to a certain incident. If that incident metric is less than zero, an incident occurred. Examples for such metrics could be the tail clearance in case of the incident tailstrike, or the vertical speed prior to touchdown with respect to hard-landing. For the runway overrun, we use the stop margin SM that is defined as follows:

$$SM = LDA - ALD \qquad (1)$$

Here, *LDA* is the Landing Distance Available (runway length) and *ALD* refers to the Actual Landing Distance. Our incident model will compute the ALD. An overrun occurs if the SM is negative (Eq. 2):

$$Flight\ Operations \begin{cases} No\ Incident & if\ SM > 0 \\ Incident & if\ SM \leq 0 \end{cases} \quad (2)$$

In order to calculate the safety margin, we use the aforementioned incident model $f_{Incident}$

$$SM = f_{Incident}(CF) \quad (3)$$

The function $f_{Incident}$ includes the relevant functional relationships between the contributing factors CF such as flight dynamics (physics), flight procedures, and aircraft systems. The output is the incident metric based on the input samples of the CF, here the stop margin.

The following Fig. 2 illustrates the concept. The red and dotted incident area equals the probability of the incident, which is unknown and has to be quantified. The aircraft symbols equal samples. If a sample is not within the incident area, no incident occurred. The distance between a sample and the incident area represents the incident metric (safety margin . The greater the distance between the incident area and the sample, the greater is the value of the incident metric. The aim of this paper is to quantify the size of the incident area that equals the incident occurrence probability by applying the subset simulation method and use its samples to estimate sensitivities of the each contributing factor.

Figure 2: Estimating Incident Probabilities

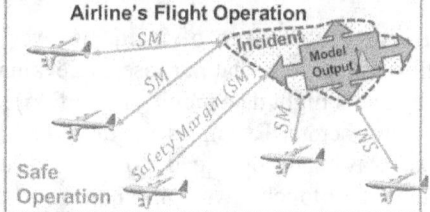

The deceleration of an aircraft can be expressed as follows by using Newton's second law of motion [5],

$$\dot{V} = \frac{1}{m}[T - D - mg \cdot \sin\gamma - \mu_F(mg \cdot \cos\gamma - L)] \quad (4)$$

where $\dot{V}$ equals the deceleration of the aircraft, m expresses the aircraft mass, g is the gravitation constant, γ equals the flight path (here it is assumed to be equal to the runway slope). The forces within this Eq. (4) are the propulsion (thrust) forces T, the aerodynamic drag D and the aerodynamic lift L. The friction coefficient during braking is a function of the runway condition and brake application (Eq. 5).

$$\mu = f(runway\ condition, brake\ pressure) \quad (5)$$

We use the following expressions to model the lift and drag:

$$L = \bar{q} \cdot S \cdot C_L = \frac{\rho}{2} V_A^2 \cdot S \cdot C_L = \frac{\rho}{2}(V_{GS} - V_W)^2 \cdot S \cdot C_L \quad (6)$$

$$D = \bar{q} \cdot S \cdot C_D = \frac{\rho}{2} V_A^2 \cdot S \cdot C_D = \frac{\rho}{2}(V_{GS} - V_W)^2 \cdot S \cdot C_D \quad (7)$$

In Eq. (6-7), $\bar{q}$ represents the dynamic pressure, S the reference area of the aircraft, C_L the aerodynamic lift coefficient and C_D the corresponding aerodynamic drag coefficient. The dynamic pressure can be expressed by using the air density ρ and the aerodynamic speed V_A.

$$\bar{q} = \frac{\rho}{2} V_A^2 \quad (8)$$

Here, we express the aerodynamic speed using the ground speed V_{GS} as well as the wind speed V_W.

$$V_A = V_{GS} - V_W \qquad (9)$$

This overrun model also incorporates operational dependencies. The touchdown behaviour of pilots heavily depends on the runway length. Specifically, it depends on the difference between LDA and the Required Landing Distance (RLD). The RLD can be obtained from aircraft manuals. The expected value of the touchdown distance $\mu_{Touchdown}$, i.e. the distance from the runway threshold until the aircraft touches the runway, can be obtained in feet by the following relationship [6]:

$$\mu_{Touchdown} = \begin{cases} 12.5\delta + 1300 & if\ \delta < 55 \\ 2000 & if\ \delta \geq 55 \end{cases} \qquad (10)$$

Here the buffer is computed as follows:

$$\delta = \frac{LDA - LDR}{LDA} \qquad (11)$$

Eq. 10 shows that the smaller the buffer, i.e. the more critical a landing in terms of a runway overrun, the smaller the touchdown distance. However, as shown in [6], the standard deviation of the touchdown distance remains the same. If the buffer is greater than 55, the mean value for the touchdown distances does not change anymore.

2.3. Contributing Factors

In order to estimate the occurrence probabilities of incidents, we have to describe the statistics (distributions) of their contributing factors. For that purpose, we evaluate the operational data that can be obtained from an airline, e.g. by one's flight data monitoring (FDM) system. We distinguish between continuous and discrete contributing factors. Examples for discrete contributing factors are: the flap configuration for landing, or the runway condition (e.g. dry or wet). In contrast, examples for continuous contributing factors are: landing mass, or touchdown point. For the continuous ones, we fit continuous probability distributions to the data. Hence, we have a probabilistic model of each contributing factor.

As shown in the Fig. 1, many contributing factors vary during flight operation or even exceed airline-specific limits. Examples for such violations are tailwind components, late touchdowns, etc. As mentioned above, the incidents are usually the result of combined "extreme" contributing factors, such as landing out of the touchdown zone, late start of braking, or low runway friction. We account for this fact, by taking these "extreme" deviations from the nominal values into account. During the fitting process, we also consider that the probability distribution does not only have to represent values around the data's mean value but also describe the data in their boundary areas (left and right tail). This means, that the occurrence of such extreme values of contributing factors, such as the before mentioned touchdown distances or low friction coefficients are not underestimated. Due to the large amount of data, the following figure presents our algorithm approach to identify the distribution type that fits to the data best (Fig. 3).

Figure 3: Fitting Algorithm

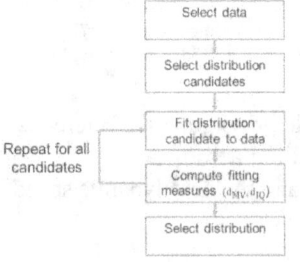

Depending on the domain Ω of each contributing factor, a set of distribution candidates is selected. Hence, we make sure that our probabilistic model for each contributing factor is valid. For example, values of the friction coefficients can only be positive and are in the range of zero and one, i.e. Ω = [0; ∞). Then we fit each of the distribution candidates to the data of the current contributing factor and calculate its goodness of fit. For that purpose, we use two measures, as used in Ref. [7], the first measure is the Integrated Quadratic Distance (IQD) which is defined as follows,

$$d_{IQ}(F,G) = \int_{-\infty}^{\infty} (F(t) - G(t))^2 w(t)\, dt \tag{12}$$

where G is the empirical distribution function (obtained from the data) and F the cumulative distribution function of the distribution candidate. In principle, a weighting function *w(t)* can be included that allows us to emphasize certain areas of interest, e.g. the tails of a distribution. The integration is adapted to the valid domain of the contributing factor. Recalling the friction coefficient example, the domain would be changed to Ω = [0; ∞). In order to counteract an over-fitting due to the weighting function, we also evaluate a second measure d_{MV} that is called the Mean Value Divergence.

$$d_{MV}(F,G) = \left[\mu_F - \frac{1}{k}\sum_{i=1}^{k} y_i\right]^2 \tag{13}$$

Here, μ_F is the mean value obtained by the distribution candidate F. This measure equals the difference between μ_F and the empirical mean value based on the data. This ensures that also the first moment (i.e. mean value) of the distribution candidates fit closely to the empirical mean value.

The following figures (Fig. 4a, b) show an example of fitting various distribution types to the measured touchdown distances. The touchdown distance is defined as the distance between the runway threshold (begin of the runway) and the point at which the aircraft touches the ground. The fitting was performed without any weighting function. For this example, the Gamma distribution fits the data best.

Figure 4: Fitting Example

a) Probability density function 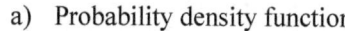   b) Cumulative distriubtion function

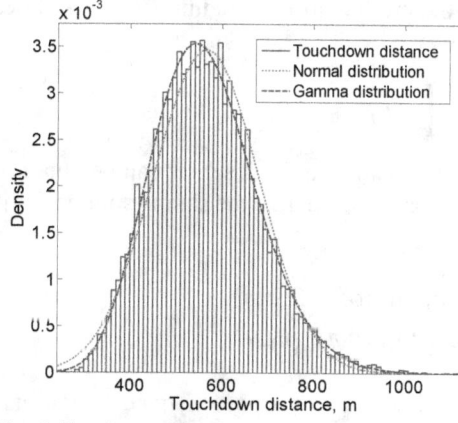 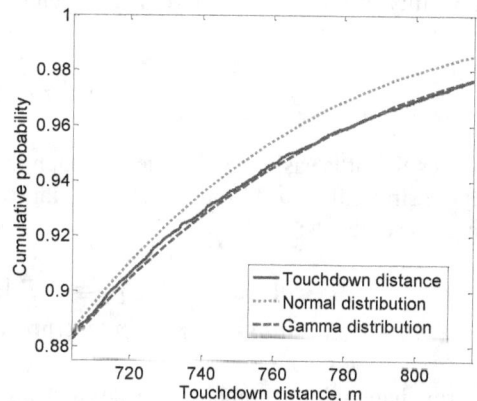

The following table shows the measures obtained for the five best fitting distributions of the touchdown distance that are shown in Fig. 4.

Table 1: Fitting Measures

Rank	Distribution Type	Integrated Quadratic Distance	Mean Value Divergence
1	Gamma	0.005	< 10e-4
2	Generalized Extreme Value	0.0176	0.0202
3	Nakagami	0.0260	0.0362
4	Lognormal	0.0392	0.0142
5	Birnbaum-Saunders	0.0397	< 10e-4

The fitting of the Normal (also known as Gaussian) distribution as shown in Fig. 4 was ranked as number ten, with an IQD of 0.1044. In addition, Figure 4b shows that large values of touchdown points (long-landings) are underestimated even if such long landings can be observed within the data.

3. SUBSET SIMULATION

3.1. General Principle

As described in Sec. 2, we want to estimate the occurrence probabilities of incidents p_I, here of the runway overrun. As shown in Fig. 2, the probability p_I equals the size of the incident, and $\pi(\Theta)$ equals the probability density function of our incident metric SM. Hence, we can write

$$P(SM < 0) = \mathrm{E}[p_I] = \iint_{Flight\ Operation} \mathbb{I}(\theta)\pi(\theta)\,d\theta = \iint_{Incident} \pi(\theta)\,d\theta \tag{14}$$

With the indicator function $\mathbb{I}$

$$\mathbb{I}(\theta) = \begin{cases} 1 & \text{if incident} \\ 0 & \text{otherwise} \end{cases} \tag{15}$$

In most of the cases, the probability density function of the incident metric is unknown and we have to estimate p_I. A possible and straightforward method to obtain an estimate $\hat{p}_I$ for the probability is the Monte-Carlo approach, as follows.

$$p_I \approx \frac{1}{N}\sum_{i=1}^{N}\mathbb{I}(\theta_i) = \hat{p}_I \tag{16}$$

However, for rare events, with small occurrence probabilities, the number of required samples increases. In order to reduce the number of samples, we apply the subset simulation method [4]. The idea is to express the failure domain as a subset of multiple larger failure domains. If the intermediate failure domain is chosen properly, the intermediate probability p_j can be large (e.g. $p_j = 0.1$).

$$\mathbb{R}^d = F_0 \supset F_1 \supset \cdots \supset F_m = F \tag{17}$$

The probability of the system failure then is determined as the product of the conditional probabilities of each subset.

$$p_I = \prod_{j=1}^{m} P(F_j|F_{j-1}) = \prod_{j=1}^{m} p_j \tag{18}$$

This ensures that it is easier for samples to reach the incident domain. The first subset can be obtained by applying straightforward Monte-Carlo. A Markov Chain is used to generate the samples for each the following subset levels:

$$\begin{aligned}\text{Subset } j = 1 \quad & p_1 = P(F_1|F_0)\ (Plain\ Monte-Carlo) \\ \text{Subset } j \geq 2 \quad & p_j\ estimation\ according\ to\ \pi(\cdot|F_{j-1})\end{aligned} \tag{19}$$

The intermediate failure domains are defined adaptively [4]. For that purpose a fixed conditional probability $p_j = p_0$ is defined (here: p_0 equals 0.2). Based on the conditional probability the number of samples in the subset failure domain is determined as:

$$n_F = N \cdot p_0 \tag{20}$$

In other words, an incident (failure) would occur if the SM is not less than zero but less than a certain threshold value y_i. So if the stop margin of a sample is less than the intermediate threshold this sample lies in the intermediate failure domain.

$$\theta_{n+1} = \begin{cases} SM < y_i & \text{intermediate failure domain} \\ SM > y_i & \text{with the probability } 1 - a(\theta_n, y) \end{cases} \tag{21}$$

3.2. Metropolis Algorithm

In the original subset simulation method, the authors used the Metropolis Hastings algorithm [4] in order to create a Markov Chain. Due to the fact, that we have to consider multiple contributing factors with different domains and different distribution types, we perform the sampling in the standard normal space. This simplifies the sampling since we do not have to consider for multiple proposal distributions. In order to generate new samples, we can use the original Metropolis algorithm [8].

This means that we can use a symmetric proposal distribution that further simplifies the sampling. Given a sample θ_n in the failure domain of subset j, we draw a sample candidate y according to $q(y|\theta_i)$. The next step is to compute the acceptance ration a:

$$a = (\theta_n|y) = min\left\{1, \frac{\pi(y)}{\pi(\theta_n)}\frac{q(\theta_n|y)}{q(y|\theta_n)}\right\} \quad (22)$$

Then we assign new sample θ_{n+1} as follows

$$\theta_{n+1} = \begin{cases} y & \text{with the probability } a(\theta_n, y) \\ \theta_n & \text{with the probability } 1 - a(\theta_n, y) \end{cases} \quad (23)$$

Now, we have to evaluate if a new sample θ_{n+1} lies in the failure of the current subset, otherwise we have to reject this new sample.

$$\theta_{n+1} = \begin{cases} \theta_n & \text{with the probability } a(\theta_n, y) \\ \theta_n & \text{with the probability } 1 - a(\theta_n, y) \end{cases} \quad (24)$$

We apply the steps from Eq. (22) until Eq. (24) until we have obtained N samples which are distributed according to $\pi(\cdot|F_{j-1})$ and lie within the next failure domain. In this paper, we used the uniform distribution as the proposal distribution q with a spread of 1.4.

4. DESIGN POINT AND SENSITIVITIES

4.1. Method

The design point DP is the most likely point in the failure domain, i.e. the distance from the design point to the origin in the standard normal space is the shortest. The following Fig. 5 illustrates the concept.

Figure 5: Design Point in U-Space

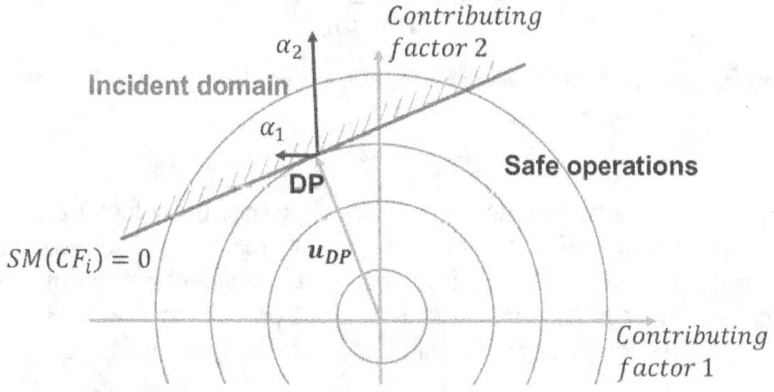

Here, the DP is the shortest distance of the failure domain to the origin of the U-space. For illustration, only two contributing factors are shown. The sensitivities are indicated by the component of the vector u_{DP}.

In order to approximate the DP we use the samples obtained from the subset simulation. Therefore, we look for all samples that lie in the failure domain. Then we transform the samples into the standard normal space using the following transformation:

$$u = U^{-1}(F(x)) \tag{25}$$

Here, u represents the sample in the standard normal space (U-space), U is the cumulative distribution function (CDF) of the standard normal distribution, and F equals the CDF of the contributing factor with its value x. The following figure (Fig. 6) shows the samples of the wind component vs. the stop margin in the X-space as well as the corresponding samples in the U-space.

Figure 6: Wind versus Stop Margin

a) X-space   b) Standard normal space

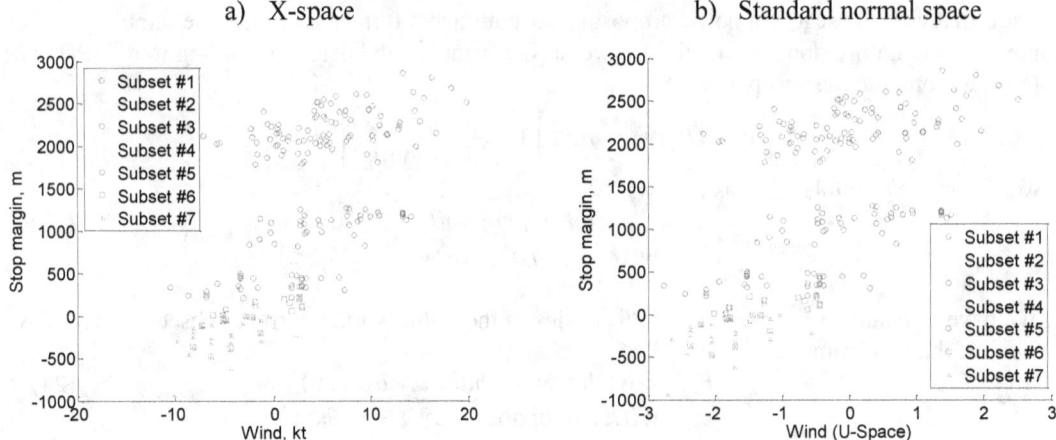

Each sample is a vector with n elements corresponding to n contributing factors. We use these samples and compute their distance to the origin of U-space by computing their norm.

$$\|u\| = \sqrt{u_1^2 + \cdots + u_n^2} \tag{26}$$

The sample with the smallest norm equals the design point u_{DP}. It has the greatest joint probability of the contributing factors [9].

$$u_{DP} = \operatorname{argmin}\|u\| \tag{27}$$

By applying the First-Order Reliability method (FORM), we are able to estimate the sensitivities for each contributing factor. The sensitivities are obtained by normalizing the negative gradient vector:

$$\boldsymbol{\alpha} = -\frac{\nabla G(u_{DP})}{\|\nabla G(u_{DP})\|} \tag{28}$$

The relationship between the sensitivity vector $\boldsymbol{\alpha}$, the design point u_{DP}, and the reliability index β

$$u_{DP} = \boldsymbol{\alpha} \cdot \beta \tag{29}$$

By solving Eq. (29) for the sensitivity vector $\boldsymbol{\alpha}$, we obtain the sensitivity factor α_i for each contributing factor. The sensitivity factor takes values from -1 until 1. The greater the absolute value of the sensitivity factor of a contributing factor, the greater its influence on the incident probability. A sensitivity factor close to zero indicates almost no impact on the incident probability. If the sensitivity factor of a contributing factor is positive, the incident probability increases with higher values of the contributing factors and vice versa.

4.2. Application Example

Now, we tag the contributing factors with the following stakeholders (flight operation, ground operation, training, and environment) and/or postholder (human, maintenance, environment, and organization). This allows us to quantify the contribution of stakeholders to the incident runway overrun. For the runway-overrun example, we allocate the following stakeholders to the following contributing factors (Table 2):

Table 2: Postholder Allocation

Contributing Factor	Postholder		
	Environment (ENV)	Training (TRA)	Flight Operation (FOPS)
Headwind	X	X	
Landing weight			X
Air pressure (QNH)	X		
Temperature	X		
Approach speed deviation	X	X	
Time of spoiler deployment	X		
Start of braking		X	
Reverser deployment		X	
Touchdown distance	X	X	

Due to the fact that we did not have any operational flight available, our analysis is based on artificial flight data. This means, that we generated samples of fictive flights performed to various airports by one aircraft type. So, the values for the sensitivities of the contributing factors are based on artificial distributions. The distribution type for each of the contributing factor is shown in the appendix A.1.

For this application example, we used 130000 samples out of which 4259 samples fell in the incident domain. The following Fig. 7 shows the norm of each sample that lies in the failure domain sorted by their length. The norm of the vector of the design point u_{DP} equals 5.49. However, multiple samples had the same length, and these samples were not concentrated at single point but spread (widely) within the domain. By using the sample with the lowest distance to the origin of the standard normal space some values for the sensitivity of some contributing factors were not reasonable from a physical point of view. Therefore, we included more samples. We used hundred samples with the lowest value of the norm.

Consequently, we obtain for each contributing factor one hundred values for its sensitivity. Figure 8 shows a histogram of sensitivities that are calculated by using the one hundred samples closest to the origin of the standard normal space. The values vary within a range of -0.366 to 0.0045.

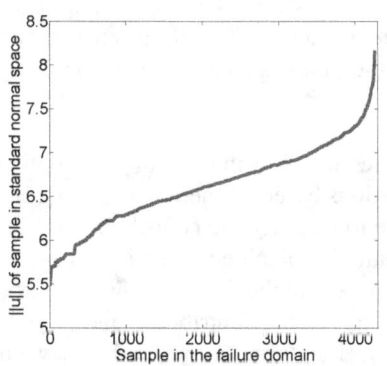

Figure 7: Norm of u-Vectors

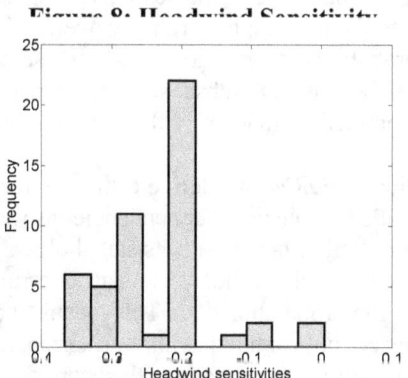

Figure 8: Headwind Sensitivity

Table 3 shows the average sensitivity of each contributing factor.

Table 3: Sensitivities

Contributing Factor	Sensitivities
Headwind	-0.2531
Landing weight	0.1710
Air pressure (QNH)	0.0071
Temperature	0.0852
Approach speed deviation	0.1848
Time of spoiler deployment	0.0235
Start of braking	0.2462
Reverser deployment	0.0258
Touchdown distance	0.8347

These sensitivities values are reasonable. For example, the headwind's sensitivity is negative that means smaller values for the headwind increase the likelihood of an overrun. This is plausible as a negative headwind equals tailwind that increases the kinetic energy of the aircraft at touchdown, which has to be absorbed. In contrast, the approach speed deviation has a positive sensitivity that means the greater the approach speed the greater the distance the aircraft requires to stop. Due to the fact, that the runway length in our analysis was quite low, the touchdown distance in this scenario is the main driver.

In order to calculate the individual sensitivity α_j of a stakeholder j, we sum up the absolute values of each contributing factor that belongs to the individual contributing factor CF_j.

$$\alpha_j = \sum_i |\alpha_i(CF_j)| \tag{30}$$

The following table shows the result of the contribution of each stakeholder.

Table 4: Sensitivities

Stakeholder j	Sensitivity α_j	Contribution
TRA	1.7671	44.2%
ENV	1.5788	49.5%
FOPS	0.2257	6.3%

In Table 4, the stakeholder contribution SC_j to runway overrun is computed as

$$SC_j = \frac{\alpha_j}{\sum_i \alpha_{i,j}} \tag{31}$$

The stakeholder ENV contributes most to the incident type runway overrun, based on the contributing factors being analysed.

5. SUMMARY AND CONCLUSION

This paper studies the quantification of stakeholder contribution to runway overrun. Therefore, a model tailored to the incident type runway overrun is presented. Then, based on data from a fictive airline, we fitted probability distributions that represent the contributing factors. The occurrence probability is estimated by using the subset simulation method. We used the samples that we obtained through subset simulation method to identify the design point.

By applying FORM, we derive estimates for the sensitivities for each individual contributing factors that include the functional dependencies as well as the deviations based on their distribution. By tagging each contributing factor with its stakeholders, we were able to quantify the contribution of the various stakeholders to the incident runway overrun. However, all the obtained sensitivities are based on artificial flight operation data. This means, that dependencies within the data, which might exist within an airline are not captured at all. Furthermore, the distributions of each contributing factor might not be close to reality. Nevertheless, this paper shows that, if such information would be available, the stakeholder's contribution can be quantified.

However, we also found out, that using only the sample with the lowest distance to the origin of the standard normal space does not necessarily provide sensitivities that are reasonable, especially from a physical point of view. Therefore, we used multiple samples that are close to the origin of the standard normal space and calculated sensitivities using an average. The results are much more reasonable and are in line with the physical and probabilistic knowledge of the contributing factors and their functional relationship. Furthermore, it also becomes clear, that the stakeholder contribution heavily relies on a proper allocation of the stakeholder to the contributing factors and all relevant factors have to be taken into account.

Consequently, the next step is to apply this approach to data obtained from a real flight operation of an airline. Then, the allocation of stakeholder might change as well as additional stakeholder needs to be included. In addition, other contributing factors, such as flap setting or runway condition have to be taken into account.

Appendix
A.1

Table A.1 Distribution of the Contributing Factors

Contributing Factor	Distribution Type	Distribution Parameter
Headwind	Generalized Extreme Value	$k = 0.0243$, $\sigma = 3.4585$, $\mu = -5.4383$
Landing weight	Generalized Extreme Value	$k = -0.6716$, $\sigma = 4.7701$, $\mu = 67.4481$
Air pressure (QNH)	Burr	$\alpha = 1025.38$, $c = 235.38$, $k = 4.38$
Temperature	Weibull	$a = 21.6713$, $b = 3.7698$
Approach speed deviation	t Location-Scale	$\mu = 0.0207$, $\sigma = 3.1942$, $v = 36.6177$
Time of spoiler deployment	Generalized Extreme Value	$k = -0.0633$, $\sigma = 0.4288$, $\mu = 0.5848$
Start of braking	Generalized Extreme Value	$k = -0.0007$, $\sigma = 1.0430$, $\mu = 1.9173$
Reverser deployment	Generalized Extreme Value	$k = -0.0123$, $\sigma = 0.8448$, $\mu = 3.6858$

References

[1] EASA. Annual Safety Review 2012. European Aviation Safety Agency, 2013.

[2] G.W.H. van Es. K. Tritschler, M. Tauss. Development of a Landing Overrun Risk Index. NLR, 2009

[3] G.W.H. van Es. Running out of runway. Analysis of 35 years of landing overrun accidents. NLR, 2005

[4] S-K. Au, J. L. Beck. Estimation of small failure probabilities in high dimensions by subset simulation. Prob. Eng. Mech. 2001; 16; 263-77

[5] L. Drees, F. Holzapfel, Determining and Quantifying Hazard Chains and their Contribution to Incident Probabilities in Flight Operation, AIAA Modeling and Simulation Technologies Conference, Minneapolis, MN, Aug. 13-16, 2012 AIAA-2012-4855

[6] M. Wendt, P. Fabregas. Characterization of Landings Deep and Long Landings and of the Runway Excursion Risk. Presentation at the EASA EOFDM Conference 2012.

[7] T. L. Thorarinsdottir, T. Gneiting, N. Gissibl. Using proper divergence functions to evaluate climate models. 2013

[8] N. Metropolis; A.W. Rosenbluth, M.N. Rosenbluth, A.H. Teller, Equations of state calculations by fast computing machines. Journal of Chemical Physics (6): 1087-109

[9] S. Amatya, Y. Honjo, Use of Subset Simulations to Determine Design Point in Reliability Analysis, 3rd International ASRANet Colloquium, 10-12th July 2006, Glasgow, UU

International Space Station End-of-Life Probabilistic Risk Assessment

Gary Duncan*
ARES Technical Services, Houston, TX, USA

Abstract: Although there are ongoing efforts to extend the ISS life cycle through 2028, the International Space Station (ISS) end-of-life (EOL) cycle is currently scheduled for 2020. The EOL for the ISS will require de-orbiting the ISS. This will be the largest manmade object ever to be de-orbited, therefore safely de-orbiting the station will be a very complex problem. This process is being planned by NASA and its international partners. Numerous factors will need to be considered to accomplish this such as target corridors, orbits, altitude, drag, maneuvering capabilities, debris mapping etc. The ISS EOL Probabilistic Risk Assessment (PRA) will play a part in this process by estimating the reliability of the hardware supplying the maneuvering capabilities. The PRA will model the probability of failure of the systems supplying and controlling the thrust needed to aid in the de-orbit maneuvering.

Keywords: PRA, ESD, ISS, Footprint, End State

1. INTRODUCTION

When the ISS is de-orbited it will be the largest manmade object to ever reenter the earth's atmosphere. This presents numerous technical challenges as well as logistical problems to the ISS international partners. The target risk to the ground population is less than 1 in 10,000 for a single Progress option. (Note that the estimated energy for a loss of life incident is an impact to a person of about the energy of a baseball pitched by a major league pitcher or about 130 joules). This paper presents the structure of planned re-entry using a single Russian Progress M cargo vehicle and the ISS Service Module (SM) for re-entry maneuvering. This plan is being reevaluated and will probably be revised to use three Progress M vehicles to supply re-entry maneuvering; however, the structure and methodology will be similar. Because of the sensitive nature of the material, quantitative values will not be presented here.

There are many driving factors that will play into the successful re-entry of the ISS. These include vehicle inclination, altitude at start of re-entry sequence, first interface with atmosphere, and breakup sequence. Experiments are being performed to collect data on these parameters. Because of these factors there will be uncertainty in the plan even if the hardware functions nominally. The PRA team's role in this process is to model the probability that the hardware will successfully provide the thrust necessary to support the planned re-entry.

2. RE-ENTRY PLAN

The re-entry plan is still in a preliminary planning stage. The current plan is outlined below:

- Planned ISS EOL will begin with natural drag from 400km to final operations starting at 200km. This phase will last over a year with only phasing burns to nearly 200km with a circular orbit.
- ISS will be reduced to a 3-man crew.
- Two decision points in the final week will allow holds, if needed, with the same intended ground track available 4 days and 1 day later than first planned.
- Numerous consecutive-orbit preparatory burns will shape the final orbit using Progress rendezvous and docking (R&D) thrusters.

- Ballistic/altitude plan leading into the final burn will support an optimized combination of abort capability and projected footprint that is completely contained in the ocean.
- Final burn will be a combination of (up to) three Progress M vehicles, and propulsion enhancements selected below. This analysis considers a single Progress case.
    - SM main engine
    - Cross-flow of aft Progress resupply tanks to aft Progress main engine
    - Burn time extension of Progress R&D engines
- ISS will be in an aerodynamically-trimmed configuration with its center of pressure behind the center of drag, and minimum exposure of solar arrays to RAM pressures.

3. APPROACH

Industry standard event tree and fault tree methodology will be used to model the hardware systems required for successful ISS re-entry. Although many of the systems required are already modeled in the ISS PRA Progress R&D and ISS SM models, there are still modeling challenges ahead. Some of the hardware, such as thrusters, may be required to operate outside of their design parameters. In these cases, the use of the current failure rate data, developed for nominal operations, will need to be reassessed. Test data could help fill in the gaps between data acquired for nominal cases and data to evaluate a system challenged beyond its design constraints.

4. ANALYSIS

Figure 1: ISS EOL Re-entry Burn Sequence, Single Progress Option

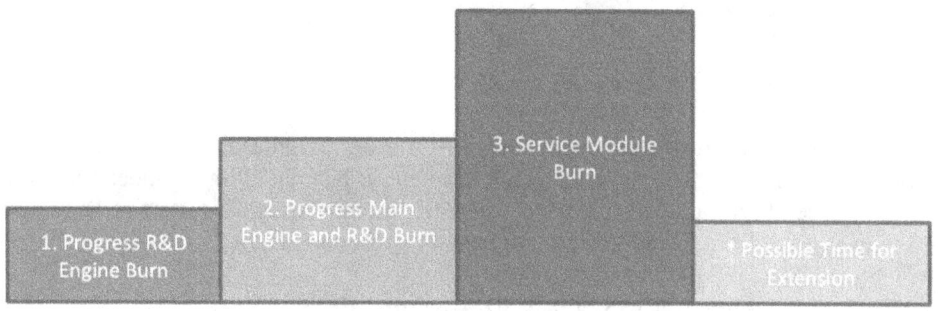

* Optional Time Extension

Burn 1 utilizes Progress R&D Engines, Burn 2 utilizes Progress R&D and Main Engines, Burn 3 uses the ISS SM engines, and Burn 4 is a contingency burn which could be used to make up failures that occur during the first three burns. Event Sequence Diagrams (ESDs) (Figures 2 through 8) were developed for the planned re-entry scenarios.

The single Progress planned re-entry will rely on the burn sequence presented in Figure 1. This plan, once initiated, can accommodate up to two abort/restart sequences as needed on subsequent orbits. The ESDs could result in the five possible end states listed below:

- P_n = Nominal Re-entry
- P_r = Random Re-entry
- P_p = Pseudo Random
- P_l = Large Under-Burn
- P_s = Small Under-Burn

Figure 2: ISS EOL ESD De-Orbit Attempt 1

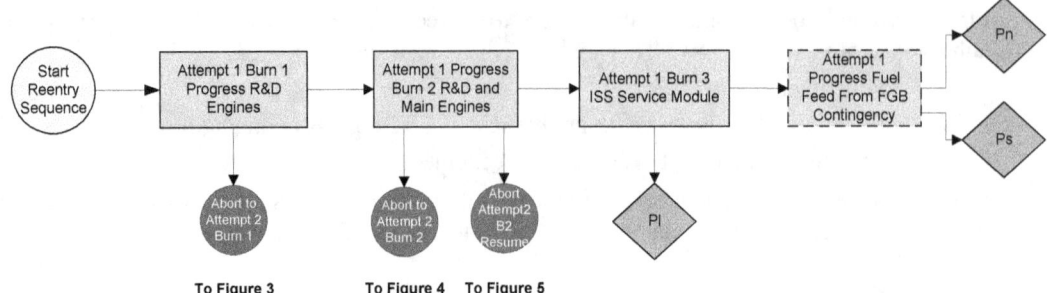

Figure 2 presents the initial re-entry attempt. There are 3 paths that result in re-entry: Nominal, Small Under-Burn, and Large Under-Burn. Three abort paths are included: abort to Attempt 2 with a restart at Burn 1, abort to Attempt 2 with a restart at Burn 2, and abort to Attempt 2 by resuming Burn 2.

Figure 3: ISS EOL ESD Attempt 2 Restart Burn 1

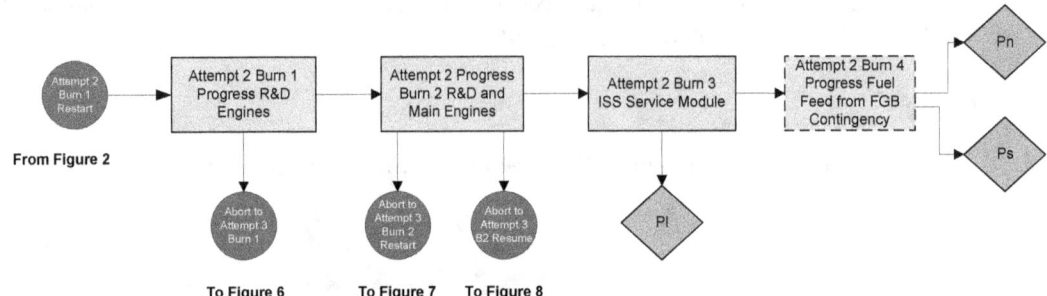

Figure 3 is the first of three abort trees from Attempt 1; this tree assumes the burn sequence will restart from the beginning. There are 3 paths that result in re-entry: Nominal, Small Under-Burn, and Large Under-Burn. Three abort paths are included: abort to Attempt 3 with a restart at Burn 1, abort to Attempt 3 with a restart Burn 2, and abort to Attempt 3 by resuming Burn 2.

Figure 4: ISS EOL ESD Attempt 2 Resume Burn 2

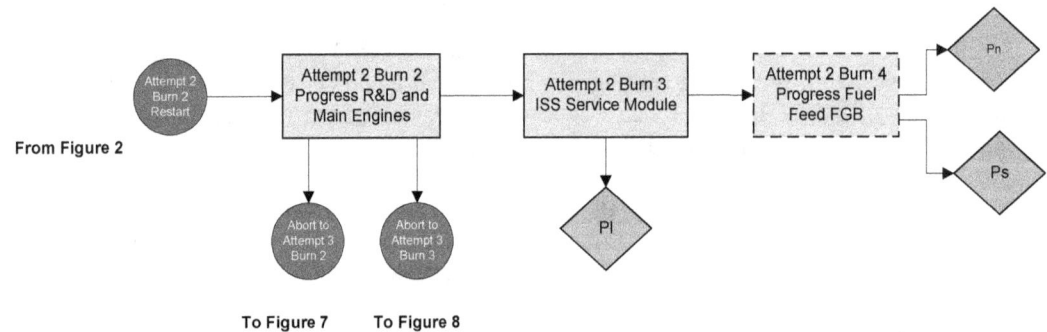

Figure 4 is the second of the aborts possible from Attempt 1 and assumes Burn 1 completed and so this burn will restart from the beginning of Burn 2. There are 3 paths that result in re-entry: Nominal, Small Under-Burn, and Large Under-Burn. Two abort paths are possible: abort to Attempt 3 with a restart at Burn 2, and abort to Attempt 3 with a restart at Burn 3.

Figure 5: ISS EOL ESD Attempt 3 Resume Burn 2

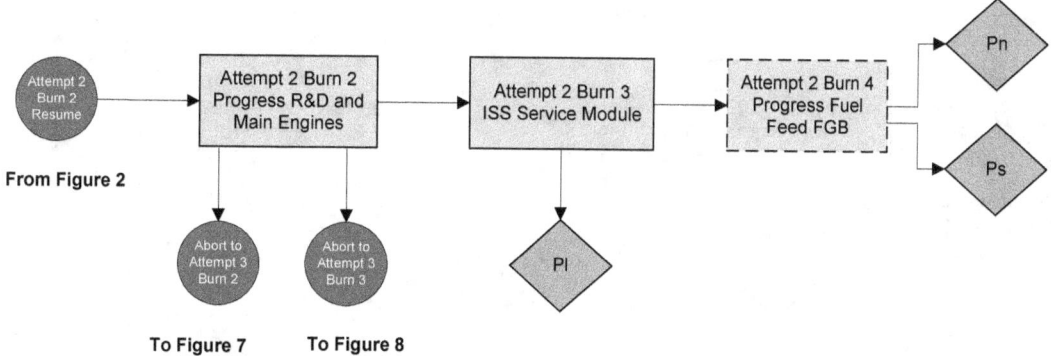

Figure 5 is the third of the aborts possible from Attempt 1 and assumes Burn 2 was aborted while in progress. There are 3 paths that result in re-entry: Nominal, Small Under-Burn, and Large Under-Burn. Two abort paths are possible: abort to Attempt 3 with a restart at Burn 2, and abort to Attempt 3 by resuming Burn 2.

Figures 6, 7, and 8 model Attempt 3, which is the final attempt to complete the process – there are no abort paths possible. At this point Nominal Re-entry can still be achieved.

Figure 6 – Attempt 3 ISS EOL ESD Restart Burn 1

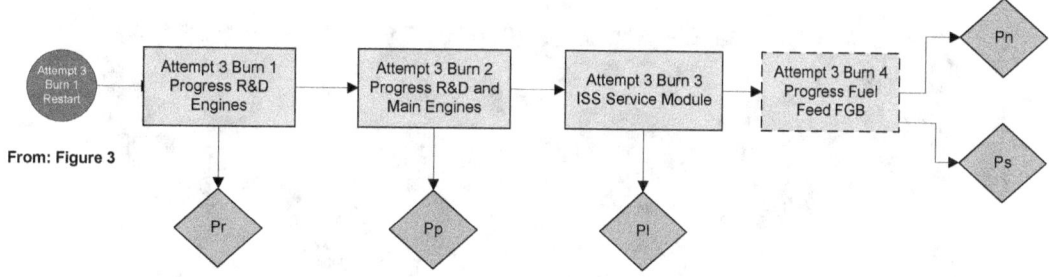

Figure 7 – ISS EOL ESD Attempt 3 Restart Burn 2

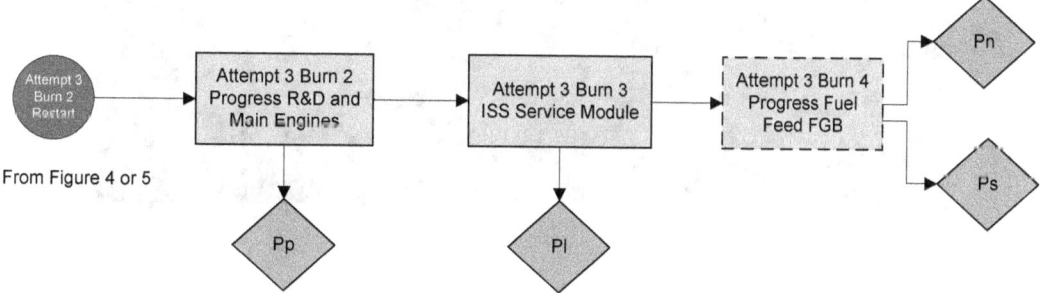

Figure 8 – ISS EOL ESD Attempt 3 Restart Burn 3

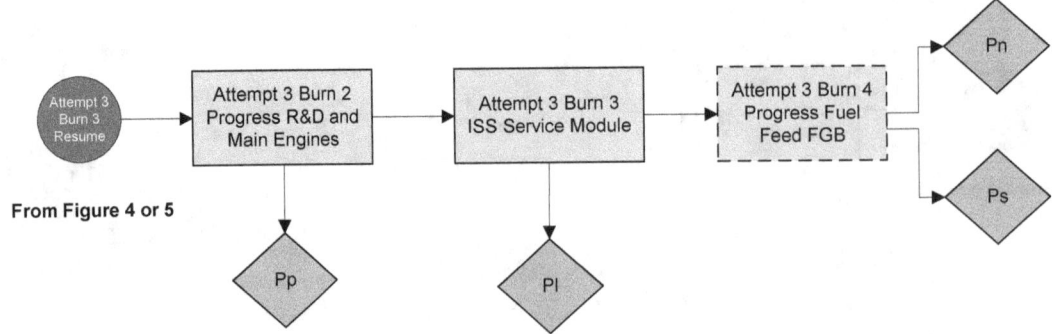

The end states from the model are P_n (Nominal Re-entry), P_r (Random Re-entry), P_p (Pseudo Random), P_l (Large Under-Burn) and P_s (Small Under-Burn), and will be binned into the following Footprints.

- Nominal Footprint
- Ocean Footprint
- Populated Footprint

Figure 9 shows the potential footprints for the start of the re-entry process.

Figure 9 – ISS Re-entry Footprints

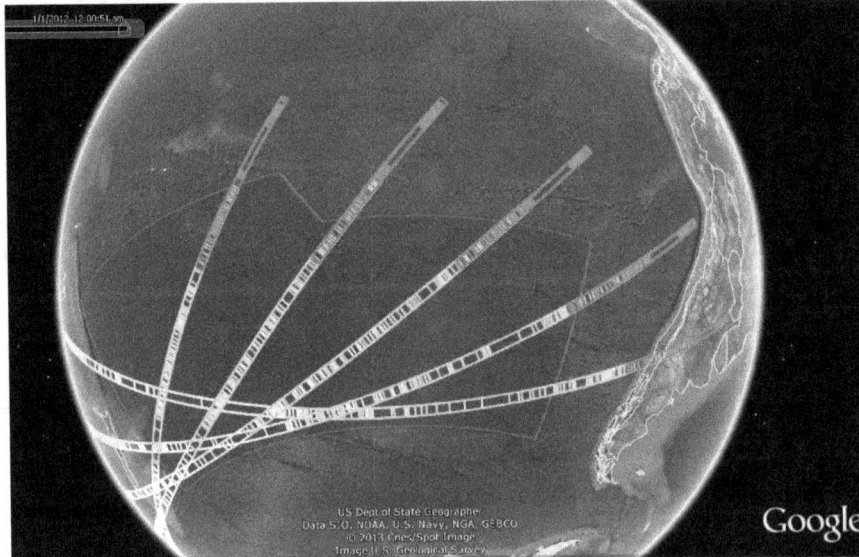

Probabilistic Safety Assessment and Management PSAM 12, June 2014, Honolulu, Hawaii
*gary.w.duncan@nasa.gov

Figure 10 is the type of chart that will be used to present the PRA results, once completed.

Figure 10 – Example: Probability of ISS Re-entry Footprints

5. METHODOLOGY

ISS PRA will utilize event tree/fault tree methodology. Some of the challenges facing the ISS PRA team are discussed in this section.

Any analyst that has attempted to model common cause using Multiple Greek Letter or Alpha Models knows it gets convoluted after about four components. The Progress has 28 thrusters and the ISS SM has 32 thrusters, whose redundancy and similar design make them susceptible to common cause failures. The Global Alpha Model (as described in NUREG/CR-5485) [1] can be used to represent the system common cause contribution, but NUREG/CR-5496 [2] supplies global alpha parameters for groups only up to size six. Because of the large number of redundant thrusters on each vehicle, regression is used to determine parameter values for groups of size larger than six. An additional challenge is that thruster failures must occur in specific combinations in order to fail the propulsion system; not all failure groups of a certain size are critical. The calculation of common cause will become even more complicated if the ISS program opts to use the thrusters on three Progress vehicles in addition to the SM to de-orbit the ISS vehicle. The methodology that will be used to model common cause failures of the thrusters required to de-orbit the ISS has already been used to model common cause failure of thrusters on the ISS Visiting Vehicles and is described in "*Modeling Common Cause Failures of Thrusters on ISS Visiting Vehicles*" [3].

Fault trees are based on Boolean algebra failures and therefore failures are generally bounded as failed or operational – this makes modeling difficult in an area where reduced performance issues can occur. The challenge is mapping thrusters under performance issues into bins and establishing how these under-performance issues will stack up if they occur in sequence. There is also difficulty in dealing with hard thruster failures in cases where the thruster vectoring can be equaled or approximated by other thruster combinations.

There are current Progress models accounting for functional failures during R&D operations; however, these models focus on motion control and include propulsion and supporting control systems and power systems. The current models are in the process of being refined for this re-entry analysis. The command and control of the overall de-orbit vehicle configuration, whichever configuration is selected (single Progress with SM or three Progress) will also pose a challenge. Once the burn sequence begins, the de-orbit maneuver becomes software-dependent, and the current model does not account for software reliability.

6. CONCLUSION

The ISS EOL and resulting vehicle de-orbit will be a highly publicized event involving many nations – those invested in the safe completion of the ISS Program, and those potentially affected by the footprint of the ISS debris following re-entry. The ISS EOL PRA will contribute to the planning and safe execution of this event by capturing the probability of hardware failures associated with ISS EOL. The PRA will be used as one of the components of the overall plan, to assist in ISS Program decisions regarding vehicle configuration as measured against the likelihood of success. The ISS EOL PRA package will be integrated into the overall risk plan for ISS EOL re-entry, and may be used by the agency to accept any residual risk (complete with uncertainty bounds) to the general population.

References

[1]   U.S. Nuclear Regulatory Commission, *Guidelines on Modeling Common-Cause Failures in Probabilistic Risk Assessment, NUREG/CR-5485*, November 1998.

[2]   U.S. Nuclear Regulatory Commission, *CCF Parameter Estimations, 2009 Update to NUREG/CR-5496*, January 2011.

[3]   Haught, Megan, *Modeling Common Cause Failures of Thrusters on ISS Visiting Vehicles*, Probabilistic Safety Assessment and Management (PSAM), Honolulu, Hawaii, June 2014.

MCSS Based Numerical Simulation for Reliability Evaluation of Repairable System in NPP

Daochuan Ge[a,b,*], Ruoxing Zhang[b], Qiang Chou[b], Yanhua Yang[a]

[a] School of Nuclear Science and Engineering, Shanghai Jiao Tong University, Shanghai, China
[b] Software Development Center, State Nuclear Power Technology Corporation, Beijing, China

Abstract: The quantitative analyses of Nuclear Power Plant (NPP)'s repairable systems are conventionally Markov-based methods. The thing is, systems' state space grows exponentially with the increase of basic events, which makes the problem hard or even impossible to solve. In addition, the maintenance /test activities are frequently imposed on some safety-critical components, which make the Markov based approach unavailable. In this paper, a new numerical simulation approach based on MCSS (Minimal Cut Sequence Set) is proposed, which can get over the shortcomings of the conventional Markov method. Two typical cases are analyzed and results indicate that the new approach is correct as well as feasible.

Keywords: Numerical Simulation, Reliability, Minimal Cut Sequence Set, Repairable System

1. INTRODUCTION

After Fukushima nuclear accident, more and more countries focus their attention on NPP's safety, especially the reliability of safety-critical systems. The real-life safety-critical systems often have sequence- and function-dependent failure behaviours. For the description of these dynamic failure behaviors, traditional static fault tree is unfeasible. To overcome the shortcomings of the conventional static fault tree method, some researchers [1] introduced a few new dynamic gates, such as PAND, SPARE, FEDP and SEQ into static fault tree, i.e., DFT. Compared with previous static fault tree, the DFT greatly extends modeling capacities. Considering the intuitiveness and compactness of DFT, NPPs often adopt DFT to model safety-critical system's failure mechanism. The commonly-used methods for quantifying a DFT are mainly based on Markov approaches [2,3,4] or multi-integration approaches [5,6,7]. Unfortunately, each of these methods has its own shortcomings: For the Markov-based approach, it requires the time-to-failure/time-to-repair of components follows exponential distribution. In addition, the approach may confront the notorious problem of "state space explosion"; as to the multi-integration -based approach, although it avoids the problem of "state space explosion", it is only applicable for non-repairable systems. Given that the components in NPP system are usually repairable and their failure and repair time are not exponent, the methods mentioned above are unavailable. To solve these problems, some researchers proposed a Monte Carlo Simulation-based method [8,9]. This Monte Carlo Simulation method is based on the failure behaviours of DFT gates. As to simple dynamic gates, this method is easy to implement. However, when dynamic gates are highly-coupled and complex, this method usually becomes hard to carry out.

In this paper, a MCSS-based numerical simulation method is presented, which is applicable for any complex DFT and easy to implement. Results show this method is feasible and correct.

The remainder of this paper is organized as follows: Section 2 reviews some related concepts including unavailability, Minimal Cut Sequence Set, etc; Section 3 presents our proposed method. Section 4 provides two cases studies to validate our proposed method. Section 5 gives final conclusions.

2. RELATED CONCEPTS

---

[*] Corresponding author: Phone: +86-10-18817554483; Fax: +86-10-58197250
E-mail: gdch-2008@163.com

2.1. System's Unavailability

Suppose a repairable unit that is put into working at time t=0. As the unit fails, a repair activity is implemented to restore the function of the failed unit. The state of the unit at time t is defined by a state variable:

$$X(t) = \begin{cases} 1 & \text{if the unit is working at time } t \\ 0 & \text{otherwise} \end{cases} \quad (1)$$

Then the reliability of a repairable unit may be measured by the availability of the unit at time t:

$$A(t) = P_r(X(t) = 1) \quad (2)$$

Sometimes, $A(t)$ is referred to be as the point availability. Note if the unit is not repaired, then we can get: $A(t)=R(t)$. Where $R(t)$ is the reliability of a non-repairable unit at time t. Similarly, we can define the unavailability of a non-repairable component at time t as the probability that the unit is not in working state at time t:

$$\overline{A(t)} = 1 - A(t) = 1 - P_r(X(t) = 1) \quad (3)$$

In NPP, we are more interested in the average or mission availability $A(0, t)$ in time interval $(0, t)$, which is defined as:

$$A_{av}(0,t) = \frac{1}{t} \int_0^t A(t) dt$$

The average availability $A_{av}(0, t)$ can be interpreted as the mean proportion of the time interval $(0, t)$ where the unit is able to operate. Suppose a repairable unit that starts to work at time $t=0$. Whenever the unit fails, it is repaired to an "as good as new" state or substituted by a new one. Then a sequence diagram of up-times (life times) U_1, U_2 ... and down-times (outage times) D_1, D_2 ... appearing alternately is obtained as shown in Fig.1.

Fig.1: alternate state-time of a repairable unit

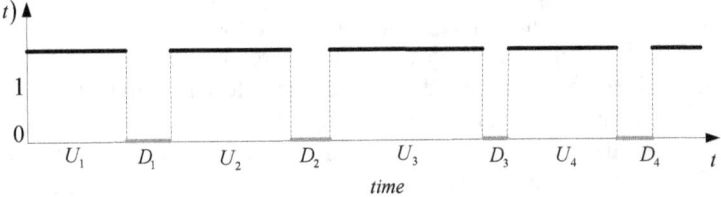

In this paper, we suppose the up-times $U_1, U_2, \cdots$ are independent and identical distributed and $D_1, D_2, \cdots$ are independent and identical distributed as well. In addition, we also suppose $U_i + D_i$ for $i=1, 2, \cdots$ are independent. Assume a unit has just finished n repair, then the unit's up-times $U_1, U_2, \cdots, U_n$ and down-times $D_1, D_2, \ldots, D_n$ are obtained. When the $n \to +\infty$, then the average availability of the unit can be expressed as:

$$A_{av} = \frac{\lim_{n \to \infty} \sum_1^n U_i}{\lim_{n \to \infty} \sum_1^n U_i + \lim_{n \to \infty} \sum_1^n D_i} \quad (4)$$

Then the average unavailability of the unit can be written as:

$$\overline{A_{av}} = 1 - A_{av} = \frac{\lim_{n \to \infty} \sum_1^n U_i}{\lim_{n \to \infty} \sum_1^n U_i + \lim_{n \to \infty} \sum_1^n D_i} \quad (5)$$

2.2. Minimal Cut Sequence Set

As to a system modelled by DFT, the occurrence of the top event (system failure) not only depends on the combination of basic events but also depends on their failure sequences. Thus traditional minimal cut set is not able to describe this sequential failure behaviour correctly. To solve this problem, Tang et al [10] presents a concept of Minimal Cut Sequence (MCS) that expresses the minimal failure sequence that causes the occurrence of the top event of a DFT. The original expression of a minimal cut sequence comprises several capitals denoting a failure of a basic event and several temporal connecting symbols "→" which is used to express the failure sequence, i.e., the left event fails before the right one. However in a real-life system's DFT, the failure behaviours of the basic events may be not the same: Some components are initially powered on; some components may be initially powered on just at a reduced power; and some others may be originally in a standby state without any power. In this paper, to distinguish the failure behaviours of the basic events, four special symbols are introduced: "X" denotes the component X being initially powered on at a full energy and fails randomly; "$_X^0Y$" indicates the component Y as a cold spare of X is initially unpowered and fails after X; "$_X^1Y$" expresses the component Y as a warm spare of X and fails after X; "$_X^{\alpha}Y$" shows the component Y as a warm spare of X and fails before X. For an illustrative purpose, an example is given in Fig.2.

Fig.2: An Illustrative Example

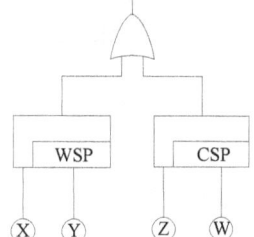

For the system's top event is connected by the logic OR of the two dynamic gates, i.e., WSP gate and CSP gate, according to the failure behaviours of dynamic gates mentioned in [11], the minimal cut sequences of the system failure are obtained as { $X \to {}_X^1Y, {}_X^{\alpha}Y \to X, Z \to {}_Z^0W$ }. For a DFT, it may have more than one MCS and all these MCSs compose an aggregate, i.e., Minimal Cut Sequence Set (MCSS). Since the occurrence of each MCS leads to the failure of a system, the MCSS captures the complete information about a system's failure. Note that whether a system is repairable or not, the corresponding MCSS is unique. Therefore we can get the MCSS of a repairable system using the approaches developed for non-repairable systems [12,13].

2.3. Basic Events' Failure Behaviors

The failure behaviour of a basic event refers to the randomness of its failing. As mentioned above, basic events involved in a DFT may have different failure behaviours. According to the failure behaviours of the basic events, we classify the basic events into three categories: random basic events, semi-random basic events and decided basic events. As we know the failure time of a basic event is completely random during its mission time. Note that a component's mission time doesn't always equal the system's mission time. For example, the mission time of a cold spare is always dependent on the primary component.

In general, the occurrences (failure behaviours) of components providing the main functions during system mission period are considered to be random basic events. And the occurrences of components supplying standby function are considered semi-random basic events. In most cases, the entire spare components except hot spares are semi-random basic events, and the remains are the random basic events. In NPP, some components' function failure is caused either by a random event (random failure event) or by a decided event. The decided events here refer to the regular maintenance/test activities imposed on the safety-critical components to improve the reliability. However, when the safety-critical components are forced outage for the regular maintenance/test activities, the risk of the system will increase. In this paper, the decided basic event is supposed to be a virtual component. To reflect the influence of the decided basic events to a component' function, this paper developed a new function dependent dynamic gate. To differ from the traditional FDEP gate, this new dynamic gate is called Decided Function Dependent gate (DFDEP) where the trigger events are the decided events. When the decided events occur, the dependent events are forced outage. The DFDEP gate is shown in Fig.3. The symbols $\zeta_1 \ldots \zeta_n$ represent the decided events such as maintenance activity, test activity, etc. and the $E_1 \ldots E_n$ means the dependent events. In general, the classifications of the basic events are listed in Table 1.

Table 1: Classification of the Basic Events

Category	Basic Events	Symbols
Random basic events	The basic events under AND, PAND, OR gate; The basic events under SPARE gate representing the primary components;	X
Semi-Random basic events	The basic events under PAND gate denoting the standby components;	$_x^0Y, _x^1Y, _x^1Y$
Decided basic event	Virtual events expressing a series of maintenance, test, etc, activities.	$_x\zeta$

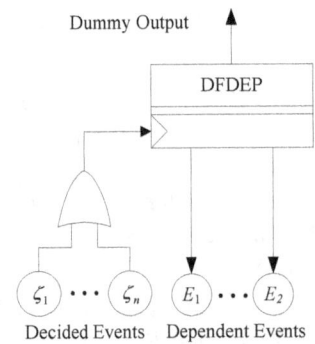

Fig.3: A Decided Function Dependent gate

3. NUMERICAL SIMULATION FOR THE MCSS

3.1. Numerical Simulation for the Failure Behavior of a Basic Event

As to a repairable component, its failure behavior can be simulated by a Monte Carlo-based approach. It is known the time-to-failure and time-to-repair of a component are only determined by their respective Cumulative Probability Distributions (CDFs). Consider the CDFs of a component's time-to-failure and time-to-repair are $F(x)$, $G(x)$, and then the time-to-failure T_f and the time-to-repair T_r are obtained by the following expressions:

$$\begin{cases} T_f = F^{-1}(\varepsilon) \\ T_r = G^{-1}(\eta) \end{cases} \quad (6)$$

Where the ε, η are uniform random numbers generated by any standard random number generators. For the random basic events being active initially, its running state in system's mission time can be simulated directly using Eq. (6). However, for the semi-random basic event, when there is no demand, it will keep up in standby state or may be in a failed state due to on-shelf failure. Therefore the failure behaviors of the semi-random events are relatively complicated. For the cold spares, they never fail during the standby states. Considering the outage time (time-to-repair) of the primary component is the mission time of the cold spare, the failure behavior can be simulated using Eq. (6) during this mission time. As to the warm spare, the situation is even more complex. The failure rates of a warm spare in standby state and in working state are not the same. In other words, the warm spares have two CDFs of the time-to-failure in different states. Generally speaking, the failure rate of a warm spare in working state is higher than that in standby state. When the primary component is staying in a working period (time-to-failure), the failure behavior of the standby component is simulated by the Eq. (6) with one time-to-failure CDF. Similarly, when the primary component is staying in an outage period (time-to-repair), the failure behavior of the standby component is simulated by the Eq. (6) with the other time-to-failure CDF. Apparently the failure behaviors of the semi-random basic events are dependent on but not affecting the random basic events. Finally, as to the decided events, the occurrences time of these events can be obtained directly from the scheduled maintenance/test management documents.

Considering the correlation between the basic events' failure behaviors, the simulation precedence is required as: random basic events →decided events →semi-random basic events.

3.2. Numerical Simulation for the MCSS

After simulating the basic events' failure behaviors, the numerical simulation of the system's MCSS is carried out. If a MCS occurs at some point, the system is considered to be failed. For demonstration purpose, an illustrative example is shown in Fig.4.

Fig.4: An Illustrative Example

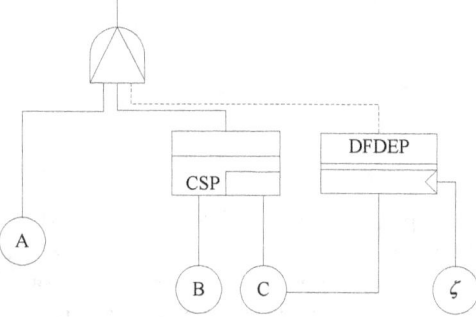

Based on the temporal rules mentioned in [12,13], the MCSS of the example system is: $\{A \rightarrow B \rightarrow {}_B^0 C,\ A \rightarrow B \rightarrow {}_c\zeta\}$. Where the symbol ${}_c\zeta$ represents a series of regular maintenance activities imposed on component C. Therefore, the system has two failure scenarios: the scenario 1 is $A \rightarrow B \rightarrow {}_B^0 C$ and the scenario 2 is $A \rightarrow B \rightarrow {}_c\zeta$. The state-time of the example system (MCSS) is depicted in Fig.5.

Fig.5: The State-time of The Example System's MCSS

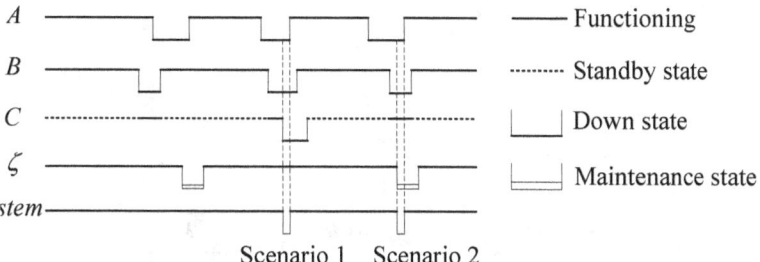

For generating the system state time diagram, all components state time profiles, including virtual components (test/maintenance activities), involved in every MCS are compared. The system will fall in a failed state if all the components contained in a MCS failed in a pre-assigned sequence (usually from left to right), as shown in Fig.5 (first and second scenarios). In the scenario 1, the active component (A) failed followed by the second component (B), and then followed by the third component (C), the system is identified as failure since the failure sequence meets $A \rightarrow B \rightarrow {}_B^0 C$. As to the scenario 2, although the standby component (C) is functionally available during the repair period of the primary component (B), it is forced outage for the imposed maintenance activity, and the system is still considered to be failed since the failure order meets $A \rightarrow B \rightarrow {}_c\zeta$.

3.3. Calculation the System Reliability Indexes

In NPP, we are interested in system average unavailability and unreliability. To obtain these reliability indexes, the total outage time (t_o^i) and time to first failure (t_f^i) of the system in a simulation are recorded. Let $\varphi(t)$ be the system state variable and the logic value of the variable is defined as:

$$\varphi(t)=\begin{cases}1, & t<T\\0, & t\geq T\end{cases}$$

Where, T is the mission time of the system.

Then, the system average unavailability in the mission period is evaluated as:

$$\overline{A_{av}}=\frac{\sum_{i=1}^{N}t_o^i}{NT} \qquad (7)$$

Where, the N is the simulation number. And the system unreliabilityy R_s during the mission time T can be calculated using the follow equation.

$$R=\frac{\sum_{i=1}^{N}\varphi(t_f^i)}{N} \qquad (8)$$

4. CASE STUDY

4.1. Case Study 1

In this section, a case study is presented to validate the proposed method. The case is excerpted from an I&C (Instrument and Controller) system in one Chinese NPP. The simplified DFT model of this system is shown in Fig.6. And every capital letter represents a component. In I&C system, every component is repairable. The components failure and repair parameters are listed in Table 2.

Fig.6: Simplified DFT of an I&C Controller System

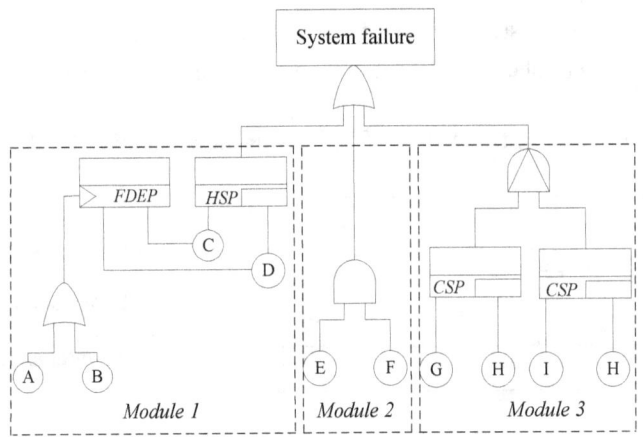

Table 2: Components failure and repair parameters

Component	Failure rate(h)	Repair rate(h)	Component	Failure rate (h)	Repair rate(h)
A	1.0E-4	0.25	F	5.0E-3	4.00
B	5.0E-4	1.20	G	1.4E-3	2.00
C	1.0E-3	1.50	H	2.0E-4	0.50
D	1.5E-3	1.00	I	2.5E-3	3.00
E	5.0E-3	5.00	-	-	-

The MCSS of the DFT is {A, B, C→D, D→C, E→F,F→E, I→G→H,G→H→I}. For the mission time 10^4h, the unavailability of I&C system calculated by our proposed method is 8.18E-4. For validation purpose, the Markov-based approach is adopted as a benchmark. To reduce the system state space, the system is divided into three independent sub-modules via modularization. Each sub-module, denoted with the dotted box, is solved by the Markov approach. Then the results of the three sub-modules are integrated to obtain the system unavailability. As applying the Markov-based approach, the unavailability of the system is obtained 8.20E-4, which is accorded with ours. In addition, the unavailability time and first time to failure distribution for the system is shown in Fig.7-8, respectively.

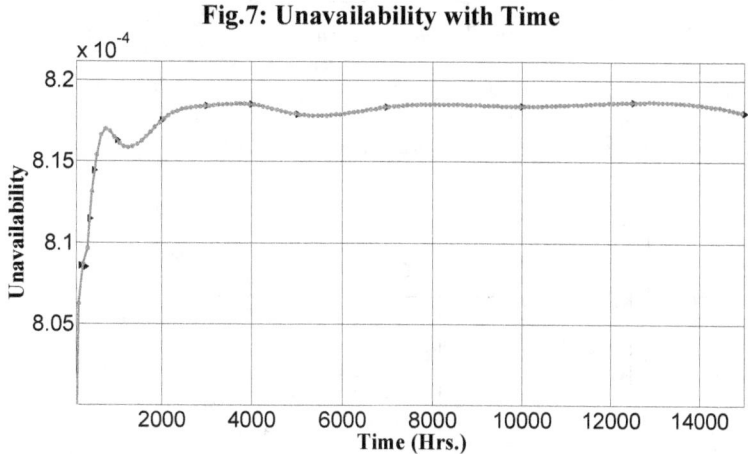

Fig.7: Unavailability with Time

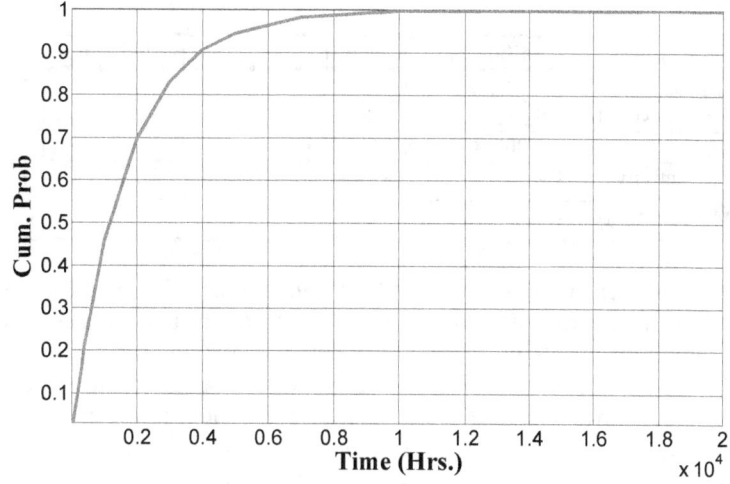

Fig.8: First Time to Failure Distribution

4.2. Case Study 2

For further validation purpose, another case with complex failure behaviors is studied. The case is the electrical power supply system of typical NPP. The system contains three subsystems: the Grid supply subsystem known as Class IV supply is the main power which feeds all the load; The diesel generator subsystem, known as Class III supply, as the standby power of the Class IV supply is providing the emergency power in the absence of the primary power; The sensing & control subsystem is used to trigger the redundant diesel generator once detecting the failure of Grid supply system. To ensure the reliability of the electrical power supply system, the redundant diesel generator is forced outage for regular test/maintenance. Therefore the system has two failure scenarios: The Grid supply subsystem fails, and then redundant diesel generator fails or is unavailable for test or maintenance outage; the sensing & control subsystem fails before the primary diesel generator fails and it makes the standby power be not triggered. The top event (station blackout) of the system modeled by DFT is shown in Fig.9. The component failure and maintenance information is listed in Table 3.

Fig.9: Dynamic Fault Tree Model for the Station Blackout Accident

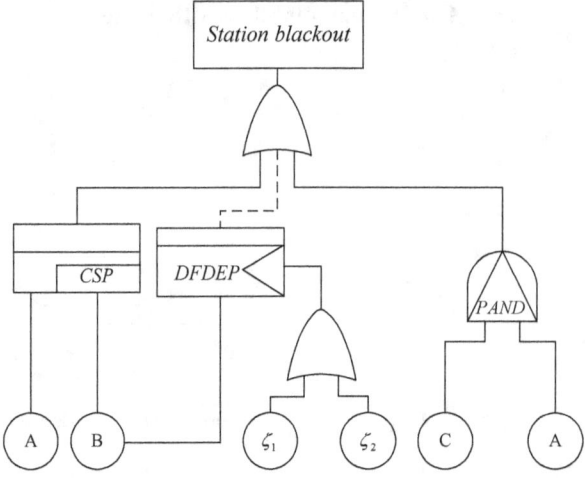

Table 3: component failure and maintenance information

component	description	Failure rate	Repair rate	Test period/time	Maint. period/time
A	Grid supply	2.34E-4	2.590	-	-
B	Standby supply	5.33E-4	8.695E-2	-	-
C	Sensor	1.00E-4	2.500E-1	-	-
ζ_1	Test activity	-	-	168/8.33E-2	-
ζ_2	Maint. activity	-	-	-	216/8

For a general dynamic repairable system, the DFT can be quantified by a Markov-based approach. However, in this case, it becomes unavailable since the test and maintenance activities are decided events. Hence an approximate solving strategy is adopted as: the unavailability of CSP gate (Q_{CSP}) is approximately obtained by the unavailability of Grid supply being multiplied the unavailability of standby supply, and the unavailability Q of the standby component is solved by the equation: $Q = [1-(1-e^{-\lambda T})/\lambda T] + [f_m T_m] + [\lambda T_r] + [\tau/T]$ suggested in IAEA P-4 [14], where λ is failure rate, T is test interval, f_m is frequency of preventive maintenance, T_m is duration of maintenance, T_r is repair time, τ is test duration; As to the unavailability of PAND gate (Q_{PAND}), it is can be solved by the conventional Markov approach. Then the approximate solution of the system unavailability (Q_{syst}) is calculated by the following equation: $Q_{syst} = Q_{CSP} + Q_{PAND} - Q_{CSP} \cdot Q_{PAND}$. For the mission time $10^4 h$, the unavailability of the system is 3.89e-7 using the approach mentioned above.

At last, the unavailability of the electrical power supply system is calculated by our proposed simulation approach. The system MCSS is {C→A, A→B, A→ζ_1, ζ_1→A, A→ζ_2, ζ_2→A}, then the system unavailability is finally simulated as 3.87e-7 with 10^7 simulation numbers. Obviously, the result obtained using our method is in good agreement with that calculated by the approximate solving strategy. In addition, the unavailability-time and outage-time distribution of the system are obtained as shown in Fig.10-11.

Table.10: Unavailability with Time

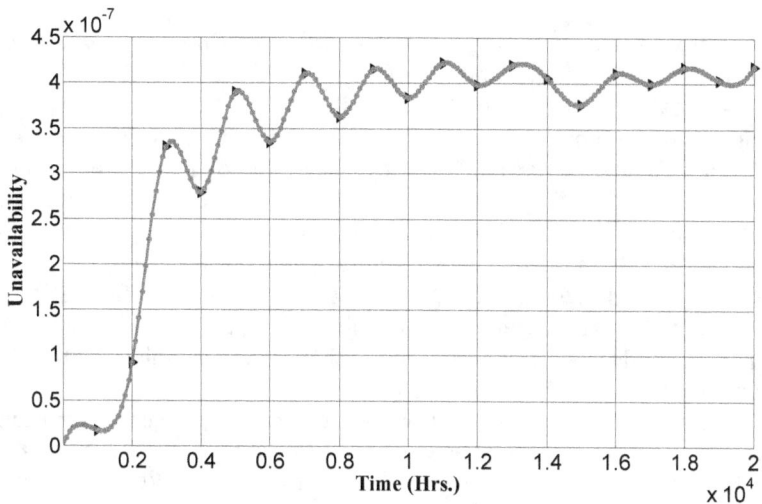

Table.11: Outage Time Distribution

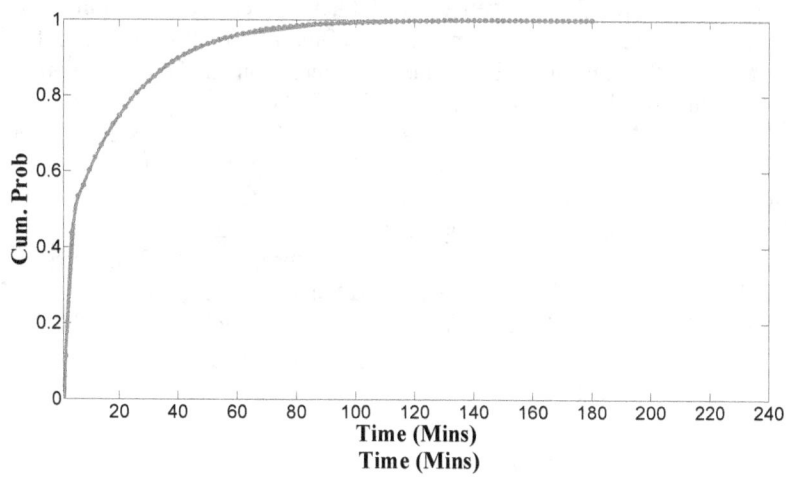

5. CONCLUSION

In this paper, we propose an efficient numerical simulation approach for evaluating the reliability of repairable system in NPP. By contrast to the existing approaches, such as Markov model, multi-integration model, our proposed approach has no limitation in the size of DFT, exponential components time-to-failure and time-to-repair distributions. Moreover the proposed approach is applicable to the system components with scheduled tests and maintenance activities. The results show this simulation method is correct. Although it is intensively computational for the top event with a small occurrence probability, it is a valuable approach to be studied with the rapid development of computer technology.

References

[1] Dugan JB, Bavuso SJ, and Boyd MA. "*Dynamic fault-tree models for fault –tolerant computer systems*". IEEE Transaction on Reliability, 41(3), pp. 363-377, (1992).
[2] Alam M, Al-Saggaf UM. "Quantitative reliability evaluation of repairable phased-mission systems using Markov approach", IEEE Transaction on Reliability, R-35(5), pp. 498-503, (1986).
[3] Dugan JB, Bavuso SJ, and Boyd MA. "Fault trees and Markov models for reliability analysis for fault tolerant systems", Reliability Engineering and System Safety, 39(3), pp. 291-307, (1993).
[4] Dugan JB, Sullivan KJ, Coppit D. Developing a low-cost high-quality software tool for dynamic fault-tree analysis. IEEE Transaction on Reliability, 49(1), pp. 49-59, (2000).
[5] Long W, Sao Y, and Horigome M. "Quantification of sequential failure logic for fault tree analysis", Reliability Engineering and System Safety, 67(3), pp. 269-274, (2000).
[6] Amari SV, Dill G, and Howald E. "A new approach to solve dynamic fault tree." In Proc. Annu. Reliab. Maintainability Symp., pp. 1-7, (2003).
[7] Liu D, Zhang C, Xing W, Li R and Li H. "Quantification of Cut Sequence Set for Fault Tree Analysis", HPCC2007, Lecture Notes in Computer Science, Springer-Verlag, pp. 755-765, (2007).
[8] Durga Rao K, Gopika V, Sanyasi Rao VVS, Kushwaha HS, Verma AK, and Srividya A. "Dynamic fault tree analysis using Monte Carlo simulation in probabilistic safety assessment" , Reliability Engineering and System Safety, 94(4), pp. 872-883, (2009).
[9] Zhang P, Chan KW. "Reliability Evaluation of Phasor Measurement Unit Using Monte Carlo Dynamic Fault Tree Method", IEEE Transaction on Smart Grid, 3(3), pp. 1235-1243, (2012).
[10] Tang Z, Dugan JB. "Minimal Cut Set/Sequence Generation for Dynamic Fault Tree", In Proc. Annu. Reliab. Maintainability Symp., pp. 207-213, (2004).
[11] Merle G, Roussel J.-M, Lesage J.-J. "Dynamic Fault Tree Analysis Based On the Structure Function", In Proc. Annu. Reliab. Maintainability Symp., pp. 1-6, (2011).
[12] Liu D, Xing W, Zhang C, Li R, and Li H, "Cut Sequence Set Generation for Fault Tree Analysis." in Proc. Lecture Notes in Computer Science, pp. 592-603 (2007).
[13] Merle G, Roussel J.-M, Lesage J.-J. "Algebraic determination of the structure function of Dynamic Fault Trees", Reliability Engineering and System Safety, 96(2), pp. 267-277, (2011).
[14] Procedure for conducting probabilistic safety assessment of nuclear power plants (level 1). Safety series no. 50-p-4. International Atomic Energy Agency, 1992.

Design for Reliability of Complex System with Limited Failure Data; Case Study of a Horizontal Drilling Equipment

Morteza Soleimani[a*], Mohammad Pourgol-Mohammad[b]
[a]Tabriz University, Tabriz, Iran
[b]Sahand University of Technology, Tabriz, Iran

Abstract: In this paper, a methodology is developed for reliability evaluation of electromechanical systems. The method is applicable in early design phase where there is only limited failure data available. When experimental failure data is scarce, generic failure data are searched from some related reliability data banks. In this method, Reliability Block Diagrams (RBD) is used for modeling the system reliability. Monte Carlo Simulation technique is employed to simulate the system for reliability and availability calculation. Current methodology contains the reliability importance analysis and reliability allocation to optimize the reliability. Evaluating reliability of complex systems in reverse engineering (competitive) design phase is one of the applications of this method. As a case study, a horizontal drilling equipment is used for assessment of the proposed method. According to the results, motor sub-system and hydraulic sub-system are the critical elements from reliability point of view. A comparison of the results is done with the results of reliability evaluation for a system with more failure and maintenance data available. Benchmark of the results indicates the effectiveness and performance quality of presented method for reliability evaluating of systems.

Keywords: Reliability, Monte Carlo Simulation, Design Phase, Reliability Allocation.

1. INTRODUCTION

Today's competitive world and increasing customer demand for highly reliable products, makes reliability engineering more challenging task. The reliability analysis has grown at a rapid rate as tracked in the literature on the subject. Design for reliability is an important research area, specifically in the early design phase of the product development. Mainly, the reliability analysis takes place at the end of design process, after determination of structure layout. The role of the analysis is to verify whether the reliability of the structure satisfies the demanded reliability. However, at the end of the design process, it is costly, or there is not enough time available to introduce major changes in the structure. Therefore, the results of the analysis have little influence on design. If reliability analysis applied during the conceptual design phase, its impact will be more remarkable on the design process producing high quality items. Results in more reliable and less expensive structure; a structure that is reliable in concept is less expensive than a structure that is not reliable in concept, Even with improvement in a later phase of the design process.
During the recent years, the requirement of modern technology, especially the complex systems used in the industry, leads to a growth in the amount of researches around the reliability base design. Avontuur [1-2], emphasises the importance of reliability analysis in the conceptual design phase. He demonstrates that it is possible to improve a design by applying reliability analysis in the conceptual design phase, reveals how to quantify failure and unavailability in cost, and compare them with investment cost to improve the reliability. Al-kheer [3] developed a reliability-based design approach by integrating the randomness of tillage forces into the design analysis of tillage machines, aiming at achieving reliable machines. The proposed approach was based on the uncertainty analysis of basic random variables and the failure probability of tillage machines. For this purpose, two reliability methods, namely the Monte Carlo simulation technique and the first-order reliability methods were used. Halloran [4] presented a case study for the early design reliability prediction method (EDRPM) to calculate function and component failure rate distributions during the design process such that components and design alternatives can be

selectively eliminated. The output of this method is a set of design alternatives that has a reliability value at or greater than a pre-set reliability goal.

In the most of the recent reliability base design researches, experimental data was used as the main source of the component reliability data; also a part of a system (for example electrical or mechanical part) was studied and hybrid electromechanical systems were not analysed. Furthermore, these studies are not comprehensive to include the various reliability calculations such as: reliability importance, reliability allocation, availability and uncertainty analysis in its results.

In this research, a reliability-based design methodology was developed for electromechanical systems. It overcomes the drawbacks of other reliability evaluation approaches which are not suitable for complex systems with availability of limited failure data for their components. Additionally, this methodology is integrated to include the several reliability estimation. Also two distributions are used to model the components reliability to contain different phases of life-cycle. Reliability evaluation of complex systems in reverse engineering (competitive) design phase, is one of the applications of the presented method.

This paper demonstrates a case study to support the proposed method as an early design reliability tool. In section 2, new method is summarized and illustrates its steps. Section 3 introduces the case study and demonstrates the results. The final section provides a conclusion and future work.

2. STRUCTURE OF THE METHODOLOGY

In this research a methodology is developed for reliability evaluation of electromechanical systems.
The new methods flowchart is shown in figure 1. This flowchart involves eight steps and applies for the case study in the following.

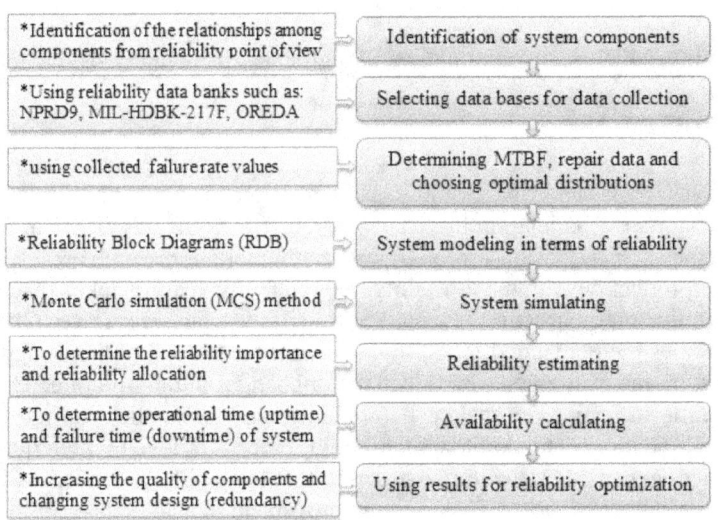

Figure 1: The new methods flowchart as an early design reliability tool

In the first step, subsystems and components of a system are identified and also the their functional relationships are determined. In the next step, the system components maintenance and failure data are collected from some data bases like MIL-HDBK-217F [5], OREDA [6] and NPRD-95 [7]. As mentioned above, this method is applicable in early design phase when there is only limited failure data. This method works well in the lack of suitable methods for reliability evaluation of complex system.. Also, expert judgement is used for specific components failure estimation for which there is no generic failure data. Since, there is a possibility for an error occurrence in data selecting for components failure, uncertainty is evaluated for the obtained results. In this method, Monte Carlo technique is utilized for system simulating. In modelling of the system, Weibull and exponential distributions are used because of their capability for modelling components reliability in different phases of life-cycle (specially Weibull distribution for wear-out phase). In subsequent steps of this method, the estimation is done for reliability and availability value. Also reliability importance and reliability allocation is calculated for reliability optimization. In this research and for the case study, the failure of the selected components (even a simple headlight) leads to system failure and system operation breakdown.

2.1. Reliability data type

There are some reliability data types; including experimental failure data, maintenance data and classified generic reliability data. For a specific system, experimental and maintenance data are collected from its maintenance information during its life cycle. Generic reliability data for a system are collected from similar systems information or classified generic reliability data that are collected from valid data bases. Classified generic reliability data is a suitable source when there is only limited maintenance or experimental failure data for a system. Usually these reliability data are applied in systems design phase.
In this paper, MIL-HDBK-217F, OREDA and NPRD-95 was used as the primary source of component reliability data.

2.2. Monte Carlo Simulation

Monte Carlo analysis is a powerful sampling tool for modeling the reliability of systems. Monte Carlo simulation method uses statistics to mathematically model a real-life process and then estimate the likelihood of possible outcomes. Before performing a Monte Carlo simulation, two category must be determined, one of them is the statistical distribution of the failure and repair processes and other is the logic function of the system, in other word, the relationship of the systems components and sub-systems should be determined [8]. Monte Carlo technique, take samples from cumulative distribution function (CDF) due to its suitability for taking weighted samples from more dense area. Figure 2 [9] illustrates the Monte Carlo simulation.

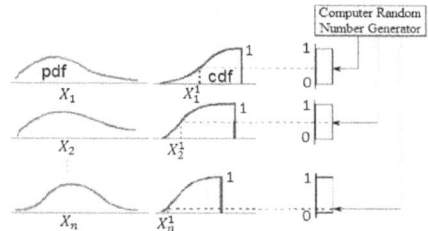

Figure 2: The Monte Carlo computer procedure

3. SIMULATION AND RESULTS FOR A CASE STADY

A Horizontal Drilling Equipment is considered in the reverse engineering stage, as a case study for evaluating the present method. Weibull and exponential distribution are selected in modelling of the system and all results are calculated for this two distributions. As mentioned earlier, RDB is method of choice for reliability modelling of the system. Figure 3 shows this modelling that created from ReliaSoft Blocksim8 software [10]. Each blocks in this model consists of several subsystems and components. After the system simulation, reliability parameters are calculated and analysed.

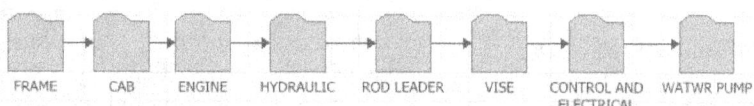

Figure 3: horizontal drilling equipment dividing for modelling

3.1. Reliability

Figure 4 shows the reliability estimation results of drilling equipment for two distribution. Results show that in the lower time the reliability value of system with Exponential distribution is less than system reliability value with Weibull distribution. This is for the value of the Shape parameter (β) that assumes 2 with expert assumption for modelling components reliability in wear-out phase.

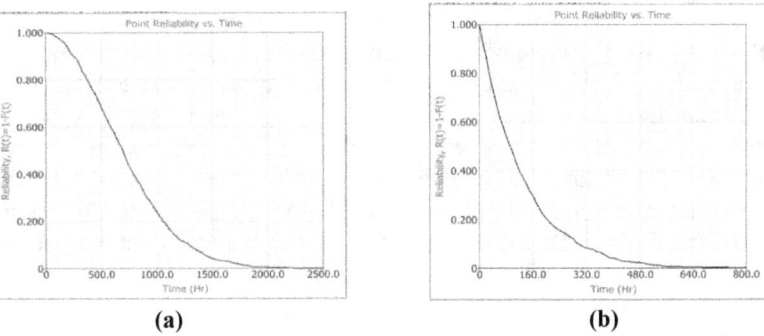

(a)   (b)

Figure 4: Reliability diagrams of drilling equipment with Weibull (a) and Exponential (b) distribution

Failure rate for Weibull distribution in lower time of life-cycle is very small and optimistic but not reasonable. Notwithstanding, values of the failure rate for Exponential distribution in the same life-cycle is reasonable. The results is reverse for Weibull and Exponential distribution in upper life of life-cycle. Therefore with choosing the Exponential distribution for lower time of life-cycle and Weibull distribution for upper time of life-cycle we can consider the system with Bathtub Curve behaviour. Figure 5 shows the failure rate estimation results for drilling equipment and its subsystems in Weibull and Exponential distributions.

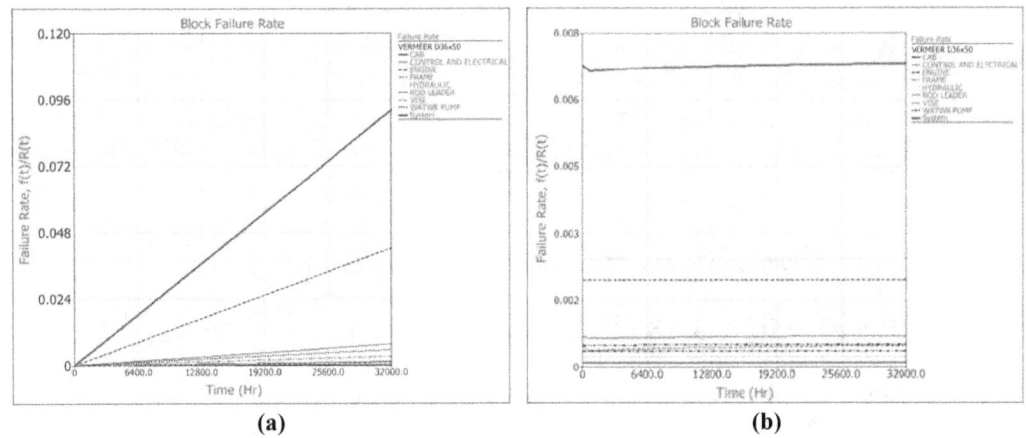

(a)   (b)

Figure 5: Failure Rate diagrams of drilling equipment and its subsystems with Weibull (a) and Exponential (b) distribution

3.2. Reliability Importance

Reliability evaluation of a system depends on its components contribution. In a system with a series structure that has several subsystems, the subsystem with the lowest reliability value, has the highest effect on the system reliability. Therefore, a small variation in the subsystem with the lowest reliability value leads to a great variation in system reliability value. Whereas, variation in other elements doesn't have as much influence as on the system reliability value. The calculation of reliability importance value for a component of a system with 'n' element is defined as the following mathematical formulae:

$$IR_i = \frac{\partial R_s}{\partial R_i} \qquad (1)$$

In this equation, R_s stands for the reliability of system and R_i is reliability of a component. In this case study, cab subsystem has the maximum reliability value for both Exponential and Weibull distribution. And the minimum is for hydraulic subsystem and motor subsystem for Exponential and Weibull distribution respectively. Therefore, occurrence of failure in hydraulic and motor subsystems is more susceptible. Furthermore, among all components of the system, motor starting has maximum failure rate and reliability importance. So with increase in the quality of component in this subsystems or design change (for example redundancy) reliability of system can be improved.

3.3. Reliability Allocation

Reliability allocation is an important step in system design. It allows determination of the reliability of constituent subsystems and components so as to obtain a targeted overall system reliability. Since 1950s [11], several studies have been devoted to this problem and decent number of researches were devoted on this subject. But no general method has been proposed to solve the reliability allocation problem satisfactorily. This situation is due to increasing complexity of current systems and necessity of considering multiple constraints such as cost, weight, and component obstruction among others. An overview is recently published of the methods developed during the past 3 decades for solving various reliability optimization problems [12-13].

Aeronautical Radio Incorporated (ARINC) [14] technique is one of the well-known reliability allocation type that performs based on weighting factors to subsystems of a series structure system. In this method, weighting factors for a subsystem is equal to division of the failure rate of the subsystem to the sum of all subsystems failure rates of system. In this research, ARINC technique is used to achieve the results of reliability allocation. Table 1 shows the results of reliability allocation for subsystems of drilling equipment with Weibull distribution. For this system, 0.95 is considered as a target reliability for 2000 hours (that is equal to 1.25 functioning years for drilling equipment). It should be noted that, these results are obtained for 95% of confidence level.

Table 1: Initial reliability and target reliability for subsystems of drilling equipment with Weibull distribution

Subsystem	Reliability importance (2000 hours)	Initial reliability (2000 hours)	Weighting factors	Target reliability (2000 hours)
Frame	0.004	0.830	0.032	0.998
Cab	0.003	0.996	0.001	0.999
Engine	0.045	0.071	0.461	0.976
Hydraulic	0.023	0.142	0.341	0.983
Rod Loader	0.005	0.626	0.082	0.996
Vise	0.003	0.995	0.007	0.999
Control & Electrical	0.004	0.923	0.014	0.999
Water Pump	0.005	0.704	0.061	0.997
The whole system	-	0.003	-	0.95

3.4. Availability

Availability is defined as the probability that a repairable system is operating satisfactorily at any random point in life-cycle time. In other words, the probability that a product or system is operational. The availability is formulated as follows that "u" is uptime and "d" is downtime of the system.

$$A = u/(u + d) \qquad (2)$$

In a repairable system, because of renewal process in components, the value of system reliability is not a good metrics for decision making about the system. Therefore, in repairable systems analysis, availability measure is used that is a combination of reliability and maintainability parameters [14].

Average availability value for drilling equipment in 32000 hours (that is equal to 20 functioning years for drilling equipment) is about 0.95. Figure 5 shows the average availability diagram for drilling equipment for two distribution. Results show that in a repairable system reliability value is a good parameter only for determining the reliability importance of components. However, the reliability analysis is a beneficial factor for repairable systems.

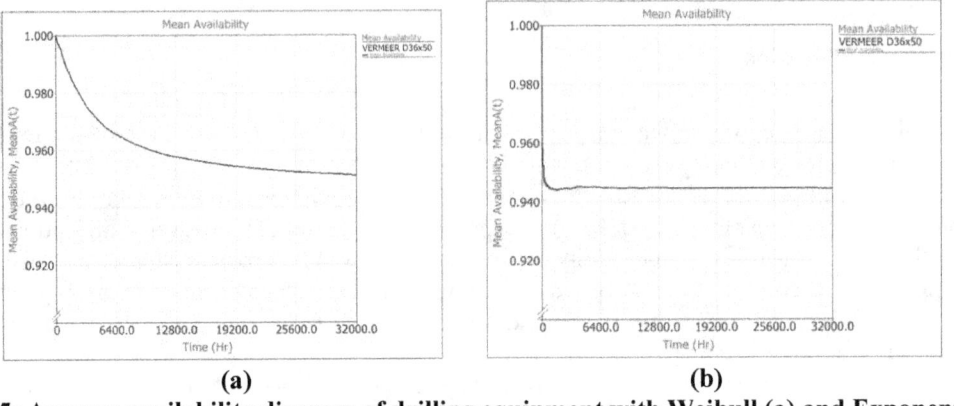

Figure 5: Average availability diagram of drilling equipment with Weibull (a) and Exponential (b) distribution

3.5. Uncertainty

Uncertainty is a measure of the "goodness" of an estimate. Without such a measure, it is impossible to judge how closely the estimated value relates or represents the reality. Uncertainty arises primarily due to lack of reliable information [14]. Engineering systems are often associated with vast amounts of data. This data can be used to estimate the performance of the system, consequences of system failure, and its risks [9,15].

In this research uncertainty analysis is applied for reliability and availability results. Figure 7 illustrates the average, upper bound and lower bound for availability and reliability of drilling equipment with Weibull distribution. This results is obtained for 95% two-sided confidence bound.

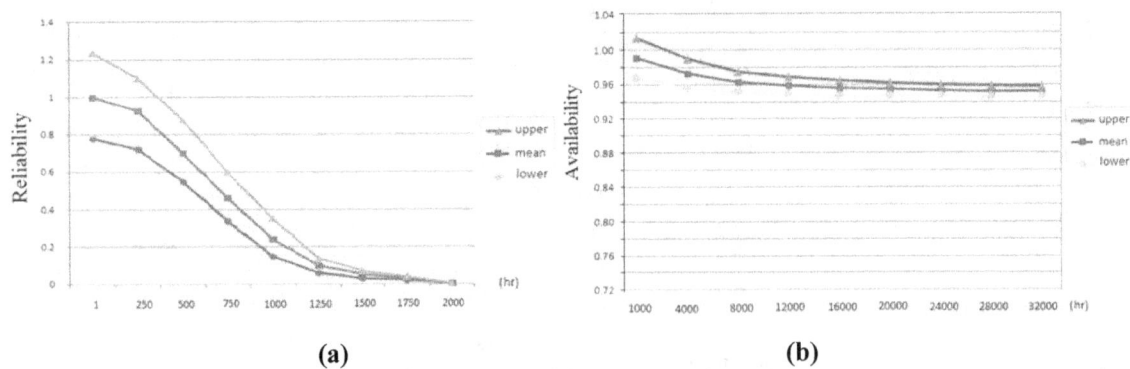

Figure 7: Uncertainty result for reliability (a) and average availability (b) of drilling equipment for 95% two-sided confidence bound with Weibull distribution

3.6. Benchmark Test

For validation of presented methodology, a benchmark research has been done with the results of similar projects. The results of a project that is about the reliability evaluation of mining equipment (Dump Tracks) in a copper mine [16], is selected for comparison with the results of current methodology. The reason behind this selection is the existence of similar subsystems and work conditions of Dump Tracks and Drilling Equipment. This project has an advantage for comparison and that is experimental data for Dump Tracks components failures. Table 2 shows the Drilling Equipment and Dump Tracks reliability value in different life-cycle time. Values in table 2 indicates the approximate equal results for both systems.

Table 2: Comparison of Drilling Equipment and Dump Tracks Reliability Value

Time (hours)	Reliability of Drilling Equipment	Reliability of Dump Tracks
0	1	1
50	0.7	0.55
500	0.029	0.0023
1000	0.001	5E-6
3000	5.9E-10	1.75E-17

4. CONCLUSION

In this paper, a reliability-based design methodology for electromechanical systems was developed to overcome the drawbacks of other reliability evaluation approaches which are not suitable for complex systems having limited failure data for their components. This method is applicable in early design phase when there is only limited failure data. So reliability evaluation of complex systems in reverse engineering design phase, is one of the applications of the presented method. The main steps of this approach were presented and applied for a drilling equipment as a case study. Reliability and availability parameters are calculated for this case study. All calculations are computed for Weibull and Exponential distribution. Comparing the various results show that the Exponential distribution is suitable for lower phase of life-cycle and Weibull distribution for upper phase. Reliability importance analysis illustrates that hydraulic and motor subsystems are the critical elements in terms of reliability. In addition, among all components of the system, motor starter has maximum failure rate and reliability importance. So with increase in the quality of components in the subsystems or changing of design (for example redundancy), reliability of system is improved. At the end, a benchmark of the result of this research with similar projects indicates the effectiveness and performance of presented method for reliability evaluation of systems existing in design phase with limited failure data for its component.

References

[1] Avontuur G. C, Van der Werff K. "An implementation of reliability analysis in conceptual design phase of drive trains", Reliability Engineering and System Safety journal, April ,3,155-165, (2001).
[2] Avontuur GC. Reliability Analysis in Mechanical Engineering Design. Delft, The Netherland: Delft University Press, (2000).
[3] A. Abo Al-kheer , A. El-Hami, M.G. Kharmanda, A.M. Mouazen, "Reliability-based design for soil tillage machines", Journal of Terramechanics, July, 48, 57-64, (2011).
[4] Bryan M. O'Halloran, Chris Hoyle, Robert B. Stone, Irem Y. Tumer. "The early design reliability prediction method", ASME, November, 9-15, (2012).
[5] Handbook MIL-HDBK 217F. 1991 "Reliability prediction of electronic equipment" Revision F.

[6] Participants O. OREDA Offshore Reliability Data Handbook. 4th edition, DNV, PO Box,(2002).
[7] G. C. William Denson, William Crowell, Amy Clark, Paul Jaworski. , 1994 "Nonelectric Parts Reliability Data" vol. 2, D. o. Defense, Ed. 1995, Rome.
[8] Dirk P. Kroese and etal., 2011. Handbook of Monte Carlo Methods. John Wiley & Sons, INC.
[9] Modarres M., 2006. Risk analysis in engineering: techniques, tools, and trends, CRC.
[10] Blocksim 8 User's Guide; ReliaSoft Corporation; www.reliasoft.com , 2012.
[11] Modarres M, Kaminskiy M, Krivstov V. , 2010 " Reliability Engineering & Risk Analysis" a practical guide: CRC: second edition.
[12] W. Kuo and V. R. Prasad, "An annotated overview of systems reliability optimization, "IEEE Trans. Rel., vol. R-49, no. 2, pp. 176–187, (2000).
[13] W. Kuo, V. R. Prasad, F. A. Tillman, and C. L. Hwang, Optimal Reliability Design: Cambridge University Press, (2001).
[14] Dodson B, Nolan D. 1999 ,"Reliability Engineering Handbook" CRC.
[15] Sanchez A, Carlos S, Martorell S, Villanueva JF, "Addressing imperfect maintenance modeling uncertainty in unavailability and cost based optimization" *Reliability Engineering & System Safety* 94, (2009).
[16] Amin Moniri Morad, Mohammad Pourgol-Mohammad, Javad Sattarvand, "Reliability-Centered Maintenance for Off-Highway Truck: Case Study of Sungun Copper Mine Operation Equipment" ASME, IMECE2013-66355, (2013).

Analyzing Simulation-Based PRA Data Through Clustering: a BWR Station Blackout Case Study

Dan Maljovec[a]*, Shusen Liu[a], Bei Wang[a], Valerio Pascucci[a], Peer-Timo Bremer[b], Diego Mandelli[c], and Curtis Smith[c]

[a]SCI Institute, University of Utah, Salt Lake City, USA
[b]Lawrence Livermore National Laboratory, Livermore, USA
[c]Idaho National Laboratory, Idaho Falls, USA

Abstract: Dynamic probabilistic risk assessment (DPRA) methodologies couple system simulator codes (e.g., RELAP, MELCOR) with simulation controller codes (e.g., RAVEN, ADAPT). Whereas system simulator codes accurately model system dynamics deterministically, simulation controller codes introduce both deterministic (e.g., system control logic, operating procedures) and stochastic (e.g., component failures, parameter uncertainties) elements into the simulation. Typically, a DPRA is performed by 1) sampling values of a set of parameters from the uncertainty space of interest (using the simulation controller codes), and 2) simulating the system behavior for that specific set of parameter values (using the system simulator codes). For complex systems, one of the major challenges in using DPRA methodologies is to analyze the large amount of information (i.e., large number of scenarios) generated, where clustering techniques are typically employed to allow users to better organize and interpret the data. In this paper, we focus on the analysis of a nuclear simulation dataset that is part of the risk-informed safety margin characterization (RISMC) boiling water reactor (BWR) station blackout (SBO) case study. We apply a software tool that provides the domain experts with an interactive analysis and visualization environment for understanding the structures of such high-dimensional nuclear simulation datasets. Our tool encodes traditional and topology-based clustering techniques, where the latter partitions the data points into clusters based on their uniform gradient flow behavior. We demonstrate through our case study that both types of clustering techniques complement each other in bringing enhanced structural understanding of the data.

Keywords: PRA, computational topology, clustering, high-dimensional analysis

1. INTRODUCTION

A recent trend in the nuclear engineering field is the implementation of computationally-intensive codes for the design and safety analysis of nuclear power plants. In particular, the new generation of system analysis codes aims to embrace phenomena such as thermo-hydraulic, structural behavior, system dynamics, human behavior, as well as uncertainty quantification and sensitivity analysis associated with these phenomena. The use of dynamic probabilistic risk assessment (DPRA) methodologies allows a systematic approach to uncertainty quantification.

DPRA methodologies account for possible coupling between triggered or stochastic events through explicit consideration of the time element in system evolution, often through the use of dynamic system simulators. Such methodologies are commonly needed when the system has multiple failure modes, control loops, processes, software/hardware components, or human interactions. A DPRA is typically performed by 1) sampling values of a set of parameters from the uncertainty space of interest (using the simulation controller codes), and 2) simulating the system behavior for that specific set of parameter values (using the system simulator codes).

---

* Corresponding author, maljovec@cs.utah.edu

Due to the intrinsically high level of detail within such a process, one would need to handle large amounts of data generated within the simulation [9]. In [7] we have presented a framework that visualizes high-dimensional scalar functions through a topological segmentation of its input surfaces. The input of such a high-dimensional scalar function arises from the set of n uncertain parameters $x_1, x_2, ..., x_n$, whereas the output originates from some safety-related outcomes, such as maximum core temperature of each simulation. Our topological tools aim to reconstruct the topological structure of such a function, i.e., the response surface, in the high-dimensional space. We further explore the topology-based clusterings that lie beneath such a framework for DPRA datasets [6].

In this paper, we focus on the analysis of a particular nuclear simulation dataset based upon our previously developed analysis and visualization framework [6, 7]. The dataset is part of the risk-informed safety margin characterization (RISMC) boiling water reactor (BWR) station blackout (SBO) case study [8]. We enrich our tool by combining traditional and topology-based (hierarchical) clustering, as well as dimensionality reduction (DR) techniques. We demonstrate through our case study that both types of clustering techniques complement each other in bringing enhanced structural understanding of the data. In particular, the topology-based clustering helps highlight key features of the data that are otherwise hidden using the traditional techniques.

BWR system. The system considered in our case study is a generic BWR power plant with Mark I containment. The three main structures are: 1) the reactor pressure vessel (RPV), a pressurized vessel that contains the reactor core; 2) the primary containment including the dry well (DW) that houses the RPV and circulation pumps; and 3) the pressure suppression pool (PSP), also known as the wet well. The PSP is a large torus-shaped container that contains a large amount of water (almost 1 M gallons of fresh water) and is used in specific situations as an ultimate heat sink. The original BWR Mark I includes a large number of systems, but for the scope of this report and for the case study considered, we use a smaller subset of systems that includes the RPV level control systems, the RPV pressure control systems, the cooling water inventory, and the AC power system, which consists of two power grids, emergency diesel generators (DGs), and battery systems for the instrumentation and control systems.

The RPV level control systems provide manual and automatic control of the water level within the RPV and consist of two components, the reactor core isolation cooling (RCIC) and the high pressure core injection (HPCI). The RCIC provides high-pressure injection of water from the CST to the RPV. Water flow is provided by a turbine-driven pump that takes steam from the main steam line and discharges it to the suppression pool. The HPCI is similar, but allows much greater water flow rates. The RPV pressure control systems provide manual and automatic control of the RPV internal pressure and consist of a set of safety relief valves (SRV), safety valves, and the automatic depressurization system (ADS). The SRVs are DC-powered valves that control and limit the RPV pressure, and the ADS is a separate set of relief valves that are employed in order to depressurize the RPV. The cooling water inventory includes the condensate storage tank (CST), the PSP, and the fire water system. The CST in the considered plant is a 375 Kgal fresh water reservoir that can be used to cool the reactor. The PSP contains a large amount of fresh water that is relied upon as an ultimate heat sink when AC power is lost. Water from the fire water system can be injected into the RPV when other water injection systems are disabled and when the RPV is depressurized.

SBO scenario. The analysis considered is a **BWR Mk. I** system during a loss of offsite power (LOOP) event followed by loss of the diesel generators (DGs), i.e., station blackout (SBO). In more detail, at time $t = 0$, LOOP condition occurs due to an external event (i.e., the offsite power lines are damaged). The LOOP alarm triggers the following events: a successful scram of the reactor is performed by the operators; MSIVs are successfully closed, isolating the primary containment from the turbine building; emergency DGs successfully start keeping the AC power busses energized; DC systems (i.e., batteries) are functional; and the decay heat generated by the core is removed from the pressure vessel through the residual heat removal system.

Next, the SBO condition occurs due to internal failure, which results in the failure of the DGs. As a result of the loss of external power, removal of decay heat is impeded. Reactor operators start the SBO emergency procedures and perform the following: RPV level control using RCIC or HPCI, RPV pressure control using SRVs, and containment monitoring (both dry well and PSP). At this point, plant staff start recovery operations to bring back on-line the DGs while the recovery of the off-site power grid is underway as well. Due to heavy usage, battery power can be depleted. When this happens, all remaining control systems are off-line, causing the reactor core to heat until the maximum temperature limit for the clad is reached: a core damage (CD) condition occurs. If DC power is still available and one of three conditions is met (i.e., failure of both RCIC and HPCI; HCTL limits have been reached; RPV water level becomes too low), then the reactor operators activate the ADS in order to depressurize the RPV and allow fire water injection, if available. When AC power is recovered, through successful restart/repair of DGs or off-site power, RHR can be employed to keep the reactor core cool.

2. BACKGROUND

Dimensionality reduction (DR) and traditional hierarchical clustering are widely used techniques for analyzing structures of high-dimensional data. To extend the existing framework we have developed in [6, 7], we employ a visualization system that utilizes both dimensionality reduction and clustering, where dimensionality reduction constructs a mapping for the clustering results for intuitive visual analysis. We begin with a brief description of DR and traditional hierarchical clustering and then focus on the topology-based clustering, which may be unfamiliar to nonspecialists.

Dimensionality reduction. DR techniques [1], such as principal component analysis (PCA), multi-dimensional scaling (MDS), and Isomap, are common tools for analyzing high-dimensional data by constructing its low-dimensional representation. Since direct visualization of high-dimensional data is extremely challenging, we would like to obtain some intuition regarding the structure of the data through its low-dimensional embedding. Such embeddings are typically constructed in 2D or 3D spaces for visualization purposes. We have integrated a number of DR techniques into our system. For the purpose of our study, we use primarily PCA, a linear DR technique, in the analysis due to its simplicity and computational efficiency. However, using DR alone as a black box solution in the analysis suffers a major limitation. That is, the results of DR could be hard to interpret as a certain amount of structural information has been lost during the DR process. Therefore, we try to impose structural context onto the embeddings by combining DR results with a clustering obtained from the original high-dimensional data.

Traditional hierarchical clustering. Clustering groups the data in such a way that points are more similar to those in the same cluster than to those outside the cluster. There are numerous criteria for defining what constitutes a cluster, which are based on density, distribution, distance, connectivity, etc. In our current analysis, we choose average-linkage hierarchical clustering [2] (among others available in the system). Such a clustering technique is based on point-wise connectivities where points are considered more related to nearby points than points that are farther away. Starting from individual points as their own clusters, this technique builds a dendrogram from the bottom up, merging clusters with nearby clusters. In our system, we do not need to specify the number of clusters we are looking for; instead we interactively expand or collapse different levels of clustering in the hierarchy during the analysis.

Approximated Morse-Smale complex and topology-based hierarchical clustering. We consider an alternative method for clustering high-dimensional data based on the concept of the Morse-Smale complex. We give a brief overview of these concepts. See [6, 7] for details. The Morse-Smale complex is a type of topological structure that serves as a structural summary of a given high-dimensional scalar function. We consider a scalar function $f : \mathbb{X} \to \mathbb{R}$ defined over a finite set of points $\mathbb{X}$ in $\mathbb{R}^n$. The approximated Morse-Smale complex, at its finest level, partitions the points in $\mathbb{X}$ based on their uniform gradient

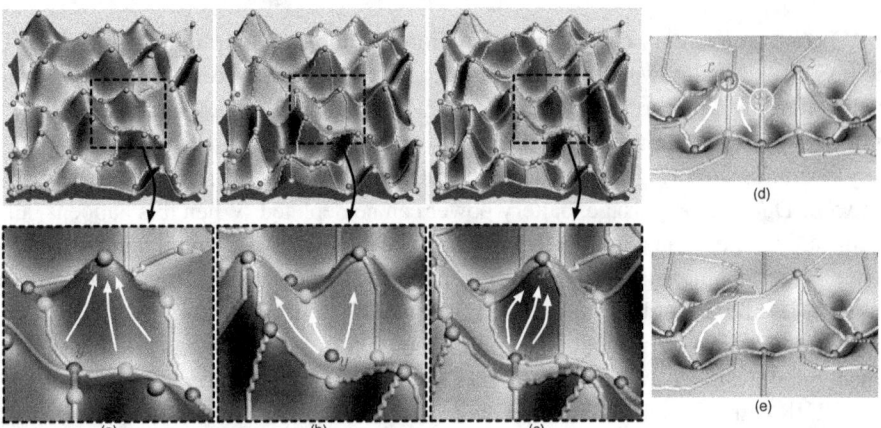

Figure 1: For a height function defined on a 2D domain (where maxima, minima, and saddles are colored red, blue, and green, respectively): (a) For each point in the brown region, the gradient flow (white arrow) ends at the same maxima x; (b) For each point in the green region, the gradient flow starts at the same minimum y; (c) For each point in the blue (Morse-Smale) cluster, the gradient flow begins and ends at the same maximum-minimum (i.e., (x,y)) pair. To illustrate merging of clusters based on persistence simplification, in (d), the left peak at the maximum x is considered less important topologically than its nearby peak at maximum z, since x is lower. Therefore, at a certain scale, we would like to represent this feature as a single peak instead of two separate peaks, as shown in (e), by redirecting gradient flow (white arrow) that originally terminates at x to terminate at z. In this way, we simplify the function by removing (canceling) the local maximum x with its nearby saddle y. On the cluster level, the clusters (i.e., decompositions of the domain separated by edges connecting the saddles and extrema) surrounding the left peak x are merged into clusters surrounding the right peak z. Figure reproduced from [6].

behavior. First, points in $\mathbb{X}$ are connected with a neighborhood graph (e.g., k-nearest-neighbor (KNN) graph). Second, the steepest ascending edge adjacent to a given point is used to estimate the gradient flow at the point. All points with no neighbors of higher/lower values are considered local maxima/minima. Finally, points are clustered based on the unique minimum-maximum pair from which their gradient flows start and end. A topology-based clustering at the finest level for a height function defined on a 2D domain is illustrated in Figure 1(a)-(b). We can then merge clusters based on persistence simplification [3], where less (topologically) significant clusters are merged into more significant ones. We avoid the technical details here but simply illustrate such a process in Figure 1(d)-(e).

Topological skeleton obtained through DR. Given a topology-based clustering at a fixed scale, we further our analysis by computing a collection of summary curves that serves as the topological skeleton of the data in the visual space. We follow a three-step process, as detailed in [4]: 1) perform inverse regression with data in each cluster and obtain a 1D curve embedded in $\mathbb{R}^n$; 2) project the curves in $\mathbb{R}^n$ to a curve in the visual space using PCA [5], and 3) align the curves in the visual space to meet at their shared extrema to maintain the coherency of the extracted structure. The resulting topological skeleton serves as a structural summary of the data, and it is visualized to encode information, as illustrated in Figure 2. Finally, the topological skeleton can also be visualized based on the cluster labels. In addition, we distinguish the clusters based on configurations of their input dimensions through a collection of inverse-coordinate plots. Suppose we use a point sampling of the same 2D height function in Figure 2. The above process is illustrated in Figure 3. For more details of the visualization pipeline as well as additional views, see [4, 6, 7].

3. CASE STUDY DATASET

An ensemble of 4997 transient simulations has been generated using classical Monte-Carlo sampling of seven input parameters, among which 833 scenarios resulted in system failure (the core temperature

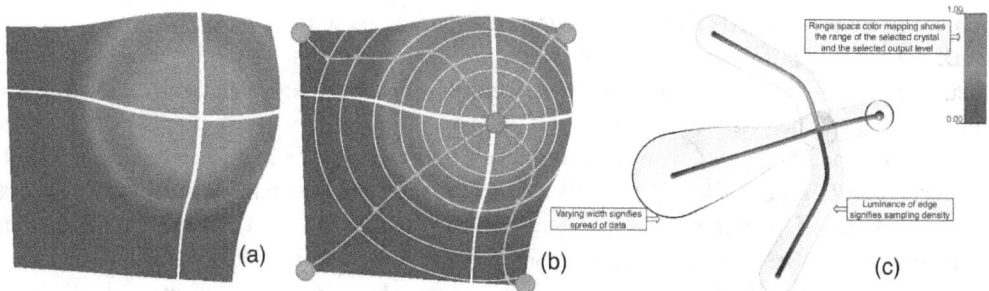

Figure 2: An illustrative example of our visualization of a topological skeleton extracted from a 2D height function: (a) the surface is first segmented into clusters of uniform gradient flow; (b) then each level set (white line) is averaged to a single point and consecutive level sets are connected to form a curve per cluster (orange curves); and (c) finally the resulting topological skeleton is visualized. Each summary curve in the visual space corresponds to a cluster of the original high-dimensional data. In the visualization, the color of each curve signifies the average value of each level set, and a transparent region encloses a given curve, where its width represents a direction-independent estimate of the spread of data and the luminance of its boundary edges signifies the sampling density.

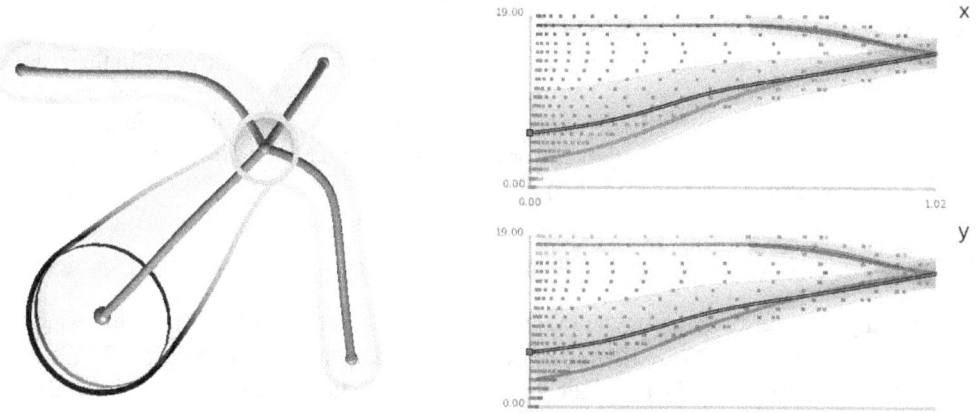

Figure 3: Left: topological skeleton colored by cluster labels. Right: inverse coordinate plots. Data points are visualized by their cluster labels, and summary curves are projected. For the inverse coordinate plots, the horizontal axis represents the output dimension (e.g., height values), and each vertical axis represents an input dimension (e.g., x or y coordinates of the domain). The projected summary curve in each inverse-coordinate plot gives the average value (of the input dimension of interest) at each level set and uses a dimension-specific standard deviation for the width of the transparent region.

reached the clad failure temperature threshold of 2200 F), whereas the rest of the 4164 scenarios ended up in system success (AC power is recovered or the firewater becomes available if the RPV is depressurized). Each simulation includes information regarding the timing of various recovery attempts (e.g., cooling recovery, fire water, etc.) and component failures (e.g., battery life exhausted, a safety relief valve gets stuck open, etc.). The seven input parameters are listed below, as they are the only uncertain parameters under consideration.

- **FailureTimeDG**: Failure time of the DGs corresponding to the time of the SBO event.
- **ACPowerRecoveryTime**: min{Recovery time of DGs, off-site power recovery time}. The minimum of these two times will determine when the simulation is considered recovered.
- **SRVStuckOpenTime**: The time when an SRV is stuck in the open position.
- **CoolingFailtoRunTime**: max{HPCI failure time, RCIC failure time}. As long as one of these systems is functioning, the reactor is being actively cooled, so it is important to understand when

both systems have failed. Thus, we take the maximum of these two times.
- **ADSactivationTimeDelay**: The time when the operator manually depressurizes the RPV by activating the ADS system. This parameter measures the time delay from the HCTL event, not the time from 0 to when ADS is activated.
- **firewaterTime**: As an emergency action, when RPV pressure is below 150 psi, plant staff can connect the fire water system to the RPV to cool the core and maintain an adequate water level.
- **ExtendedECCSOperation**: Battery life combined with extended ECCS operation. That is, operators may extend RCIC/HPCI and SRV control even after the batteries have been depleted. They manually control RCIC/HPCI by acting on the steam inlet valve of the turbine and/or supply DC power to the SRVs through spare batteries.

All of the above time-related parameters are measured from the time of the SBO event (in seconds), which is the FailureTimeDG, with the exception of FailureTimeDG, which is measured from the LOOP event, and the ADSactivationTimeDelay, which is measured from the time of the HCTL event. The output variables obtained from the simulations are: 1) **maxCladTemp**, which is the maximum clad temperature reached during the entire course of the simulation; and 2) **simulationEndTime**, which for failure cases represents the time to reach the failure temperature of 2200 F. We study the topology of scalar functions with each of these outputs as the scalar value in isolation. The above data is pre-processed with a Z-score standardization, whereby values V of each dimension are recomputed as $V - \text{mean}(V)/\text{std}(V)$; therefore all input parameters have the same mean (0) and standard deviation (1) but may vary in their ranges.

In this study, the domain scientists are interested in what combination of conditions (in the form of input simulation parameters) can cause potential reactor failure (i.e., nuclear meltdown witnessed by maximum core temperature exceeding a threshold value).

4. RESULTS

We provide analysis under both traditional (Section 4.1) and topological hierarchical clustering (Section 4.2) using the 7D input data. For each subsection, we consider two separate cases. In the first case, referred to as the **All Scenarios Case**, we analyze all 4997 simulations, using maximum clad temperature as the observed output variable. Note that in this case, all failure cases have the same output variable of 2200 F. In the second case, referred to as the **Failure Scenarios Case**, we focus on clustering of the 833 failure scenarios. Since the maximum clad temperature does not vary for these cases, we treat the end simulation time as the output variable. We give a comprehensive picture by providing comparisons among the two clustering techniques and discuss the benefits and limitations inherent in each approach.

4.1. Traditional Clustering

For traditional clustering, we map the data into eight dimensional space by considering the seven input variables and the output variable, maximum clad temperature. We start our analysis by applying PCA to reduce the eight dimensional data to its two dimensional embedding for direct visual analysis.

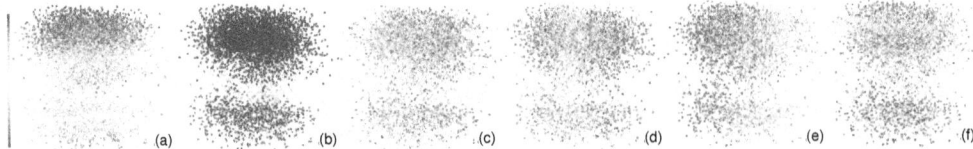

Figure 4: PCA embedding for the 8D dataset under the **All Scenarios Case**. The dimensions shown exhibit relatively strong correlation patterns within the embedding. We use a spectral colormap where red/blue represents low/high value. (a) ACPowerRecoveryTime; (b) maxCladTemp; (c) CoolingFailToRunTime; (d) firewaterTime; (e) SRVStuckOpenTime; (f) ExtendedECCSOperation.

All Scenarios Case. To study the distribution/variation of each dimension with respect to the embedding, we first color the points according to each dimension, as illustrated in Figure 4. All the dimensions shown exhibit a certain amount of visual correlation within the embedding, except for the omitted dimensions, ADSActivationTimeDelay and FailureTimeDG. It is important to note that a vertical or horizontal pattern of variation corresponds to the variance of the dimension. That is, a larger variance corresponds to a more noticeable pattern, which is likely due to the fact PCA is inherently optimized for capturing dominant directions of maximum variance. In Figure 4(b), there appear to be only a few data points with a moderate maxCladTemp as the top portion of the embedding is dominated by success scenarios characterized by low MaxCladTemp values (in red), and the bottom portion of the data consists of mostly failure scenarios characterized by high (constant) MaxCladTemp (in blue). It is therefore obvious that maxCladTemp separates the success from failure scenarios in the embedding. This claim can be further validated by coloring the points with known labels of success/failure. In Figure 4(a), ACPowerRecoveryTime varies smoothly within both the success and failure scenarios, but it does not serve as a differentiating factor between the successes and failures. Furthermore, in Figure 4(f), relatively high ExtendedECCSOperation time can be observed among all the success scenarios, so we suspect that a long extended ECCS operation time is a main contributing factor for stable system recovery. However, ExtendedECCSOperation is likely not a sufficient condition to separate successes from failures as there are a few points with high ExtendedECCSOperation values within the lower half of the embedding (i.e., failures scenarios). In Figure 4(c)-(e), the remaining three dimensions vary orthogonally with respect to maxCladTemp. This observation implies that these dimensions have less impact on the outcomes of the simulation, which are characterized by variations in maxCladTemp.

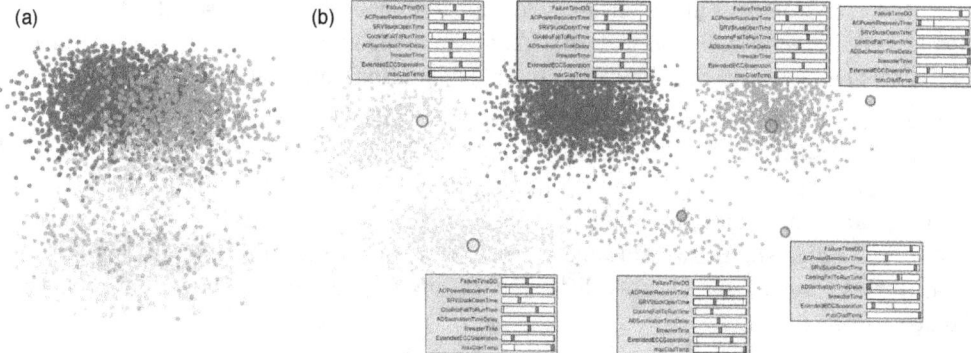

Figure 5: (a) 2D embedding of the data colored by cluster labels. (b) In order to provide a more clear view for the clusters, we provide a separate illustration of each individual cluster and its summary statistics.

In addition, combined with traditional hierarchical clustering, our analysis framework enables us to color the points in the embedding based on cluster labels. Furthermore, the tool also visualizes the statistical summary of each dimension for points within each cluster (enclosed in a box next to the clustered points). In the statistical summary of a given cluster, each row represents a dimension of the data, where the yellow bar corresponds to its min-max range and the red marker indicates its mean value across all points in the cluster. With these summaries across all clusters, we can quickly compare and investigate the defining characteristics of each cluster at a glance.

During the interactive exploration of the embedding, we apply cluster expansions recursively to study the data from coarse to fine resolutions. At the coarsest level, the data is split into two clusters, where the upper cluster contains exclusively success scenarios, and the lower cluster contains all failure scenarios and a few number of successes (via validations by known labels of success/failure). We subdivide these clusters by applying a few steps of cluster expansion. We then arrive at a level in the clustering hierarchy that consists of seven clusters, as shown in Figure 5. Four of the top clusters decompose all of the success scenarios (top half of the embedding). The extremely small purple cluster likely consists of outliers in the data, since its points share extremely low ACPowerRecoverTime and maxCladTemp. These points correspond to the success scenarios where AC power is recovered very quickly and clad temperature never increases drastically. Although the blue and cyan clusters share similar statistical summaries across most dimensions, ACPowerRecoveryTime seems to be the most likely factor that differentiate these two clusters. The differentiating factor between the orange cluster and the other three clusters is its late SRVStuckOpenTime. On the other hand, three of the bottom clusters partition primarily the failure cases. The dark green cluster again contains the outliers and its points share extremely late SRVStuckOpenTime and firewaterTime. These correspond to the failure scenarios where all SRVs operate correctly for a long time and the firewater is injected very late, not in time to avoid the core damage from overheating. The light green and pink clusters differ mostly in ExtendedECCSOperation and CoolingFaillToRunTime. The green cluster is concentrated with data points exhibiting lower ExtendedECCSOperation and higher CoolingFailToRunTime compared to the pink cluster. Therefore, differentiating clusters based on variations across different dimensions allows the user to organize and interpret the trends in scenario evolution and risk contributors for each scenario.

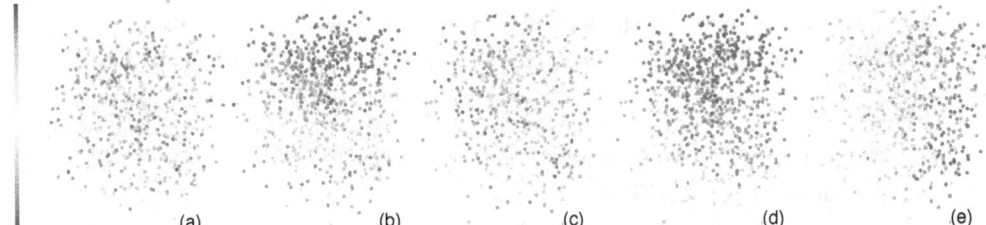

Figure 6: PCA embedding for the 8D dataset under the **Failure Scenarios Case**. The dimensions shown exhibit relatively strong correlation patterns within the embedding. (a) CoolingFailToRunTime; (b) ExtendedECCSOperation; (c) FirewaterTime; (d) simulationEndTime; (e) SRVStuckOpenTime.

Failure Scenarios Case. Once again, we color the points in the embedding for all failure scenarios, as illustrated in Figure 6. There are clear variations among points in the embedding under ExtendedECCSOperation, firewaterTime, and SRVStuckOpenTime. FirewaterTime and SRVstuckOpenTime vary along the horizontal direction, whereas ExtendedECCSOperation varies vertically. We also notice that very few points exist with a high simulationEndTime among all the failure scenarios. Comparing this case with the **All Scenarios Case**, it is much more difficult to obtain insights from the original data based on this visualization alone.

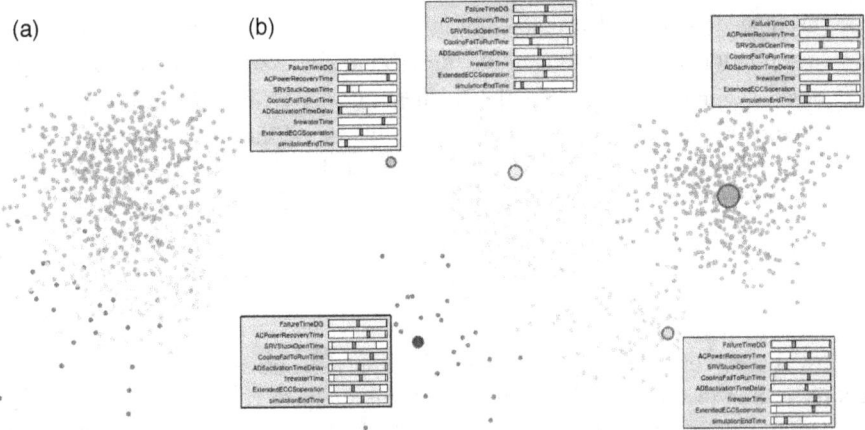

Figure 7: (a) 2D embedding of the data colored by cluster labels. (b) A separate illustration of individual clusters and their summary statistics.

Using clustering expansion, we arrive at a level of the hierarchy where five clusters are presented in the data (Figure 7). The purple cluster contains outliers that share late ACPowerRecoveryTime and

CoolingFailToRunTime. In this focused analysis of all the failure scenarios (without the interference from the dominating dimension MaxCladTemp), we obtain various insights regarding separations of clusters. For example, both green and red clusters consist of points with low simulationEndTime. The differentiating factors here are the CoolingFaillToRunTime and ExtendedECCSOperation.

4.2. Topology-Based Clustering

For topology-based clustering, we map the data into a seven dimensional scalar function, where its input includes the seven input parameters of the simulation, and its output corresponds to maxCladTemp for the **All Scenarios Case**, and EndSimulationTime for the **Failure Scenarios Case**.

All Scenarios Case. After careful analysis of the clustering hierarchy, we focus on a level consisting of four clusters. Figures 8 and 9 summarize our results. In Figure 8, three of the clusters share a common global maximum, whereas the fourth cluster (cyan) consists of points exhibiting low MaxCladTemp values, which correspond to success scenarios. Here we study the conditions that lead to distinct local minima, that is, the different parameter settings that yield stable success scenarios, by focusing on the behavior of the projected summary curves in the inverse coordinate plots of Figure 8(right).

Recall the vertical axis of each inverse coordinate plot is labeled by one input parameter, and the horizontal axis corresponds to maxCladTemp. Since we study conditions that lead to minimal values of maxCladTemp, we focus on the left side of the horizontal axis of each plot, which corresponds to low values of maxCladTemp. The local minimum that belongs to the pink cluster exhibits an early ACPowerRecoveryTime, a late firewaterTime, and an early ExtendedECCSOperation time. The local minimum of the blue cluster, on the other hand, has a late ACPowerRecoveryTime, a very early firewaterTime, an early ADSActivationTimeDelay, and a late ExtendedECCSOperation time. The third local minimum, shared by the green and cyan clusters, has a moderate firewaterTime paired with an early ACPowerRecoveryTime and a late ExtendedECCSOperation time. The input parameters that seem to be irrelevant in differentiating these clusters are the FailureTimeDG, the CoolingFailToRunTime, and the SRVStuckOpenTime. This last observation seems well aligned with the observations we have made in Figure 4, where we see that there is no visual correlation between the maxCladTemp and the FailureTimeDG (therefore we omit the plot for FailureTimeDG), and that the CoolingFailToRunTime and SRVStuckOpenTime are orthogonal in variations to the maxCladTemp in the PCA embeddings. The new information we obtain from topology-based clustering is that the firewaterTime does play a role in differentiating the pink, green, and blue clusters, as we see clear separation among the left end points of all three summary curves in its inverse coordinate plot.

Therefore, from a safety analysis perspective, we observe that, in order to assure a low value of maximum clad temperature, the high pressure injection system needs to be available for a long time for scenarios to remain system successes. On the other hand, the failure time of DGs (FailureTimeDG, initial time of the SBO condition) does not play a relevant role in guaranteeing a low value of max clad temperature. For the pink cluster in (Figure 8), an early AC recovery time guarantees system success even for early values of SRVstuckOpenTime, ExtendedECCSoperation time, and late firewaterTime. This means, even in the case of an early RPV depressurization (i.e., SRV stuck open), the core heating rate is slow enough that an early AC recovery time guarantees low values of max clad temperature.

For comparison, we could use the same 2D embedding from Section 4.1 and color the points based on the topological clustering results. This analysis is illustrated in Figure 9. Here we see four clusters colored in green, cyan, purple, and blue, respectively. This type of visualization gives less information regarding the correlations of input variables with respect to the output. On the other hand, for each cluster, it provides a more compact summary statistics of each input dimension.

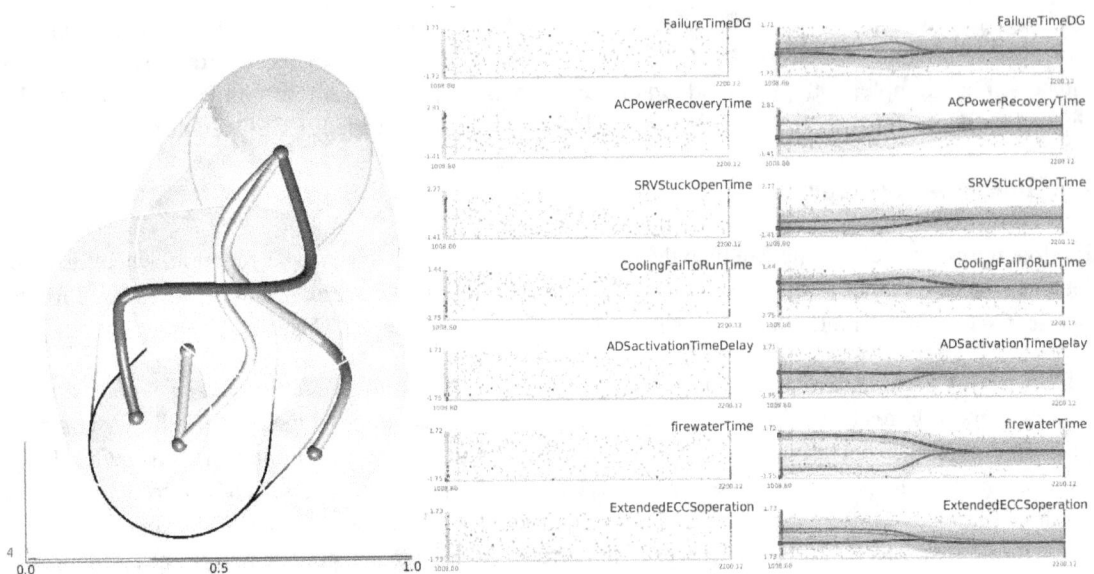

Figure 8: Left: the topological skeleton of all 4997 scenarios. Middle and right: inverse coordinate plots, where points (middle) along with summary curves (right) are colored by cluster labels.

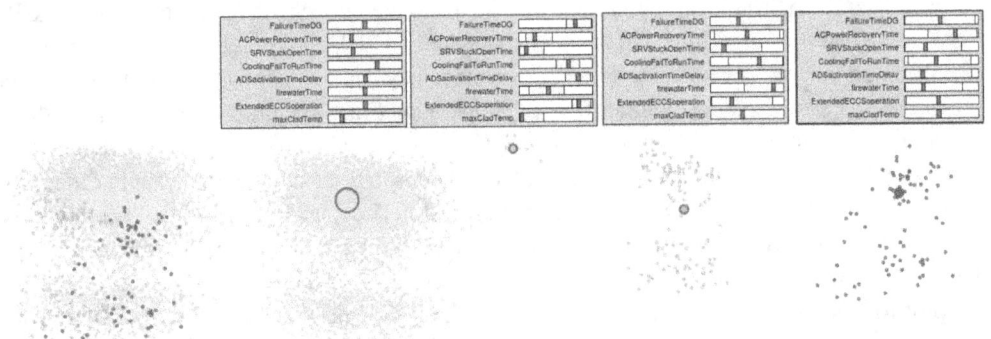

Figure 9: Left: 2D embedding of the data colored by topology-based clustering labels. Right: a separate illustration of individual clusters and their summary statistics with respect to the input dimensions.

Failure Scenarios Case. In this case, we consider only failure scenarios and use simulationEndTime, that is, the time to reach the failure temperature of 2200 F, as the output variable. We obtain a topology-based clustering that consists of four clusters. Results are shown in Figure 10 and Figure 11. In Figure 10(left), four clusters share a global minimum, characterized by a simulationEndTime of 434.82 seconds. There are four distinct local maxima. One interpretation is to look at the local maxima as independent, near-success scenarios, as they represent within their own cluster, the latest time to reach the failure states (e.g., when the simulations terminate). In other words, the temperature for each of these local maxima scenarios grows slowly during the simulation, thereby allowing a longer simulation time.

From a safety analysis perspective, we are interested in understanding the conditions under which we have a late core damage event. Recall in the inverse coordinate plots of Figure 10(right) that the horizontal axis corresponds to the simulationEndTime. Therefore we focus our analysis on the right side of the horizontal axis, where a long simulation corresponds to a late core damage event. For the green cluster in Figure 10(right), as expected, a driving factor to reach a late core damage is a high value of ECCS operation. This observation implies that it is preferable to keep the RPV pressurized as long as possible and maintain high pressure cooling, instead of activating the ADS system and obtaining cooling through the FW system. Also note for the green cluster that a late core damage is also correlated with a late ACPowerRecoveryTime. For all scenarios contained in the purple cluster, we notice that the latest core damage within the cluster is reached for high values of FailureTimeDG, since a large quantity of heat has been discharged before reaching the SBO condition. On the contrary, for the red cluster, the latest core damage within the cluster occurs when a small quantity of heat has been rejected from the core following reactor scram (i.e., low value of FailureTimeDG) and late failure of the high pressure core cooling system (i.e., high value of CoolingFailToRunTime). In summary, for all clusters, a late failure of the high pressure core cooling system and a late ACPowerRecoveryTime are always needed in order to guarantee a late core damage condition. In addition, FailureTimeDG when coupled with the firewaterTime also plays a relevant role in understanding the conditions for reaching late core damage.

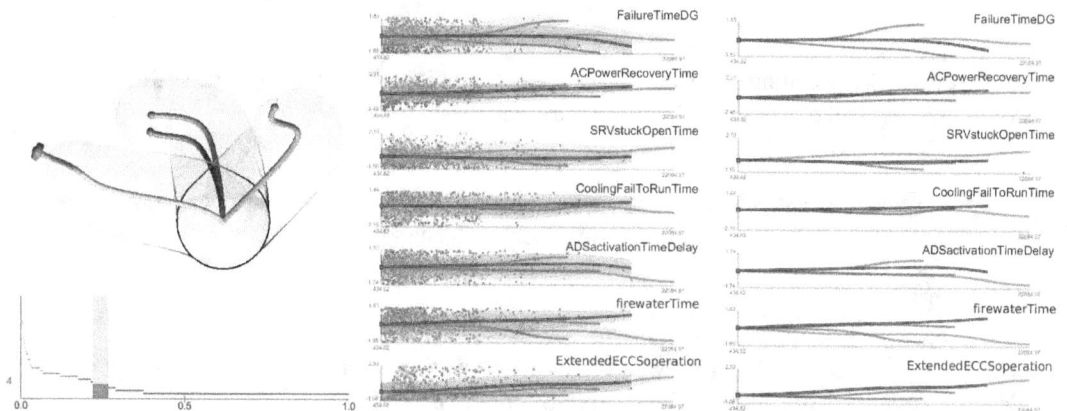

Figure 10: Left: topological skeleton of all failure scenarios. Inverse coordinate plots with (middle) and without (right) points projection. Points and summary curves are colored by cluster labels.

For comparison, as before, we color points in their 2D embedding based on the topological clustering results, as shown in Figure 11. We are able to see how the clusters differ in terms of the statistical summaries of the input dimensions. However, the information regarding how the output variable varies among the clusters remains hidden. For example in Figure 11, ACPowerRecoveryTime varies in its range and mean value across the four clusters; however, the inverse coordinate plot in Figure 10 reveals that such an input parameter is not a differentiating factor across the four clusters at the local maxima. As a matter of fact, the summary curves of this parameter overlap significantly in its inverse coordinate plot.

5. CONCLUSION

We investigate the use of both traditional and topology-based hierarchical clustering in conjunction with dimensionality reduction techniques on DPRA datasets. We provide the domain scientist with an analysis and visualization tool for obtaining insights about system responses under the simulated accident scenarios. We focus on a dataset that simulates the response of a BWR system during an SBO accident scenario. We obtain such a dataset by performing a series of simulations where, for each simulation run, we randomly change timing and sequencing of events. We would like to identify how timing or sequencing of these events affects the maximum core temperature.

We have observed that a traditional clustering combined with DR is adequate to distinguish failure scenarios with success scenarios, and to group points with similar parameter settings. On the other hand, topology-based clustering captures information regarding how input parameters are correlated with the output, and how input parameters settings help differentiate local extrema (i.e., local maxima or minima)

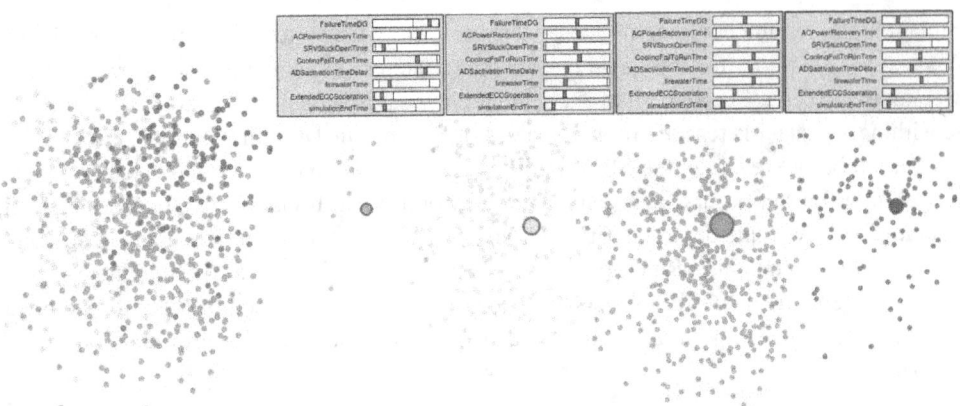

Figure 11: Left: 2D embedding of the data colored by topology-based clustering labels. Right: a separate illustration of individual clusters and their summary statistics.

of the output. We believe that pairwise comparisons and validations of both types of clustering techniques complement each other in bringing enhanced structural understanding of the data.

REFERENCES

[1] M. A. Carreira-Perpinan, "A review of dimension reduction techniques", *Department of Computer Science. University of Sheffield. Tech. Rep. CS-96-09*, pp. 1–69, 1997.

[2] D. Defays, "An efficient algorithm for a complete link method", *The Computer Journal*, vol. 20, no. 4, pp. 364–366, 1977.

[3] H. Edelsbrunner, D. Letscher, and A. J. Zomorodian, "Topological persistence and simplification", *Discrete and Computational Geometry*, vol. 28, pp. 511–533, 2002.

[4] S. Gerber, P.-T. Bremer, V. Pascucci, and R. Whitaker, "Visual exploration of high dimensional scalar functions", *IEEE Transactions on Visualization and Computer Graphics*, vol. 16, pp. 1271–1280, 2010.

[5] I. T. Jolliffe, *Principal Component Analysis*. Springer-Verlag, 1986.

[6] D. Maljovec, B. Wang, D. Mandelli, P.-T. Bremer, and V. Pascucci, "Analyze dynamic probabilistic risk assessment data through topology-based clustering", *International Topical Meeting on Probabilistic Safety Assessment and Analysis (PSA)*, 2013.

[7] D. Maljovec, B. Wang, V. Pascucci, P.-T. Bremer, M. Pernice, D. Mandelli, and R. Nourgaliev, "Exploration of high-dimensional scalar function for nuclear reactor safety analysis and visualization.", *International Conference on Mathematics and Computational Methods Applied to Nuclear Science & Engineering*, 2013.

[8] D. Mandelli, C. Smith, T. Riley, J. Schroeder, C. Rabiti, A. Alfonsi, J. Nielsen, D. Maljovec, B. Wang, and V. Pascucci, "Support and modeling for the boiling water reactor station black out case study using relap and raven", Idaho National Laboratory (INL), Tech. Rep. INL EXT-13-30203, 2013.

[9] D. Mandelli, A. Yilmaz, T. Aldemir, K. Metzroth, and R. Denning, "Scenario clustering and dynamic probabilistic risk assessment", *Reliability Engineering & System Safety*, vol. 115, pp. 146–160, 2013.

Quantification of MCS with BDD, Accuracy and Inclusion of Success in the Calculation – the RiskSpectrum MCS BDD Algorithm

Wei Wang[a], Ola Bäckström[a], and Pavel Krcal[ab]
[a] Lloyd's Register Consulting, Stockholm, Sweden
[b] Uppsala University, Uppsala, Sweden

Abstract: A quantification of a PSA can be performed through different techniques, of which the Minimal Cut Set (MCS) generation technique and Binary Decision Diagrams (BDD) are the most well known. There is only one advantage with the MCS approach compared to the BDD approach - calculation time, or rather, the capability to always solve the problem. In most cases the MCS approach is fully sufficient. But as the number of high probability events increases, e.g. due to seismic risk assessments, more accurate methods may be necessary. In some applications, a relevant numerical treatment of success in event trees may also be required to avoid overly conservative results.

We discuss the quantification algorithm in RiskSpectrum MCS BDD, especially with regard to success in event trees. A BDD for both the failure and success parts of a sequence can be generated separately - and thereafter, the BDD structures can be combined. Under some conditions, this calculation will yield exactly the same result as if a complete BDD for both the failure and success parts was generated. Properties of the algorithm are also demonstrated on several examples including a large size PSA.

Keywords: PSA, Binary Decision Diagrams, Minimal Cut Set List Quantification

1. INTRODUCTION

The PSA models are of increasing size and complexity, and they are used for an increasing number of applications. In most cases the standard calculation methods (e.g. minimal cut set upper bound) are fully sufficient. But as the number of high probability events is increasing, e.g. due to seismic risk assessments, more accurate methods may be necessary. In some of the applications, a relevant numerical treatment of success in event trees may also be required to avoid overly conservative results.

Bryant's Binary Decision Diagrams [9] present an efficient data structure for encoding of Boolean functions. Their main industrial applications lie in the area of computer aided design of electronic circuits and formal verification. Since their introduction to fault tree analysis [5,6] they stand as a challenge to traditional cut set based approaches. The BDD approach would yield the exact result unaffected by cutoff truncation, first order approximation and imperfect negation treatment. In spite of a large body of work, see e.g. [7,10,11], it has not been possible to solve current PSA models with BDD because of their size – it is therefore not a realistic approach today.

A less ambitious approach uses a variant of BDDs, ZBDDs [4,8], for a partial solution of the problem, for quantification of a coherent structure (e.g., MCS lists produced by traditional cut set generation methods). ZBDDs serve as a compact representation of the cut set list in question, but the quantification procedure results in the same value as rare event approximation.

The calculation algorithm in RiskSpectrum MCS BDD was briefly described at PSAM9 [1], and this paper will discuss the quantification in more detail, especially with regard to success in event trees. We apply BDD technique to quantify cut set lists with two aims:
- Avoid conservative results of the first order approximation for important events
- Allow exact negation treatment for success in sequences

A characteristic of a BDD is the possibility to calculate a structure given the event combinations that have, or have not, happened. This characteristic is discussed and its use within the algorithm is presented. A BDD for both the failure and success parts of a sequence can be generated separately - and thereafter, the BDD structures can be combined. Under some conditions, this calculation will yield exactly the same result as if a complete BDD for both failure and success part was generated.

The structure of the paper is as follows. First, we briefly introduce BDD background. Section 3 presents the MCS BDD algorithm together with some heuristics applied during BDD generation. Then we describe properties of the algorithm and argue for its correctness. Finally, some of the properties are demonstrated in Section 5 on example problems (including a full scale PSA study).

2. BACKGROUND

The classic MCS quantification methods like minimal cut set upper bound or 1st, 2nd or 3rd approximation are effective and efficient, but they also have some obvious drawbacks, such as overestimation of the top results, especially with some high probability events. Another problem is the treatment of success in sequence and consequence analysis.

The MCS BDD algorithm has therefore been developed as a part of the RiskSpectrum software package, which gives a better approach to quantify top results from MCS list.

The MCS BDD algorithm is derived from Shannon non-intersect decomposition,

$$f = xf_1 + \bar{x}f_0$$

where x is one of decision variables. The functions f_1 and f_0 are Boolean functions evaluated at $x=1$ and $x=0$ respectively.

Minimal cut sets can be converted to a group of mutually exclusive variables by non-intersect operation, then the top result is obtained by the summation of the probabilities of these variables.

Example 1: MCS list {A, BC, D}.

By no-intersect decomposition, the Boolean expression of the MCS list can be converted as follows:

$$Top = A + BC + D = A + \bar{A}D + \bar{A}\bar{D}BC$$

The top event probability can be calculated directly:

$$P(Top) = P(A) + P(\bar{A}D) + P(\bar{A}\bar{D}BC)$$

BDD is a natural non-intersect structure, in which all the paths from the root to the terminal "1" (failure state) nodes make the whole set of mutually exclusive minimal cut set variables. If we succeed to build the BDD structure for the target MCS list, it means that we can get a non-intersect form of the MCS list, from which the exact top result could be retrieved easily.

The disjoint paths through the BDD structure in Figure 1 are: A, $\bar{A}D$, $\bar{A}\bar{D}BC$.

3. GENERATING BDD FROM A MCS LIST

The MCS BDD algorithm uses a pivotal decomposition method to construct the BDD structure from the MCS list. By continually picking up pivotal elements (decision variables) from the MCS list, both failure and success branches are being further developed. When all the branches have reached "1" (failure state) or "0" (success state), the BDD structure is generated completely.

Figure 1. BDD structure in Example 1

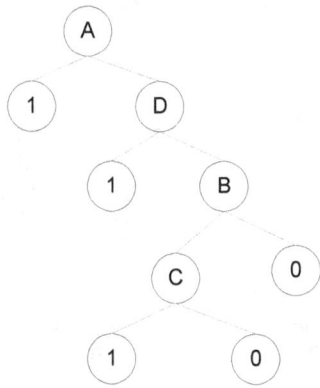

The key process of building the BDD structure is to select the pivotal element from the remaining MCS list and append it to the branch of the current pivotal node under construction. A backtracking has to be performed before selecting the next pivotal element in order to get all the elements along the backward path; the MCS list can then be minimized against the states of these pivotal elements.

The purpose of minimization is to reduce the size of the cut set list that has to be maintained during run-time. However, instead of doing a complete cut set minimization for each pivotal element, which is actually time consuming, we do a limited minimization to simply remove those events and cut sets with fixed states. If any cut set that contains the current pivotal element has got a fixed state after minimization or there is no cut set remaining in the current MCS list, it indicates that the branch has reached a terminal state ("1" or "0"); otherwise a new pivotal element shall be picked up from those events in the current MCS list whose states have not been decided yet, and appended to the target branch of the current pivotal node.

Take Example 1 to illuminate the BDD building process. First we choose A as a pivotal element, its failure branch leads directly to terminal state "1" after minimization, while its success branch is still to be developed. Then we need to select the next pivotal element from the remaining MCS list (with A succeeded) to extend the success branch of A; followed by selecting D as the new pivotal element, and the state of the failure branch can be decided to terminal state "1" by a propagation of the remaining MCS list (with D failed, A succeeded), while its success branch is still left to be further developed. In the same way, we can add another two pivotal events B and C to the structure. When all paths are terminated by "1" or "0", the building process has come to an end.

3.1 Function Event Success in a MCS List

First, we describe the way in which we store information about success in minimal cut set lists. We assume that the cut set list is obtained from analyzing one or more sequences of one or more linked event trees. All function events that succeed along a sequence are analyzed separately in the following way. We generate a minimal cut set list where each cut set fails at least one function event which succeeded. This means that these cut sets cannot be a part of the sequence solution. We label this minimal cut set list with the set of successful function events which were used to generate it and call it, together with the label, a *success module*.

We note that a success module uniquely identifies a sequence. This is obvious from the fact that each sequence for one or more linked event trees is uniquely defined by enumerating function events that succeed (or, that fail).

The algorithm for cut set generation will add the corresponding success module to each cut set generated from a sequence with successful function events.

3.2 MCS Cutoff Term

In order to reduce the amount of computation to a practical scale, in the MCS BDD algorithm it is possible to define a threshold on the cut set list so that those parts of cut sets that are below the threshold shall be evaluated with ordinary MCS quantification. The algorithm then would focus on the other part of the MCS list to quantify the result exactly, because this part will make the greatest contribution to the accuracy of the top result.

The results of the two parts shall be summed up with the minimum cut set upper bound method to ensure a conservative top result. If the threshold is set to 0%, the algorithm will build the BDD structure based on the whole MCS list.

3.2 MCS Grouping

Frequency events, usually as initiating events, and success modules are considered naturally mutually exclusive. It is possible to split the MCS list into different groups which are mutually exclusive to each other. This can greatly reduce the scale of the problem: the top result then can be calculated by a direct accumulation of the results of each separate group.

In the MCS BDD algorithm we first group the MCS list with different frequency events, and for each such group, we then split the cut sets into sub-groups with different success modules. The input to the algorithm for BDD generation is a cut set list in which all cut sets contain the same frequency event and the same success module. We assume that the frequency event and the success module are already removed from the cut sets.

In the MCS BDD algorithm, instead of building a complete BDD for the whole sequence, we build one main BDD structure for the sequence MCS list without success modules, as well as another BDD for the success module itself. In a later stage, the success module BDD is appended and merged to the main BDD to get a full representation of the logic of the sequence.

3.3 Pivotal element ordering

The BDD is generated from the MCS list by successively adding events from the MCS list. The order of pivotal elements makes a great impact on the size of the BDD structure. The problem of deciding whether there is an ordering of pivotal elements resulting in a BDD of a given size is NP-complete [3]. Therefore, it is unlikely that a polynomial algorithm for finding the best order of pivotal elements in order to obtain the optimal BDD structure exists. In the MCS BDD algorithm, instead of using a predefined fixed order of these events, we use a dynamic method to select new pivotal elements from the remaining MCS list with some basic principles, e.g. number of event occurrences, cut set order, which allow a more concise BDD structure to be obtained.

3.4 BDD modules

The MCS-BDD algorithm will also try to group events together in modules to simplify the BDD structure. The modules are detected and created during runtime based on the actual MCS list. Each time when selecting a new pivotal element, it will check whether it is possible to find any modules. Modules always have a higher priority when selecting a pivotal element.

3.5 Approximate treatment

When a pivotal element is chosen the MCS list has to be treated and minimized based on that failure or success. There are basically two different ways of generating the BDD structure in the MCS BDD algorithm:
- Exact treatment
- Approximate treatment

The exact treatment represents the normal way of building a BDD structure from an MCS list. This means that an event A is considered to be either failed or successful, and then apply that on the MCS list.

The approximate treatment means that when the MCS is analyzed and an event is failed then only MCSs where the event is included are considered further (in principle, the same way as ZBBDs [3,4] are built). The reason for this is that all the MCSs not containing the event will be a part of the success branch. Hence, with a slight modification of the quantification of the structure it is possible to have a very good approximation of the solution that normally is very efficient in reducing the structure.

Take Example 2 to illuminate the approximate treatment:

Example 2: MCS list {AB, BC, DE}

The complete BDD structure from exact treatment is shown in Figure 2 (the pivotal elements in the order of A, B, C, DE as a module).

Figure 2. BDD of Example 2

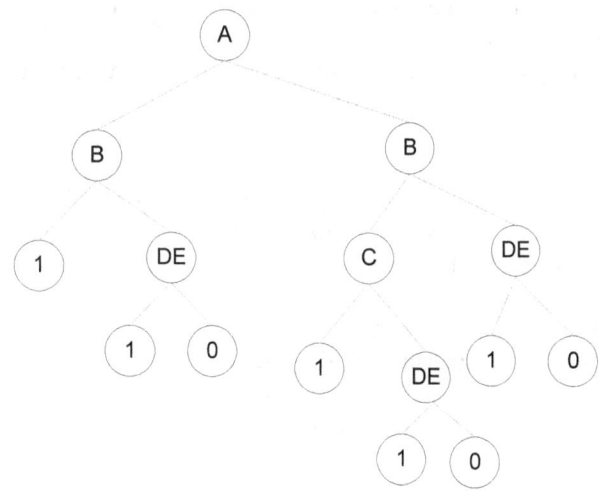

Because the pivotal element A is only included in the cut set {AB}, the remaining two cut sets should be in the success branch of A. With approximate treatment, the above BDD structure can be simplified as shown in Figure 3.

The new BDD structure still keeps the information of all the minimal cut sets, and the failed branch of A is simplified. As can be understood from the above the quantification of the approximate structure must be different. The quantification of the BDD structure will in this case be:

P(Top) = P(A) * [P(MCS including A) + P(MCS not including A) - P(MCS including A) * P(MCS not including A)] + (1 - P(A)) * P(MCS not including A)

Figure 3. BDD of Example 2 with approximate treatment

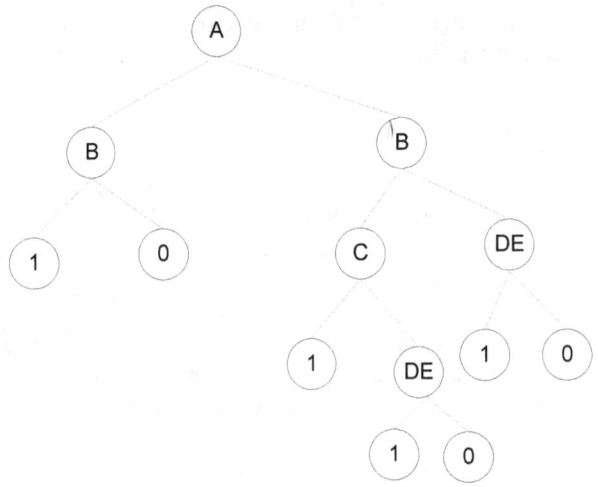

This can be simplified as:
P(Top) = P(A) * P(MCS including A) + (1- P(A) * P(MCS including A)) * P(MCS not including A).

When this approximate treatment is used the following conditions apply:
- The approximate treatment can never be used when the event is negated in the cut set list.
- The approximate treatment is conservative for coherent structures, since dependencies between MCS are not treated exact. The conservative error is however very small if the pivotal element has low probability.
- When the two parts of the cut set list both contain events that occur in the success module, exact treatment has to be used or we could switch to other more conservative quantification methods, e.g. ZBDD.

The use of the approximate treatment is optional and is actuated by two different triggers:
- Fussel-Vesely importance above a specified level
- Unavailability above a specified level

Fussel-Vesely importance is defined as a measure of the contribution a basic event makes to the top event probability or frequency. It is calculated using the formula:

$$FV_E = \frac{\sum Q_{\overline{A}_E}}{Q_{Top}},$$

where each $Q_{\overline{A}_E}$ is the probability value of the cut set $\overline{A}_E$ containing the element E.

4. CORRECTNESS

First, note that the MCS list (without success modules) and success modules are both generated from coherent fault trees. Fault tree models in RiskSpectrum can contain negated gates. These gates are not interpreted according to standard rules of Boolean algebra, expressing that a failure occurs if some subsystems succeed. Not-logic is treated as a syntactic shortcut for cut set removal during the fault tree analysis [2].

4.1 Correctness of Quantification

We build a BDD from the minimal cut set list in the standard way. We use events from success modules as pivot elements first before all other events. These events are treated exactly. When we have processed all of these basic events then we can use both exact and approximate way of building BDD for the remaining basic events.

Finally, we merge this structure with the success module BDD. Because we have built nodes for basic events from success modules in the exact way, we can do this in a way that allows for exact quantification of success modules.

Claim 1. The MCS BDD algorithm quantifies success modules exactly.

If all nodes in the BDD are treated exactly then the quantification procedure returns the exact probability of the MCS list, including exact quantification of the success modules. This means that the whole MCS BDD algorithm handles also the non-coherent part, as if cut sets with success modules were expanded into prime implicants.

The approximate treatment might lead to overestimating the actual probability of the failure part of the BDD. Both BDD substructures of a node which is treated approximately represent coherent structures (note that success modules and these substructures are disjoint). Because of this, the term "- P(MCS including A) * P(MCS not including A) " in the quantification formula for approximate treatment will never lead to an under-approximation. On the other hand, it will give a more precise result than the standard ZBDD quantification.

Claim 2. The MCS BDD algorithm is always conservative and it is exact if all nodes are treated exactly.

The conservatism of the top result will be determined by the way we select basic events for exact or approximate treatment. Because we select most important events for exact treatment, the over-estimation will be negligible in most cases.

4.1 Effects of Cutoff

The whole event/fault tree analysis algorithm applies cutoffs to keep the problem in a manageable size. There are three types of cutoff:
- Cutoff during cut set generation of the failure part
- Cutoff during generation of success modules
- Cutoff before building the MCS BDD structure

The first cutoff type has the same effect as for the standard MCS based event/fault tree analysis. It decreases the top value probability and potentially makes the result non-conservative. It is up to the PSA analyst to set this cutoff correctly so that the cutoff error remains acceptably small (in the best case, negligible with respect to the calculated top value).

The second type of cutoff is always conservative. It removes cut sets from success modules. By doing so, the actual probability of the success module in question grows. It might also result in removing all cut sets containing a particular basic event. Then this event might be treated approximately in the MCS BDD algorithm. But, this leads again only to an over-approximation of the final probability. This means that we can use greater cutoff values for success modules without the risk of underestimating the top probability.

The last cutoff type is also conservative. It excludes some cut sets from the MCS BDD treatment and quantifies them by the minimal cut set upper bound method. This results in an over-approximation and the total probability for these cut sets is always greater then (or, equal to if all BDD nodes are treated approximately) the result of the MCS BDD algorithm without any cutoff would be.

This brings us to the following conclusion.

Claim 3. If the analysis algorithm does not apply any cutoff and if the MCS BDD generation process always uses the exact treatment of basic events then it returns the exact top value by the means of MCS BDD.

4.4. Importance with MCS BDD

The importance factor that suffers most from over-approximations in the standard MCS list quantification is the Risk Increase Factor (RIF). The main problem is that simply setting the probability of a basic event to one might leave the MCS list with many non-minimal cut sets. The minimal cut set upper bound algorithm counts in also all of these non-minimal cut sets. This results in overly high RIF values. If the MCS BDD algorithm treats the basic event for which we calculate RIF exactly then there will be no over-approximation coming from what would correspond to non-minimal cut sets in the regular cut set list quantification. If we use approximate treatment then we always get a result which is at least as good as from the standard MCS list quantification with minimal cut set upper bound.

5. EXAMPLES

The use of the MCS BDD algorithm is demonstrated in the following examples. The first example uses a full scale PSA model, which is slightly modified so that it is not producing results that are applicable for any plant. The problem, size and numbers, are still representative of the type of problem we want to demonstrate. Figure 4 shows the analyzed event tree.

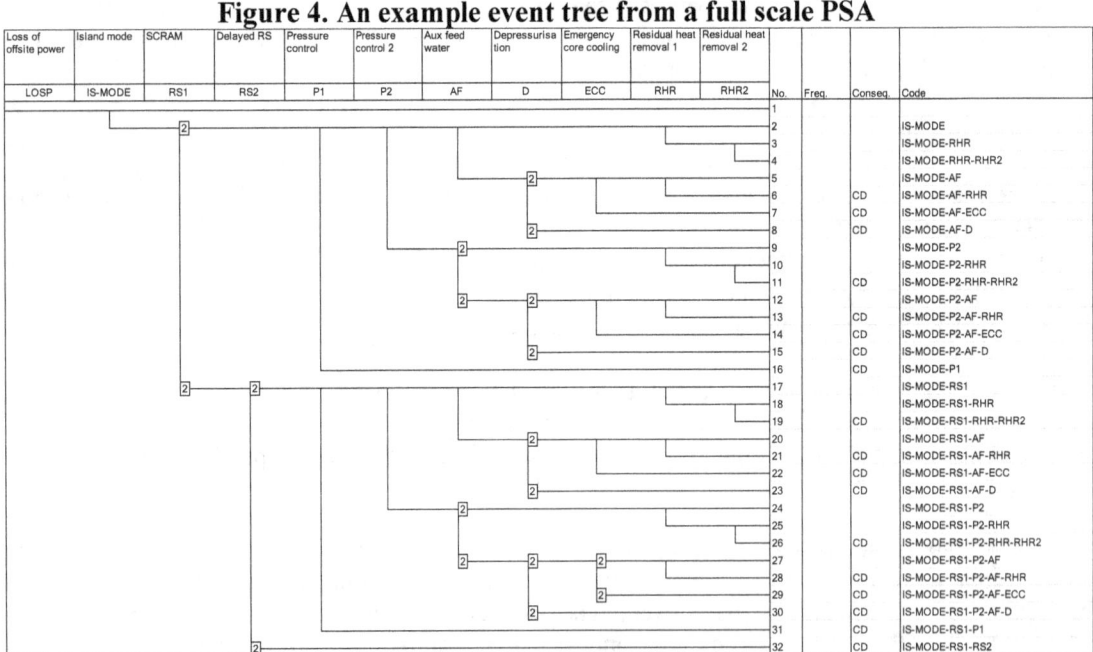

Figure 4. An example event tree from a full scale PSA

The initiating event is loss of offsite power and its frequency is 0.204/a. This is a PSA level 1, using large fault trees and small event trees. Since it is a PSA level 1, the probabilities for the top events (function events) in the event tree are typically low.

In RiskSpectrum PSA, there are several quantification methods that may be used. In this example, we are going to compare Logical ET Success, Logical and Simple Quantitative and MCS BDD (including success modules).

In short,

- Logical ET success means to logically ensure that the MCS that are generated cannot be in two different sequences. There is no consideration of the success path.
- The term *logical and simple quantitative* means, in addition to the above, that the successes are quantified as a FT analysis case ($P_{failure}$), and then the success probability is 1-$P_{failure}$.
- MCS BDD including success modules generates the MCS using the logical and simple quantitative method, but the success modules are appended to the BDDs that are generated for the failed part.

The example is run to illustrate the results from the different methods and to demonstrate that the MCS BDD method is scalable and applicable to real size problems. The number of MCS generated for the sequences spans from a few MCSs up to around 100 000 MCSs, including around 4000 events.

The example is, for each sequence, using 99,95% of the top frequency as basis for the BDD, and exact BDD treatment for events that have a FV larger than 1E-2 or a probability greater than 1E-2. The results are generated based on an MCS cut off defined at 1E-12. The core damage frequency is approximately 1,3E-5. This means that individual sequences close to cutoff will not be correct – but they will on the other hand have no impact on the overall results.

The comparison of individual sequences is shown in Table 1.

Table 1. A comparison of three different quantification algorithms

	MCS BDD	Log Simp Q	Logical ET Success
LOSP:0001	1,51E-01	1,51E-01	2,04E-01
LOSP:0002	5,13E-02	5,13E-02	5,31E-02
LOSP:0003	1,24E-06	1,26E-06	1,45E-06
LOSP:0004	1,33E-07	1,35E-07	1,40E-07
LOSP:0005	1,03E-05	1,06E-05	1,10E-05
LOSP:0006	4,18E-09	4,28E-09	4,44E-09
LOSP:0007	8,25E-06	1,11E-05	1,15E-05
LOSP:0008	2,12E-06	2,14E-06	2,22E-06
LOSP:0009	1,77E-03	1,83E-03	1,84E-03
LOSP:0010	3,54E-08	3,63E-08	4,05E-08
LOSP:0011	3,71E-09	3,78E-09	3,79E-09
LOSP:0012	4,99E-06	5,56E-06	5,59E-06
LOSP:0013	5,17E-09	5,50E-09	5,52E-09
LOSP:0014	5,81E-07	8,04E-07	8,06E-07
LOSP:0015	1,46E-07	1,52E-07	1,52E-07
LOSP:0016	5,49E-10	5,69E-10	5,70E-10
LOSP:0017	1,13E-04	1,15E-04	1,78E-04
LOSP:0018	4,86E-09	5,00E-09	8,70E-09
LOSP:0019	5,06E-10	5,07E-10	7,89E-10
LOSP:0020	5,85E-05	7,17E-05	7,42E-05
LOSP:0021	6,07E-08	6,36E-08	6,57E-08

LOSP:0022	2,36E-08	3,75E-08	3,88E-08
LOSP:0023	1,29E-07	1,42E-07	1,47E-07
LOSP:0024	3,58E-06	3,67E-06	5,55E-06
LOSP:0025	8,64E-11	8,75E-11	1,60E-10
LOSP:0026			
LOSP:0027	2,11E-06	2,55E-06	2,55E-06
LOSP:0028	6,10E-11	6,18E-11	6,19E-11
LOSP:0029	1,90E-09	2,34E-09	2,34E-09
LOSP:0030	1,45E-09	1,73E-09	1,73E-09
LOSP:0031			
LOSP:0032	1,26E-06	1,36E-06	1,36E-06
	2,04E-01	2,04E-01	2,59E-01

It can be seen from the table above, which presents all sequences, that both MCS BDD and logical and simple quantitative methods produce results that sum up to the initiating event frequency. It can also be noticed that MCS BDD always produces lower results than logical and simple quantitative, which in turn produces lower results than logical ET success.

The requirement to produce results that sum up to the initiating event frequency is that success treatment is included in the evaluation. There are some sequences that differ significantly between MCS BDD and logical and simple quantitative. The results of the MCS BDD are always lower (or equal), which is expected since the quantification of the success module is conservative in the simple quantitative approach.

If an analysis is set up to quantify the results for a group of sequences, the results can simply be summed for the MCS BDD and the simple quantitative results (or to define a consequence analysis). For logical ET success, the situation is different, since success is not included and hence some MCSs may be the same, or non-minimal between sequences. Therefore, the logical ET success result has to be calculated as a consequence analysis case.

The results for the CD with the different settings are:
- Logical ET success: 1,31E-5
- Logical and simple quantitative: 1,58E-5
- MCS BDD: 1,26E-5

This example demonstrates that:
- The simple quantitative approach produces slightly conservative results, but it does give a reasonable estimate of individual sequence results.
- The logical ET success will produce conservative estimates for sequences in which success probability has a significant impact, and it does produce good approximations for normal cases where core damage is studied.
- The MCS BDD approach produces very good estimates of sequence results, including successes and it does also produce results that are accurate for a group of sequences.

The advantage with the MCS BDD is both the accuracy in the quantification of high probability events and the calculation of success. The other example illustrates the impact of success treatment and it is compared to the logical and simple quantitative and logical ET success approaches.

The example is illustrated in Figure 5, together with the correct results. Sequence 1 is the combination -(A+B), Sequence 2 will not produce any results. Sequence 3 is represented by (A-B) and (B-A) and Sequence 4 is (AB).

Figure 5. A simple event tree with dependencies between function events

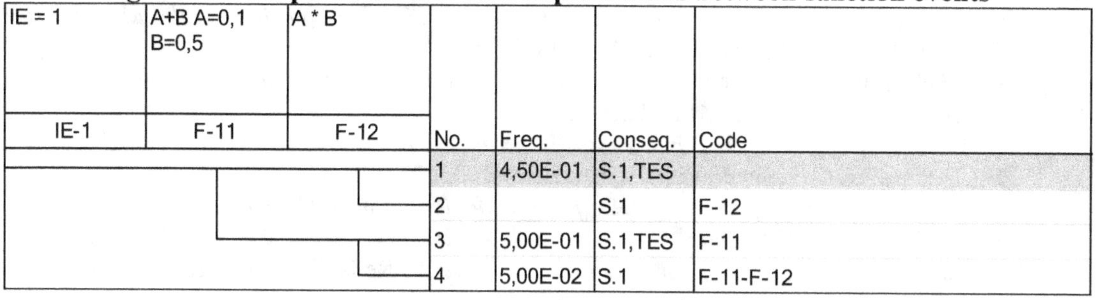

IE = 1	A+B A=0,1 B=0,5	A * B				
IE-1	F-11	F-12	No.	Freq.	Conseq.	Code
			1	4,50E-01	S.1,TES	
			2		S.1	F-12
			3	5,00E-01	S.1,TES	F-11
			4	5,00E-02	S.1	F-11-F-12

Table 2 presents the results for the different quantification methods:

Table 2. Comparison results for the simple event tree

	Log ET	Log Simple Quant	MCS BDD
Seq1	1	0,45	0,45
Seq2	0	0	0
Seq3	0,55	0,5249	0,5
Seq4	0,05	0,05	0,05

As can be seen, the logical and simple quantitative is a good approximation, but when there are dependencies between the failed part and the successful part of the MCS the results are overestimated. This is illustrated by Sequence 3, where the logical and simple quantitative produce following MCSs:
A –F12
B –F12

F12 is represented by the MCS
AB

Hence, the probability P(-F12) is calculated as $1-P_{F12} = 1 - P_A P_B = 0,95$

The problem is that there are dependencies between the failed part and the success part. For example, the MCS A-(AB) is always conservative, since this should be reduced to A(-B). The success module should be quantified depending on each MCS, but this would be very time consuming.

The MCS BDD, where the BDD for the success is appended/merged with the failed structure will be considered. In the MCS BDD the correct results will therefore be retrieved.

6. CONCLUSIONS

The MCS BDD algorithm presents a new way to quantify MCS lists in RiskSpectrum. Its main advantages are improved accuracy over the first or third order approximation and exact quantification of the part of the MCS list which represents event tree success. Especially, the new treatment of success enables a more accurate quantification in presence of function events with high probability. Heuristics implemented in the algorithm, e.g., combination of exact and approximate treatment of BDD nodes, make the algorithm scalable while always giving a conservative result. On the other hand, the algorithm has capability of returning the exact top event probability/frequency. If there are no cutoffs used and all BDD nodes are treated exactly then the new method returns the same result as a complete BDD algorithm.

References

[1] O. Bäckström and D. Ying, "*A presentation of the MCS BDD algorithm in the RiskSpectrum Software Package*", In Proc. of PSAM9, Hong Kong, 2008.

[2] O. Bäckström and P. Krcal, "*A Treatment of Not Logic in Fault Tree and Event Tree Analysis*", In Proc. of PSAM11, Helsinki, 2012.

[3] B. Bollig and I. Wegener, "*Improving the Variable Ordering of OBDDs is NP-Complete*", IEEE Trans. on Computers, Vol. 45(9), pp. 993-1002, 1996.

[4] S.-I. Minato, "*Zero-Suppressed BDDs for Set Manipulation in Combinatorial Problems*", In Proc. of DAC'93, 1993.

[5] A. Rauzy, "*New Algorithm for Fault Tree Analysis*", Reliability Engineering and System Safety, Vol. 40, pp. 203-211, 1993.

[6] O. Coudert and J.-C. Madre, "*Fault Tree Analysis: 10^{20} Prime Implicants and Beyond*", In Proc. of Annual Rel. and Maint. Symp. 1993, Atlanta, 1993.

[7] A. Rauzy, "*Binary Decision Diagrams for Reliability Studies*", Handbook of Performability Engineering, pp. 381-396, 2008. ISBN 978-1-84800-130-5.

[8] W. S. Jung et al., "*A Fast BDD Algorithm for Large Coherent Fault Trees Analysis*", Reliability Engineering and System Safety, Vol. 83, pp. 369-374, 2004.

[9] R. Bryant. "*Graph Based Algorithms for Boolean Function Manipulation*", IEEE Transactions on Computers, Vol. 35(8), pp. 677-691, 1986.

[10] O. Nusbaumer, "*Analytical Solutions of Linked Fault Tree Probabilistic Risk Assessments using Binary Decision Diagrams with Emphasis on Nuclear Safety Applications*", PhD Thesis, ETH No. 17286, 2007.

[11] R. Remenyte and J. D. Andrews, "*Qualitative analysis of complex modularized fault trees using binary decision diagrams*". Journal of Risk and Reliability, Vol. 220 (1), pp. 45-53, 2006.

Developing a New Fire PRA Framework by Integrating Probabilistic Risk Assessment with a Fire Simulation Module

Tatsuya Sakurahara[a]*, Seyed A. Reihani[a], Zahra Mohagheh[a], Mark Brandyberry[a], Ernie Kee[b], David Johnson[c], Shawn Rodgers[d], and Mary Anne Billings[d]

[a] The University of Illinois at Urbana-Champaign, Urbana, IL, USA
[b] YK.risk, LLC, Bay City, TX, USA
[c] ABS Consulting Inc., Irvine, CA, USA
[d] South Texas Project Nuclear Operating Company, Wadsworth, TX, USA

Abstract: Recently, the fire protection programs at nuclear power plants have been transitioned to a risk-informed approach utilizing Fire Probabilistic Risk Assessment (Fire PRA). One of the main limitations of the current methodology is that it is not capable of adequately accounting for the dynamic behavior and effects of fire due to its reliance on the classical PRA methodology (i.e., Event Trees and Fault Trees). As a solution for this limitation, in this paper we propose an integrated framework for Fire PRA. This method falls midway between a classical and a fully dynamic PRA with respect to the utilization of simulation techniques. In the integrated framework, some of the fire-related Fault Trees are replaced with a *Fire Simulation Module (FSM)*, which is linked to a plant-specific PRA model. The FSM is composed of simulation-based physical models for fire initiation, progression, and post-fire failure. Moreover, FSM includes the uncertainty propagation in the physical models and input parameters. These features will reduce the unnecessary conservativeness in the current Fire PRA methodology by modeling the underlying physical phenomena and by considering the dynamic interactions among them.

Keywords: Nuclear Power Plants, Fire PRA, Integrated PRA Framework

1. BACKGROUND

After a fire event at the Browns Ferry nuclear power plant (NPP) in 1975 [1], fire protection began to be recognized as one of the important elements of nuclear safety. Conventionally, the fire protection program (FPP) at NPPs has been implemented by a deterministic approach based on Title 10 of the Code of Federal Regulations, Part 50, Section 48 (10 CFR 50.48) [2], and Appendix R [3]. In 2004, the U.S. NRC revised 10 CFR 50.48 to allow licensees to voluntarily shift to risk-informed fire protection under NFPA 805 [4]. As of 2012, 47 reactors in the U.S. (out of a total of 104 reactors operating in the country) plan to shift (or are in the process of shifting) to the risk-informed and performance-based (RI-PB) approach [5].

1.1. Deterministic vs. Risk-informed Fire Protection Program

The transition from a deterministic to a risk-informed approach in the fire protection program at NPPs should be regarded as part of the general effort by the U.S. NRC and the nuclear industry to expand the use of the probabilistic risk assessment (PRA) technique for the improvement of safety at NPPs. According to U.S. NRC PRA Policy Statements [6], the usage of PRA in the nuclear safety arena contributes to (i) decision-making that is enhanced by the use of PRA insights, (ii) more efficient use of resources, and (iii) reduction in unnecessary burdens on licensees. In order to benefit from these advantages, PRA should be used "to reduce unnecessary conservatism" and the PRA output "should be as realistic as possible" [6].

In the context of fire protection at NPPs, the benefits from PRA as compared to those of the deterministic approach are summarized as follows [5]. First, the nuclear operators can obtain a better understanding

---

* sakurah2@illinois.edu

of plant risk by quantitatively identifying the risk-significant fire compartment and event sequences. It will allow nuclear operators to effectively allocate their limited resources to risk-significant factors, and allow them flexibilities in areas that have been assessed as insignificant with respect to plant risk. Second, the licensing conditions and requirements become less complicated. If a licensee transitions to risk-informed FPP under NFPA 805 [7], the licensee can obtain fire protection licensing amendments based on a single standard, NFPA 805. This is considerably simpler than the conventional deterministic FPP, where licensing conditions are subject to a number of guidance documents, communications, and regulatory issue summaries. Transition to a risk-informed FPP can eliminate a significant resource burden, both on the U.S. NRC and the nuclear industry, that is caused by complicated exemption and deviation approval processes in the fire protection program under the deterministic approach [4]. These will help the NRC staff and plant operators focus their available resources on risk-significant issues. Third, according to some licensees and experts, plant safety has actually been improved through the process of transition to fire PRA by the extensive fire analyses and modifications [5]. Due to these benefits, the risk-informed approach has the potential to lead to more effective fire protection at NPPs, outperforming the traditional deterministic approach.

1.2. Limitation in Current Fire PRA Methodology

The current methodology for Fire PRA, established by NUREG/CR-6850, was issued in 2005 [8,9]. NUREG/CR-6850 provides state-of-the-art methods, tools, and data for Fire PRA at operating NPPs. This document aims at consolidating the existing state-of-the-art methods and technical bases into one standard methodology. In addition, the new methods were developed in several areas such as a post-fire plant-safe shutdown response model, fire event data and fire frequency analysis, detection and suppression analysis, circuit analysis, Human Reliability Analysis (HRA), etc.

Despite these advancements, it has been recently recognized that the current Fire PRA methodology, established in NUREG/CR-6850, has some limitations. The reported limitations in the literature include (i) an unexpectedly high transition cost reported by pilot plants [5], (ii) lack of human resources familiar with fire modeling [5], and (iii) overly conservative assumptions and unrealistic output [5,10]. Among them, from the technical point of view, the limitation that is frequently pointed out is conservatism [5] [8,10]. The overly conservative assumptions are considered in two areas: First, the data necessary to develop the model of fire phenomena and damage to equipment are insufficient. As one example, NUREG/CR-6850 [8] itself states that spurious actuation likelihood caused by cable damage, which has been derived from expert elicitation, is considered to be generally conservative, while the extent of conservatism remains unidentified. The second cause of overly conservative output is the fact that fire damage and operator response are modeled in a static context [8]. According to NUREG/CR-6850 [8], although the impact of this assumption is judged to be conservative for most fire scenarios, the extent of conservatism has not been identified either qualitatively or quantitatively.

The overly conservative output can result in the misallocation of available resources to areas that have been evaluated as risk-significant by current Fire PRA, in spite of actually being insignificant with respect to actual risk [5]. Thus, in order to reduce the plant risk more effectively, it is necessary to improve the current methodology to reduce the unnecessary conservatism and to produce an "as realistic as possible" result. To achieve this, modeling based on physical simulation techniques should play a key role. A simulation-based model can contribute to overcoming the causes of unnecessary conservatism in the current methodology by modeling time-dependent fire phenomena and damage [8], and by compensating for the lack of data on failure probabilities of equipment and human reliability [11].

2. INTEGRATED FIRE PRA FRAMEWORK

Although a simulation-based/dynamic PRA is the ideal goal for nuclear power risk analysis, it is, on a short-term basis, impractical and can be quite costly. Currently, NPPs utilize classical PRAs and changing them to fully simulation-based PRAs would require significant time and resources. In this paper, we are proposing an *integrated* framework that, with respect to modeling techniques, stands between classical PRA and simulation-based/dynamic PRA (See Figure 1). In other words, the classical

Figure 1: Modeling state of the integrated framework with respect to PRA evolution.

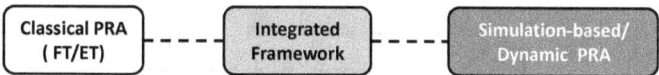

PRA of the plant would be used, but the fire phenomena would be modeled in a simulation-based module (separate from PRA) and the module would then be linked to the classical PRA of the plants. The goal is not to translate the fire phenomena into a fault tree (FT) and event tree (ET) context, but instead to model them in a simulation-based environment up to the last point of events of interface with the plant-specific PRA. Similar *integrated* approaches have been developed, by several of the authors of the current paper, for the incorporation of the effects of organizational factors into PRA [12] and for the risk-informed resolution of Generic Safety Issue 191 (GSI-191) [13].

The major features of the proposed *integrated* Fire PRA approach (Figure 2) are:

(1) *Plant-specific PRA Module* composed of ET/FT used in the classical plant-specific PRA. This module will be explained in Section 2.1

(2) *Fire Simulation Module (FSM)* which includes the simulation and uncertainty quantification of realistic phenomenological models for fire initiation, dynamic progression of fire effects, and post-fire damage .The elements of FSM will be explained in Section 2.2 ,

(3) *Input Module* (explained in Section 2.3), which provides the required input for the *Fire Simulation Module*.

The FSM replaces some part of FTs in the current PRA methodology and produces the conditional probabilities of basic events that are input to *Plana-Specific PRA Module*. Using this *integrated* approach would allow us to "simulate" the fire phenomena and would create the possibilities of (a) advancing quantification of dynamic interactions, (b) more adequate depiction of contextual factors (e.g., physical factors and human performances), and (c) advancing propagation of uncertainties involved in the physical phenomena. These three features would lead to more "realistic" modeling of fire that, ultimately, could reduce the unnecessary conservatisms. Another advantage of the *integrated* approach is that it is a step toward having a fully simulation-based PRA. When, and if, NPPs are ready to switch to simulation-based PRAs, the *FSM* developed in this work would be an appropriate engine for their PRAs.

This research project consists of two phases. In Phase I, we will develop a *FSM* by integrating the existing experimental, statistical, and physical models (related to fire initiation, progression, and post-fire failure modeling) that have been proposed as a result of research activity over the past three decades. In this process, we will use the simulation-based model everywhere possible. In addition to those models used in the current Fire PRA methodology [9], we will search for models having the potential for applicability as proposed in existing literature in both the nuclear and non-nuclear arenas. Then, the *FSM* will be connected to the *Plant-specific PRA Module*. The connection provides the conditional failure probability distribution of interface of basic events computed by *FSM* to the *Plant-specific PRA Module*. The target risk metric (i.e., core damage frequency) is calculated using the *Plant-specific PRA Module*.

After obtaining the primary results from Phase I, which would contribute to the development of a dynamic and realistic fire risk analysis, for Phase II of the project, we would advance some of the physical models in the FSM, especially in areas currently dominated by statistical and expert elicitation methods. This would, thereby, create an even more accurate quantification of risk.

Figure 2: An integrated Fire PRA framework for a hypothetical fire-induced fault tree resulting in a SBLOCA.

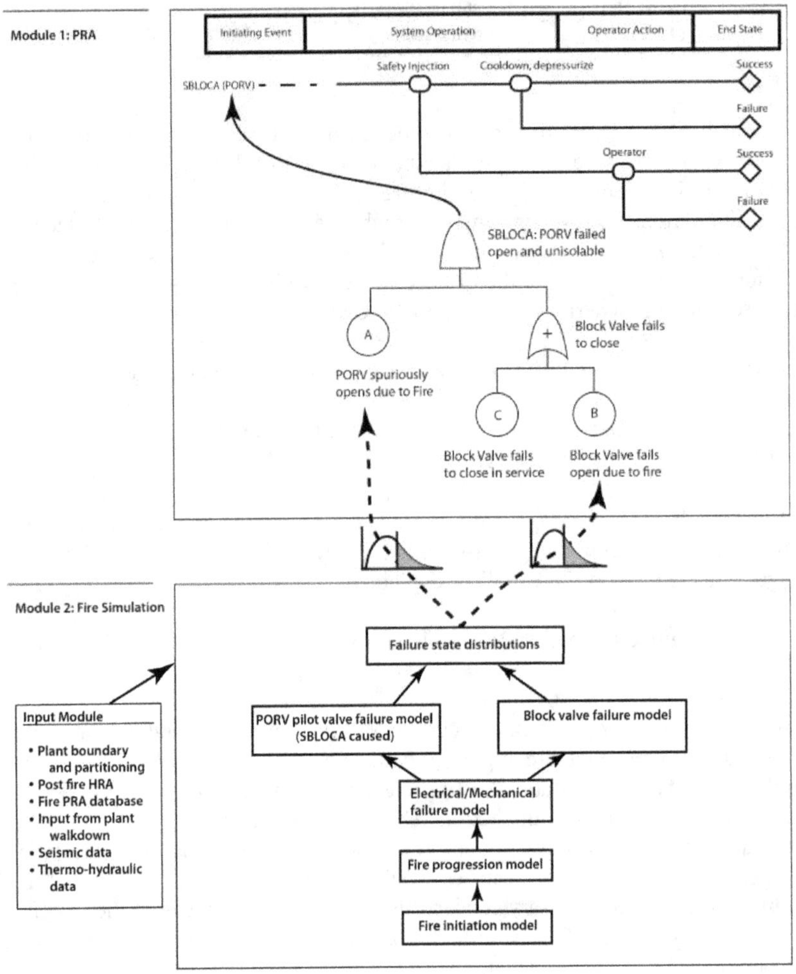

The primary contributions from Phase I of this project would be to:

(1) Change quantification techniques of the Fire risk analysis for NPPs from FT/ET to a simulation-based approach in a *FSM*: The plan is to extract the majority of fire-related fault trees from PRA and, instead, cover them in a simulation-based context. Our purpose is to integrate the existing time-dependent physical fire models into a simulation module so that their dynamic interactions can be more adequately covered.

(2) Propagate uncertainty in the *FSM*: The *FSM* integrates the physical phenomena and propagates uncertainty in physical models and input parameters from fire initiation to potential core damage precursors. Uncertainty propagation can be accomplished by sampling the input parameters assuming that the input parameters are random variables with epistemic distributions derived from historical data, experimental data, expert elicitation, physics, or a combination of these sources. These values would be propagated through the *FSM* to provide an estimator of a key performance measure, such as the probability of a subsystem failure, which is passed to the *Plant-specific PRA Module*.

(3) Link the *FSM* to the *Plant-specific PRA Module:* In Phase I, the purpose is to use the existing experimental and physical models and to mainly focus on integrating them into a simulation module and, ultimately, into the *integrated* framework. However, in Phase II of this research project, if necessary, some of the elements of the *integrated* framework, such as the physical models that apply to fire initiation and target damage, will be advanced.

Next sections explain the elements of the modules of Figure 2 and map them into the tasks of the current Fire PRA methodology in NUREG/CR-6850 [9]. Note that all elements of the *integrated* framework are not necessarily mapped into the current methodology. This is because the proposed framework is simulation-based and the fire-induced phenomena are stated in the classical PRA language using FT/ET. We will also summarize several simulation-based approaches from literature that are candidates for sub-modules of the *Fire Simulation Module*.

2.1. *Plant-specific PRA Module* (Module 1 in Figure 2)

1.A) Identifying the dominant fire-induced PRA scenarios:

First, we would need to determine on which fire-related scenarios we would develop the *FSM*. We will categorize all the possible fire-related scenarios into several groups based on the similarity in the failure modes of basic events and then develop the *FSM* for each group of scenarios. This step will be related to Task 1 (Plant Boundary & Partitioning) and Task 2 (Fire PRA Component Selection) defined in [9]. A hypothetical example would be a fire in an area that would cause the pressurizer Pilot-Operated Relief Valve (PORV) to spuriously open and would result in a Small Break Loss of Coolant Accident (SBLOCA) until the block valve could be closed (Figure 2).

1.B) Identifying the key basic events of interface between PRA and the *FSM*:

Another element would focus on finding the basic events of the PRA, which are the interfaces between the *Plant-specific PRA Module* and the *FSM*. For example, in this simplified fire-induced scenario (Figure 2), the basic events A and B are the interfaces. This step is similar to Task 4 (Qualitative Screening) as defined in NUREG/CR-6850 [9].

1.C) Calculating total risk:

The ultimate risk for the plant will be calculated by a *Plant-specific PRA Module* (Module 1) using the conditional probabilities calculated by the *FSM* for the basic events of interface. For instance, the core damage frequency (CDF) is mathematically expressed by

$$\text{CDF} = \sum_i f_i \left(\sum_j P_{j|i} \left(\sum_k P_{CD:k|i,j} \right) \right) \tag{1}$$

where f_i denotes the frequency of postulated fire i, $P_{j|i}$ denotes the conditional probability that the postulated fire causes damage to equipment j, and $P_{CD:k|i,j}$ denotes the conditional probability that the operator fails to recover the plant which results in core damage given the postulated fire i and fire-induced damage to the equipment j [9] [14]. In our *integrated* framework, f_i and $P_{j|i}$ are estimated in *FSM* using the mechanistic simulation-based models to the full extent, while $P_{CD:k|i,j}$ is provided in the *Input Module*, based on the post-fire HRA.

This step relates to Task 14 (Fire Risk Quantification), Task 15 (Uncertainty & Sensitivity Analysis), and Task 16 (Fire PRA Documentation) in NUREG/CR-6850 [9].

2.2. *Fire Simulation Module* (Module 2)

The main elements of the *FSM* (in Figure 2) include: (3.2.A) Fire initiation model, (3.2.B) Fire progression model, (3.2.C) Post-fire failure model, and (3.2.D) Failure state distributions. The purpose of this module is to integrate the models of the physical phenomena leading to basic events of interface with the *Plant-specific PRA Module*, and to propagate uncertainties in these physical models and input parameters in order to estimate the probability distribution of the basic events of interface (e.g., A and B in Figure 2).

We have reviewed the literature concerned with the models of fire physical phenomena that were published after the issuance of NUREG/CR-6850. Those models are categorized into three groups:

- Category I. Deterministic simulation of fire physical phenomena based on deterministic physical equations (e.g., Fire Dynamic Simulation [15] and CFAST [16]).
- Category II. Probabilistic model of fire physical phenomena (e.g., [17])
- Category III. Integration of deterministic simulation of fire physical phenomena with probabilistic model (e.g., coupling of CFAST with Monte Carlo simulation [18,19]).

The models grouped in Category III [18,19] are conceptually the most similar to the *FSM* being developed in this research. However, the *FSM* differs from Category III models in several aspects. The main differences are in the following:

a) They are not linked to a Plant-specific PRA model. Their target outputs include the cumulative distribution function of physical variables (e.g., maximum heat release rate, maximum temperature of cables [19]) and the failure probability of individual electrical cable caused by fire-induced environmental conditions [18], rather than the total risk (e.g., core damage frequency) or the fire-induced failure probability of safety-related systems.

b) Their scope is limited to one or two fire scenarios. Both references [18] and [19] compute the scenario of electrical cable fire inside the cable tunnels. They do not account for any other type of fire scenarios such as fuel tank fire and battery fire.

c) They only deal with intermediate phases of fire events (i.e., fire ignition, fire progression, and post-fire failure). For instance, both references [18] and [19] do not take into account the fire ignition process and its probability since they focus on the conditional probability or fire-induced environment given a certain fire ignition. In addition, their scope is limited to the cable-level analysis, and the post-fire failure (e.g., spurious operation of equipment) is not considered [18,19].

In the following, the conceptual design of each sub-module in *FSM* is summarized as well as the current result of the literature review on the models of fire physical phenomena that are candidates for sub-modules.

2.2.1. Fire Initiation model:

This step is related to Task 2 (Fire PRA Compartment Selection), Task 3 (Fire PRA Cable Selection), and Task 6 (Fire Ignition Frequencies) in NUREG/CR-6850 [9]. Industrial fires can involve transient or in-situ combustibles. They can be initiated either by human error or, more frequently, from an electrical short. Electrical system design normally incorporates a fire suppression system for the rapid termination of the fire progression by eliminating the energy source (high resistance current flow). Even if the energy source is sustained, there is normally very little combustible material available to sustain the burn. Of course, there are still some systems in power reactors that are designed with a substantial amount of combustible material. Examples are fuel tanks for diesel generators, oil-cooled transformers, hydrogen-cooled generators, water treatment systems with large quantities of acid and base chemicals that could be a high energy source if they were to come into contact with an electrical short. Each of these systems would be amenable to uncertainty quantification in an accident progression represented by linked engineering models of fire spread, suppression, and energy/combustible material sources.

Bayesian updating has been used to estimate the generic fire ignition frequencies for use in Fire PRA (e.g., [9] and [20]). Because the current highly data-driven methodology is simple and conservative (due to assumed prior distributions), the resultant fire frequency data can be too conservative for a realistic modeling of the frequency of fire initiation. As a specific example, consequential events with an assumed prior frequency of 1E-03 or 1E-02 would be significantly affected by the absence of experience in the lifetime of a plant. We anticipate advancing the initiating event modeling beyond its current statistical/experimental state (e.g., Section 17; Appendix A of [21]) by physically modeling the initiation and growth phases of initiating events. This allows physical insights into the assumptions currently made for fire initiation. Research into physical modeling of cabinet fires, effects on solid state equipment, and cable tray propagation modeling (advancing the state as compared to [21], Section 11), will bring the state of Fire PRA closer to a true best-estimate, risk-management-capable framework.

As far as the authors know, the recent efforts for obtaining more accurate fire ignition frequency in the area of nuclear engineering have been mainly directed toward improving the methodology of statistical estimates with some modifications (e.g., trend analysis and choice of prior distributions in [21], while incorporating newer fire event data and considering between-plant variability using the Hierarchical Bayes approach [20]).

2.2.2. Fire Progression model:

The fire progression model in the proposed *integrated* framework is related to Task 8 (Scoping Fire Modeling) and Task 11 (Detailed Fire Modeling) defined in NUREG/CR-6850 [9]. For a more risk-informed (RI)/performance-based (PB) method of fire protection in NPPs, the NRC has verified and validated five mechanistic fire simulation codes, cited in NUREG-1824 [22]: Fire Dynamics Tools (FDT) [23], Fire-Induced Vulnerability Evaluation Revision 1 (FIVE-Rev1) [24], Consolidated Model of Fire Growth and Smoke Transport (CFAST) [16], MAGIC [25], and Fire Dynamics Simulator (FDS) [15]. NUREG-1934 [26] provides guidance on the appropriate selection and application of the models in the RI/PB approach. These simulation codes would be mainly used in our *FSM* to predict the behavior of fire and fire-induced effects (e.g. thermal radiation, high temperature gas, smoke density) on equipment and electrical cables. Table 1 summarizes characteristics of the models that are delineated in NUREG-1934 [26] and the literature that investigated the application of these models. The type of fire scenario to be analyzed and characteristics of the models will determine the type of the fire model. For instance, the FDS code is capable of simulating the fire progression and fire-induced environmental conditions by accounting for the complex geometrical configuration of a fire compartment and complex vent conditions [26]. Recently, the application of the FDS code [15] to a few fire scenarios has been reported by several authors [27-29]. Y. M. Ferng et al. [27] modeled the burning behavior of electrical control cables using the FDS code [15] and compared the outputs of the model with experimental results. They reported that the predicted transient profile of the heat release rate (HRR) was in agreement with experimental data; the maximum value of HRR agreed within 5 %, and the time of maximum HRR agreed within 10 %, while the qualitative shape of the profile (i.e., after a peak, HRR decreased and reached a low steady value) was well reproduced. In addition, S. Qiang et al. [28] simulated a fire event in the fuel tank room for an emergency diesel generator using the FDS code. They were able to simulate the qualitative interaction between the fire source and fire suppression by sprinklers such as the decay of the HRR time-profile and the decrease of 3-D temperature distribution around the fire source after the sprinkler was actuated. Although their work should be validated quantitatively, this example suggests that fire simulation models can be applied in current studies to analyze the effectiveness of fire detection and suppression. These applications of mechanistic fire simulation codes are grouped in Category I as defined at the beginning of this section.

Table 1: Main characteristics of five fire models verified and validated in NUREG-1824 [22].

Model	Category	Advantages	Limitations	Reference
FDT	Algebraic model	Low computational cost; Suitable for a comprehensive sensitivity study;	Not fully considering physical mechanism; Only applicable to steady-state fires or simply defined transient fires; Verified and validated for limited application range	
FIVE-Rev1				
CFAST	Zone model	Low computational cost; Suitable for a comprehensive sensitivity study; Verified and validated for wide range of use;	Larger errors with increasing deviation from a rectangular enclosure; Difficulty in treating horizontal flow paths;	[18] [30] [31]
MAGIC				[32]
FDS	Computational Fluid Dynamics Model	Applicable to complex configuration and vent conditions; Verified and validated for wide range of use;	Large effort to produce input files and post-processing of outputs; Long computational time;	[27] [28] [29] [33]

Several authors have studied the simulation-based modeling of electrical cable failure induced by fire using the THIEF model [34] (Category I), finite-element method [35] (Category I), and the combination of Monte Carlo simulation and CFAST [18] or FDS [19] (Category II). These methods enable us to obtain the time-profile of an electrical cable failure probability during a fire event with the consideration of a time delay caused by the heat transfer process from surrounding hot gas or cable surface to insulator inside the cable. The first two physical models (THIEF model and finite-element method), categorized in Category I, can be used in our *FSM* to simulate the time-dependent temperature distribution inside the electrical cable and to develop their conditional degradation probability. As input data, these models use the outputs from mechanistic fire progression codes such as surrounding environmental conditions (e.g., temperature, heat radiation from fire source, gas composition). Besides, the combination of the sampling method and the fire simulation code (e.g., Monte Carlo sampling and CFAST code [18] or FDS [19], categorized in Category III, is capable of simulating both the fire progression and its effect on targets within one computational framework. The clear advantage of this method is that the time-dependent probability distribution of target damage induced by fire is obtained by random sampling of input variables to the fire simulation code. In other words, this combinational method allows us to quantify and propagate the uncertainties that arise from input parameters directly through failure probability.

However, as mentioned in Table 1, the use of a detailed mechanistic fire model, especially the CFD model, is very resource-intensive in input generation, simulation, and output analysis, even with the current availability of multicore and cluster computing. Typically, the two-zone model (e.g., CFAST and MAGIC) is able to produce the solution in seconds to minutes, while the CFD model produces the corresponding answer in days to weeks [26]. Therefore, algebraic and zone models would be used in our *FSM* where they are adequate in terms of their applicability and accuracy. Also, for the fire scenarios where the algebraic or zone models are not applicable, the other alternative approach for computational cost reduction is to develop a response surface model [36] constructed from simulation output and use it as a part of the input data to *FSM*.

In Phase II of our research project, in order to have a realistic model in the *FSM*, additional "experimental tests", similar to experiments performed in [37-41], may be required to provide the supporting data for the fire simulation codes (e.g., Fire Dynamics Simulator; [15]) and the required analysis (e.g., electric cable degradation analysis). For instance, regarding the fire-induced electric cable damage, although several test programs [42,43] have been reported, we may still need to perform some additional experiments in order to obtain the degradation data under the plant-specific conditions (thermal exposure, cable type and cable arrangement).

2.2.3. Post-fire failure model:

Figure 2 shows a simplified example of the heat from a fire (for example, the pressurizer heater wiring) resulting in a malfunction of the PORV so that it fails and becomes stuck open. At this point, the event is a Small Break Loss of Coolant Accident (SBLOCA). Located near the PORV are the motor-operated block valves that are normally used to terminate this event by closing off the relief path for the respective PORV (usually, two valves are supplied). The same fire could damage either the block valve motor or its wiring to the extent that it would be inoperable. In Figure 2, post-fire failure models correspond to the "Electrical/Mechanical failure model", the "Block valve failure model", and the "PORV failure model". These are related to Task 9 (Detailed Circuit Failure Analysis) of NUREG/CR-6850 [9].

In the current Fire PRA methodology, the likelihood of fire-induced damage to an electric circuit is estimated based on a formulation developed through an expert elicitation process [9,44]. This methodology attempts to account for physical configurations based on limited experimental evidence [37-41], but is too general for a realistic Fire PRA since it cannot take into account plant-specific factors such as the specific shape and layout of electric cables. The inherent uncertainties in knowledge about specific cable placement, condition, configuration, etc. need to be treated in a more detailed manner than is currently possible.

There are several areas where the advancement in mechanistic modeling will benefit by predicting more accurate estimates of the likelihood of post-fire failure. One example of these areas is the calculation of spurious actuation probabilities. In the current methodology [9,45], these probabilities are calculated using either (i) the failure mode probability estimate table derived from expert elicitation [44], or (ii) a reverse-engineered formula from the fire test data [46], with a deterministic assumption (i.e., in cases where the cables of concern are dependent, the likelihood of spurious actuation should be determined by the first cable failure). As pointed out in NUREG/CR-6850 [8], additional consideration of the circuit failure mode likelihood values is needed. Mechanistic modeling will enable us to capture the elements which have not been taken into account in the current methodology, such as the intermediate modes of cable faulting between spurious actuation and fuse blow [44,45] and the effect of multiple cable failures on the spurious actuation [8,45].

2.2.4. Failure state distributions:

This step is related to Task 10 (Circuit Failure Mode & Likelihood Analysis) and Task 15 (Uncertainty & Sensitivity Analysis) in NUREG/CR-6850 [9]. Advanced uncertainty quantification and propagation techniques would be added to the physical models (in the *FSM*) in order to estimate the "probabilities" of basic events of interfaces and then, these probabilities would be incorporated into the Plant-specific PRA. Similar uncertainty propagation has been undertaken in the risk-informed resolution of GSI-191 [13]. Also, recent advances in uncertainty propagation techniques for physical modeling in the computational sciences can provide methodologies in order to circumvent the "curse of dimensionality" in uncertainty quantification. In the models used inside our *FSM*, there are uncertainties associated with the input variables (e.g., mass stoichiometric ratio of air to fuel, heat of combustion) and those associated with the sub-models and correlations (e.g., a sub-model to simulate the flame height) used in those simulation codes [17]. Research in the tailoring of existing uncertainty analysis techniques for application to the proposed *integrated* approach for fire risk analysis would be needed.

2.3. Input Module

The elements of the *FSM* would need certain input data. As Figure 2 shows, the *Input Module* includes:
- Plant boundary and partitioning (related to Task 1; Plant Boundary & Partitioning in NUREG/CR-6850 [9]).
- Post-fire HRA (related to Task 12; Post-Fire Screening in NUREG/CR-6850 [9]).
- Fire PRA database containing (i) plant partitioning and fire compartment designation, (ii) plant cable and raceway data (e.g., type and configuration of electrical cables and their relationship with safety equipment), and (iii) fire PRA equipment. It also contains the input parameters to mechanistic models (e.g., physical and chemical properties of electrical cable).
- Input from plant walk downs (e.g., ventilation features, possibly connected compartments).
- Seismic data (related to Task 13; Seismic-Fire Interaction in NUREG/CR-6850 [9])

Plant boundary and partitioning would be inputs to the "fire initiation model". Post-fire HRA would provide input to both fire initiation and fire progression models. HRA data will be also needed for the *Plant-Specific PRA Module*. Having a simulation-based fire module would create the possibility of advancing post-fire HRAs (i.e., using simulation-based HRA models). The Fire PRA database and input from plant walk downs would also be required for fire initiation and progression models. The data from a seismic model would be required to consider the potential seismic-fire interaction in both initiation and progression models.

In addition, as mentioned in the previous section, the generalized models developed based on the output from simulation-based models would be provided in the *Input Module*. The concept is that, instead of directly integrating complex simulation-based models into our *FSM*, we would develop a generalized model such as response surface model or correlation model (e.g., [36]) and provide those generalized models as input data.

3. CONCLUSION

The risk-informed fire protection program, when compared to the traditional deterministic approach, has a potential for more effective fire protection at NPPs. One of the widely recognized drawbacks of current Fire PRA methodology is the limitation in accounting for dynamic aspects of fire phenomena since it uses the classical PRA methodology based on ET/FT. As a solution to this limitation, we propose the integrated framework Fire PRA to model the dynamic phenomena of fire events without moving to a totally simulation-based or dynamic PRA.

In the integrated framework (Figure 2), some of fire-related FTs are replaced with a *FSM* separated from the plant-specific PRA model (i.e., ETs and FTs are extracted from the current PRA model). This FSM contains simulation-based realistic physical models for fire initiation, dynamic progression of fire effects, and post-fire failure. In addition, the uncertainties from the physical model and input parameters would be propagated through the module by sampling of the inputs. The output of the *FSM*, namely the distributions of conditional failure probability for interface basic events, will be imported to the *Plant-Specific PRA Module* and be used to calculate the risk metrics (i.e., core damage frequency).

This research project consists of two phases. In Phase I, we focus on integrating the *existing* physical models of fire events into the *FSM*. The main contributions from Phase I include: (i) change of quantification methods of fire-related risk from ET/FT to simulation-based techniques, (ii) propagation of uncertainties by randomly sampling the input parameters, and (iii) linkage of the *Fire Simulation Module* to the *Plant-specific PRA Module*. In Phase 2, after obtaining the results from Phase I, we would refine some of the physical models to create a more realistic quantification of fire risk.

In future work, for Phase I of the research project, we will continue building the *FSM* by integrating the existing models of fire physical phenomena. The first step in building the module is to identify (and categorize) the dominant fire-induced scenarios and determine the corresponding interface basic events. Then, the *FSM* will be developed for each category of interface basic event. In addition to the models used in the current methodology (NUREG/CR-6850), the models proposed in recent literature, in both the nuclear and non-nuclear arenas, will be considered as candidates for the elements of this module. We plan to complete the construction of the *Fire Simulation Module* and obtain primary outputs by the end of the current year. Then, we will proceed to uncertainty propagation, and verification and validation of the *Fire Simulation Module*, followed by connecting it to the Plant-specific PRA Module and importance measure analysis.

References

[1] NUREG-0050, "Recommendations Related to Browns Ferry Fire," U.S. Nuclear Regulatory Commission, Washington, D.C., 1976.

[2] "U.S. Code of Federal Regulations, Title10 (Energy), Part 50 (Domestic Licensing of Production and Utilization Facilities), Section 48 (Fire Protection)," Office of the Federal Register, Washington, D. C.

[3] "U.S. Code of Federal Regulations, Title 10 (Energy), Appendix R to Part 50 (Fire Protection Program for Nuclear Power Facilities Operating Prior to January 1, 1979)," Office of the Federal Register, Washington, D.C.

[4] "Voluntary fire protection requirements for light water reactors: adoption of NFPA 805 as a risk-informed, performance-based alternative (69 FR 33536)," Federal Registers, 2004.

[5] "Nuclear Regulatory Commission - Oversight and Status of Implementing a Risk-Informed Approach to Fire Safety (GAO-13-8)," United States Government Accountability Office (GAO), 2012.

[6] U.S. Nuclear Regulatory Commission, "Use of Probabilistic Risk Assessment Methods in Nuclear Regulatory Activities; Final Policy Statement," 1995.

[7] National Fire Protection Association (NFPA), "NFPA 805 Performance-Based Standard for Fire Protection for Light Water Reactor Electric Generating Plants (2001 Edition)," 2001.

[8] EPRI 1011989 and NUREG/CR-6850, "EPRI/NRC-RES Fire PRA Methodology for Nuclear Power Facilities: Volume 1: Summary and Overview," Electric Power Research Institute (EPRI) and U.S. Nuclear Regulatory Commission, Office of Nuclear Regulatory Research (RES), 2005.

[9] EPRI 1011989 and NUREG/CR-6850, "EPRI/NRC-RES Fire PRA Methodology for Nuclear Power Facilities: Volume 2: Detailed Methodology," Electric Power Research Institute (EPRI) and U.S. Nuclear Regulatory Commission, Office of Nuclear Regulatory Research (RES), 2005.

[10] J. Chapman, "Seeking Realism in Fire PRA," *ANS PSA 2013 International Topical Meeting on Probabilistic Safety Assessment and Analysis*, Columbia, SC, 2013.

[11] U. Schneider, "Introduction to fire safety in nuclear power plants," *Nuclear Engineering and Design*, vol. 125, pp. 289-295, 1991.

[12] Z. Mohaghegh, A. Mosleh, and R. Kazemi, "Incorporating Organizational Factors into Probabilistic Risk Assessment (PRA) of Complex Socio-technical Systems: A Hybrid Technique Formalization," *Reliability Engineering and System Safety*, vol. 94, no. 5, pp. 1000-1018, 2009.

[13] Z. Mohaghegh, E. Kee, S. Reihani, R. Kazemi, D. Johnson, R. Grantom, K. Fleming, T. Sande, B. Letellier, G. Zigler, D. Morton, J. Tejada, K. Howe, J. Leavitt, Y. Hassan, R. Vaghetto, S. Lee and S. Blossom, "Risk-Informed Resolution of Generic Safety Issue 191," *ANS PSA 2013 International Topical Meeting on Probabilistic Safety Assessment and Analysis*, 2013.

[14] G. Vinod, R. Saraf, A. Ghosh, H. Kushwaha and P. Sharma, "Insights from fire PSA for enhancing NPP safety," *Nuclear Engineering and Design*, vol. 238, pp. 2359-2368, 2008.

[15] K. McGrattan, B. Klein, S. Hostikka, and J. Floyd, "Fire Dynamic Simulator (Version 5) User's Guide," National Institute of Standards and Technology, 2007.

[16] W. Jones, R. Peacock, G. Fomey and P. Reneke, "CFAST - Consilidated Model of Fire Growth and Smoke Transport (Version 6) Technical Reference Guide," U.S. National Institute of Standards and Technology, 2009.

[17] V. Ontiveros, A. Cartillier and M. Modarres, "An Integratd Methodology for Assessing Fire Simulation Code Uncertainty," *Nuclear Science and Engineering,* vol. 166, pp. 179-201, 2010.

[18] S. Hostikka and O. Keski-Rahkonen, "Probabilistic simulation of fire simulation," *Nuclear Engineering and Design,* vol. 224, pp. 301-311, 2003.

[19] A. Matala and S. Hostikka, "Probabilistic simulation of cable performance and water based protection in cable tunnel fires," *Nuclear Engineering and Design,* vol. 241, pp. 5263-5274, 2011.

[20] R. Wachowiak, "An Improved Methodological Approach for Estimating Fire Ignition Frequencies," EPRI 1022994, 2011.

[21] K. Canavan, "Fire PRA Methods Enhancements: Additions, Clarifications, and Refinements to EPRI 1019189," EPRI 1016735, 2008.

[22] NUREG-1824 and EPRI 1011999, "Verification and Validation of Selected Fire Models for Nuclear Power Plant Applications, Volume 1: Main Report," U.S. Regulatory Commission, Office of Nuclear Regulatory Research, and Electric Power Research Institute (EPRI), 2007.

[23] NUREG-1805, "FDT: Fire Dynamics Tools (FDT): Quantitative Fire Hazard Analysis Methods for the U.S. Nuclear Regulatory Commission Fire Protection Inspection Program, Final Report," U. S. Nuclear Regulatory Commission, Office of Nuclear Reactor Regulation, Washington, DC, 2004.

[24] EPRI 1002981, "Fire Modeling Guide for Nuclear Power Plant Applications", Electric Power Research Institute (EPRI), Palo Alto, CA, 2002.

[25] Gay, L., "User Guide of MAGIC Software Version 4.1.1, EdF HI82/04/022/B," Electricité de France, 2005.

[26] NUREG-1934 and EPRI 1023259, "Nuclear Power Plant Fire Modeling Analysis Guideline (NPP FIRE MAG)," U.S. Nuclear Regulatory Commission, Office of Nuclear Regulatory Research (RES), and Electric Power Research Institute (EPRI).

[27] Y. Ferng and C. Liu, "Investigating the burning characteristics of electric cables used in the nuclear power plant by way of 3-D transient FDS code," *Nuclear Engineering and Design,* vol. 241, pp. 88-94, 2011.

[28] S. Qiang, M. Rongyi, L. Juan, Z. Jiaxu, Z. Chunming and C. Jianshe, "Numerical Simulation Study on Sprinkler Control Effect in UBS Fuel Tank Room of Nuclear Power Plants," *Procedia Engineering,* vol. 43, pp. 276-281, 2012.

[29] S. Suard, M. Forestier and S. Vaux, "Toward predictive simulations of pool fires in mechanically ventilated compartments," *Fire Safety Journal,* vol. 61, pp. 54-64, 2013.

[30] S. Arshi, M. Nematollahi and K. Sepanloo, "Coupling CFAST fire modeling and SAPHIRE probabilistic assessment software for internal fire safety evaluation of a typical TRIGA research reactor," *Reliability Engineering and System Safety,* vol. 95, pp. 166-172, 2010.

[31] Y.-H. Lee, J.-H. Kim and J.-E. Yang, "Application of the CFAST zone model to the Fire PSA," *Nuclear Engineering and Design,* vol. 240, pp. 3571-3576, 2010.

[32] L. Gay, B. Sapa and F. Nmira, "MAGIC and Code_Saturne developments and simulations for mechanically ventilated compartment fires," *Fire Safety Journal,* vol. 62, pp. 161-172, 2013.

[33] Y.-M. Ferng and C.-H. Liu, "Numerically investigating fire suppression mechanisms for the water mist with various droplet sizes through FDS code," *Nuclear Engineering and Design,* vol. 241, pp. 3142-3148, 2011.

[34] G. Valbuena and M. Modarres, "Development of probabilistic models to estimate fire-induced cable damage at nuclear power plants," *Nuclear Engineering and Design,* vol. 239, pp. 1113-1127, 2009.

[35] A. Stein, E. Sparrow and J. Gorman, "Numerical simulation of cables in widespread use in the nuclear power industry subjected to fire," *Fire Safety Journal,* vol. 53, pp. 28-34, 2012.

[36] M. Brandyberry and G. Apostolakis, "Response Surface Approximation of a Fire Risk Analysis Computer Code," *Reliability Engineering and System Safety,* vol. 29, pp. 153-184, 1990.

[37] A. Stein, E. Sparrow and J. Gorman, "Numerical simulation of cables in widespread use in the nuclear power industry subjected to fire," *Fire Safety Journal,* vol. 53, pp. 28-34, 2012.

[38] M. Brandyberry and G. Apostolakis, "Response Surface Approximation of a Fire Risk Analysis Computer Code," *Reliability Engineering and System Safety,* vol. 29, pp. 153-184, 1990.

[39] NUREG-2128, "Electrical Cable Test Results and Analysis During Fire Exposure (ELECTRA-FIRE) – Draft Report for Comment," U.S. Nuclear Regulatory Commission, Office of Nuclear Regulatory Research (RES), 2012.

[40] NUREG/CR-7010, vol. 1, "Cable Heat Release, Ignition, and Spread in Tray Installations During Fire (CHRISTIFIRE) Phase 1: Horizontal Trays," U.S. Nuclear Regulatory Commission, Office of Nuclear Regulatory Research (RES), 2012.

[41] NUREG/CR-7100, "Direct Current Electrical Shorting in Response to Exposure Fire (DESIREE-Fire): Test Results," U.S. Nuclear Regulatory Commission, Office of Nuclear Regulatory Research (RES), 2012.

[42] NUREG/CR-7102, "Kerite Analysis in Thermal Environment of FIRE (KATE-Fire): Test Results," U.S. Nuclear Regulatory Commission, Office of Nuclear Regulatory Research (RES), 2011

[43] NUREG/CR-7150, BNL-NUREG-98204-2012, and EPRI 1026424, "Joint Assessment of Cable Damage and Quantification of Effects from Fire (JACQUE-FIRE) Volume 1," U.S. Nuclear Regulatory Commission, Office of Nuclearar Regulatory Research (RES), and the Electric Power Research Institute (EPRI), 2012.

[44] NUREG/CR-6776, "Cable Insulation Resistance Measurements Made During Cable Fire Tests," U.S. Nuclear Regulatory Commission, Office of Nuclear Regulatory Research (RES), 2002.

[45] NUREG/CR-6931, "Cable Response to Live Fire (CAROLFIRE) Volume 2: Cable Fire Response Data for Fire Model Improvement," U.S. Nuclear Regulatory Commission, Office of Nuclear Regulatory Research (RES), 2008.

[46] R. Kassawara, "Spurious Actuation of Electrical Circuits Due to Cable Fires: Results of an Expert Elicitation," Electric Power Research Institute (EPRI), EPRI 1006961, 2002.

Lessons Learned from the US HRA Empirical Study

Huafei Liao[a*], John Forester[a,b], Vinh N. Dang[c], Andreas Bye[d], Erasmia Lois[e], Y. James Chang[e]

[a] Sandia National Laboratories, Albuquerque, NM, USA
[b] Idaho National Laboratory, Idaho Falls, ID, USA
[c] Paul Scherrer Institute, Villigen PSI, Switzerland
[d] OECD Halden Reactor Project, Institute for Energy Technology, IFE, Halden, Norway
[e] U.S. Nuclear Regulatory Commission, Washington, DC, USA

Abstract: The US Human Reliability Analysis (HRA) Empirical Study (referred to as the US Study in the article) was conducted to confirm and expand on the insights developed from the International HRA Empirical Study (referred to as the International Study). Similar to the International Study, the US Study evaluated the performance of different HRA methods by comparing method predictions to actual crew performance in simulated accident scenarios conducted in a US nuclear power plant (NPP) simulator. In addition to identification of some new HRA and method related issues, the study design of the US Study allowed insights to be obtained on some issues that were not addressed in the International Study. In particular, because multiple HRA teams applied each method in the US Study, comparing their analyses and predictions allowed separation of analyst effects from method effects and allowed conclusions to be drawn on aspects of methods that are susceptible to different application or usage by different analysts that may lead to differences in results. The findings serve as a strong basis for improving the consistency and robustness of HRA, which in turn facilitates identification of mechanisms for improving operating crew performance in NPPs.

Keywords: HRA, simulator data, nuclear power plant, performance shaping factor.

1. INTRODUCTION

As an effort to improve the robustness of human reliability analysis (HRA), the US Nuclear Regulatory Commission (NRC) participated in and supported the International HRA Empirical Study [1-4] (referred to as the International Study hereafter), in which HRA predictions of different analysts and methods were compared to observed crew performance data at Halden Reactor Project's HAMMLAB (HAlden huMan-Machine LABoratory) simulator facilities. Many HRA methods were tested in the study; however since, with one exception, there was one HRA team applying each method, this limited our ability to make decisive conclusions concerning how the HRA analysts' applications of a given method could contribute to the variability in HRA results and how the methods themselves contributed to the variability.

In contrast, the HRA methods evaluated in the US HRA Empirical Study [5-7] (referred to as the US Study hereafter) were applied by multiple HRA teams. This allowed us to obtain new insights on factors contributing to variability in HRA results as well as strengths and weaknesses of the chosen HRA methods. Another notable difference from the International Study was that the HRA analyst teams in the US Study were able to visit the plant, observe a crew in a simulator training scenario, and interview training personnel to collect information needed to perform HRA. Third, the US Study was performed on a US nuclear power plant (NPP) training simulator, which allowed us, to some extent, to evaluate the generalizability of Halden human performance studies to US applications.

In the following sections, the study methodology, simulator data, and predictive quantitative results will be presented. Then, the findings on the contributing factors to variability in HRA results will be discussed.

---

* Corresponding author.
 Email address: hnliao@sandia.gov

2. STUDY METHODOLOGY OVERVIEW

The US Study capitalized on the design and methodology of the International Study. It focused on control room personnel actions required in the response to initiating events typically modelled nuclear power plant (NPP) probabilistic risk assessments (PRAs). Three scenarios were developed and five human failure events (HFEs) were defined (see [5] for detailed description of scenarios and HFEs). Scenario 1 was a total loss of feedwater (LOFW) followed by a steam generator tube rupture (SGTR), for which three HFEs (HFEs 1A, 1B, and 1C) were defined. Scenario 2 was a loss of component cooling water (CCW) and reactor cooling pump (RCP) sealwater, for which one HFE (HFE 2A) was defined. Scenario 3 was an SGTR scenario without further complications, in which one HFE (HFE 3A) was defined. HFE definitions were based on the definitions of similar HFEs from real plant PRAs and were defined on a functional level (i.e., "fails to perform X before Y" or "fails to perform X within t minutes"). In some cases the HFEs were defined with stricter success criteria than many HFEs in standard PRA scenarios. The reason for this is that although the HFE success criteria used in the study should relate to those commonly used in PRA/HRA, for the purposes of this study, it was important that they be clearly observable in the simulated scenarios.

Four crews of five licensed operators from a participating US NPP simulated the scenarios on the plant full-scope pressurized water reactor (PWR) training simulator. Their performance data were analyzed and described in the following three ways.

- Performance on the HFE related actions expressed in operational terms ("operational descriptions");
- Assessment of the performance shaping factors (PSFs) (main drivers) for each action;
- Number of crews failing to meet the success criteria for each action and an assessment of the difficulty of the action

Nine HRA teams participated in the US Study and each team applied an HRA method to predict performance relative to the HFEs defined in the study. Two teams used ATHEANA (A Technique for Human Event Analysis) [8], two teams used SPAR-H (Standardized Plant Analysis Risk-Human Reliability Analysis) [9], two teams used ASEP ((Accident Sequence Evaluation Program Human Reliability Analysis Procedure) [10], two teams used the EPRI HRA Methodology (implemented with the HRA Calculator version 4.1.1) [11] and used CBDT (Cause-Based Decision Tree) [12] and HCR/ORE [12] for the diagnosis portion of the response and THERP (Technique for Human Error Rate Prediction) [13] for response execution, and one team used a hybrid CBDT+THERP+ASEP method with similarities to the EPRI HRA method. The predictions were compared to crew performance data to assess the predictive power of the method applications and examine aspects such as traceability, method guidance, and insights for error reduction.

Predictive power was assessed in terms of qualitative predictive power and quantitative predictive power. Qualitative predictive power was evaluated based on a comparison of predicted operational expressions and observed operational descriptions, as well as a comparison of predicted PSFs and observed performance drivers. Quantitative predictive power was assessed in terms of the absolute values of the HEPs predicted by each HRA team and the ranking of the HFEs based on the magnitude of the predicted HEPs. Given the small sample of observations, the accuracy of the predicted HEPs is difficult to assess. Therefore, although the quantitative predictive power was assessed to the extent possible in light of the small number of data points, method assessment was conducted primarily from a qualitative analysis perspective. In practice the assessment of quantitative predictive power were performed as a prelude to analysis of qualitative issues.

As mentioned above, the HRA methods in the US Study were applied by multiple HRA analyst teams. Therefore, the predictions of the different teams using the same method were compared to separate analyst effects from method effects to the extent possible and gain insights on the factors that can contribute to variability in HRA results. The intra-method comparison focused on the differences in qualitative predictions, quantitative results, analysis approaches and assumptions.

3. EMPIRICAL AND PREDICTIVE QUANTITATIVE RESULTS

Simulator data were collected for four HFEs. The crew failure rates and HFE difficulty ranking are listed in Table 1. The difficulty ranking of the HFEs was determined in terms of the crew failure rates and the challenges experienced by the crews in diagnosing plant status and executing manual actions to bring the plant to a stable state.

Table 1. Crew Failure Rates and HFE Difficulty Ranking

HFE	Failure Rate	Difficulty Ranking
HFE 2A	4/4	1 (Very difficult)
HFE 1C	3/4	2 (Difficult)
HFE 1A	0/4	3 (Fairly difficult to difficult)
HFE 3A	0/3	4 (Easy)

* No data were collected for HFE 1B.

The mean HEPs predicted with each method are presented in Figure 1 alongside the Bayesian uncertainty bounds derived from the simulator data with a non-informative prior (Jeffrey's prior). The HFEs are ordered by their difficulty ranking on the horizontal-axes, and the HEPs are presented on the vertical-axes in a logarithmic scale. The following observations are made from the HEP curves (see [5-6] for more detailed discussion).

- Overall, most methods identified the correct HFE difficulty ranking with a few exceptions in terms of the relation between HFEs 2A and 1C.
- For most HFEs, there is about one order of magnitude or less difference across the HRA analyst teams using the same method, especially if we consider the HEP of HFE 1C predicted by the CBDT & HCR/ORE Team 2 was caused by a misunderstanding of the HFE definition.
- Across the four methods, it seems that ASEP, ATHEANA, and CBDT & HCR/ORE produced relatively more consistent quantitative results than the SPAR-H.
- Many teams underestimated the most difficult HFE 2A. This seems to be caused either by insufficient qualitative analysis to understand the scenario dynamics, or by inappropriate assumptions or interpretations based on information obtained from the interviews with plant personnel.
- Although all of the HRA analyst teams concluded that HFE 3A was the easiest, there is significant variability in the HEPs with a couple of estimates being much lower than the other HEPs. This seems to indicate that there is a lack of consensus in HRA in terms of what the baseline HEP for generally good conditions should be.

It could be argued that with the exception of HFE 3A, there was less variability in the predicted HEPs across the methods in the US Study compared to the International Study. In addition, the difference in the within method HEP predictions was less than might have been expected based on general HRA performance in the International Study and on the couple cases where similar methods were used. That is, one might argue that the analysts using the same methods in the US Study did a relatively good job in many cases and often corresponded relatively well in their predicted difficulty rankings of the HFEs. Potential reasons for these results include the following:

- There may have been some learning effects between the International Study and the US Study. Some of the HRA analysts participated in both studies and most participants in the US Study were familiar with the results of the International Study. Thus, the lessons learned may have improved the HRA team's applications in that they had a better idea of what they needed to do to perform a better analysis with the method they were using. There also appears to be some evidence of the learning effects between the different phases of the International Study.

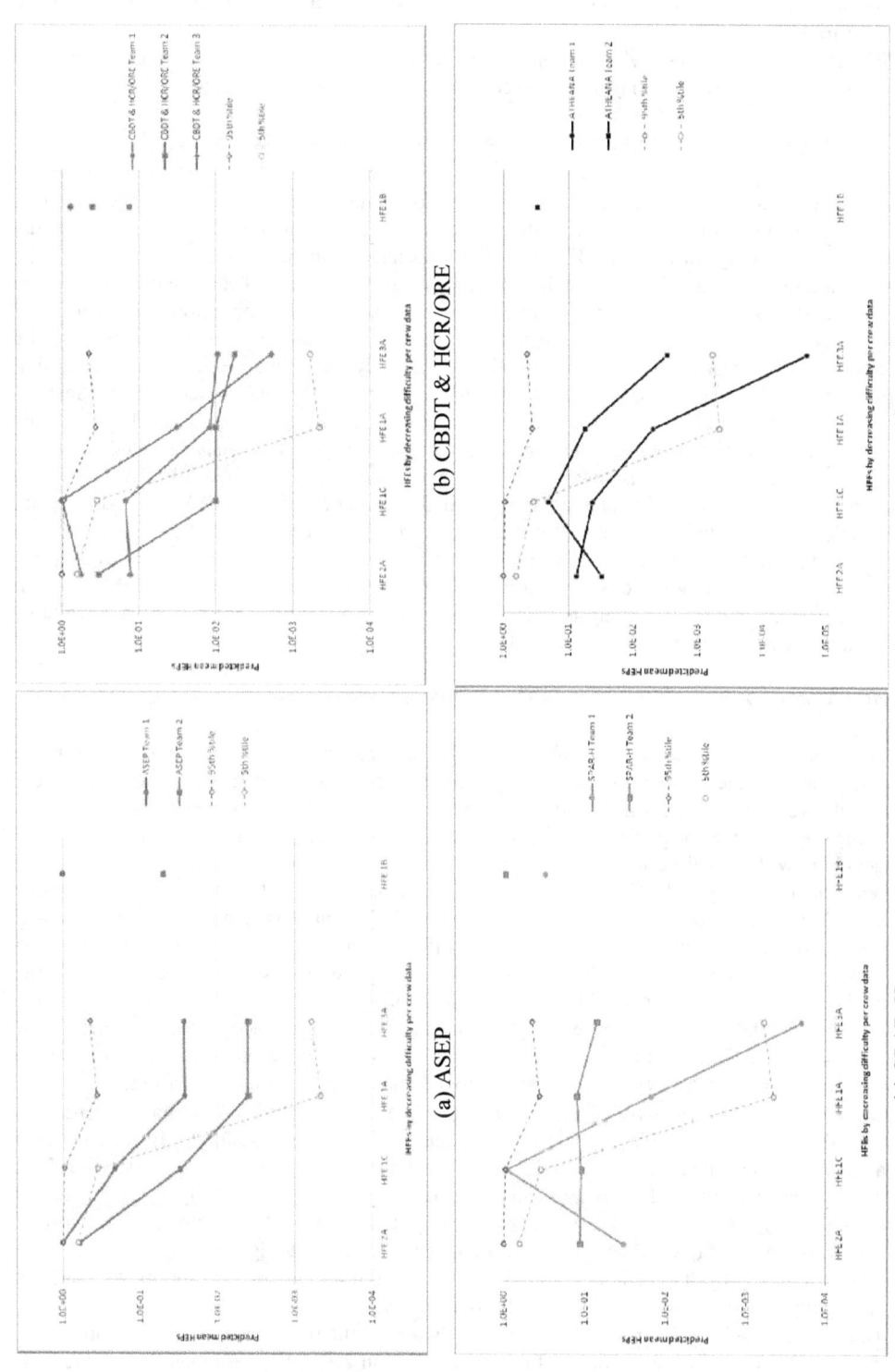

Figure 1. Predicted Mean HEPs by HRA Methods with Bayesian Uncertainty Bounds

- It is also possible that the HRA teams were, in fact, better at predicting the performance of US crews on a US simulator than the performance of foreign crews. That is, their previous experience with how US crews work and interact with procedures etc. may have facilitated their performance.
- Similarly, their ability to visit the plant, interact with plant personnel, and obtain the usual HRA related information may also have contributed to improved performance.

4. FACTORS CONTRIBUTING TO HRA PREDICTIVE DIFFERENCES

The International Study identified that significant variability can occur in the results of applying different HRA methods for the same HFE due to the differences and limitations in the methods' technical and methodological bases [1-4] and in the associated guidance provided for applying the methods. In addition, although it appeared that the analysts themselves could contribute to differences in results (analyst driven factors), the design of the International Study did not allow clear separation of method effects from analyst effects because there was only one case where two teams used the same method. With at least two teams per method, the US Study was able to identify that in addition to inherent method-driven factors, analyst-driven factors and the interactions between the analyst-driven factors and method-driven factors can also cause significant variability in the HRA results. Overall, the US Study revealed that a major source of variability across analysts using the same method was analyst decisions about how to apply various aspects of the method. Analysts are often called upon to make decisions in their analyses, and the guidance of the HRA methods are not sufficient or specific enough, so that analysts may, to some degree, have to rely on their own subjective interpretation of the guidance. Moreover, the methods sometimes allow analysts to apply different analysis or modeling options without providing clear criteria for when to use the different options. The factors leading to analysts' subjectivity and thus contributing to HRA predictive differences are discussed below.

4.1. Factors Contributing to Differences in Qualitative Analysis

- *Differences in qualitative analysis approaches, scope and depth.* One reason for the differences is the degree to which a specific method framework specifies or guides the qualitative analysis process. Some methods (e.g., SPAR-H) do not provide specific or complete guidance for performing the qualitative analysis, and thus HRA analysts are left to decide how they will perform the qualitative analysis and its level of detail. Although some other methods (e.g., ATHEANA) support the search and treatment of a more comprehensive set of performance drivers at a more detailed level, the guidance is somewhat open-ended and not always well-structured for translating the information into HEPs in a consistent manner; thus, to some extent leaving the level of detail up to the analysts. Another reason is the analysts' level of effort devoted to their qualitative analyses. Some analysts may undertake a detailed qualitative analysis by even going beyond method guidance as they see necessary. Although more detailed qualitative analyses tend to lead to a better understanding of scenario dynamics, it does not necessarily lead to improved quantitative predictive performance.
- *Differences in task decomposition approaches.* Most HRA methods do not have a consistent approach or provide much guidance for task analysis and decomposition. Insufficient task analysis to understand scenario conditions, procedural interactions, and key manual actions may cause analysts to fail to appreciate the task complexity, especially for complicated scenarios. This can lead to different task groupings and dependency modeling, which, in turn, can affect quantitative results. It can also cause analysts to ignore cognitive activities involved in step-by-step actions. In addition, the level of task decomposition may impact the application and traceability of the quantitative analysis.
- *Timeline analysis.* Timing is either explicitly or implicitly considered in various HRA methods; hence the uncertainties in timing analysis can affect HEP estimates. For methods strongly based on time reliability correlations (TRCs) (e.g., ASEP and HCR/ORE), HEPs are estimated as a function of the time available for operators to respond to accident scenarios and

the HFEs of interest. Considering the characteristics of TRCs, the HEPs can be sensitive to timing analysis results, particularly for the HFEs that are time critical, that is, small differences in time windows can lead to large differences in HEPs. For some methods that are not TRC-based (e.g., SPAR-H), the availability of time is often considered as a PSF. The differences in the rating of this PSF can also lead to variability in HEPs, which can be as large as that obtained with TRCs. The variability in timing analysis can occur because: (1) analysts need to make decisions on which procedural steps to include in the timing analysis; (2) some methods do not provide specific guidance on assessing the values of timing points (e.g., how to assess the time required for an action); and (3) analysts' judgment is needed to account for the impact of distractions, delays, and competing task demands on timing analysis.

- *Difficulty in understanding and treating complexity.* HRA methods differ in their ability to deal with complexity. It can be difficult for the analysts to identify and treat the issues in complex scenarios that are beyond the scope of a given method. Although narrative-based methods (e.g., ATHEANA) appear to present some advantages over PSF-based methods in addressing complex HFEs, there are some aspects that are still left to the analysts. While the analysts can always go beyond the methods in qualitative analysis for HFEs that are broader than the scope of the factors explicitly treated by a given method, such an extended analysis is subject to variability because the analysts will determine the appropriate scope of performance drivers without a common basis for this judgment. Additionally, as mentioned above, the results of this extended analysis may not be easily incorporated into the quantification portion of the method to produce improved quantitative predictions.

4.2. Factors Contributing to Differences in Quantitative Analysis

- *Judgment about when to credit recovery.* Although some methods provide an option to credit error recovery, it is often the analysts' decisions whether or not to consider it in an HRA application and whether the criteria are adequately met. Some analysts may choose not to credit recovery just for a preference toward a more conservative result. Different biases between analysts toward obtaining conservative results can obviously lead to variability in results. This type of variability is not necessarily a problem as long as the bases for the decisions are made clear and analysts have the opportunity to follow up on the estimates and do more detailed and realistic analysis for important contributors.

- *Attempt at compensation for inadequate range of PSFs.* It has been observed in the US Study that in some situations the PSF-based methods were inadequate in coping with some HFE-specific performance drivers identified in the crew data simply because they were not addressed by the method. As a result, some analysts had to compensate for such limitations by stretching the method to fit the situation based on their experience. This led to quantitative differences in HRA results between the applications. This finding suggests that to be able to reliably predict performance, the PSFs need to have a sufficient coverage. Nevertheless, as was also shown and discussed in the International Study, the US Study showed that the PSF-based methods sometimes produced reasonable HEPs without identifying all performance drivers, particularly for the easy HFEs. It could not be determined whether this reflects an inherent characteristic of the methods or whether it was just a coincidental effect.

- *Decision on PSF selection and rating.* For some methods (e.g., SPAR-H, ASEP and CBDT), judgment about the relevance of a particular factor and the specific level of that factor in a given scenario must be made, and for others (e.g., ATHEANA) the analyst must determine what factors are present and characterize them, including the strength of their impact. Definition overlap between PSFs and inadequate guidance on determining the PSF status can cause variations in analysts' interpretation of the scope of the PSFs and in the ratings assigned to the PSF for a given issue or performance condition. This underscores the importance of addressing such issues for PSF-based methods.

- *Inadequate coupling between qualitative analysis and quantification model.* A broad qualitative analysis in evaluating likely crew performance does not necessarily lead to HEPs that are consistent with the observed crew performance. For some methods, the guidance on quantification of the impact of PSFs on crew performance is limited, and to varying degree left to analysts' judgment, particularly when the analysts' qualitative analysis goes beyond the method guidance. In addition, it seems that not all HRA methods cover an adequate range of PSFs to predict operating crew performance for all circumstances; as a result, analysts may have to rely on their judgment to decide how to integrate the role of factors not explicitly covered by a method in HEP estimation, which can obviously lead to variability in results. A good tie or dovetailing between the qualitative analysis and the quantification approach is important for consistency in the results of HRA applications within methods.

- *Efforts to compensate for poor treatment of diagnosis.* Methods that strongly rely on TRCs to quantify diagnosis (e.g., ASEP) appear to be poorly equipped to address the difficulties in operators' cognitive activities. In addition, it does not appear that all the HRA methods in the study are well equipped to address the full scope of cognitive activities related to operators' overall response to the scenarios. This can lead analysts to attempt to compensate for the method's shortcomings by doing more qualitative analysis than is directed by the method and trying to incorporate the information into the quantification approach by adjusting the method based on their own experience, which can introduce variability in results. Moreover, when the methods do address diagnosis, analysts tend to focus on operators' cognitive activities in understanding the plant situation to decide the appropriate response plan (i.e., initial diagnosis). The cognitive activities in executing the response plan are typically omitted. This has significant effects when the response plan is more complex than simple skill-of-the-craft. The study results have showed that inadequate consideration of operators' cognitive challenges in working through procedures can lead to failure in identifying important performance drivers and result in underestimations of HEPs. Although analysts may compensate for methods' inadequacy in diagnosis treatment based on their experience, it can lead to quantitative differences in HRA results.

4.3. Factors Related to HRA Practices

- *Different levels of reliance on interview information.* Information from interviews with plant personnel was used for several purposes, such as estimating the time required for diagnosis and execution and evaluation of training and experience. Most of the time, the information provided valuable insights for analysts to understand the scenario dynamics. However, the analysts differed in the extent to which the interview information was used in their analyses. Some analysts tended to rely on input directly from the interviews while others tended to rely on their own analysis and judgment with interview information as a supplement. The US Study has shown that over-reliance on trainer or operator opinions can sometimes negatively impact HRA results. Furthermore, it also seems to suggest that the differences in plant personnel's opinions may increase with the increase of scenario complexity. Obtaining consensus among multiple trainers or operators and/or more detailed qualitative analysis may help reduce the effects.
- *Different approaches to plant personnel interview.* It was observed that the HRA analysts differed in the scope and level of detail of the qualitative scenario analysis conducted before the interviews. They also differed in their interview techniques. Some analysts focused on questions regarding whether operators would take some actions. In contrast, some analysts asked the plant personnel to do a detailed talk-through together with them and focused on questions regarding timelines and the interactions between operators and procedures. The US Study seemed to indicate that sufficient preparation before interviews and the talk-through based interview technique would provide a better understanding of scenario dynamics and complexity.
- *Reasonableness check of HEPs.* The HEPs resulting from the application of each HRA method in the US Study were assessed with respect to their relative values (rank order of HFEs by failure probabilities) and the overall differentiation among these probabilities. Overall, the HEPs showed reasonable differentiation. However, in some cases, the HEPs for the two most difficult HFEs (2A and 1C) were not consistent with the difficulty ranking and/or fell into a narrow range. Particularly, the HEPs of SPAR-H Team 2 did not show differentiation expected from their qualitative analysis. This suggested that the team did not

check the reasonableness of the obtained probabilities or performed an inadequate check in view of their qualitative findings. This is especially remarkable for their analysis of HFE 3A.

Although HEPs are often checked for reasonableness in external reviews of the PRA/HRAs, where each individual HEP cannot be reviewed in detail and emphasis is placed on the relative values of the HEPs, there appears to be little documented guidance on how to perform reasonableness checks. Some of the factors to be considered in terms of similarity and levels of challenge include:

- Available time (time window)
- Decision complexity, basic vs. complex scenarios (number of issues, need to prioritize)
- Task complexity, number of tasks, need for manual control, fine-tuning, adjustment
- Number of issues, adverse performance shaping factors, and failure modes identified for the HFE

Comparing the related HFEs (for the same tasks performed in different scenarios) or HFEs with similar performance conditions typically leads analysts to review their differences in HEPs to determine whether the HEPs correspond to their qualitative analysis. For the HRA teams that did perform such checks, the identified discrepancies between HEP results and qualitative expectations would lead them to review the quantification and in some cases adjust the quantification of the HFEs. In summary, the development of guidance for reasonableness checks would help to promote a structured review of HRA results that emphasizes the consistency between qualitative findings and quantification results.

5. FINDINGS ON HRA GENERAL ISSUES

There was significant agreement in the findings between the International and US Studies. The conclusions from both studies about HRA in general and the identified needed improvements are summarized below.

- *Consideration of cognitive activities.* Both studies agreed that the consideration of operator cognitive activities is an important contributor to the adequacy of HRA predictions. It is especially beneficial to understand the difficulties in operators' assessing the situations and/or making new response plans in complex scenarios. However, the US Study has also revealed that even when diagnosis is explicitly considered, the methods still show some limitations in the ability to assess crews' cognitive activities in order to adequately support the understanding and identification of failure mechanisms and the HEP quantifications, particularly for the more difficult HFEs.
- *Explicit guidance and framework to support structured and consistent qualitative analysis.* While a good qualitative analysis is a relative strength of some methods (e.g., ATHEANA), one conclusion from both studies is that qualitative analysis is a shared weakness across all methods. The variability across HRA applications is not unexpected given the differences in the technical bases and methodologies of the methods. However, it seems that the variability also has its root in the fact that the methods do not provide sufficient guidance or an explicit framework for analysts to conduct a structured and consistent qualitative analysis. This is clearly evidenced by the variability in timing analysis, HFE decomposition, the range of factors considered, and the treatment of complexity.
- *Method improvement and extended qualitative analysis to treatment of complexity.* There is a need for method improvement to cover a broader scope of performance drivers in all methods. Given that complex scenarios normally involve relatively more cognitive challenges than easy scenarios, one priority of method improvement should focus on providing means and frameworks for analysts to identify and characterize contextual factors and mechanisms that can cause failures at the cognitive level and provide a structured and systematic way to incorporate the information into the quantification process. However, it should be realized that HRA applications will always rely to some extent on analysts' experience and expertise.

Nevertheless, the goal should be to provide as much structure and guidance as possible to support analysts at differing levels of expertise.
- *Coherent coupling between qualitative analysis and quantitative model.* As discussed above, extended qualitative analysis can help analysts uncover scenario-specific performance drivers. However, it may not necessarily lead to appropriate HEP estimates in all cases because of the difficulties in translating qualitative analysis into HEP impact, especially for complex scenarios.
- *Adherence to good practices with improved guidance.* Improved guidance is needed for performing plant visits and personnel interviews. Additionally, there is little documented guidance on how to perform reasonableness checks on HEPs and the consistency between qualitative findings and quantitative results in terms of performance conditions.

6. ACHIEVEMENTS AND OVERALL CONCLUSIONS

The US Study and the International Study are two large-scale systematic data collection and HRA method evaluation efforts. The achievements from the studies are summarized below:

- The US Study and the International Study demonstrated the feasibility of using simulator data to evaluate HRA methods. The methodological tools developed in the studies, such as (1) the development of the experimental design, focusing on evaluating HRA methods, (2) the methodology for collecting simulator data, (3) the methodology for analyzing simulator data for PRA-type of scenarios and tasks, and (4) the methodology for data-to-method comparisons, were tailored to HRA needs and are proving to be very useful achievements. They have also demonstrated that important information on HRA and HRA methods can be obtained without using impractically large numbers of operating crews and scenarios.
- The studies have shown that simulator data are highly useful for HRA studies. Although simulator data was used as the empirical basis against HRA methods' predictions, the promising results from this study encourage and promote the use of simulator data in the future, as well as encouraging analysts to use it in different ways. The potential of using and aggregating empirical simulator results from multiple studies to strengthen the empirical basis for both method assessment and extending the scope of methods to address some of the identified shortcomings. In summary, while there are other sources of HRA data, this study reinforced the relevance of simulator data for HRA in general. We also saw similar crew performance in the US Study as in the International Study, indicating that the results from the HAMMLAB simulator are applicable to the human performance in NPPs in other countries.
- The scenarios developed in the studies are similar to those modelled in PRA and represent difficulty levels from basic to highly complex. They can be used as standard scenarios for other HRA benchmarking studies. Complex scenarios can be used to determine whether an HRA method may lead to HEP underestimation. Basic scenarios can be used to establish a baseline performance. The difficulty levels can be used to test whether a method can produce HEPs with appropriate differentiation.

Acknowledgements

The authors gratefully acknowledge the contributions of Helena Broberg and Salvatore Massaiu, the Halden Reactor Project, Bruce Hallbert and Tommy Morgan, Idaho National Laboratory, and Amy D'Agostino, USNRC for major parts of the experimental work done in the project. The work of the nine HRA teams has of course been of invaluable importance, as was that of the additional assessment team members, Alysia Bone, USNRC, Katrina Groth, Sandia National Laboratories, and Stuart Lewis, Electric Power Research Institute. Very special thanks goes to the US nuclear power plant that supported the study with their training simulator, operating crews, instructor support in designing the scenarios, and multiple staff supporting the data collection and analysis. The plant support is obviously a major contribution to supporting the improvement of HRA and the safety of nuclear power plants.

This study is a collaborative effort of the Joint Programme of the OECD Halden Reactor Project, the U.S. Nuclear Regulatory Commission (USNRC), the Swiss Federal Nuclear Inspectorate (DIS-Vertrag Nr. 82610) and the U.S. Electric Power Research Institute. In addition, parts of this work were performed at Sandia National Laboratories and Idaho National Laboratory (INL) with funding from the USNRC. Sandia is a multi-program laboratory operated by Sandia Corporation, a Lockheed Martin Company, for the United States Department of Energy under Contract DE-AC04-94AL85000. INL is a multiprogram laboratory operated by Battelle Energy Alliance LLC, for the United States Department of Energy under Contract DE-AC07-05ID14517. The opinions expressed in this paper are those of the authors and not those of the USNRC or of the authors' organizations.

References

[1] E. Lois, V.N. Dang, J. Forester, H. Broberg, S. Massaiu, M. Hildebrandt, P.Ø. Braarud, G. Parry, J. Julius, R. Boring, I. Männistö, and A. Bye. "*International HRA Empirical Study—Phase 1 Report: Description of Overall Approach and Pilot Phase Results from Comparing HRA Methods to Simulator Data. NUREG/IA-0216, Vol. 1,*" US Nuclear Regulatory Commission, 2009, Washington, DC.

[2] A. Bye, E. Lois, V.N. Dang, G. Parry, J. Forester, S. Massaiu, M. Hildebrandt, P.Ø. Braarud, H. Broberg, J. Julius, I. Männistö, and P. Neslson. "*International HRA Empirical Study—Phase 2 Report: Results from Comparing HRA Method Predictions to Simulator Data from SGTR Scenarios. NUREG/IA-0216, Vol. 2,*" US Nuclear Regulatory Commission, 2011, Washington, DC.

[3] V.N. Dang, J. Forester, R. Boring, H. Broberg, S. Massaiu, J. Julius, I. Männistö, H. Liao, P. Neslson, E. Lois, and A. Bye. "*The International HRA Empirical Study - Phase 3 Report: Results from Comparing HRA Methods Predictions to HAMMLAB Simulator Data on LOFW Scenarios. NUREG/IA-0216, Vol. 3,*" US Nuclear Regulatory Commission, 2011, Washington, DC.

[4] J. Forester, V.N. Dang, A. Bye, E. Lois, S. Massaiu, H. Broberg, P. Broberg, R. Boring, I. Männistö, H. Liao, J. Julius, G. Parry, and P. Neslson. "*The International HRA Empirical Study— Final Report: Lessons Learned from Comparing HRA Methods Predictions to HAMMLAB Simulator Data NUREG-2127,*" US Nuclear Regulatory Commission, 2013, Washington, DC.

[5] J. Forester, H. Liao, V.N. Dang, A. Bye, M. Presley, J. Marble, H. Broberg, M. Hildebrandt, E. Lois, B. Hallbert, and T. Morgan. "*The US HRA Empirical Study – Assessment of HRA Method Predictions against Operating Crew Performance on a US Nuclear Power Plant Simulator. NUREG-2156,*" US Nuclear Regulatory Commission, 2014, Washington, DC.

[6] J. Marble, H. Liao, J. Forester, A. Bye, V.N. Dang, M. Presley, and E. Lois. "*Results and Insights Derived from the Intra-Method Comparisons of the US Empirical HRA Study,*" Proc. PSAM11&ESREL2012, June 25-29, 2012, Helsinki, Finland.

[7] A. Bye, V.N. Dang, J. Forester, M. Hildebrandt, J. Marble, H. Liao, and E. Lois. "*Overview and First Results of the US Empirical HRA Study,*" Proceedings of the 11th International Probabilistic Safety Assessment and Management Conference, June 25-29, 2012, Helsinki, Finland.

[8] Technical Basis and Implementation Guidelines for A Technique for Human Event Analysis (ATHEANA), NUREG-1624, Rev. 1, US Nuclear Regulatory Commission, Washington, D.C., May 2000.

[9] D. Gertman, H. Blackman, J. Marble, J. Byers, L. Haney, and C. Smith. "*The SPAR-H Human Reliability Analysis Method NUREG/CR-6883,*" U.S. Nuclear Regulatory Commission, 2005, Washington, D.C.

[10] A.D. Swain. "*Accident Sequence Evaluation Program Human Reliability Analysis Procedure. NUREG/CR-4772/SAND86-1996,*" Sandia National Laboratories for the U.S. Nuclear Regulatory Commission, 1987, Washington, D.C.

[11] J. Julius, J. Grobbelaar, D. Spiegel, and F. Rahn. "*The EPRI HRA Calculator® User's Manual, Version 3.0, Product ID #1008238,*" Electric Power Research Institute, 2005, Palo Alto, CA.

[12] G. Parry. et al. "*An Approach to the Analysis of Operator Actions in PRA, EPRI TR-100259,*" Electric Power Research Institute, 1992, Palo Alto, CA.

[13] A.D. Swain and H. E. Guttman. "*Handbook of Human Reliability Analysis with Emphasis on Nuclear Power Plant Applications NUREG/CR-1278-F,*" U.S. Nuclear Regulatory Commission, 1983, Washington, D.C.

Extracting Human Reliability Information from Data Collected at Different Simulators: A Feasibility Test on Real Data

Salvatore Massaiu
OECD Halden Reactor Project, Halden, Norway

Abstract: This paper presents a feasibility test on extracting HRA-relevant information form data collected at different plant/simulators. Newly proposed methodologies for HRA simulator-data collection are trying to overcome the aggregation and generalization problems that stranded previous attempts at the creation of HRA data banks. Common to the different methodologies is that they insist on the need to precisely characterize the performance conditions. The difference is on the type of information they aim to collect, some focus on failure probabilities, other on situational influences on the performance of the join human-machine system. This paper investigates whether further information of use for HRA, like timing and performance variability, could be added to and extracted from such databases. The data used in this test derive from three simulator experiments. Two experiments were conducted at the Halden Human-Machine Laboratory, while the third at a training simulator at a U.S. nuclear power plant. All together 23 crews of licensed operators from four plants in two countries participated, and ten emergency scenarios were run. The test considers the data as a subset from a larger database, selected by a HRA user as relevant for the target application. The test shows that it is possible to extract three types of HRA-relevant data from records obtained at different simulators and plants: mean times of actions and diagnoses, response-time variability for critical actions, and standardized margins-to-failure information. This paper shows the feasibility of including and re-using traditional types of HRA data in newly proposed approaches to HRA database construction.

Keywords: HRA, Full-scale simulation, HRA Data, HRA database.

1. INTRODUCTION

During the eighties several simulator experiments were organized in order to provide empirical data for development and validation of Human Reliability Analysis (HRA) methods [1]. However, large-scale efforts directed at gathering HRA data were limited and mainly time reliability correlation data were acquired [2]. In the nineties the HRA discipline underwent a radical critique and new methods and practices were advanced, but without new systematic evaluations against empirical data. Efforts to build HRA databases were made at that time (e.g., the NUCLARR and HERA databases sponsored by the U.S. Nuclear Regulatory Commission) but are now discontinued. The result is that nowadays no publicly accessible data banks of HRA data exist.

There is general agreement on the need for collecting and sharing human performance data for probabilistic safety analysis (PSA) but no standardized HRA data collection approach exists. There are conceptual challenges relating to the generalizability of the data from the plant, scenario and task were they are acquired to other plants, and even to other scenarios and tasks at the same plant. There are disagreements on the classification of the observations (e.g., error types, characterization of the context). And there are practical impediments relating to proprietary and sensitivity issues.

These challenges are known and several international cooperative activities are addressing them in a renewed call for collection and exchange of HRA data. The Nuclear Energy Agency/Committee on the Safety of Nuclear Installations/R(2008)9 report [3] proposes a standardized HRA data collection approach for plant training simulators and provides a generic framework for collection and exchange of human performance data. The NRC has developed the Scenario Authoring, Characterization, and Debriefing Application (SACADA), a methodology and a software tool for creating a HRA database of human performance data from plants' routine training simulations [4,5], and a HRA Data Collection project is underway with a participating nuclear power plant (NPP). Electricité de France and the Institute for Energy Technology in Norway/Halden Reactor Project have started work on a classification system for Emergency Operating Systems (EOS), for allowing exchange and re-use of HRA data and information collected at different facilities [6].

This paper builds on these efforts by taking a look at real simulator data from different plant/studies and trying out ways of aggregating the data for possible reuse. The paper capitalizes on data from three dedicated HRA experiments that together constitute a limited but workable sample of observations, resembling a selected set of records from a larger HRA database. In this way reuse of data collected at different plants (the main challenge that have historically hindered previous efforts at creating HRA databases) is investigated.

2. TRADITIONAL HRA DATA COLLECTION

As soon as realistic simulations of control room operation became feasible, simulator studies became the natural arena for seeking nominal error probabilities (i.e., the probability that a given action will be performed erroneously when the task is not influenced by plant and situation specific factors). The idea was that from sets of scenarios containing the same human action, and from large samples of operators, it could be meaningful to at least count frequencies of the unsafe instances of the human actions. In this way, the reliability of the 'human component' could be estimated in basically the same manner as conventional reliability does for technical components.

Incorporating simulator data into probability safety analysis (PSA) and HRA in such a direct way proved unfeasible. When quantification was sought at the level of actions modeled in the PSA event trees, such as depressurizing the primary system or isolating a ruptured steam generator, two problems arose. First, it was noted that an extremely high number of sessions and crews (sample size) would be needed to observe these kinds of failures. For instance, Dougherty [7] refers to simulator studies performed in the late 1980's where, out of 1600 simulator opportunities, zero 'significant deviations' were recorded. Second, it was not trivial to define nominal conditions, whether similar actions observed in different scenarios and conditions could be combined for the calculation of the failure rates. Moray [8], who studied human performance for informing HRA, summarized the problem in the following way: "The attempt to find a single number is an attempt to establish a context-free universal fact about human performance. No such thing exists."

The importance of the situation for human performance is recognized by several research directions in the fields of psychology and human factors, such as situated cognition [9], ecological psychology [10, 11] and cognitive system engineering [12,13], to name a few. Vicente, working within the Rasmussen, Goldstein and Hollnagel research tradition developed at the RISØ laboratory in Denmark, has termed the importance of the situation for human performance "context-conditioned variability" [14]. The impact of context-conditioned variability to HRA is that the parallel with component reliability where the context is more or less irrelevant and statistical data are available, cannot be fully maintained for the calculation of the reliability of the human component. HRA data are context laden, and therefore information about the context need to be captured.

One attempt at avoiding the sample and the context-conditioned variability problems in one single move was to collect human error probabilities for lower-level operator's tasks, i.e., tasks having a more defined context and higher failure probabilities (and therefore failure would be observed in smaller samples). This was the strategy followed by Swain and Guttman in THERP [15]: "Our general approach is to divide human behavior in a system into small units, find data (or use expert judgment to obtain appropriate estimates) that fit these subdivisions and then recombine them to derive estimates of error probabilities for the tasks performed in NPPs".

The fact that control room operation is performed by a team and the observation that most errors on small tasks were in fact corrected by other team members, or self corrected when the operators had enough time [16, 17], required the introduction of new concepts: recovery and dependency. Recovery has to consider the possibility that errors on one or several sub-tasks in a sequence could be detected and corrected in time (before an irreversible state is reached). Dependency accounts for the possibility that failing a task might influence the probability of failing another task later in the sequence. Recovery and dependence are in a sense ways of taking into account the team and dynamic aspects of accident operation without explicit models of team cognition and behavior.

Quantifying the building blocks of the tasks represented in the PSA, instead of these directly, could address the sample size problem, and data were provided to validate and extend decompositional methods like THERP. On the other hand, context-conditioned variability issues remained, and impaired the generalizability of the data. The standards of performance that define the failure of the basic actions refer to the given plant technology, the actual procedures, the organization of work, as well as other plant and situation specific aspects. This is an obstacle to direct generalizability of the failure data to other tasks, scenarios, and plants, with the additional drawback that the data might obsolete even at the same plant when the specific features are changed. In general, by collecting error data defined as behavior that diverges from plant-specific standards, specific contexts are imposed on the empirical material so that contextual differences need to be accounted for when re-using the data: for instance, if the procedures' steps are taken as standard of correct performance, data on the number of deviations form a step might be unusable if the step is modified or deleted. Even for error data on smaller bits of operator performance the context needed to be accounted for.

The following points summarize the traditional approaches to HRA data collection:
1. Data are collected on human errors for NPP tasks in nominal contexts.
2. 'Core damage' failures in nominal contexts are rarely observed in training or research simulators, therefore failure data for smaller tasks are more often obtained in the data collections.
3. The results cannot easily be generalized to different plants, scenarios and tasks, since the context of performance for the source data is not sufficiently and systematically described.

The same considerations apply to time reliability data, in fact the most collected type of traditional HRA data [2]. Response times can be collected for both high level tasks and for small units of behavior, and may be used to estimate probability of non-response curves, i.e., probabilities of human errors as a function of time available. The problems of generalizing to different tasks and contexts are similar for time reliability data as for human errors for required tasks.

3. CURRENT APPROACHES TO HRA DATA COLLECTION

Conscious of the limitations of the previous attempts to create HRA data banks, as well as the needs of second-generation HRA methods and emerging applications (e.g. HRA for not-at-power, external events) new approaches are being developed and tested for collecting, storing and exchanging HRA data. The SACADA [4], EOS [6] and the "Methodology for Conducting Simulator Experiments for HRA" developed at the Idaho National Lab [18], although from different theoretical perspectives and aiming at different end uses of the data in the HRA process, have all abandoned the notion of a nominal context. Instead they agree on the need to precisely characterize the performance conditions, in order to overcome the generalizability problems that have jammed the old attempts at the creation of HRA data banks.

This has important implications for the methodology used to generate and code human reliability data. Scenarios and contexts should be representative of the needs for PRA and the different performance conditions that operating crews may encounter. For subsequent use of the data it is therefore necessary to ensure that important information on the conditions of performance accompanies the data in order to prevent its potential aggregation with other data that reflect dissimilar conditions. This requires describing, with specificity, the characteristics of the join human-machine system (e.g. the plant, interfaces, procedures, crews, conduct of operations), and the kinds of cognitive and human performance demands present in the simulated scenarios that are used for data collection. It also requires such characteristics to be described consistently, following an approach that may be standardized, as well as applying associated, accepted models of human performance and accident operation. Characterizing plant, crews, scenarios and contextual conditions systematically is also an important part of understanding the data, not only the means of identifying and selecting data for future use and study.

The SACADA and EOS approaches to HRA data collection employs different ways to solve the limitation of classical HRA data where it comes to generalizing the data. The type of data and end-use considered by the two approaches are also different. SACADA collects success and failure data on tasks performed in simulator training (hence most actions are at a lower level of granularity compared to the Human Failure Events (HFEs) typically modeled in PSA event trees). Failure rates are obtained and re-used by matching tasks on their cognitive demand profile, i.e., a type of cognitive task analysis of the simulated tasks, assuming similar human error probabilities for similar cognitive demand profiles.

The EOS method focuses on reusing information in rich qualitative task analyses. The plant and simulation settings are systematically profiled in terms of the characteristics of the join human-machine system, i.e. the ensemble of crew, interface, and procedures, by describing the system according to standardized categories, e.g., the crew composition, the prescribed way of working, the communication style, the procedures type. The EOS will then help assessing the similarity and differences of the systems (plants) and consequently the relevance of the empirical material to the target application. The EOS approach is more focused on reusing qualitative information for the purposes of second-generation HRA methods. The EOS approach also considers other HRA analyses as data (knowledge) that can be reused, as they contain information and reasoning about possible system failure elaborated by other experts.

In these approaches types of information that are still widely used by HRA are left out. One example is timing information. Another is procedure progression variability. In this paper we concentrate on showing how timing information and performance variability information could also be collected in HRA simulator experiments and reused via data-banks, by benefitting form the context-of-performance characterizations provided, e.g., by the SACADA and EOS methods.

2. TEST DATA

The data analyzed in this paper derive from experiments that were originally arranged for evaluating aspects of HRA practice, including assessing the predictive capabilities of commonly used HRA methods [19], intra-method consistency [20], and for investigating teamwork and procedure following in complex emergency operation [20, 21]. The studies provided useful insights to the specific research questions addressed, but it became apparent that there was no agreed methodology to extend the use of the collected data beyond the specific scopes that motivated their generation, as for instance recording the results in central data bank. This is a typical situation when it comes to the use of research simulators for collecting HRA data. It is in fact not uncommon for individual plants to use the results of simulator observations to support aspects of their plant-specific PSA. Yet, collected data are rarely, if ever, used to verify or validate HRA methods, or to improve the HRA discipline at large.

The data used in this test derive from three simulator experiments. Two experiments were conducted at the Halden Human-Machine Simulator, while the third at a training simulator at a U.S. nuclear power plant. All together 23 crews of licensed operators from four plants in two countries participated, and ten emergency scenarios were run. Table 1 below shows the details of the dataset used, including the experiment name and references, the simulated plant type, the number of crews in the experiment, and the number of design basis scenarios run. All crews used similar versions of the emergency operation guidelines developed by Westinghouse Owner Group. The test imagines that the data are a subset from a larger database, created according to the prescripts of modern approaches to HRA data base creation, and that therefore it was possible for the HRA user to select a sample that matched the target application.

Table 1: Sources of the data used

Experiment name	Ref. #	Simulated plant	# of crews	# of scenarios
PSF/Masking	[19]	Framatome PWR 900 MW	14	4
US training simulator	[20]	Westinghouse PWR	4	3
HRA-2011	[21, 22]	Westinghouse PWR	5	3

4. RESULTS

Three types of information have been extracted from the data set. The first type of information is about mean times for diagnosis and actions. The second type of information is response time distributions for critical tasks, i.e., tasks that impact the plant process, and likely the likelihood of subsequent Human Failure Events in the event sequences. The third type of information is margin-to-failure scores, a measure of human performance variability to normalize different tasks outcomes on a common scale.

4.1. Timing of Actions

Time is a factor of uppermost importance for HRA. Some approaches use time as a surrogate cause of failure and allow the time factor to incorporate the effects of most, if not all, drivers of performance. For these approaches the estimation of the timing of actions in a scenario is essential and it is not limited, as for all other HRAs, to the correct estimation of the maximum allowable times from thermohydraulic calculations (i.e. the times when irreversible state arise and the required operators' actions are not longer useful). Typically, these methods have to differentiate between the time required for the actions to be completed, the time available for formulating the correct diagnosis of the situation, the time available for recovering errors, and the time for eventual delays. All these timings are connected to each other, but in most time-reliability methods the single most important estimation is the diagnosis time, since it is often the case that the HEPs are derived from time response curves for diagnosis of abnormal events in the control room, or outside.

The HRA analysts calculate the timing of action based on the qualitative information obtained for the task analysis. An important source for timing determinations is the estimated entry and progressions in the emergency procedures set. Scenario specific factors also play a role, for instance the combination of events might create cumbersome progressions in the procedures or cause reasons for delaying important actions (e.g. tripping the reactor). HRA data collected at research and training simulator can constitute a valuable source for timing information determinations.

These points can be illustrated by reference to the International Empirical Study, where various HRA analyst where asked to predict the outcome of simulated emergency scenarios [19]. Tables 2 and 3 below summarizes the different timing estimates made for two HFEs in two steam generator tube rupture scenarios, and the assumed impact for task success. The first HFE (1A) is the identification and isolation of the ruptured steam generator in a scenario without added complications (base scenario), given a maximum available time defined by study's assessment group in 20 minutes (from rupture to isolation). The second HFE (1B) is again the identification and isolation of the ruptured steam generator, but in this case in a scenario complicated by a steam line break concurrent with the tube break and by the lack of radiation indications (partly due to the steam line isolation that follows the steam line break and partly to failures in other radiation sensors). The study defined the maximum available time for isolating the ruptured SG in this scenario in 25 minutes.

Table 2: Timing estimates for HFE1A: ruptured steam generator isolation

Team	Method	Delay	Diagnosis	Required	Available	Impact of time to failure
NRC	ASEP-THERP	7		13	20	Time as main HEP driver.
EPRI	CBDT+THERP	7			20	From RX trip, but main error cause is missing the first transition step to E-3. Some time for recovery.
INL	SPAR-H		10		20	Based on 10 minutes to reach E-3. Sufficient time if Rx is tripped manually.
NRC	SPAR-H		8-10	13-15	20	The remaining 5 to 7 minutes to manually trip the reactor are sufficient.
PSI	CESA				20	Limited time for recovery. But failure mainly due to random execution errors.
NRI	DT+ASEP				20	Time not a limiting factor for the HFE.
EDF	MERMOS				20	Lack of urgency main factor for failure. 20 min criterion seen as arbitrary.
Ringhals	HEART			16-18	20	Time from trip to isolation. Time shortage is the main HFE difficulty.
IRSN	PANAME		13-15	18-20	20	Time is tight and the main HEP driver for an easy diagnosis.
VTT	B-THERP		10-12	13-15	20	Time sufficient for HFE success.
NRC	ATHEANA	11		18-26	20	The crews might wait up to 11 min. to trip the reactor and then the time will bee too short. The allowed 20 min. considered as arbitrarily defined.
KAERI	K-HRA	5	11		20	Delay: 5 minutes to trip the reactor. Time is tight and the main driver.
NRI	CREAM				20	Adequate time but with small margin.

Table 3: Timing estimates for HFE1B: ruptured steam generator isolation in a scenario with complications

Team	Method	Delay	Diagnosis	Required	Available	Impact of time to failure
NRC	ASEP-THERP		9-12		25	Time sufficient to reach a transition to the right procedure.
EPRI	CBDT+THERP		18-20		25	Time minor contributor, even if no time for recovery.
INL	SPAR-H	>5		20	25	Time insufficient, HEP = 1.
NRC	SPAR-H				25	Limited time and limited time to recover, but rated nominal.
PSI	CESA				25	Short time but not a determining factor.
NRI	DT+ASEP				25	Short time (and unrealistic), but not main factor.
EDF	MERMOS				25	Lack of urgency as main factor together with short available time.
Ringhals	HEART				25	Shortage of time as second most important factor.
IRSN	PANAME		15-20	20-25	25	Time is short and one of two main factors.
VTT	B-THERP		15	18-25	25	Time available close to time required and other negative PSFs determine a high HEP for the HFE.
NRC	ATHEANA	15	6	18-26	25	No diagnosis in the first 15 minutes (delay) due in part to masking. The crews may simply run out of time to meet the "arbitrary" 25-minute time frame.
KAERI	K-HRA	15			25	Delay: no signs of SGTR for 15 min after Rx trip. Time is the most important factor for HFE failure.
NRI	CREAM				25	Delayed interpretation one of failure types identified.

The table shows that although the analysts received the same information regarding the scenarios, the HFEs, the crews, the simulated plant, and even detailed printouts of plant status parameters at different points in the scenarios, there was variation in the timing estimates. Furthermore, the timing evaluation had in many cases a significant impact on the quantification of the HEPs, and not only for methods that employed time-reliability curves.

4.2. Average Times

Observing simulator sessions is a good HRA practice and would have obviously benefited the analyses. Table 4 below, reporting the timing information observed in the data, shows that many analysts' time estimates were realistic, but some were not and in many cases strongly influenced quantification. For instance, there were no significant delays in tripping the reactor in the base scenario, as by some assumed. Also, it took on average double as long in the complex scenario to enter the SG isolation procedure compared to the base scenario. This information would have resulted in very different HEP estimates for some analyses (e.g. K-HRA and ATHEANA in the base scenario).

Table 4: Time responses in SGTR scenarios

SGTR to:	Base N=14		Complex N=14		U.S. crews N=8		All N=36
	Mean	Max	Mean	Max	Mean	Max	Mean
RX trip	02:23	06:29	At SGTR	-	02:11	03:39	**02:19**
E-0	At trip	07:06	At trip	01:11	At trip	-	**At trip**
Stop AFW to ruptured SG	06:44	09:41	13:27	22:42	02:52	04:43	**08:59**
Enter E-3	10:15	16:08	20:46	40:32	12:58	18:12	**15:18**
SG isolated	15:32	21:29	26:54	45:27	21:13	33:42	**21:34**

The table also reports the corresponding timing information regarding U.S. crews that run a base SGTR scenario in their training simulator and other U.S. crews that run a multiple SGTR scenario in the Halden Man Machine Laboratory (HAMMLAB). The times are in line with those observed with Swedish crews, considered the differences in scenario specifics, simulated plants, and emergency procedures details. The table also shows the averages for the entire sample: if these data where provided with standardized context of performance information they could be used as generic timing data (conservative in this case, as over half the scenarios are SGTR with complications) for cases where the analysts do not possess specific plant/scenario data, if comparable in terms of context-of-performance characteristics to target plant.

For methods that use time reliability curves, one of the most important information is the time available for diagnosis, as this can be associated with an HEP – once other contextual considerations are made (e.g. type of event, familiarity). In the case of SGTR events, the diagnosis is formally completed by entering the tube rupture isolation procedure (E-3 for Westinghouse plants). The complete sample of 36 SGTR observations across the three studies indicates an average time of about 15 minutes to enter this procedure from the initial cues associated with the tube rupture (i.e. radiation alarms or automatic reactor trip in these simulations). Again, an analyst working on a SGTR scenario for a Westinghouse PWR plant could benefit from access to such empirical information.

4.3 Response-Time Variability

A second type of data important in many HRA applications is response-time variability. For a given task in a given scenario and for a given plant, an extensive data collection could result in empirically informed non-response time distributions. If a data sample of the same size collected at different plants was available the same could not be done easily, as different plant PRA criteria, procedural criteria and other plant/simulation specific aspects would likely strongly reduce the relevant sample. However, the data from different plants can provide very useful information to HRA analyses at other plants.

Figure 1 below is based on data from the three studies collected on total loss of feedwater (TLFW) scenarios. In this event the crews have to trip the reactor as early as possible after feedwater flow to the steam generators is lost, in order to keep secondary water inventory longer in the steam generators, and hence increase the time available before the occurrence of fuel damage. This action is critical as it has direct consequences on the evolution of the plant process and likely impact the failure probabilities of the HFEs in the event sequence. Figure 1 shows the crew variability for this action and that significant delays are possible. In the specific case, the crew that tripped the reactor 51 seconds after total LOFW had less than 30 minutes available to establish Feed and Bleed before core damage, compared to about 90 minutes for the fastest crews.

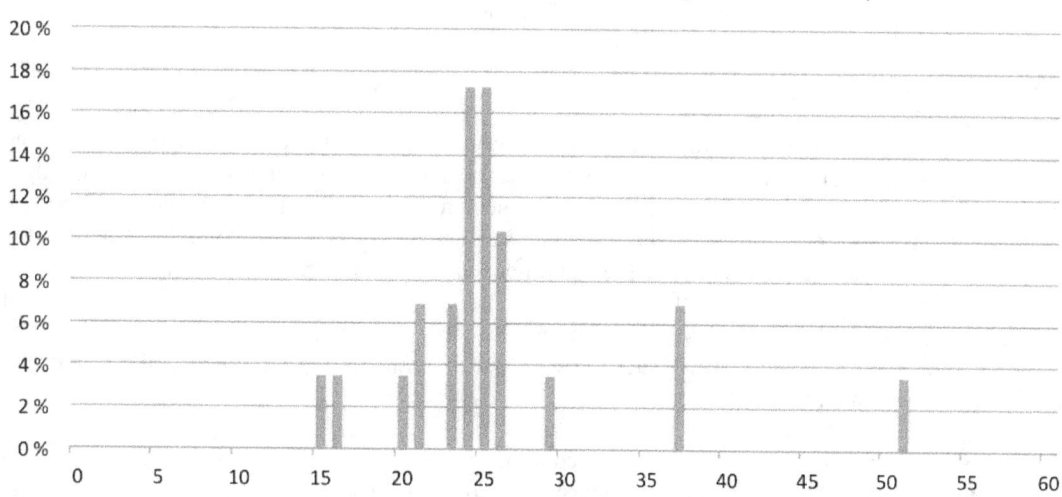

Figure 1: Seconds from TLFW to reactor trip (percent of crews) - N=29

Another example, from the same total loss of feedwater scenarios is depicted in figure 2. After the crews trip the reactor, they will be directed to the relevant emergency guidelines for the event. One of the steps in the heat sink procedure will instruct the crews to stop the reactor cooling pumps (RCPs). Also this action has consequences for the plant process and the available time to core damage: the earlier the RCPs are stopped the longer the time available. Figure 2 shows that two of 29 crews significantly delayed this action, thereby consuming available time to avoid core damage.

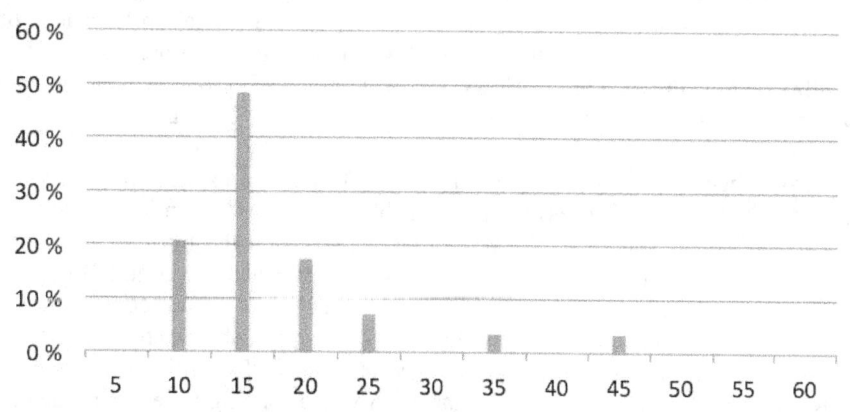

Figure 2: Minutes from reactor trip to RCPs stop (percent of crews) - N=29

Response-time distributions of the type presented here cannot obviously be used to uncritically derive HEPs or to obtain non-response time curves. They are nonetheless a valid source of empirical information, and the value of the data is not limited to analysts that do not have access to direct observation of operator performance for their analyses. For instance, the presence in the data of significant delays and their relative frequency cannot be disregarded by the fact that similar delays have not be observed at the analyzed plant. There is in fact no guarantee that the performance conditions and task demands the crews have been trained for and evaluated on in their simulator sessions were fully representative of the conditions encompassed by a PSA scenario, and of those that could arise in a real accident.

A critical aspect is therefore that the database allows access to the individual data records in order to determine the circumstances surrounding the delays (e.g. plant conditions, procedures, crew composition, training) thus allowing the user to assess the relevance to the analyzed context. Likewise, the user should be allowed full sorting and filtering access of the data, in order to obtain the most relevant sub-sample for further analysis. In this respect, coding the observations according to the EOS and SACADA classifications, would not only provide the data required for these methodologies, but also facilitate the re-use of timing information in new contexts.

4.4. Margins to Failure

A proposal for generalizing HRA data from one plant/scenario to different plants/scenarios is to score human performance on different tasks on a common scale. The notion of success/failure is already a normalized measure of human performance. However, such a binary characterization limits the ability to draw useful insights from even substantial amounts of observed trials. For example, it is possible that all crews in a simulation session complete the needed actions to mitigate a design basis event. This would indicate only successes on the HFEs and no conclusion regarding the relative difficulty of the tasks could be obtained. However, performance quality variability typically exists, and the information indicating that some crews succeed with substantial 'margin' to spare while others did not, is lost by the dichotomous measure.

For this reason Hallbert and al. [18] propose a continuous measure to characterize performance that capture issues such as available or remaining margin to failure and variability in performance amongst crews. The approach is called the "limit state concept" and includes a data analysis technique that normalizes the raw performance measurement in terms of the limit state. "This means that the raw performance measures are adjusted using their relationship to the limit state as a notionally common scale" [18, p. 40].

Figure 3 below, show the application of this data analysis approach to performance data from the three studies considered in this paper. The crew performances of four HFEs in three different design basis events have been analyzed according to the limit state concept. The limit states (i.e. success criteria for the HFEs) are defined, with some specificity and uncertainty, by the PRA of the design basis accidents. One HFE was "establish Feed and Bleed before core damage" in total loss of feedwater events. The second HFE was "isolate the ruptured steam generator before it overfills" in SGTR scenarios. The third and forth HFEs are in a RCP Seal LOCA scenario (one requiring to stop the RCPs and the other to start the positive displacement pumps before irreversible damage states).

The 73 crew performance observations of the four HFEs are scored on a scale ranging from 0 margins (i.e., failure) to 1 (i.e., maximum success margin possible), based on the timing of their actions in the scenario, the status of plant parameters at the time of actions, and the limit states (success criteria) provided by the PSA model. For instance, in the case of the HFE "Feed and Bleed", the success margin would represent the percentage of remaining available time before core damage at the time Feed and Bleed was established, based on the time the crew tripped the reactor. In this scale 0 would correspond to the action executed after the core began to melt, and 1 the action was executed at the same time as reactor trip (100 percent time available). A curve relating time form reactor trip to core damage (obtained form the PSA) is adopted for calculating the available time. Similar calculations are made for the other HFEs.

Figure 3 shows the margin to failure scores obtained by the crews relative to the time available for the four HFEs. The interpolated polynomial curve resembles typical "probability of non-response" curves for the considered time region, although the upper and lower bounds would likely be larger here, as failures and near failures were seen also with extensive time available. It should be stressed, however, that the Y-axis here represents the margin to failure scores on a linear scale and not human error probabilities on a logarithmic scale.

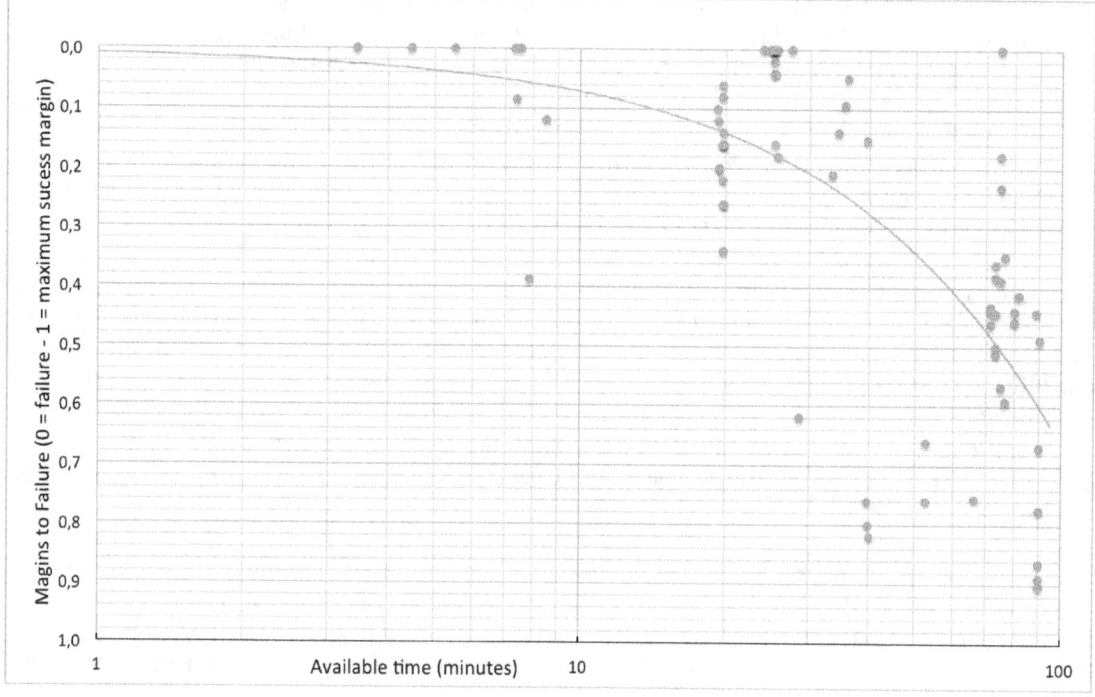

Figure 3: Margins to failure by available time - N=73

This approach permits insights to be drawn regarding three properties related to performance reliability:
1. Whether the actions meet the success criteria for the defined HFE;
2. The amount of margin available between task performance and the limit state for the task as defined by the HFE;
3. Variability among crews in performing the task(s).

The measure would also permits estimation of measures of central tendency and calculating statistics that can be used in reliability analysis.

5. CONCLUSIONS

Several international cooperative activities are addressing the need of collecting and sharing human performance data for probabilistic safety analysis (PSA). New methodologies for data collection, classification and storage have been proposed, and some are being tested. Common to the different methodologies is that they insist on the need to precisely characterize the performance conditions, in order to overcome the aggregation and generalization problems that stranded previous attempts at the creation of HRA data banks.

The paper follows on such developments and assumes that by means of these classification systems a HRA user has extracted a sample of relevant data from the database. The test investigates whether a HRA user could extract further relevant information, beyond the primary uses assumed by the methodologies. The data used in this trial are real simulator observations from three dedicated HRA experiments performed with three different simulators, and with crews from four different plants in U.S. and Europe.

The first type of information that is extracted is about timing of actions. The paper shows that average response times from the entire sample could be used as generic timing data, provided the data match the target application. Given the nature of the data, with over half the records from design basis scenarios with extra complications, such mean values could be assumed to be conservative estimates.

The second type of information is response time distributions for critical tasks, i.e., tasks that impact the plant process, and likely the likelihood of subsequent HFEs in the event sequences. This information does not allow direct derivation of HEPs or non-response time curves. Rather, it is a source of empirical information about crew variability and of the possibility and relative frequency of significant delays that need to be taken into account in the qualitative and quantitative HFE modeling. This information cannot be disregarded even by analysts that have collected performance observations on-site, since there is no guarantee that the performance conditions and task demands the crew have been trained for and evaluated on in their simulator sessions were fully representative of the conditions encompassed by a PSA scenario, and of those that could arise in a real accident.

The third type of information extracted applies a methodology for scoring human performance on different tasks on a common scale, on the lines of the "limit state" concept for generalizing HRA data. This approach permits insights to be drawn regarding three properties related to performance reliability: (1) whether the actions meet the success criteria for the defined HFE; (2) the amount of margin available between the task performance and the failure criteria for the task as defined by the HFE; and (3) variability among crews in performing the task(s). The measure would also permits estimation of measures of central tendency and calculating statistics that can be used in reliability analysis.

A critical assumption for this test is that the database allows access to the individual data records in order to determine the circumstances surrounding performance (e.g. event description, plant conditions, procedures, crew composition, training) thus allowing the user to assess the relevance to the analyzed context. Likewise, the user should be allowed full sorting and filtering access of the data, in order to obtain the most relevant sub-sample for analysis.

This paper shows that timing and standardized performance data coded according to the new classification systems for HRA simulator data collection could not only provide the data required for these methodologies, but also facilitate the extraction and re-use of other types of information of critical importance to HRA, like timing, margins-to-failure, and crew variability information. This data can be naturally collected at research simulators, while raw data collected at training simulators could be later analyzed, formatted, and stored in a publicly available data bank.

References

[1] A. J. Spurgin, "*Human Reliability Assessment Theory and Practice*," CRC Press, 2009.
[2] D. I. Gertman and H. S. Blackman, "*Human Reliability and Safety Analysis Data Handbook*," John Wiley & Sons, 1993.
[3] Nuclear Energy Agency/Committee On The Safety Of Nuclear Installations. "*HRA Data and Recommended Actions to Support the Collection and Exchange of HRA Data*," NEA, 2008, Paris.
[4] Y. J Chang and E. Lois, "*Overview of the NRC's HRA Data Program and Current Activities*", in "Proceedings of the 11th International Probabilistic Safety Assessment and Management Conference & The Annual European Safety and Reliability Conference", 25–29 June 2012, Finland.
[5] L. Criscione, S.-H. Shen, R. Nowell, R. Egli, Y. J. Chang and T. Koonce. "*Overview of Licensed Operator Simulator Training Data and Use for HRA*", in Proceedings of the 11th International Probabilistic Safety Assessment and Management Conference & The Annual European Safety and Reliability Conference. 25–29 June 2012, Finland.
[6] S. Massaiu, P. O. Braarud, and P. Le Bot. "*Including organizational and teamwork factors in HRA: the EOS approach*", in proceedings of the Enlarged Halden Program Group, 2013, Norway.

[7] E. M. Dougherty, "*Human reliability analysis--where shouldst thou turn?*", Reliability Engineering & System Safety, 29, pp. 283–299, (1990).

[8] N. Moray. "*Dougherty's dilemma and the one-sidedness of human reliability analysis (HRA)*", Reliability Engineering and System Safety, 29, pp. 337-344, (1990).

[9] P. Robbins, and M. Aydede. "*The Cambridge handbook of situated cognition,*" Cambridge University Press, 2008.

[10] R. Shaw and J. Bransford, "*Perceiving, Acting, and Knowing: Toward an Ecological Psychology,*" Lawrence Erlbaum, 1977.

[11] J. J. Gibson. "*The Ecological Approach To Visual Perception,*" Psychology Press, 1986.

[12] E. Hollnagel and D. D. Woods. "*Cognitive systems engineering: New wine in new bottles*", International Journal of Human-Computer Studies, 51(2), pp. 339–356, (1983).

[13] J. Rasmussen, A. M. Pejtersen and L. P. Goodstein, "*Cognitive systems engineering,*" Wiley, 1994.

[14] K. J. Vicente, "*Cognitive work analysis: toward safe, productive, and healthy computer-based work,*" Routledge, 1999.

[15] A. D. Swain and H. E. Guttman, "*Handbook of human reliability analysis with emphasis on nuclear power plant applications,*" Sandia National Labs-NUREG CR-1278, Nuclear Regulatory Commission, 1983, Washington DC.

[16] D. D. Woods. "*Some results on operator performance in emergency events*", In D. Whitfiled (Ed.), "*Ergonomic Problems in Process Operations,*" pp. 21–32, Institute of Chemical Engineers Symposium, 1984, Birmingham.

[17] L. Norros and P. Sammatti, "*Nuclear power plant operator errors during simulator training,*" VTT Technical Research Centre of Finland Helsinki, 1986, Finland.

[18] B. Hallbert, T. Morgan, J. Hugo and J. Oxstrand. "*Simulator Data Sharing for HRA: Tenets & Methodology for Conducting Simulator Experiments for HRA - Data Generation and Analysis Focused on the Limit-State Concept,*" NRC Technical report INL/EXT-12-26327 (Draft), Nuclear Regulatory Agency, 2012, Washington DC.

[19] J. Forester, V. N. Dang, A. Bye, E. Lois, S. Massaiu, H. Broberg, P. Ø. Braarud, R. Boring, I. Männistö, H. Liao, J. Julius, G. Parry and P. Nelson, "*The International HRA Empirical Study – Final Report – Lessons Learned from Comparing HRA Methods Predictions to HAMMLAB Simulator Data*", HPR-373, OECD Halden Reactor Project, 2013, Norway.

[20] J. Marble, H. Liao, M. Presley, J. Forester, A. Bye, V. Dang and E. Lois. "*Results and Insights Derived from the Intra-Method Comparisons of the US HRA Empirical Study*", in proceeding of 11th International Probabilistic Safety Assessment and Management Conference and the Annual European Safety and Reliability Conference 2012 (PSAM11 ESREL 2012), (2012).

[21] S. Massaiu, and L. Holmgren, "*Complex diagnosis of multiple steam generators leaks: Preliminary results from the HRA-2011 experiment*", in proceedings of the Enlarged Halden Program Group, 2013, Norway.

[22] S. Massaiu, and L. Holmgren, "*An Empirical Investigation of Team Decision-Making with Emergency Procedures*", paper presented at the American Nuclear Society's 2013 Risk Management Embedded Topical Meeting: Risk Management for Complex Socio-technical Systems, (2013).

Simplified Human Reliability Analysis Process for Emergency Mitigation Equipment (EME) Deployment

Don E. MacLeod[a]*, Gareth W. Parry[a], Barry D. Sloane[a], Paul Lawrence[b], Eliseo M. Chan[c], and Alexander V. Trifanov[d]

[a] ERIN Engineering and Research, Inc., Walnut Creek, USA
[b] Ontario Power Generation, Inc., Pickering, Canada
[c] Bruce Power, Toronto, Canada
[d] Kinectrics, Inc., Pickering, Canada

Abstract: For a variety of different reasons, it is becoming more common for nuclear power plants to incorporate the use of portable equipment, for example, mobile diesel pumps or power generators, in their accident mitigation strategies. In order for the Probabilistic Risk Assessment (PRA) to reflect the as-built, as-operated plant, it is necessary to include these capabilities in the model. However, current Human Reliability Analysis (HRA) methodologies that are commonly used in the nuclear power industry are not designed to accommodate the evaluation of some of the tasks associated with the use of portable equipment, such as retrieving equipment and making temporary power and pipe connections.

This paper proposes a method for estimating the component of the human error probability (HEP) associated with the deployment of portable equipment. The other components of the HEP, such as the failure to identify the need to initiate portable equipment deployment, can be addressed with existing methodologies and are not addressed by this approach. This approach is intended for application to a variety of hazard risk assessments, including internal events, internal flooding, high winds, internal fires, external flooding, and seismic events.

Keywords: HRA, Portable, Temporary, External Events.

1. INTRODUCTION

For a variety of different reasons, it is becoming more common for nuclear power plants to incorporate the use of portable equipment (referred to in this paper as Emergency Mitigation Equipment (EME)) in their accident mitigation strategies. In order for the Probabilistic Risk Assessment (PRA) to reflect the as-built, as-operated plant, it is necessary to include these capabilities in the model. Current Human Reliability Analysis (HRA) methodologies that are commonly used in the nuclear power industry are not designed to accommodate the evaluation of some of the tasks associated with the use of such portable equipment, such as retrieving equipment and making temporary power and pipe connections. Some methodologies, such as ATHEANA [1] and FLIM [2], are, in principle, capable of assessing these types of failures, but require an extensive coordination of resources, rely heavily on expert judgment, and have limited exposure in the industry.

This paper proposes a method for estimating the human failure probability associated with deploying EME. This approach is intended for application to a variety of hazard risk assessments, specifically internal events, internal flooding, high winds, internal fires, external flooding, and seismic events. Further, since the approach is intended for application to a range of methods of EME deployment, it is generic in nature and therefore tends to be conservative. While this methodology was developed to address the conditions that were anticipated to be the most relevant to EME deployment, application of the methodology has thus far been limited to work performed by Ontario Power Generation and Bruce Power to support the modeling of EME in their PRAs. As with other developing PRA techniques, it is expected that experience with the process will identify areas for further refinement and enhancement.

---
* demacleod@erineng.com

2 SCOPE

The intent of this paper is to document a methodology for evaluating the failure probability associated with the retrieval, transportation and installation of the EME, referred to in this paper as EME deployment. For this methodology, "installation" is considered to include tasks such as making temporary piping and power connections, loading a portable generator, and/or pressurizing a water system header with a portable pump, although it is expected that the specific tasks may be different for each site. The contributions associated with determining the need to initiate the EME deployment (i.e., the diagnosis and decision making) and with operating the equipment, once deployed, are not part of this methodology. The use of this methodology is only proposed for cases where there is clear guidance provided by abnormal incidents manuals (AIMs), emergency mitigating equipment guides (EMEGs), emergency operating procedures (EOPs) (or equivalent guidance) to deploy EME and that these decisions are made within the main control room or secondary control area by authorized staff (e.g., authorized nuclear operators, control room shift supervisors, shift managers). Further, the actions to initiate operation of the EME equipment, once deployed, are performed entirely within the control room or secondary control areas or using field actions required to initiate EME (e.g., opening manual valves), and these can be addressed within the scope of the hazard-specific modelling of operator response by nominal plant HRA methods. Reliability of EME hardware is also not addressed in this paper.

3. METHODOLOGY

This methodology is a simplified process that applies adjustment factors to represent the impact of performance shaping factors (PSFs) on a hazard-specific basis on a base human error probability (HEP). The impacts of each PSF are tracked in an HRA decision tree and the combined impact of all decision branches, which characterize the implementation conditions for the site being evaluated, determine the scenario-specific HEP. An example decision tree is described in Section 5.

In the event that it is necessary to evaluate EME deployment for other types of hazards, additional HRA decision trees can be developed. For consistency in a given PRA model, it is suggested that any additional HRA decision trees be developed in the same manner as those presented in this paper.

3.1. Base HEP

The base HEP is used as the starting point in the EME deployment HRA decision trees. It is considered to represent the failure probability of a deployment activity that:

- Is governed by procedures that provide all of the information required to effectively perform the task. For example,
  - It is assumed that there are step-by-step instructions for the installation tasks, such as making temporary power and piping connections, aligning valves, starting and loading a generator, and/or pressurizing water system header.
  - The level of detail associated with procedures for acquiring and transporting equipment is expected to be lower, but if the EME is loaded on a trailer that can only be towed by a specific set of trucks, the procedure should identify which vehicles are capable of supporting the task.
- The responsible plant personnel have been trained on. Determining the adequacy of training is somewhat subjective, but adequate training could be described as:
  - Having received classroom training on deployment as a prerequisite for being part of the deployment team,
  - Re-performing the training on a periodic basis has been demonstrated to be feasible under nominal conditions. For the internal events analysis, the environmental conditions in which the deployment will occur are likely similar to those

conditions in which the validation exercise was performed. It is possible that other environmental conditions may exist that were not present in the validation exercise, such as:

- Extreme cold weather
- Extreme hot weather
- Heavy rain
- Heavy snow
- Nighttime deployment

However, the base HEP is assumed to account for these factors in that it represents an average over these conditions for internal events. These are underlying random conditions that are not modeled in the PRA. These variations, such as changes in temperature from day to day could have an influence on performance, e.g., performance could degrade if the actions were being taken at very high or very low temperatures. However, since they are random with respect to when the demand could occur, and are not modeled explicitly, the HEP is characterized as being the average HEP over the spectrum of these conditions. Conditions that are correlated with the hazard being evaluated are treated explicitly in the following decision trees.

A failure probability of 1.0E-01 is assigned for this base HEP, which is consistent with a screening HEP from NUREG-1792 [3].

3.1.1 Feasibility for Specific Hazards

While the feasibility of EME deployment has been established for nominal plant conditions, it will be necessary to establish feasibility for each of the hazards in which it is credited. This includes consideration of:

Staffing: Each hazard presents different requirements on the plant and may require the performance of different activities by the available staff. For each case in which EME deployment is credited, it must be confirmed that EME deployment team personnel that are qualified to perform required duties will not be diverted to other tasks such that they would not be available to support EME deployment. Any special fitness requirements for performing deployment tasks, such as operating chainsaws (to facilitate clearing debris for example), should be considered as part of the staffing assessment.

Timing: In order to establish feasibility, it must be demonstrated that the time required to perform the deployment is less than or equal to the time available after allowing for the time to initiate the EME given successful deployment. A separate "time margin" assessment is used as part of this HRA methodology, but it is focused on determining potential credit for the recovery of errors.

Equipment and Location Accessibility: For each hazard in which the EME is credited, it must be established that the EME will not be damaged to the extent it cannot function and that it will be possible to access the equipment, transport it to the deployment area, and that it is possible to work in the deployment area. Events that could prevent this include:

o   Failure of the structure(s) that house the EME, for example:

   - Building collapse that damages the EME
   - Building collapse that prevents access to the EME,
   - Partial collapse or other damage, such as door buckling, that prevents access to the EME.

- Failures of structure(s) along the access path between the EME storage location and the point where it is to be deployed, or structural failures of the access paths,
- Obstruction of path due to debris accumulation that is beyond the capability of on-site sources to remove,
- Failures of the structure(s) where the EME is deployed (if applicable).
- Fire in an area where EME deployment activity is required: No credit is taken for EME deployment in fire scenarios where part of the activity must be performed in the same (or very nearby) location as the fire.
- Flooding in an area where EME deployment activity is required: No credit is taken for EME deployment in internal or external flooding scenarios where part of the activity must be performed in a location that is flooded.
  - If a case can be made that the quantity of water in the area would not significantly impact the deployment activity, the HRA analyst may choose to document the issue and credit EME deployment. However, due to the uncertainty in assessing the impact of flood events, it is suggested that EME not be credited when flood conditions exist in the zone where the activity is required.
  - Consideration should be given to scenarios where the EME itself may not be damaged, but access to the EME for refueling is not possible (etc.).
- For external flooding scenarios, determine if the installation of flood barriers would prevent access to equipment or transportation routes.

Safety Limits: No credit should be taken for EME deployment in conditions that exceed any safety limits established for personnel protection by the plant. For example:
No credit should be taken for the EME deployment activity for high wind events which exceed the safety limits established for plant personnel.

Communications: If EME deployment relies on communication between the deployment team and any other group, it must be verified that the communication equipment will be capable of operating in the scenario in which it is used. For example, if the communications equipment requires an antenna that would be failed in certain seismic events, then that equipment should be considered to be unavailable for those events.

Other required equipment: If any equipment is required for EME deployment that is not stored with the EME, it must be demonstrated that this additional equipment will be available and the time required to obtain it must be accounted for in the timing assessment. For example, if self-contained breathing apparatus (SCBA) or portable lighting is required, but not included with the EME, it must be demonstrated that the location of the additional equipment is known, that it can be accessed, and the deployment time must be increased to account for obtaining and using the equipment (if not already accounted for).

Other Considerations: While not technically a feasibility issue, the scenarios in which EME deployment actions are credited should be consistent with the PRA. For example, if the PRA does not credit operator actions in seismic events greater than a certain magnitude, the same limitations should apply to the EME deployment activity. Any exceptions should be documented and a basis for the exception should be provided.

3.2. Use of the decision trees

The HRA decision trees developed for this methodology should not be treated as event trees (i.e., each node does not necessarily represent an event probability). The logic for the decision node describes how the HEPs are impacted based on the paths taken for each decision node. In some cases a multiplier greater than 1 may be applied as part of the quantification, while in others, the deployment time may be doubled and its impact will be accounted for in assessing the branch points related to time margin. Each path through the decision tree will result in a separate HEP that is applicable to the conditions defined by the decision node choices.

The branches on the decision trees represent performance shaping factors that are considered to be the most critical for this activity. If other PSFs are considered relevant, additional branches could be added. For example, since some of these activities may be protracted, it might be thought that fatigue could become an issue. However, this would have already been factored into the assessment when the feasibility of the action was assessed (see Section 3.1.1).

4. ASSUMPTIONS

The following assumptions were made to support the development and initial applications of this methodology:

1. The use of EME is proceduralized, the cues that would lead to deployment are addressed in validated plant procedures that provide step-by-step guidance, and the contribution related to the diagnosis of the need for and decision to implement EME can be evaluated using an existing HRA approach.
2. The personnel responsible for deploying the EME have been trained on the procedures and are qualified to perform the necessary tasks.
3. At least one member of the on-shift deployment team has practiced deploying the EME, and therefore can help the entire team work through problems.
4. The Main Control Room operators (or personnel responsible for controlling the EME) have been trained on the operation of the EME once installed, and the procedures provide sufficient guidance that the contribution to the HEP from failure in execution can be evaluated using current HRA methodologies.
5. For multi-unit site response, the analyzed unit is the last unit to which the EME is deployed. In reality, units with more time critical conditions would likely be prioritized for deployment; however, for this revision of the methodology, no attempt has been made to credit prioritized deployment.
   - If it is desirable or necessary to account for the prioritization of EME deployment, the HRA analyst can model that decision making process and include it in the PRA using existing HRA methodologies as long as the timeline used for EME deployment is adjusted to account for the modified deployment order.
6. The self-check and independent check recovery activities are considered to be most relevant to the "installation" portion of the EME deployment activity; therefore, the time margin assessment is focused on recovering errors committed during the "installation" activity.
7. EME deployment is assumed to be performed in the absence of high radiation conditions that would affect EME deployment for one or more units.
8. In the initial applications, for seismic events with magnitudes greater 0.3g PGA, actions taken in non-seismically qualified buildings were assumed to fail, consistent with other post-seismic human response assumptions in the seismic model. (This threshold could be varied and justified based on the specific seismic model.) The exception is for the EME storage structures, which may be designed to not damage the EME when they collapse. The impacts of EME structure failure, however, should explicitly be considered in the feasibility assessment.
9. The site has developed and is in compliance with guidance that maintains the EME and plant grounds in a state that will not hinder deployment of the EME.

5. EXAMPLE DECISION TREE - HIGH WINDS

The High Winds HRA decision Tree is provided in Figure 1. The guidance for interpreting and using each of the decision tree nodes is provided below.

<u>High Wind Induced Obstruction?</u>

As identified in the "Equipment and Location Accessibility" bullet in Section 3.1.1, deployment is considered to be failed for wind events in which deployment is precluded by the effects of the hazard. Reasons for this include:

- Failure of the structure(s) that house the EME,
- Failure of the structure(s) along the access path (or the access path itself) between where the EME is stored and the point where it is deployed, which if failed, could prevent access,
- Failure of the structure(s) where the EME is deployed (if applicable),
- Obstruction of path due to debris accumulation that is beyond the capability of on-site sources to remove.

It is expected that in these cases, no additional assessment is required.

For scenarios in which the impact is not so severe as to prevent successful deployment, there could still be an impact on the deployment action. An example of this condition may be one in which an EME pump (or fire truck to be used as the EME pump) is stored under a lightweight structure that would not damage the truck when it fails (and there are no other postulated wind generated failures that would impact the truck).

Other conditions that would classify as a "high wind induced obstruction" are those in which wind speeds are adequate to introduce impediments that would require additional time to address, but would not require equipment that is not part of the EME deployment set to move (e.g., fallen branches). Some sites may store heavy equipment, such as bulldozers, with the EME to provide the capability to move larger debris. In the event that this equipment is stored with the EME or is otherwise stored such that its use would be feasible by a qualified member of the deployment team, the EME may be credited for higher severity events that may result in the deposition of larger obstructions (e.g. trees) in the areas where deployment activities are required. The determination of the potential for wind induced obstructions will require a site and scenario-specific assessment of the EME location, access path, and deployment area.

For cases with no expected wind damage, the "no" (down) branch is taken, and no time penalty is assessed.

For cases in which wind induced obstructions/impediments are expected, the "yes" (up) branch is taken and a time penalty is assigned to the deployment task. Because of the nature of the event, there is no means of determining how much of an impact the wind damage may have on the deployment task. In order to provide a means of accounting for potential setbacks, it is assumed that the deployment time for each unique impacted leg (e.g., transportation, installation) of the deployment is doubled, subject to the following:

- If the installation is in an area that is protected from the impacts of high winds (e.g., inside a structure that has not failed), it is not necessary to double the installation time ($T_{Install}$).
- Account for the impacts of increases to the T_{Trans} and $T_{Install}$ (if necessary) times for all of the units that are included in the evaluation of T_{delay} (as defined in the "Time Margin >100%" node of Appendix A) for the analyzed unit. Also, account for increases in the T_{Trans} and $T_{Install}$ times for the analyzed unit if they have not already been accounted for.

- For cases in which multiple units are assessed, there may be some common activities. For example, if a four unit site is being assessed, the timing assessment would double T_{Trans} from the EME storage area to the deployment area if the EME is used for all of the units. If different equipment needs to be deployed for the additional units it may be possible to assume that the pathway has been cleared and only the first unit's T_{Trans} would to need be doubled. This would depend on whether the transportation paths were common.

If a plant specific review demonstrates that doubling the deployment time is not appropriate for the conditions at the site, the analysis can be adjusted to employ a revised deployment time multiplier judged to be more appropriate to the situation being analyzed. The revised multiplier and basis should be documented with the HRA.

Take Action After Event?

This node addresses timing of the deployment relative to the wind event. The "yes" (up) branch corresponds to the condition in which the action can successfully be taken after the wind event has passed. The "no" branch corresponds to the condition in which the action must be taken during the high wind event to ensure success, which is considered to be a high stress condition.

Primarily, this node is most useful for events where there may be high intensity, but short lived strong winds (e.g., during a tornado).

If the up branch is taken for non-high-intensity/short lived events, a clear description of why the node is applicable must be provided.

For the "yes" (up) branch, no multiplier is applied.

For the "no" (down) branch, a multiplier of 2 is applied to represent the impact of the stress associated with having to perform the EME deployment during a high wind event. This is based on the THERP [5] Table 20-16 moderately high stress multiplier of 2, as applied to skilled personnel performing step-by-step tasks.

Wind Below Safety Limits?

This node addresses the potential for high wind events to preclude work outside the plant. An actual physical limit that would prevent work in a high wind event would be difficult to determine, but each plant/site/organization may have safety guidelines that define a wind speed threshold above which actions are not allowed in a high wind event.

For cases where the action may be taken after the high wind event, this node is a pass through (i.e., is not evaluated).

If the wind speed of the event is greater than the safety limits specified for the plant where the action is taken, then the no (down) branch is taken and the action is set to FAILURE. These are cases in which the action is required to be performed during an extreme wind event where outside action is not possible.

If the wind speed of the event is less than the safety limits specified for the plant where the action is taken, then the "yes" (up) branch is taken. A multiplier of 2 is also applied here to represent the physical difficulty of performing the task in a high wind event. The doubling of the failure probability relative to the case in which no high wind conditions exist is based on judgment.

Cases in which actions are required during a storm for which wind does not impact the reliability of the action are not considered to be high wind events.

Time Margin >100%?

This node is used to account for the potential for the deployment team to correct an error in the installation of the EME (i.e., a self-check recovery). Self-checking in this context is defined as the process of checking that, upon completion, the installation has been performed correctly. In order for the error correction to be credited, it must be demonstrated that the time margin is ≥100%. The time margin requirement of 100% is assumed in order to account for the time required to:

- Perform the initial installation (which includes actions such as making temporary piping and/or power connections, loading a generator, and/or pressurizing a water system header),
- Identify if the installation has not been performed correctly,
- Review and re-perform the installation steps, when required,

The time margin definition is borrowed from NUREG-1921 [4] and modified to address the deployment assessment:

Equation 1: Time Margin (expressed as a percentage) =

$$100 * [(T_{SW} - T_{Delay}) - (T_{Trans} + T_{Install} + T_{Exe})] / (T_{Install})$$

Where,
- T_{SW} = the system window, or the time window within which the action must be performed to achieve the function provided by the EME. For example, this time could be measured from the time the hazard impacts the plant to the time at which the EME must be delivering water to its load(s).
- T_{Delay} = time delay, or the duration of time it takes to begin initiating EME deployment for the analyzed unit, measured from the time the hazard impacts the plant. For a multi-unit site, since the analysis is for the last unit for which the EME is deployed, the time delay includes the sum of the times taken to deploy the EME for the other unit(s). Because the order of deployment is not known for multi-unit sites, it is assumed that the analyzed unit is the LAST unit to which the EME is deployed. In reality, units with conditions with more time critical conditions would likely be prioritized for deployment; however, for this revision no attempt has been made to credit prioritized deployment.
   - If it is desirable or necessary to account for the prioritization of EME deployment, the HRA analyst can model that decision making process and include it in the PRA using existing HRA methodologies as long as the timeline used for EME deployment is adjusted to account for the modified deployment order.
- T_{Trans} = the time required to transport the EME from the storage area to the area where the EME is deployed and unload any equipment that is required.
- $T_{Install}$ = the time to perform tasks such as making any necessary temporary piping and/or power connections, loading a generator, and/or pressurizing a water system header such that water is available for the load, when directed.
   - If the installation activity does not include steps that would validate EME operation such that the first opportunity to identify an error would be when the EME was required to provide power/flow to a station load, credit for error correction should not be taken unless some equivalent validation process exists and is performed at or near the time of installation.
- T_{Exe} = the time to perform the steps required to initiate water flow and/or energize electrical equipment from the time when it is directed. [Note that the failure probability of this portion of the EME implementation action is not assessed by this methodology, but the timing assessment for the deployment portion of the action is required to account for the execution time in the time margin assessment.]

Because the deployment team is attempting to correct its own error, a high dependence condition is assumed to exist between the commission of the error and the work to correct it. The THERP guidance provides equations in Table 20-17 for determining dependent failure probabilities for a range of different dependence levels. Because the equation for "high dependence" generally yields results in the 0.5 range, the self-check recovery has been assigned a failure probability of 0.5.

Independent Check Available?

An independent check is considered to be an assessment of the EME installation by a qualified member of the plant staff that was not part of the deployment activity. In order for the check to be credited, it must be demonstrated that the time margin is >=200% using Equation 1.

The time margin requirement of 200% is assumed in order to account for the time required to:

- Perform the initial installation (e.g., making temporary piping and/or power connections, loading a generator, and/or pressurizing a water system header),
- Identify if the installation has not been performed correctly,
- Review and re-perform the installation steps, when required,
- Determine that the EME is still not functioning,
- Have an independent checker trouble shoot and direct the crew to re-perform the installation, when required, to correct the error.

The benefit of concurrent verification (i.e., step review by a team member who did not perform the step, but was present when the step was performed) has not been considered in this methodology. If it is necessary to credit concurrent verification in an application to remove unnecessary conservatism, it may be used in place of independent verification provided that a basis is documented for concurrent verification failure probability.

The "yes" (up) branch is taken for cases in which independent check credit is available. A multiplier of 0.1 is used based on the probability for item 1 in THERP table 20-22. While that item is for routine tasks in normal plant conditions, it represents failure to identify errors in connections, positions of locally operated valves, and breaker positions. The application of the event specific PSF multipliers is considered to address the impact of the events on this credit.

The "no" (down) branch is taken for cases in which no independent check credit can be justified. No multiplier is applied for this branch.

6. CONCLUSION

The methodology documented in this paper is intended to provide a means of quantifying an HEP for the deployment of portable equipment for internal events and selected external events scenarios. Because hazard events, i.e., specific occurrences of the hazard, occur in different ways, the methodology is of necessity bounding in nature and therefore tends to be somewhat conservative. The results, when combined with detailed assessments of the remaining components of the human failure event (i.e., the decision to initiate deployment of portable equipment and the use of the equipment once deployed), are considered to be adequate for use in PRA applications as long as it is understood that there is considerable uncertainty associated with the numerical values.

Acknowledgements

Funding for the development of this methodology was provided jointly by OPG and Bruce Power.

Figure 1: High Wind Events

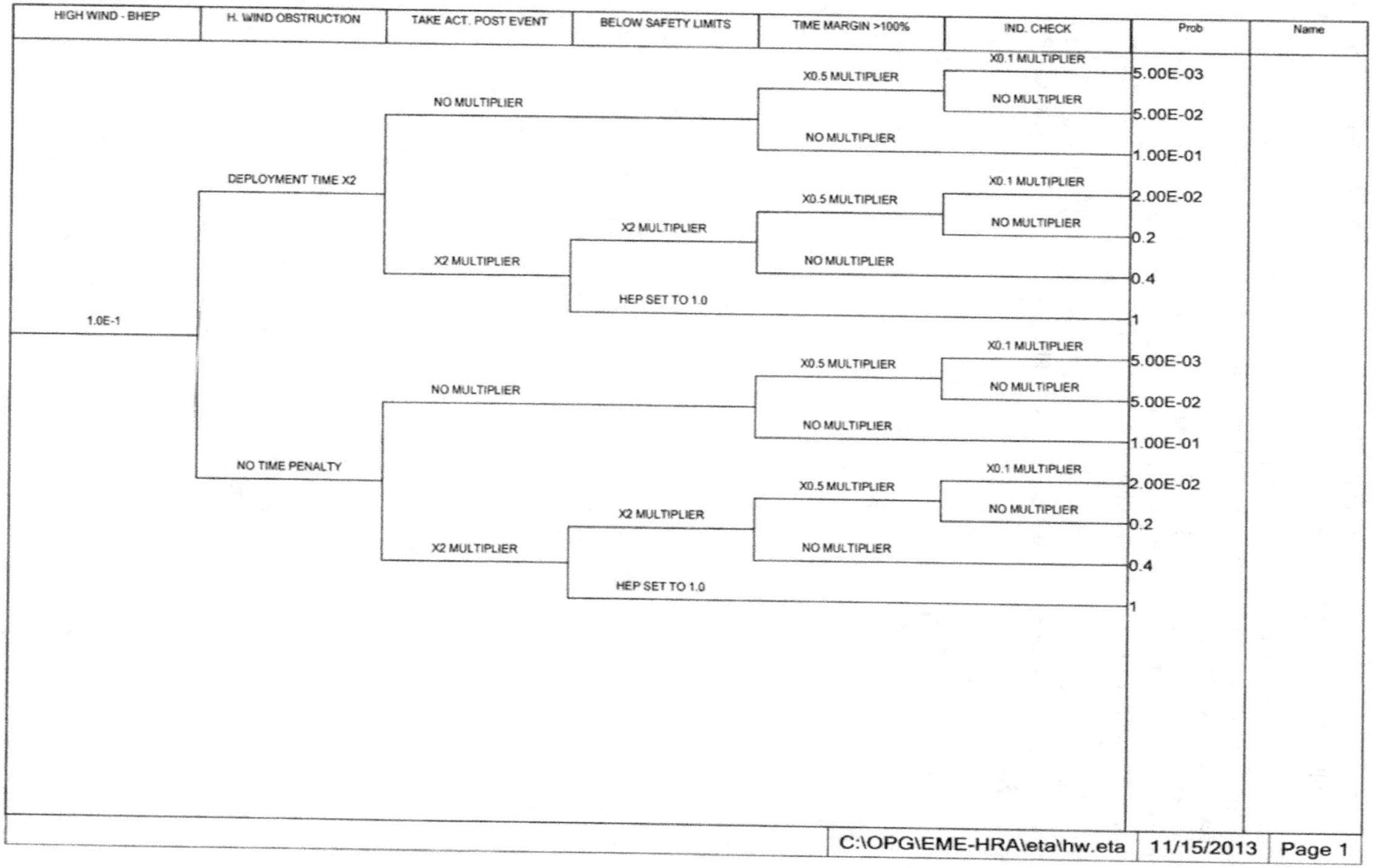

HIGH WIND - BHEP	H. WIND OBSTRUCTION	TAKE ACT. POST EVENT	BELOW SAFETY LIMITS	TIME MARGIN >100%	IND. CHECK	Prob	Name
					X0.1 MULTIPLIER	5.00E-03	
				X0.5 MULTIPLIER	NO MULTIPLIER	5.00E-02	
			NO MULTIPLIER	NO MULTIPLIER		1.00E-01	
					X0.1 MULTIPLIER	2.00E-02	
			X2 MULTIPLIER	X0.5 MULTIPLIER	NO MULTIPLIER	0.2	
		DEPLOYMENT TIME X2		NO MULTIPLIER		0.4	
			HEP SET TO 1.0			1	
					X0.1 MULTIPLIER	5.00E-03	
				X0.5 MULTIPLIER	NO MULTIPLIER	5.00E-02	
1.0E-1		NO MULTIPLIER	NO MULTIPLIER	NO MULTIPLIER		1.00E-01	
					X0.1 MULTIPLIER	2.00E-02	
	NO TIME PENALTY	X2 MULTIPLIER	X2 MULTIPLIER	X0.5 MULTIPLIER	NO MULTIPLIER	0.2	
				NO MULTIPLIER		0.4	
			HEP SET TO 1.0			1	

C:\OPG\EME-HRA\eta\hw.eta   11/15/2013   Page 1

Probabilistic Safety Assessment and Management PSAM 12, June 2014, Honolulu, Hawaii

References

[1] NRC (U.S. Nuclear Regulatory Commission), "Technical Basis and Implementation Guidelines for A Technique for Human Event Analysis (ATHEANA)", NUREG-1624, Revision 1, May, 2000.

[2] S.H. Chien A.A. Dykes, J.W. Stetkar, and D.C. Bley, "Quantification of Human Error Rates Using a SLIM-Based Approach", IEEE Fourth Conference on Human Factors and Power Plants, Monterrey, CA. June 5-9, 1988.

[3] A. Kolaczkowski et. al (U.S. Nuclear Regulatory Commission). "Good Practices for Implementing Human Reliability Analysis". NUREG-1792. April, 2005.

[4] S. Lewis. (U.S. Nuclear Regulatory Commission). "EPRI/NRC Fire Human Reliability Analysis Guidelines". NUREG-1921. July, 2012.

[5] A.D. Swain and H.E. Guttman (U.S. Nuclear Regulatory Commission). "Handbook of Human Reliability Analysis with Emphasis on Nuclear Power Plant Applications". NUREG/CR-1278. August, 1983.

Study on Operator Reliability of Digital Control System in Nuclear Power Plants Based on Boolean Network

Yanhua Zou[a,b,c], Li Zhang[a,b,c], Licao Dai[c], Pengcheng Li[c]

[a] Institute of Human Factors Engineering and Safety Management, Hunan Institute of Technology, Hengyang, China
[b] School of Nuclear Science and Technology, University of South China, Hengyang, China
[c] Human Factor Institute, University of South China, Hengyang, China

Abstract: The current human reliability analysis method of analyzing system operator's reliability, carried out from the perspective of operators themselves, is relatively static, for it hasn't taken the effect of system evolution on the operators' performance into consideration. In view of operator reliability in digital control system in nuclear power plant, this paper, based on boolean network theory, tries to explore the operators' behavior in the dynamic logic process of system evolution, aiming at finding out the dynamic evolution process of human-system interaction. A new technique, called the semi-tensor product of matrices, can convert the logical systems into standard discrete-time dynamic systems, and then the discrete-time linear equation and reliability analysis model are established. Data collected from simulation experiments carried out in full-size simulator in LingDong Nuclear Power Plant is found to be in consistence with the operator reliability model constructed before.

Keywords: Boolean network, Digital control system, Reliability analysis, Semi-tensor product of matrices.

1. INTRODUCTION

With the development of science and technology, the safety and efficiency of system and equipment have been improving, but the reliability of human-machine system has been depending on man. According to statistics, over 60% of fatal casualties and over 80% of serious casualties at home and abroad are due to human errors [1,2]. The serious consequences caused by operators have been fully realized after the accidents at the Chernobyl nuclear power plant and the America Three Mile Island nuclear power plant accidents. Therefore, research on the relationship between operators and system, on qualitative analysis and quantitative assessment on human operations, has become increasingly important in the engineering field [3].

Ever since digital control system was adopted in nuclear power plant, operators have been enjoying the conveniences it has brought about, in the meantime, they have also been facing risks of operation reliability caused by enormous and centralized information. In the main control room of a digital control system, the central display of system alarming, parameters and pictures has formed a keyhole effect with enormous information and limited display [4,5], for the operators, since in a traditional control room, the operators can take in everything at a glance while in the main control room of a digital control system, they have to use a computer to carry out interface management tasks to find information promptly and efficiently. This shows that the adoption off digital technology has brought some new risks for operators, and whether the reliability of them can meet the safe and economic requirements has become one of the urgent problems a plant has to solve.

The method of Fault Tree Analysis and Event Tree Analysis are two of the common ways in probabilistic safety assessment. However, in case of accidents at a nuclear power plant, the response of the system or the behavior of an operator changes with the process of the accident, so an operator's behavior at the next time node is closely related to the situation of the system and the operation at the previous time node. The traditional methods of Fault Tree Analysis and Event Tree Analysis are static analysis technology based on Boolean Logic [6], without taking the dynamic development between man and system into full consideration, so it is of significance to probe into the dynamic relationship and applying it to reliability analysis both in theoretical research and industrial application.

On the basis of Boolean network, now a powerful tool in system control, Cheng Daizhan put forward a new method of matrix calculation--- semi-tensor product of matrix [7,8], with which logic variable can be expressed as vector form, and logic function as multiple linear mapping form. However, under algebraic expression, Boolean network equations, having all the information of Boolean network, are represented by general discrete-time linear equations. With the method, Boolean network equations can be established to analyze the operations in the digital control system at a nuclear power plant by determining the relationship between operations with data collected through analog experiment and video analysis. The 2nd section of this paper is about basic knowledge of semi-tensor product of matrix, some basic properties needed in derivation, matrix expression of logic and Boolean network model. The 3rd section is about the establishment of Boolean network of operations, introducing the obtainment of experiment data and the specific process of model construction. The 4th section is conclusion and discussion, analyzing on experiment results and discussing future work.

2. PRELIMINARIES

First, we give some notations for the statement ease.

1) $D_k := \{0, \frac{1}{k-1}, ..., \frac{k-2}{k-1}\}, k \geq 2; D := D_2 = \{0,1\}$.

2) Let δ_n^i be the i th column of the identity matrix I_n.

3) $f : D^n \to D$ are logical functions.

4) $\Delta_n : \Delta_n = \{\delta_n^i | i = 1, 2, ..., n\}$, when $n = 2$, simply use $\Delta := \Delta_2$.

5) Denote by $COL(A)$ the set of columns of A.

6) Assume a matrix $M = [\delta_n^{i_1}, \delta_n^{i_2}, ..., \delta_n^{i_s}] \in M_{n \times s}$, its columns, $COL(M) \subset \Delta_n$. We call M a logical matrix, and simply denote it as $M = \delta_n[i_1, i_2, ..., i_s]$.

7) $\otimes$ is Kronecher product.

2.1. Definition and Proposition of the Semi-tensor Product of Matrices [7,8,9,10]

Definition (1): (i) Let $X = [x_1, ..., x_s]$ be a row vector, $Y = [y_1, ..., y_t]^T$ be a column vector.
①: If $s = t \times n$. Then

$$\langle X, Y \rangle_L := \sum_{k=1}^{t} X^k y_k \in \mathbb{R}^n \tag{1}$$

Where $X = [X^1, ..., X^t], X^i \in \mathbb{R}^n, i = 1, ..., t$. We call $\langle X, Y \rangle_L$ a semi-tensor product.
②: If $t = s \times n$. Then

$$\langle X, Y \rangle_L := (\langle Y^T, X^T \rangle_L)^T \in \mathbb{R}^n \tag{2}$$

$\langle X, Y \rangle_L$ also called a semi-tensor product.

(ii) Assume $M \in M_{m \times n}, N \in M_{p \times q}$, if n is the divisor of p or p is the divisor of n. We call $C = M \ltimes N$ is the semi-tensor product of M and N.
If C is composed of $m \times q$ blocks, $C = (C^{ij})$, meanwhile

$$C^{ij} = \langle M^i, N_j \rangle_L, i = 1, ..., m, j = 1, ..., q. \tag{3}$$

Where M^i is a row of M, N_j is a column of N.

Throughout this paper, the matrix product is assumed to be the semi-tensor product. In the following, the symbol $\ltimes$ is omitted.

The semi-tensor product has the following properties.

Proposition (1): (i) If $A \in M_{m \times np}, B \in M_{p \times q}$. Then
$$A \ltimes B = A(B \otimes I_n). \tag{4}$$
(ii) If $A \in M_{m \times n}, B \in M_{np \times q}$. Then
$$A \ltimes B = (A \otimes I_p)B. \tag{5}$$
Proposition (2): Let $X \in R^m, Y \in R^n$ be two columns. Then
$$W_{[m,n]} \ltimes X \ltimes Y = Y \ltimes X; \tag{6}$$
$$W_{[n,m]} \ltimes Y \ltimes X = X \ltimes Y. \tag{7}$$
Where $W_{[m,n]}$ is an $mn \times mn$ matrix, called the swap matrix.

Proposition (3): Let $x \in \Delta$. Then
$$x^2 = M_r x \tag{8}$$
Where $M_r = \delta_4[1,4]$ is called the power-reducing matrix.

2.2. Matrices Expression of Logic [7,8,9,10]

A logical variable means a proposition. When the proposition is true, we say that the logical variable takes value "T" or "1", and when it is false, the logical variable takes value "F" or "0". In classical logic a logical variable can only take values from $\{0,1\}$. We note

$$T := 1 \equiv \begin{bmatrix} 1 \\ 0 \end{bmatrix}; \quad F := 0 \equiv \begin{bmatrix} 0 \\ 1 \end{bmatrix}; \tag{9}$$

Four fundamental operators usually be used are Conjunction $P \wedge Q$, Disjunction $P \vee Q$, Conditional $P \rightarrow Q$, and Biconditional $P \leftrightarrow Q$. A conventional way to depict the values of an operator is using a table, called the truth table. We can have truth table for "conjunction", "disjunction", "conditional", and "biconditional", respectively as in Table 1.

Table 1: Truth Table

p	q	$p \wedge q$	$p \rightarrow q$	$p \leftrightarrow q$	$p \vee q$
1	1	1	1	1	1
1	0	0	0	0	1
0	1	0	1	0	1
0	0	0	1	1	0

Definition (2): Let σ be an r-ary logical operator. $M_\sigma \in M_{2 \times 2^r}$ is called the structure matrix of σ, if in the vector form we have
$$\sigma(p_1,...p_r) = M_\sigma \ltimes p_1 \ltimes p_2 \ltimes ... \ltimes p_r = M_\sigma p_1...p_r \tag{10}$$

Theorem (1): Assume $f(x_1,...,x_n)$ is a logical function, and in the vector form we have $f: \Delta_{2^n} \rightarrow \Delta$. Then there exists a unique logical matrix M_f, called the structure matrix of f, such that following equation(11) holds.
$$f(x_1,...,x_n) = M_f x. \tag{11}$$

Where $x = \ltimes_{i=1}^{n} x_i$.

Using Theorem (1), the structure matrix of four fundamental operators are obtained as
$$M_\wedge := M_c = \delta_2[1,2,2,2];$$
$$M_\vee := M_d = \delta_2[1,1,1,2];$$
$$M_\rightarrow := M_i = \delta_2[1,2,1,1];$$
$$M_\leftrightarrow := M_e = \delta_2[1,2,2,1].$$

2.3. Boolean Network

Boolean network was firstly introduced by Kaufman to formulate the cell networks. Then, it becomes a powerful tool in describing, analyzing, and simulating the cell networks, and also be used as models of some complex systems such as neural networks. A Boolean network is a directed network graph, consists of a set of nodes, and a set of edges.

Definition (3): A Boolean network is a set of nodes $x_1, x_2, ..., x_n$, which interact with each other in a synchronous manner. At each given time $t = 0, 1, 2, ...$ a node has only one of two different values: 1 or 0. Thus the network can be described by a set of equations:

$$\begin{cases} x_1(t+1) = f_1(x_1(t),...,x_n(t)) \\ x_2(t+1) = f_2(x_1(t),...,x_n(t)) \\ ...... \\ x_n(t+1) = f_n(x_1(t),...,x_n(t)) \end{cases} \quad (12)$$

Where $f_i : D^n \rightarrow D, i = 1, ..., n$ are n-ary logical functions, $x_i(t) \in D$ are state variables.

3. MODEL CONSTRUCTION

In digital control system, an operator's work involves monitoring, situation assessment, response planning and response implementation [11]. Suppose $x_1(t), x_2(t), x_3(t), x_4(t)$ · $x_i(i=1,2,3,4) \in D$ represent the operations at "t" (a certain time), $x_1(t), x_2(t), x_3(t), x_4(t)$ represent monitoring, situation assessment, response planning and response implementation respectively, 1 indicates that an operator takes some action, while 0 no action. For instance, if at a certain time "t", $x_1(t), x_2(t), x_3(t), x_4(t)$ has the values $(1,1,0,0)$, that means the operator take the actions of monitoring and situation assessment at this moment.

Thus, the operator's behavior at any moment can be expressed in a four-dimensional array, the evolution of the operator's behavior is equal to that of the array. With this abstract method, the operator's behavior at different time can be arranged using the method of moment, so as to analyze the dynamic process of the operator's behavior in the evolution of the system.

3.1. Experiment Data Source and Explanation

The experiment data of this paper are from the Steam Generator Tube Rupture experiment in Lingdong Nuclear Power Plant carried out on full-scale simulator. The reason why SGTR is chosen is that it is a typical problem of reliability related to operations after initial accident at a nuclear power plant, and they are crucial human's operations to be considered in PSA analysis [12]. 8 sets of data are obtained after observation and analysis:

1) $t=0$, $(0,0,1,0)$; $t=1$, $(0,0,0,1)$; $t=2$, $(1,0,0,0)$; $t=3$, $(0,1,0,0)$; $t=4$, $(1,0,0,1)$; $t=5$, $(1,1,0,0)$; $t=6$, $(1,1,1,1)$;
2) $t=0$, $(1,0,1,0)$; $t=1$, $(0,1,0,1)$; $t=2$, $(1,0,0,1)$;
3) $t=0$, $(0,1,1,1)$; $t=1$, $(1,0,0,1)$; $t=2$, $(1,1,1,1)$;
4) $t=0$, $(0,0,1,1)$; $t=1$, $(1,1,0,1)$; $t=2$, $(1,1,1,1)$;
5) $t=0$, $(1,1,1,0)$; $t=1$, $(1,1,1,1)$;
6) $t=0$, $(0,1,1,0)$; $t=1$, $(1,0,0,1)$;
7) $t=0$, $(1,0,1,1)$; $t=1$, $(1,1,0,1)$;
8) $t=0$, $(0,0,0,0)$; $t=1$, $(0,0,0,0)$;

As 17 experiment data are needed to determine a four-noded Boolean network model, 8 different time nodes are chosen as starting observation points to avoid being special, and 24 data are obtained.

In the above 8 sets of data, all the time nodes are discrete, but the time nodes of each set of experiment data are successive. The interval is the time needed by an operator to change his operation from one moment to another. As there are additional accidents planned in the experiment, four operation models happen simultaneously at some time nodes. As for the 8th set of data, it indicates that when an operator takes no action, the situation will be better, and the same will happen at next time node.

3.2. Algebraic Form of Boolean Network

From equation (12), the key to build the dynamic relationship between these four variables is to determine the four logic function.

Define $x(t) = \ltimes_{i=1}^{4} x_i(t)$, from equation (11) and (12), we have

$$\begin{cases} x_1(t+1) = M_1 x(t) \\ x_2(t+1) = M_2 x(t) \\ x_3(t+1) = M_3 x(t) \\ x_4(t+1) = M_4 x(t) \end{cases} \quad (13)$$

Where $M_i \in M_{2 \times 2^r}$, called the structure matrix of f_i. Equation (13) is called the component-wise algebraic form of (12).

Then, $$x(t+1) = \ltimes_{i=1}^{4} x_i(t+1) = M_1 x(t) M_2 x(t) M_3 x(t) M_4 x(t) \quad (14)$$

Refer to [7,8,10], (13) can further be converted as
$$x(t+1) = Lx(t) \quad (15)$$
Where $L \in L_{2^n \times 2^n}$ is called the transition matrix of the system. Equation (15) is called the algebraic form of (12).

Refer to [8], it was proved that (12),(13),(15) are equivalent to each other. While building model (12) directly seems much more difficult, we prefer to construct model (13) or (15) by calculating the structure matrix of M_i or the transition matrix L.

Next step, we use the experimental data collected in section 3.1 to construct model (13) and (15).

3.3. Dynamic Model Construction of An Operator's Performance

For the first experimental data, in the vector form [8,13], we have

$$(0,0,1,0) = X^1(0) = \delta_2[2,2,1,2], \text{ and } x^1(0) = \delta_2^2 \ltimes \delta_2^2 \ltimes \delta_2^1 \ltimes \delta_2^2 = \delta_{16}^{14}.$$

Similarly, we can calculate
$$x^1(1) = \delta_{16}^{15} \; ; \; x^1(2) = \delta_{16}^{8} \; ; \; x^1(3) = \delta_{16}^{12} \; ; \; x^1(4) = \delta_{16}^{7} \; ; \; x^1(5) = \delta_{16}^{4} \; ; \; x^1(6) = \delta_{16}^{1}.$$

The following proposition can help us to determine the column of the transition matrix L.

Proposition (4)[13]: If $x(t) = \delta_{2^n}^i$ and $x(t+1) = \delta_{2^n}^j$, the i th column of the transition matrix L is
$$Col_i(L) = \delta_{2^n}^j \qquad (16)$$

Using the Proposition, it is known that
$$Col_{14}(L) = \delta_{16}^{15}; \; Col_{15}(L) = \delta_{16}^{8}; \; Col_{8}(L) = \delta_{16}^{12};$$
$$Col_{12}(L) = \delta_{16}^{7}; \; Col_{7}(L) = \delta_{16}^{4}; \; Col_{4}(L) = \delta_{16}^{1}.$$

The 6 columns of L have been determined.

Using the same procedure to the other groups of data, certain values of column of L can be figured out.

Finally, we can obtain
$$L = \delta_{16}[1,1,1,1,3,11,4,12,3,7,7,7,3,15,8,16] \qquad (17)$$

Refer to [13], the corresponding retrievers are
$$S_1^4 = \delta_2[1,1,1,1,1,1,1,1,2,2,2,2,2,2,2,2];$$
$$S_2^4 = \delta_2[1,1,1,1,2,2,2,2,1,1,1,1,2,2,2,2];$$
$$S_3^4 = \delta_2[1,1,2,2,1,1,2,2,1,1,2,2,1,1,2,2];$$
$$S_4^4 = \delta_2[1,2,1,2,1,2,1,2,1,2,1,2,1,2,1,2].$$

Then
$$M_1 = S_1^4 L = \delta_2[1,1,1,1,1,2,1,2,1,1,1,1,1,2,1,2];$$
$$M_2 = S_2^4 L = \delta_2[1,1,1,1,1,1,1,1,2,2,2,1,2,2,2];$$
$$M_3 = S_3^4 L = \delta_2[1,1,1,1,2,2,2,2,2,2,2,2,2,2,2,2];$$
$$M_4 = S_4^4 L = \delta_2[1,1,1,1,1,1,2,2,1,1,1,1,1,1,2,2].$$

Consider the logical expression of $x_1(t)$
$$x_1(t+1) = M_1 x(t) = \delta_2[1,1,1,1,1,2,1,2,1,1,1,1,1,2,1,2]x(t).$$

It is easy to verify that
$$M_1(M_n - I_2) = 0;$$
$$M_1 W_{[2,2]}(M_n - I_2) \neq 0;$$
$$M_1 W_{[2,4]}(M_n - I_2) = 0;$$
$$M_1 W_{[2,8]}(M_n - I_2) \neq 0.$$

So $x_1(t)$, $x_3(t)$ are fabricated variables in the dynamic equation of $x_1(t+1)$.

Setting $x_1(t) = x_3(t) = \delta_2^1$, then
$$\begin{aligned}
x_1(t+1) &= M_1 x(t) = M_1 x_1(t) x_2(t) x_3(t) x_4(t) \\
&= M_1 x_1(t) W_{[2,2]} x_3(t) x_2(t) x_4(t) \\
&= M_1 (I_2 \otimes W_{[2,2]}) x_1(t) x_3(t) x_2(t) x_4(t) \\
&= M_1 (I_2 \otimes W_{[2,2]}) (\delta_2^1)^2 x_2(t) x_4(t) \\
&= \delta_2[1,1,1,2] x_2(t) x_4(t).
\end{aligned}$$

Hence its logical expression is
$$x_1(t+1) = (\neg x_2(t)) \rightarrow x_4(t).$$

The same procedure can be used to construct the logical expression of $x_2(t)$, $x_3(t)$, $x_4(t)$. Finally, the logical expression of the dynamics of the operators' performance is obtained as

$$\begin{cases} x_1(t+1) = (\neg x_2(t)) \rightarrow x_4(t) \\ x_2(t+1) = x_1(t) \vee [x_3(t) \wedge x_4(t)] \\ x_3(t+1) = x_1(t) \wedge x_2(t) \\ x_4(t+1) = x_2(t) \vee x_3(t) \end{cases} \qquad (18)$$

Its network graph depicted in Fig.1

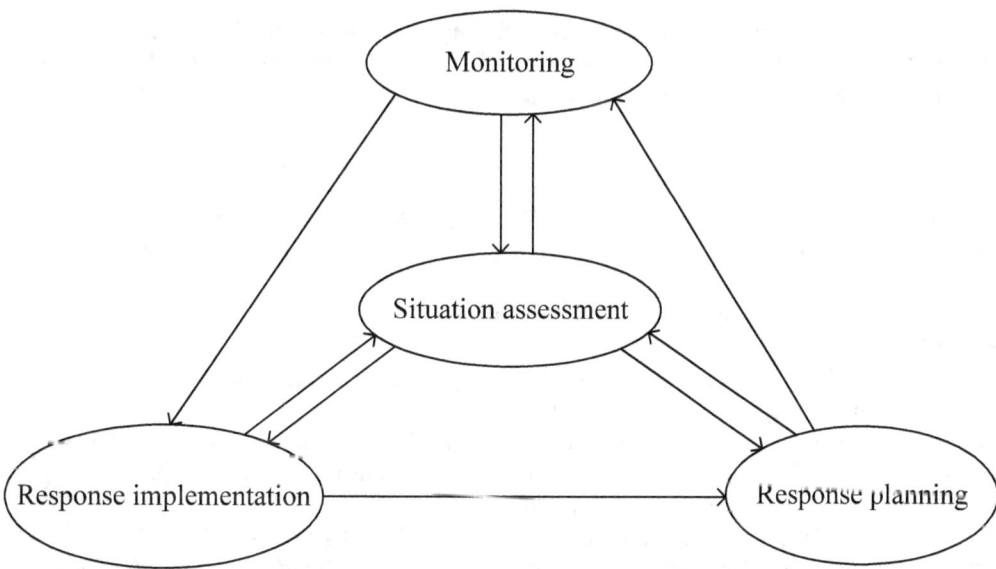

Fig.1: Network Graph

The above Boolean Network equations are the constructed dynamic model of an operator. It can be seen from equation (18) that if monitoring and situation assessment are carried out at a time node, response planning will be carried out at the next; that if situation assessment or response planning is carried out at a time node, a specific operation will be done at the next; that if monitoring and /or response planning and specific operation are carried out at one time node, situation assessment on the system will be carried out at the next; and that if situation assessment or specific operation is carried out at a time node, monitoring will be carried out at the next.

4. CONCLUSION AND DISCUSSION

In the past, analysis on human's operation was carried out through static analysis technology on the basis of Boolean algebra to work out human's error probability [3], focusing on analysis from an operator's cognitive behavior, failing to show how his operation changed with the situation of the system, seldom considering the dynamic relationship between operations at different time nodes. This paper, define the operator's behavior in the system process as a state variables, using an array of expression, the evolution of operator's behavior with process of incident could be abstract, then the logical process could be changed into algebraic expression and construct the logical equation of operations. This method has shed new light on analyzing the reliability of operations and the relationship between them, and has also shown the characteristics of some operations at a nuclear power plant.

For example, a set of data as $(1,0,0,0) \to (0,1,0,0) \to (0,0,1,0) \to (0,0,0,1)$ can be deduced from model (18), which shows that monitoring, situation assessment, response planning and response implementation, which can be overlapping, have to be carried out repeatedly at a nuclear power plant under the State Oriented Procedure(SOP).

Though the analysis method put forward in this paper is of significance in making up the inadequacy of the traditional method of static analysis on human reliability on the basis of Boolean Logic, the Boolean network models are special since they are deduced according to data obtained from one analog experiment. If different analog experiments can be carried out over and over again with this method to analyze the relationship between operations in different experiments, the general dynamic relationship between operations may be deduced in a digital control system.

Acknowledgements

This work was supported by the National Natural Science Foundation of China (Grant No.71371070, 71071051), National Natural Science Foundation for young (Grant No.71301069) and Research Project of LingDong nuclear Power Co. Ltd. (Grant No.KR70543).

References

[1] Wang Hong-de and Gao Wei. *"Study on Erroneous Operation due to Human Factor Based on Human Cognitive Reliability (HCR) Model"*, China Safety Science Journal, volume16, pp. 51-56, (2006).
[2] Zhang Li. *"Human Error Analysis and Preventives"*, Nuclear Power Engineering, volume11, pp. 91-96, (1990).
[3] Zhang Li. *"The Research on Human Reliability Analysis Technique in Probabilistic Safety Assessment"*, Atomic Energy Press, 2006, Beijing.
[4] Zhang Li, Yang Da-xin and Wang Yi-qun. *"The Effect of Information Display on Human Reliability in a Digital Control Room"*, China Safety Science Journal, volume20, pp. 81-85, (2010).
[5] Gu Pengfei, Zhang Jianbo and Sun Yongbin. *"The Human Reliability in the Accident Processing of NPP"*, Science & Technology Review, volume30, pp. 51-55, (2012).
[6] Yu Yu, Tong Jiejuan, Liu Tao, Zhao Jun and Zhang Aling. *"Accident Analysis by Phase-mission Methods in Nuclear Plant"*, Science & Technology Review, volume27, pp. 83-86, (2009).
[7] Cheng Dai-zhan, Qi Hong-sheng and Zhao Yin. "Analysis and Control of Boolean Networks: A Semi-tensor Product Approach", Acta Automatica Sinica, volume37, pp. 529-540, (2011).
[8] Cheng Dai-zhan and Qi Hong-sheng. *"Semi-tensor Product of Matrices: Theory and Applications"*, Science Press, 2007, Beijing.
[9] Cheng D. *"Matrix and Polynomial Approach to Dynamics Control Systems"*, Science Press, 2002, Beijing.
[10] Cheng D Z and Qi H S. *"A linear representation of dynamics of Boolean networks"*, IEEE Transactions on Automatic Control, volume55, pp. 2251-2258, (2010).

[11] Office of Nuclear Regulatory Research. "*Computer-Based Procedure Systems: Technical Basis and Human Factors Review Guidance*", Washington DC: US Nuclear Regulatory Commission.
[12] Yu Yuan-gao. "*Human Reliability Analysis in Steam Generator Tube Rupture Incidents*", Shanghai Jiao Tong University, 2008, Shanghai.
[13] Daizhan Cheng, Hongsheng Qi and Zhiqiang Li. "*Model Construction of Boolean Network via Observed Data*", IEEE Transactions on NeuralcNetwork, volume22, pp. 525-536, (2011).

Toward Modelling of Human Performance of Infrastructure Systems

Cen Nan[ac] and Wolfgang Kröger[b]

[a] Reliability and Risk Engineering Group (RRE), ETH Zürich, Switzerland
[b] ETH Risk Center, ETH Zürich, Switzerland
[c] Land Using Engineering Group (LUE), ETH Zürich, Switzerland

Abstract: During the last decade, research works related to modelling and simulation of infrastructure systems have primarily focused on the performance of their technical components, almost ignoring the importance of non-technical components of these systems, e.g., human operators, users. In contrast, the human operator of infrastructure systems has become an essential part for not just maintaining daily operation, but also ensuring the security and reliability of the system. Therefore, developing a modeling framework that is capable of analyzing the human performance in a comprehensive way has become crucial. The respective framework, proposed in this paper, is generic and consists of two parts: an analytical method based on the Cognitive Reliability Error Analysis Method (CREAM) for human performance assessment and an Agent-based Modeling (ABM) approach for the representation of human behaviors. This framework is a pilot work exploring possibilities of simulating human operators of infrastructure systems through advanced modeling approaches. The demonstration of the applicability of this framework using the SCADA (Supervisory Control and Data Acquisition) system as an exemplary system is also presented.

Keywords: Human Reliability Analysis, CREAM, Agent-based Modeling, Critical Infrastructure, SCADA

1. INTRODUCTION

Modern infrastructure systems, e.g., power supply, telecommunication and rail transport systems, are all large-scale, highly integrated, particularly interconnected and show complex behaviours. These systems are so vital to any country that their incapacity or destruction would have a debilitating impact on the health, safety, security, economics and social well-being [1]. The operators of these systems must continuously monitor and control them to ensure their proper operation [2]. These industrial monitor and control functions are generally implemented using an industrial control system (ICS), e.g., the SCADA system. The fundamental purpose of this type of systems is to allow its users (operators) to collect data from one or more remote facilities and send control instructions back to those facilities [3]. Most research studies on infrastructure systems, especially on this type of ICS, have taken an engineering point of view, which often underestimate the importance of their non-technical components, e.g., human operators [4, 5]. A number of studies have shown that human errors are major causes for accidents occurred in electric power, railway, aviation and maritime infrastructure sectors [6-8], highlighting the significance of examining the reliability of the human operators, which can be conducted using analytical methods and advanced modelling approaches.

2. RESEARCH STREAMS AND PROPOSED FRAMEWORK

Over the years, many Human Reliability Analysis (HRA) methods have been developed to analyse human performance in either qualitative or quantitative ways. Qualitative methods focus on the identification of events or errors, while quantitative methods focus on translating identified events/errors into Human Error Probability (HEP) [9]. The Technique for Human Error Rate Prediction (THERP), one of first generation HRA methods, is probably the most widely used technique to date [10]. THERP aims to calculate the probability of successful performance of the activities defined necessary for the accomplishment of a task. The calculations are based on pre-defined error rates (HEPs) and success is defined as the complement to the probability of making an error. Appropriate HEPs from a list around 100 factors are selected for a nominal assessment [11]. The results of the task analysis are represented graphically in a so-called HRA event tree that is a formal representation of the required sequence of actions. The use of the THERP causes limitations during human performance analysis since this method is focused on errors of omission and intends to characterize each operator action with a binary path (success or failure). Moreover, the representation of Performance Shaping Factors (PSFs) influence on human performance is quite poor and highly judgmental based on assessor's experiences [10, 12]. Success Likelihood Index Method (SLIM), another example of first generation HRA methods, is used for the purposes of evaluating the probability of a human error occurring throughout the completion of a specific task [13]. It is a decision-analytic approach, which uses expert judgment to quantify PSFs. Such factors are used to estimate a Success Likelihood Index (SLI), a form of preference index, which is calibrated against existing data to derive a final HEP. This approach is a flexible technique and able to deal with the total range of human error forms. SLIM is a subjective method and the choosing of PSFs is quite arbitrary. Another disadvantage of this approach is that there is a lack of valid calibration data [12]. A Technique for Human Event Analysis (ATHEAHA), one of second generation HRA methods, is designed to support the understanding and quantification of Human Failure Events (HFEs) [14]. This method is based on a multi-disciplinary framework that considers both human-centered factors and plant conditions creating operational causes for human-system interactions [10]. The human-centered factors and influences of plant conditions are dependent of each other, which are combined to create a situation in which the probability of making an error can be estimated. Such a situation is said to have an Error-forcing Context (EFC). The primary shortcoming of this technique lies in the fact that it is unable to produce final HEP meaning that the direct outcome of this analysis cannot be quantified [15].

CREAM (Cognitive Reliability Error Analysis Method) is one of the best known second generation HRA methods, which offers a practical approach to both performance analysis and error prediction [16]. This method presents a consistent error classification system integrating all individual, technological and organizational factors, which can be used both as a stand-alone method for accidental analysis and as part of larger design methods for interactive systems. In this method, human error is not considered to be stochastic, but shaped by different factors such as the context of the task, physical/psychological situation of the human operator, time of day, etc. One of the main features of this method is its integration of a useful cognitive model and framework that can be used in both retrospective and prospective analysis [17]. CREAM is capable of providing the estimated HEP that can be used as part of overall system analysis. Compared to other HRA methods, CREAM seems more promising as an option to assess human performance for several reasons. First, it represents a second generation HRA method with improved applicability and accuracy compared to most of the first generation methods. It is able to extend the traditional description of error modes beyond the binary categorization of success-failure and accounts explicitly for how the (performance) conditions affect the performance. Secondly, it is originally developed from the Cognitive Control Model (COCOM)* and also uses it to organize some of categories describing possible causes and effects on human action. Last but not least, CREAM can be used for performance prediction since quantified results can be provided as the final outcome. This capability especially makes the integration of the CREAM-based non-technical component model with other technical component models possible, which is a critical requirement for modelling infrastructure systems.

In recent years, a wide range of modelling approaches, e.g., Agent-based Modelling (ABM), Complex Network Theory (CNT), System Dynamic (SD), have been applied to represent technical components infrastructure systems. However, modelling efforts regarding the representation of the human behaviours remain on the adoption of classical analytical approaches, e.g., probabilistic modelling method, using a combination of fault and event tree techniques, making the analysis of the human

---

* COCOM models human performance as a set of control modes: strategic, tactical, opportunistic and scrambled and proposes a model of how transitions between these control modes.

performance in a comprehensive way particularly difficult. Furthermore, it is not an easy task to integrate this type of model with other technical system models in case all components (technical and non-technical) of an infrastructure system need to be considered. Among these approaches, the ABM seems more promising.

In this paper, a generic modelling framework is proposed and presented. The framework consists of two parts: First, an analytical method based on the CREAM for human performance assessment, which includes five working steps. In this method, a knowledge-based approach is developed in order to assess PSFs in a more efficient way. Second, an ABM approach for the representation of human behaviours. Within this approach, the human operators and uses of infrastructure systems are modelled as agents with capability of interacting with other agents, e.g., agents representing technical components. Using this advanced modelling approach, the human performance is able to be assessed and corresponding human error can be calculated in real-time dynamically based on current simulation environment, e.g., current time, simultaneous goals, etc.

3. A CLOSER LOOK AT CREAM

CREAM is derived from the method of COCOM, the purpose of which is to provide the conceptual and practical basis for developing operator performance models. In both methods, the cognition is regarded as not only an issue of processing input(s) and producing a reaction, but also an issue of the continuous revision and review of goals/intentions [18]. Therefore, the cognition should not be described as a sequence of steps, but rather a controlled use of available competence and resources [16]. The basic assumption of CREAM is that human performance is an outcome of the controlled use of competence adapted to the requirements of the situation, rather than the result of pre-determined sequences of responses to events. Four characteristic control modes are defined in the CREAM method : *scrambled control, opportunistic control, tactical control, and strategic control mode* [16]. Instead of PSFs, the method of CREAM uses CPCs (Common Performance Conditions) to determine sets of error modes and probable error causes. Total nine CPCs are proposed by Hollnagel: *adequacy of organization, working conditions, adequacy of MMI (Man-Machine Interface) and operational support, availability of procedures/plans, number of simultaneous goals, available time, time of day, adequacy of training and experience,* and *crew collaboration quality*. Various levels are also assigned to each CPC. For instance, three (CPC) levels are assigned to the CPC "working conditions": *advantageous, compatible,* and *incompatible*. The main difference between the CPCs and the PSFs is that the CPCs can be applied at the early stage of the analysis to characterize the context for the task as a whole, rather than a simplified way of adjusting probability values for each event. Therefore, the influence of CPCs is closely linked to the task analysis. Advantage working conditions such as the level "compatible" (CPC level) of "working condition" may improve the performance reliability, while disadvantage performance conditions such as the level "incompatible" may reduce the performance reliability. If the performance reliability is reduced, operators could fail more often. Relations between all nine CPC levels and their expected effects on the performance reliability can be determined based on author's general knowledge and experiences. In most first generation HRA methods, it is always assumed that PSFs are independent. This assumption raises concerns since even a cursory investigation is able to show that it is not possible that all PSFs are independent to each other. This concern has been taken into consideration by most second generation HRA methods. In the CREAM method, all the CPCs have influences on each other. For instance, the CPC "working conditions" (e.g., ambient lighting, noises from alarms, interruptions, etc) have direct impacts on both of "number of simultaneous goals" and "available time". Improved "working conditions" can be assumed to increase "available time" and decrease "number of simultaneous goals". It is very important to take these dependencies into account when applying the CREAM method (see[16] for more information).

4. FRAMEWORK PART 1-1: AN ANALYTICAL METHOD

Human error is defined as "*Any member of a set of human actions or activities that exceeds some limit of acceptability, i.e. an out of tolerance action (or failure to act) where the limits of performance are defined by the system*" in [19]. In our daily life, the human error is extremely common since everyone could commit at least some everyday. However, the human error has become a cause of great concern to the reliability of interactive infrastructure systems, since most these systems depend on the interaction with operators in order to maintain their appropriate function. A general analytical method based on the method CREAM is proposed in this chapter. To demonstrate the feasibility and applicability of the proposed framework, SCADA system is used as an exemplary system. This method can be divided into five working steps:

- Step 1: Constructing event sequence
- Step 2: Determining COCOM functions
- Step 3: Identifying most likely cognitive function failures
- Step 4: Assessing CPCs
- Step 5: Determining failure probability

In step 1, a task needs to be specified and corresponding event sequence can be constructed. In this case, a simplified task of general alarm handling is selected (see [20] for more details about introduction of the alarm handling). The overall operation of the task (task 0) involves four sequential subtasks. First, operators need to check whether or not the alarm monitor system is ready to work properly (subtask 0.1). The monitor system could include devices such as monitors, alarms, etc. Then operators start to keep checking the monitor system regularly to ensure that the new generated alarm will not be missed (subtask 0.1.1). If a new overload alarm is generated and sent by corresponding devices to the alarm monitor system, operators will be notified meaning that this identified alarm will be handled (subtask 0.1.1.1). Finally, a control command will be sent by operators (subtask 0.1.1.1.1).

In step 2, all possible COCOM functions need to be determined for each identified subtask. The model assumes that there are four basic cognitive functions: *observation*, *interpretation*, *planning*, and *execution*. Each defined typical cognitive activity can be described in terms of which combination of these four cognitive functions it requires. For example, the "monitor" activity involves "observation" as well as "interpretation". Therefore, all subtasks (cognitive activities) identified in step 1 are assigned with corresponding COCOM functions. Furthermore, it is important to determine a dominant function if the defined cognitive activity involves more than one COCOM functions. For example, subtask 0.1 (ensure the monitoring system is working) is assigned with COCOM activity "verify" that involves two COCOM cognitive functions: "observation" and "interpretation". Based on the description of the alarm handling task, this subtask involves more "observation" function and less "interpretation" function. In this case, the "observation" is the dominant COCOM function. Table 1 lists all possible cognitive functions defined for each subtask and one dominant cognitive function of each subtask is highlighted in red colour.

Table 1: Determination of cognitive functions

Subtask	Goal	Cognitive activity	Obs	Int	Plan	Exe
0.1	Ensure the alarm monitoring system is working	Verify	•	•		
0.1.1	Monitor overload alarm	Monitor	•	•		
0.1.1.1	Identify a new overload alarm	Identify		•		
0.1.1.1.1	Send command	Execute				•

Obs: observation, Int: interpretation, Plan: planning, Exe: execution

For each cognitive function, generic cognitive function failures have been defined in [16]. It is possible to use all pre-defined cognitive function failures for each cognitive activity. However, in order to make the CREAM more practical in use, one most likely cognitive function failure should be identified and used. This can be done based on the understanding and knowledge of the corresponding task in step 3. For example, three cognitive function failures can be defined for the subtask 0.1: 1) the observation of a wrong object, 2) the wrong identification made, and 3) the observation not made. According to the description of this task, it is more reasonable to assume that the possibility of missing an overload alarm is higher. Therefore, the third cognitive function failure can be identified as the most likely function failure for subtask 0.1.

Step 4, assessing CPCs, is the essential step among all 5 working steps, which is also the most challenging step. The purpose of this step is to examine and assess the CPCs under which the corresponding task is performed. Some of these CPCs can be easily assessed, e.g., the time of day (day time or night time depending on the time when the corresponding task is performed), the number of simultaneous goals, while the assessment of some CPCs can be difficult, e.g., the adequacy of organization, working conditions. In order to simplify the overall assessments of CPCs, it is necessary to assign some CPCs with a fix level. It should be noted that increasing number of CPCs with fixed levels will affect the output accuracy of the model. The effects of the CPCs on performance reliability can be quantified using the weighting factor. For instance, in the case where the expected effect is "not significant", the weighting factor is set to be 1. In the case where the expected effect is "improved", the weighting factor can be set to be less than 1 meaning that the final calculated HEP will likely be decreased. Lower weighting factor value indicates better performance. For instance, the weighting factor for level of "compatible" of CPC "working conditions" can be set to 1 and the weighting factor for level of "incompatible" can be set to 2.

To determine the CFP[†], each identified most likely cognitive function failure is firstly assigned with a nominal CFP, which can be conducted in step 5 using the information from [16]. Then, these nominal CFPs are adjusted considering the effects of the CPCs using weighting factors obtained from step 4. Table 2 lists the adjusted CFP for each subtask including best case scenario and worst case scenario.

Table 2 Adjusted CFPs for cognitive function failures

Subtask	Task step or activity	Nominal CFP	Best case scenario		Worst case scenario	
			weighting factor	adjusted CFP	weighting factor	adjusted CFP
0.1	Ensure the alarm monitoring system is working	0.07	0.2	0.014	9.6	0.672
0.1.1	Monitor overload alarm	0.07	0.2	0.014	9.6	0.672
0.1.1.1	Identify a new overload alarm	0.01	0.25	0.0025	6	0.06
0.1.1.1.1	Send the command	0.003	0.2	0.0006	9.6	0.0288

The final CFP can be obtained by choosing the maximum one from all calculated adjusted CFPs using the Equation 1:

$$CFP_{final} = max(CFP_i), i = 1, 2, ...n \quad (1)$$

Where CFP_i represents the adjusted CFP value and n represents the number of values calculated. In the case of best case scenario, three out of nine CPCs have "improved" effects on the performance reliability and none of the CPCs have a "reduced" effect ($\sum improved = 3, \sum reduced = 0$). The corresponding control mode is "Tactical" and the probability interval is from 0.001 to 0.1. The calculated final CFP, shown in Table 2, is 0.014, which falls into the interval. In the case of worst case scenario, one out of nine CPCs have an "improved" effect on the performance reliability and three of the CPCs have "reduced" effects ($\sum improved = 1, \sum reduced = 3$). The corresponding control mode is "Scrambled" and the probability interval is from 0.1 to 1. The calculated final CFP, shown in Table 2, is 0.672, which falls into the interval.

5. FRAMEWORK PART 1-2: A KNOWLEDGE-BASED APPROACH FOR CPC ASSESSMENT

As mentioned above, step 4 (assessing CPCs) is the essential step of the proposed analytical method, which are challenged by following reasons. First, it is difficult to set a numerical threshold, by which

---

[†] Within the CREAM method, the final error probability is also referred as Cognitive Failure Probability (CFP) instead of HEP.

the corresponding level can be decided. Second, the assessment depends on the knowledge and experiences related to the specific task. Furthermore, many other issues could also have direct effects on the assessment of specific CPCs. The challenges could be solved easily for assessment of some CPCs. For example, the CPC 'Time of Day" can be assessed by examining the current time of the model assuming that *"if current time is between 8 am and 20 pm, then the CPC level is set to Day Time. If not, then the CPC level is set to Night Time"*. However, it is not an easy task to assess some CPCs. For example, both the number of current simultaneous tasks and time left for operators to handle one task could have significant influences on the assessment of "available time". In order to assess this type of CPC, a knowledge-based approach using the fuzzy logic theory is proposed and developed. Fuzzy logic theory, first developed by Zadeh in [21], almost four decades ago, has emerged over last several years as a useful tool for modelling processes which are too complex or fuzzy for conventional quantitative techniques or when the available information from the process is qualitative, inexact or uncertain [10]. Fuzzy logic fills a gap between purely mathematical approaches and purely logic-based approaches. Instead of requiring accurate equations to model real-world behaviours, fuzzy logic is capable of accommodating the ambiguities of real-world human language and logic with its inference techniques. Fuzzy inference systems (FIS), developed based on fuzzy logic theory, have been successfully applied in fields such as automatic control, data classification, expert system, and decision analysis [22]. Unlike other regular mathematical systems, the FIS is related to the classes with unsharp boundaries where the output is only the matter of degrees. It is primarily about linguistic vagueness through its ability to allow an element to be a partial member of set, so that its membership value can lie between 0 and 1 [23]. Using the approach of FIS for the study of the HRA is also not a new concept. In 2006, a modelling application of CREAM methodology based on fuzzy logic technique has been developed by Konstandinidou and his colleagues [10], which can be regarded as a pilot application demonstrating the successful 'translation' of the CREAM into the language of fuzzy logic.

In order to demonstrate the applicability of the knowledge-based approach assessing CPCs, "available time" is used as an exemplary CPC. It is assumed that this CPC is mainly affected by two parameters:
- Time left: in the task analyzed using this model, each overload alarm must be handled in a predefined time period. If operators fail to process on time, the overloaded line will be disconnected automatically in order to prevent the thermal damage to the transmission line. In this task, it is assumed that the moderate overloads can be tolerated for up to 20 minutes [24, 25].
- Number of simultaneous goals: if there would be a number of simultaneous alarms, then the time to handle some of these alarms will be delayed.

Assessing "Available time" through a knowledge-based approach using a FIS can be conducted as follows:
Input:
1) *Timeleft*: the remaining time of each alarm to be handled
2) *Simgoals*: the number of simultaneous alarms that is required for operators to handle

Output: The cognitive level: "Available time": *adequate, temporarily inadequate, continuously inadequate*

Membership Functions (MF):
The MF essentially embodies all fuzziness for a particular fuzzy set [26]. The shape of membership functions used for both input and output are triangular. Three MFs are selected for both inputs, with linguistic values: "insufficient", "sufficient", and "more sufficient" for input "Timeleft" and "fewer than capacity", "match current capacity", and "more than capacity" for input "Simgoals". The range for each MF is shown in Table 3 and the graph is shown in Figure 1. It should be noted that the membership functions defined below are based on the understanding and knowledge of the analyzed task.

Table 3 Ranges of MFs for both inputs

Input	insufficient	sufficient	more sufficient
TimeLeft (min)	<10	>6 and <16	>10
Input	fewer than capacity	match current capacity	more than capacity
Simgoals	<3	>1 and <5	>3

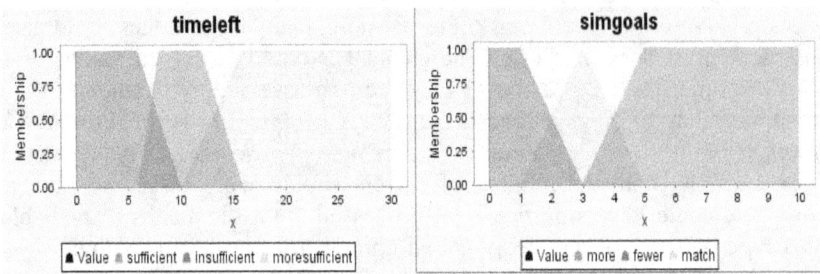

Figure 1 MF graphs of both inputs

Three output (consequence) functions are selected. The purpose of these functions is to determine the likelihood of the conclusion which is true, given a premise. The range for each MF is shown in Table 4 and MF graph is shown in Figure 2.

Table 4 The range of MF of output

Level of "Available time"	Continuously inadequate	Temporarily inadequate	Adequate
Consequence	<4	>2 and <6	>4

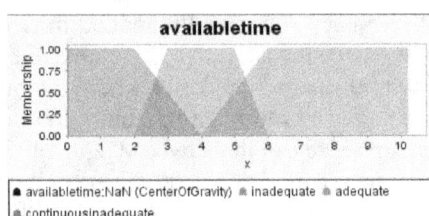

Figure 2 The graph of consequence of MFs

Rules:
Table 5 displays all fuzzy decision-making rules derived from knowledge base, developed based on the understanding and knowledge of the analyzed task. For example, the rule highlighted in the table can be read as "*If Time Left is sufficient AND the number of simultaneous goals is matching current capacity, then the level of "Available time" is set to adequate*"

Table 5 The rule table

Level of "Available time"		Number of simultaneous goals		
		Fewer than capacity	Match current capacity	More than capacity
Time left	Insufficient	Inadequate	Inadequate	Continuous inadequate
	Sufficient	Adequate	**Adequate**	Inadequate
	More sufficient	Adequate	Adequate	Inadequate

Defuzzification method:
Centre of Gravity (COG) method is implemented as the defuzzification method for combining all the consequences to make decisions, which is illustrated in the equation below. Basically this method calculates the weighted average of the centre values of the consequence membership functions (Equation 2).

$$u^{crisp} = \frac{\sum_i b_i \int \mu_{(i)}}{\sum_i \int \mu_{(i)}} \quad (2)$$

Where b_i denotes the centre of consequence membership and $\mu_{(i)}$ denotes the MF. In order to test the applicability of this knowledge-based approach, several test runs are performed. In the first test run, it is assumed that time left for operator to handle an overload alarm is 12 minutes and the number of simultaneous tasks is 2. Therefore, the inputs to the developed FIS are 12 for "Timeleft" and 2 for "Simgoals". The output of the FIS after the defuzzification is 7.24. All corresponding membership function graphs are shown in Figure 3. In this case, the level of "Available time" can be set to adequate. In the second test run, it is assumed that time left for operator to handle an overload alarm is 5 minutes and the number of simultaneous tasks is 4. Therefore, the inputs to the developed FIS are 5 for Timeleft and 4 for Simgoals. The output of the FIS after the defuzzification is 2.87. In this case, the level of "Available time" can be set to continuous inadequate.

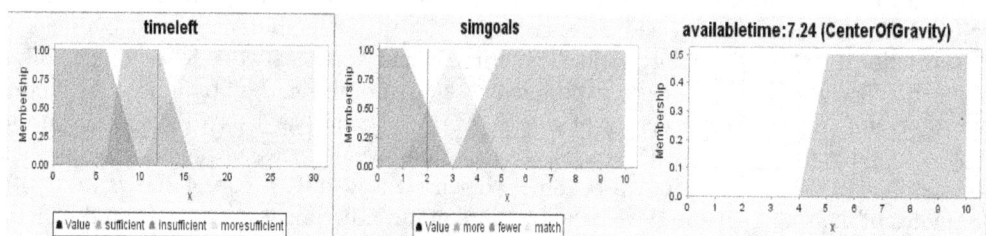

Figure 3 MF graphs of both inputs and the output for test run#1

One of advantages of integrating the approach of FIS into HRA lies in the fact that it provides a fundamentally simple way to handle complex problems without making itself exceedingly complex. It is straightforward, flexible, and easy to develop and understand. However, the FIS is a data-driven approach, meaning that the accuracy of the output depends on the quality of expert knowledge and experiences. Therefore, the membership functions, as well as developed rules, need to be carefully calibrated.

6. FRAMEWORK PART 2: MODELING HUMAN BEHAVIORS USING ABM

The ABM approach describes a whole system by its individual parts (bottom-up). Each component of the system is normally defined and modelled by an agent, capable to modify its own internal data (parameter and variable), its behaviours (function), its environment, and even adapts itself to environmental changes. An agent can be used to model both technical and non-technical components while different agents interact with each other directly or indirectly. One of the major advantages of this approach is the possibility to integrate various elements such as physical laws, Monte Carlo techniques, etc, into the overall simulation (see [27] for more details about the ABM).

In [24], a pilot human operator model is developed based on the ABM approach. The purpose of developing such a model is to assess the influences of human operator performance on the reliability of an Electricity Power Supply System (EPSS). The most critical shortcoming of this model is that the HEP is simply calculated by generating a random number between 0 and 1. Moreover, the model ignores the influences of the PSFs, restricting its applicability and accuracy. In order to overcome these shortcomings and be able to analyse the performance of the operator in a more comprehensive way, a further improved and agent-based human operator model is created using the proposed analytical method including the knowledge-based approach for CPC assessment. This model, developed as part of a SCADA model, is then integrated with an SUC (System Under Control) model in an experimental simulation platform that is built to assess interdependency-related vulnerabilities between two systems (SUC and SCADA)[‡] [28, 29]. During the simulation, if there is a request for the operator to handle an alarm, CPCs will be assessed automatically according to current simulation environment, e.g., time of day, simultaneous goals, etc, and corresponding CFP will be calculated as an input to other agents, e.g., MTU (Master Terminal Unit) agent from the SCADA model.

To simplify this assessment, it is necessary to make assumptions for following CPCs:
- working conditions (in control centre) are *compatible*
- the adequacy of organization is *efficient*
- the availability of procedures/plans is *acceptable*
- the adequacy of training and preparation is *adequate with high experience*
- the crew collaboration quality is *efficient*

[‡] It is assumed that the SUC and SCADA are parts of the EPSS.

This is the first effort to develop a human operator performance model that is capable of assessing CPCs dynamically using the ABM approach. Four CPCs are assessed and five CPCs are assumed to be fixed without further assessment due to limited data sources, which will affect the accuracy of output (CFP)[§] of this model. With the help of this model, several in-depth experiments have been developed for the identification and assessment of hidden vulnerabilities due to interdependencies, e.g., substation level single failure model experiment and small network level single failure mode experiment. The results from these experiments seem promising, highlighting the importance of human operator in the control center of the SCADA system. The lack of responses from human operators might not be the cause of failures of substation level devices, negative consequences caused by the failures of these devices could become worsen significantly. Figure 4 shows results from two case studies of the small network level single failure mode experiment. As seen from this figure, more components from both SUC and SCADA fail to function if performance of the human operator is assumed to be poor (see [28] for details and results of these experiments). This is only a pilot application demonstrating the possibility of assessing human performance using advanced modeling approaches. Motivated by these promising results, more experiments considering human operators as parts of the overall system are currently being developed, e.g., the experiment analyzing resilience related behaviors of infrastructure system.

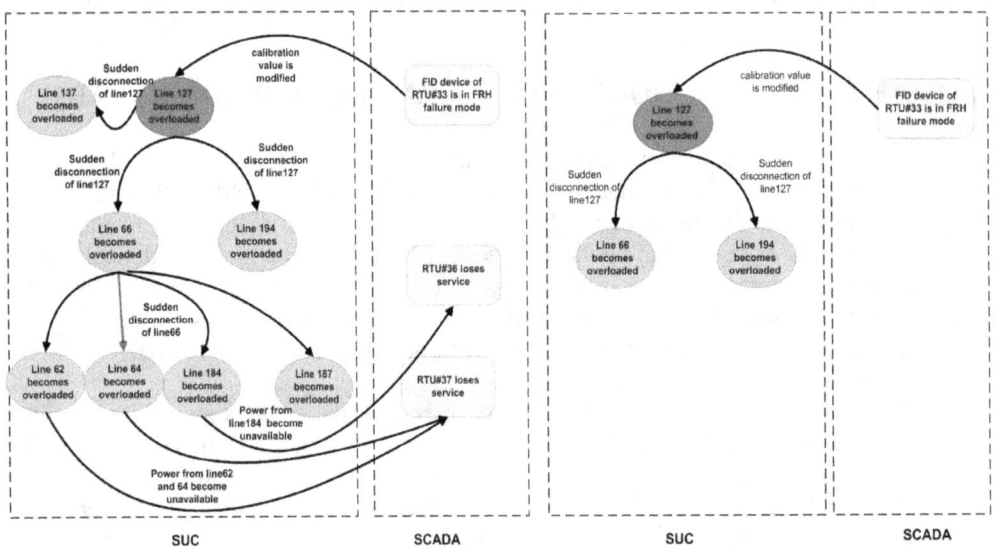

Figure 4 Affected components due to dependency between SCADA and SUC in two case studies [28]. Left: the human operator performance is assumed to be poor; Right: the human operator performance is assumed to be acceptable

7. OUTLOOK

HRA methods have been widely developed and improved during last several decades in order to provide a more applicable way to assess human performance. However, these methods are challenged by their inherent limitations, e.g., the lack of objectivity, inability to model tasks that consist of highly nested, concurrent cognitive activities, etc. These limitations hinder the possibilities of analyzing behaviors of infrastructure systems in a comprehensive way. In order to improve capability of current human operator model based on HRA method and take more contexts into consideration other than CPCs (e.g., emotion, learning ability, experiences, etc.), a conceptual agent-based hierarchy human model is proposed and currently under development, illustrated in Figure 5. This model consists of three levels. The upper level includes the components sensor and perception, which can be regarded as the input of the model. The information such as interactions with other agents, influence of environment, signal sent by technical components, predefined goals, are first processed by the sensor component. After that, the information will be further interpreted by the perception component. For

---

[§] The calculated CFP value in this case is between 0.0014 and 0.672.

example, if an alarm is received by the operators in the control room, it needs to be first received by the sensor and then interpreted by the perception in order to exact more detailed information, such as the severity of the alarm, etc. The middle level includes four components: physic status, emotion status, social status and cognition. These components contain parameters, state variables, states, rules, which can be used to determine states and behaviors of the agent. The lower level includes the components of behavior and actor, which can be regarded as the output of the model. The execution order of the agent action/behavior is determined by the component behavior, while the execution is carried out by the component actor. The information that has influence potentially on other agents/objects will be sent out by the component actor.

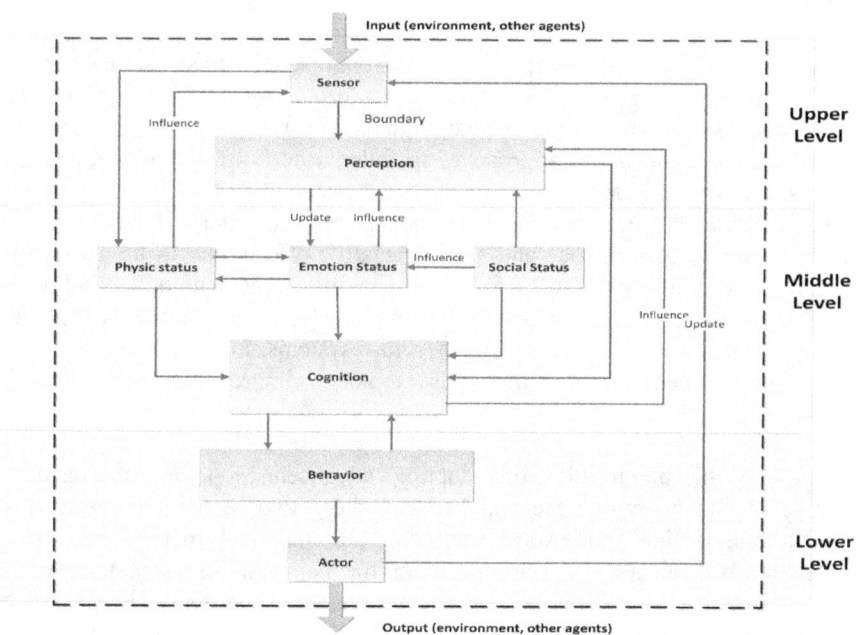

Figure 5 Overall structure of further improved agent-based human model

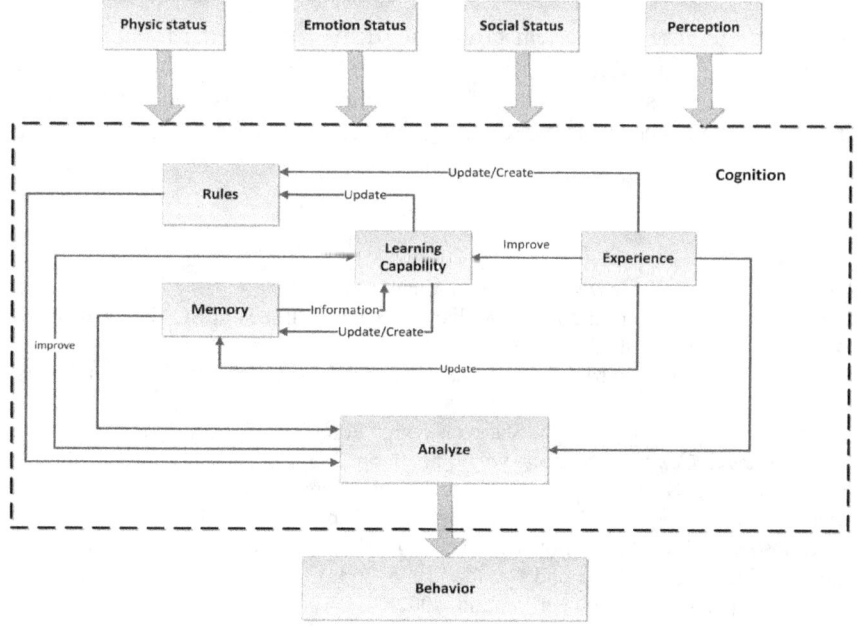

Figure 6 The structure of the cognition component

The cognition component is the most important component of the agent-based human model. The information received from outer environment and other agents will be mainly processed by this component and the corresponding behaviours will then be decided by this component. The overall structure of the cognition component is illustrated in Figure 6. Five subcomponents are included in it: rules, memory, experience, learning capability, and analyse. Compared to previous model, all the properties of an agent have been taken into account. Furthermore, this model is capable of making decisions according to previous experience, predefined rules, interaction with other agents, and outer environment.

8. CONCLUSION

Humans play an important role in the operations of vital engineered infrastructure systems. The lack of careful consideration of influences of these "non-technical components" of infrastructure systems often results in poor system performance and high costs. Their roles and impacts need to be strengthened by developing advanced approaches for human performance assessment that take more factors into consideration and focus more on the ways to analyze human behaviors efficiently in varying contexts. During the last decade, a number of research works have been developed and applied to analyze human performance and assess negative effects due to human errors, limited to errors of omission. Most of these works are based on the implementation of classical analytical approaches and seem not sufficient, which is the most critical shortcoming compared to research works focused on technical components of infrastructure systems. Full mapping of the complexity of infrastructure systems depends on continued development/ improvement of human performance models.

In order to explore the possibilities of adopting advanced modeling approaches for human performance assessment and bridge the gap between these two research communities, this paper proposes a generic modeling framework, including an analytical method for the performance assessment and an advanced modeling approach for the behavior representation, which is mainly focused on infrastructure systems, i.e. the electric power supply system. The analytical method is based on the second generation HRA method CREAM. In order to be able to assess CPCs more efficiently, a knowledge-based approach using the concept of Fuzzy Logic is proposed. The analytical method is further implemented as part of the human operator model, which is developed using the ABM approach.

The first results from the application seem promising, demonstrating the feasibility of the proposed modeling framework, not just due to its capability for representing the complexity of human performance of infrastructure systems, but also its modeling flexibility and adaptability. Thus, more application and simulation experiments based on this framework will be expected in the near future.

References

[1] W. Kröger, and E. Zio, *Vulnerable Systems*: Springer, 2011.
[2] V. M. Igure, S. A. Laughter, and R. D. Williams, "Security Issues in SCADA Networks," *Journal of Computers and Security*, vol.25, pp.498-506, 2006.
[3] S. A. Boyer, *SCADA supervisory control and data acquisition*, 3rd ed., Research Triangle Park: ISA, 2004.
[4] A. Ferscha, K. Zia, A. Riener, and A. Sharpanskykh, "Potential of Social Modelling in Socio-Technical Systems," *Procedia Computer Science*, vol. 7, no. 0, pp. 235-237, 2011.
[5] G. Baxter, and I. Sommerville, "Socio-technical systems: From design methods to systems engineering," *Interacting with Computers*, vol. 23, no. 1, pp. 4-17, 2011.
[6] C. W. Johnson, and C. M. Holloway, "A Longitudinal Analysis of the Causal Factors in Major Maritime Accidents in the USA and Canada (1996–2006)," *The Safety of Systems*, F. Redmill and T. Anderson, eds., pp. 85-104: Springer London, 2007.

[7] W. Kröger, and C. Nan, "Addressing Interdependencies of Complex Technical Networks," *Networks of Networks: The Last Frontier of Complexity*, Understanding Complex Systems G. D'Agostino and A. Scala, eds., pp. 279-309: Springer International Publishing, 2014.

[8] J. Wreathall, E. Roth, D. Bley, and J. Multer, *Human reliability analysis in support of risk assessment for positive train control*, 2003.

[9] J. Sharit, "Human Error and Human Reliability Analysis," *Handbook of Human Factors and Ergonomics, Fourth Edition*, pp. 734-800, 2012.

[10] M. Konstandinidou, Z. Nivolianitou, C. Kiranoudis, and N. Markatos, "A fuzzy modeling application of CREAM methodology for human reliability analysis," *Reliability Engineering and System Safety*, vol.91, no. 6, pp.706-716, 2006.

[11] Verein Deutscher Ingenieure, "Methods for quantitative assessment of human reliability," VDI 4006 Part 2, 2003.

[12] M. Kyriakidis, *Focal Report: A study regarding human reliability within power system control rooms*, Lab for Safety Analysis, ETH Zurich, Zurich, 2009.

[13] D. E. Embrey, P. Humphreys, E. A. Rosa, B. Kirwan, and K. Rea, *SLIM-MAUD: an approach to assessing human error probabilities using structured expert judgment. Volume I. Overview of SLIM-MAUD*, NUREG/CR-3518-Vol.1, 1984.

[14] M. Kyriakidis, *A scoping method for human performance integrity and reliability assessment in process industries*, ETH Zurich, 2009.

[15] J. Forester, A. Kolackowski, S. Cooper, D. Bley, and E. Lois, *ATHEANA User's Guide: Final Report*, U.S.Nuclear Regulatory Research, 2007.

[16] E. Hollnagel, *Cognitive Reliability and Error Analysis Method CREAM*: Elsevier, 1998.

[17] X. He, Y. Wang, Z. Shen, and X. Huang, "A simplified CREAM prospective quantification process and its application," *Reliability Engineering and System Safety*, vol.93, no. 2, pp.298-306, 2008.

[18] L. Bainbridge, *Building up behavioural complexity from a cognitive processing element*: London: University College, 1993.

[19] A. D. Swain, *Comparative evaluation of methods for human reliability analysis*, GRS-71, Institute for Reactor Safety, 1989.

[20] C. Nan, and I. Eusgeld, "Adopting HLA standard for interdependency study," *Reliability Engineering and System Safety*, vol. 96, no. 1, pp. 149-159, 2011.

[21] L. A. Zadeh, "Fuzzy logic," *Journal of IEEE Compute*, vol. 21, no. 4, pp. 83-93, 1998.

[22] R. L. Marcellus, "Evaluation of a nonstationary policy for statistical process control" pp. 89-94,1997.

[23] C. J. Harris, X. Hong, and Q. Gan, *Adaptive Modeling Estimation and Fusion from Data*, New York: Springer, 2002.

[24] M. Schläpfer, T. Kessler, and W. Kröger, "Reliability Analysis of Electric Power Systems Using an Object-oriented Hybrid Modeling Approach," in 16th power systems computation conference, Glasgow, 2008.

[25] W. R. Lachs, "Transmission-line overloads: real-time control," *IEE Proceedings*, vol. 134 (C), pp. 342-347, 1987.

[26] S. M. El-Shal, and A. S. Morris, "A fuzzy expert system for fault detection in statistical process control of industrial processes," *Systems, Man, and Cybernetics, Part C: Applications and Reviews, IEEE Transactions on*, vol. 30, no. 2, pp. 281-289, 2000.

[27] A. Tolk, and A. M. Uhrmacher, "Agents: Agenthood, Agent Architectures, and Agent Taxonomies," *Agent-Directed Simulation and Systems Engineering*, pp. 73-109: Wiley-VCH Verlag GmbH & Co. KGaA, 2010.

[28] C. Nan, I. Eusgeld, and W. Kröger, "Analyzing vulnerabilities between SCADA system and SUC due to interdependencies," *Reliability Engineering and System Safety*, vol.113, no. 0, pp. 76-93, 2013.

[29] C. Nan, W. Kröger, and P. Probst, "Exploring critical infrastructure interdependency by hybrid simulation approach," *Advances in Safety, Reliability and Risk Management*, pp. 2483-2491: CRC Press, 2011.

A Bayesian Network Model for Accidental Oil Outflow in Double Hull Oil Product Tanker Collisions

Floris Goerlandt [a,1] and Jakub Montewka [a]

[a] Aalto University, Department of Applied Mechanics, Marine Technology, Research Group on Maritime Risk and Safety, P.O. Box 15300, FI-00076 AALTO, Finland

Abstract: This paper proposes a Bayesian belief network (BBN) model for the estimation of accidental oil outflow in a ship-ship collision where a product tanker is struck. The intended application area for this model is maritime traffic risk assessment, i.e. in a setting in which the uncertainty regarding the specific vessel characteristics is high. The BBN combines a model for linking relevant variables of the impact scenario to the damage extent with a model for estimating the tank layouts based on limited information regarding the ship, as typically available from data from the Automatic Information System (AIS). The damage extent model, formulated as a logistic regression model and based on a mechanical engineering model for the coupled inner-outer dynamics problem of two colliding ships, is implemented in a discretized version in the BBN. The model for estimating the tank layout is applied for a representative set of product tankers typically operating in the Baltic Sea area. The methodology for constructing the BBN is discussed and results are shown.

Keywords: Oil spill, product tanker, Bayesian Belief Network, consequence model, risk assessment

1. INTRODUCTION

Ship-ship collisions are low-probability, high-consequence events which may have a devastating effect on the natural environment in case the struck vessel is an oil carrying tanker. In maritime traffic risk analysis, tanker spills thus are an important object of study.

Several methodologies have been proposed to determine the probability of tanker collisions occurring in a given sea area [1]–[4]. These typically provide a set of scenarios under which vessels encounter each other. These scenarios contain vessel related information such as the main dimensions, sailing speed and encounter angle. These methods are typically based on data from the Automatic Information System (AIS), which is a system where navigational parameters are transmitted from ships to one another and to shore stations, providing a rich source for the study of vessel movements. AIS data does not contain information concerning the ship masses, loading conditions or specific hull shapes. There is thus a high degree of uncertainty related to the vessel characteristics obtained from AIS data.

For the evaluation of the consequences, a link is required between the encounter conditions and the conditions at impact. In particular, collision evasive action prior to collision may change the vessel speeds and the impact angle compared to the encounter angle. For the impact scenario calculations, the impact location along the struck vessel's hull is needed as well. Several approaches have been proposed for impact scenario modeling [5], [6], and while the influence of the assumptions governing the link between encounter and impact conditions on the probability of hull breach is significant, the phenomenon is not well understood and involves high uncertainty [7].

In light of this, for the estimation of accidental oil outflow, a model is needed which can easily account for such uncertain conditions, which is why a BBN approach is selected. A number of models has been presented for tanker oil outflow in collision accidents. Przywarty [8] reports on a simple oil spill model based on the analysis of accident statistics. Montewka et al. [9] proposed a model based on a generic methodology presented by the International Maritime Organization [10]. Smailys and

---

[1] Corresponding author: floris.goerlandt@aalto.fi

Česnauskis [11] presented a more generic method for the determination of the oil outflow when limited information is available about the tanker. Their method only allows for determination of the outflow if the damage extent is known, i.e. there is no link to the impact conditions. van de Wiel and van Dorp [12] have presented a regression model for the evaluation of the damage extent and accidental oil outflow conditional to the impact conditions. Their model has been applied in maritime traffic risk assessment [3] but has the limitation that a predefined tanker layout is assumed, based on the cases presented by the National Research Council [13]. Sormunen et al. [14] present a regression model for damage extent to chemical tankers.

This paper presents a Bayesian network model for the estimation of accidental outflow for product tankers, i.e. tankers with deadweight in the range of 10000 to 60000 tonnes [15]. Such vessels are among the most common in the Baltic Sea region, which presents a pragmatic reason for this limitation in this work. The Bayesian network is learned from calculated spill sizes in a large set of damage cases for a large set of tanker layouts. The impact conditions are linked to the damages extents based on the regression equations presented in van de Wiel and van Dorp [12]. The estimation of tank arrangement and cargo tank volumes is based on a dataset of tankers which operate in the Baltic Sea, to which the procedure proposed by Smailys and Česnauskis [11] is applied for determining bulkhead locations and cargo tank volumes.

The resulting Bayesian network model is primarily meant for application in maritime traffic risk assessment, where relatively limited information about the vessels is available and the uncertainty about the impact conditions is significant. It is developed to be compatible with BBN models estimating oil spill related clean-up costs [16], oil combating [17], [18] and environmental impacts of oil spills [19].

This paper is organized as follows: in Section 2, the underlying model rationale in terms of mechanical engineering models is presented and necessary equations for linking impact scenarios to damage extents are given. Section 3 addresses the applied methodology to estimate the tank volumes and bulkhead location based on the limited data of actual tankers. Section 4 presents the methodology for constructing the Bayesian oil outflow model. An example application is shown in Section 5, illustrating the utility of accounting for uncertainty related to the impact conditions. Section 6 concludes.

2. DEFORMATION ENERGY AND DAMAGE EXTENT

2.1. Ship collision damage: phenomenon and model selection

A ship-ship collision is a complex, highly non-linear phenomenon which can be understood as a coupling of two dynamic processes. First, there is the dynamic process of two ship-shaped bodies coming in contact, resulting in a redistribution of kinetic energy and its conversion into deformation energy. The available deformation energy leads to damage to the hulls of both vessels. This process is commonly referred to as "outer dynamics" [20]. Second, there is the dynamic process of elastic and plastic deformation of the steel structures due to applied contact pressure, referred to as "inner dynamics" [20].

A number of models has been proposed to determine the available deformation energy and the extent of structural damage in a ship-ship collision, see Pedersen [21] for an extensive review. One of the few methods explicitly accounting for the coupling of outer and inner dynamics is the SIMCOL model reported by Brown and Chen [22]. This model is a three degree of freedom time-domain simulation model where vessel motion and hull deformation are tracked, from which the resulting damage length and depth can be determined. The method has been applied to evaluate the environmental performance of four selected tanker designs: two single hull and two double hull (DH) tankers of various sizes [13], for which a large set of damage calculations has been performed. The relevant parameters of these damage cases has been transformed in a statistical model based on polynomial linear and binary logistic regression by van de Wiel and van Dorp [12], linking the impact scenario variables to the damage extent and the probability of hull rupture. While more advanced collision energy and structural response models exist [21], this model is suitable as a basis for our purposes. The equations provided in the following section are implemented in the BBN.

2.2. Collision damage extent conditional to given impact scenario

The polynomial regression model by van de Wiel and van Dorp [12] uses a set of predictor variables to link the impact scenario variables to the longitudinal and transversal damage extents. These predictor variables are representative of the impact scenario. An impact scenario can be described through the vessel masses m_1 and m_2, the vessel speeds v_1 and v_2, the impact angle φ, the relative damage location l and the striking ship's bow half-entrance angle η, see Figure 1. An additional variable is used as a scaling factor between the results of the small and the large tankers given in the set of damage cases [13]. This variable is set as the vessel length L or the vessel width B depending on whether longitudinal or transversal damage extents are calculated.

Figure 1: Impact scenario variable definition

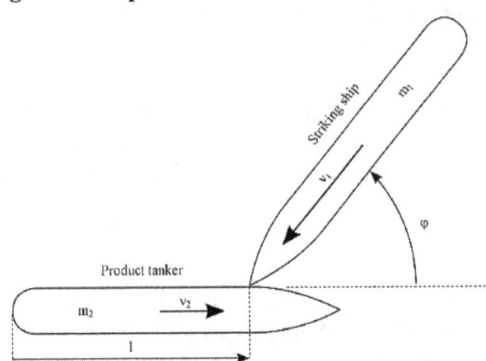

As predictor variables, dimensionless variables x_i are applied as follows:

$$\begin{cases} x_1 = 1 - \exp\left(-\frac{e_{k,p}}{\beta_p}\right)^{\alpha_p} \\ x_2 = 1 - \exp\left(-\frac{e_{k,t}}{\beta_t}\right)^{\alpha_t} \\ x_3 = Beta\left(l^* + \frac{1}{2} \mid 1.25, 1.45\right) - \\ \quad Beta\left(-l^* + \frac{1}{2} \mid 1.25, 1.45\right) \\ x_4 = CDF(\eta) \\ x_5 = CDF(L) \text{ or } CDF(B) \end{cases}$$ (Eq. 1)

where $e_{k,p}$ and $e_{k,t}$ are respectively the perpendicular and tangential collision kinetic energy, l^* the relative impact location with reference to midship and α_p, β_p, α_t and β_t parameters of a Weibull distribution for the predictor variables involving respectively the perpendicular and tangential kinetic energy. These are given in Table 1, along with the values for the empirical CDF of the bow half entrance angle η and the empirical CDF(L) and CDF(B). We write:

$$l^* = \left| l - \frac{1}{2} \right|$$ (Eq. 2)

$$e_{k,p} = \frac{1}{2}(m_1 + m_2)(v_1 \sin(\varphi))^2$$ (Eq. 3)

$$e_{k,t} = \frac{1}{2}(m_1 + m_2)(v_2 + v_1 \cos(\varphi))^2$$ (Eq. 4)

Table 1: Coefficients and parameters in predictor variables x_1, x_2 and x_4, from [12]

Parameter	Value	η [deg]	CDF(η) [-]	L [m]	CDF(L) [-]	B [m]	CDF(B) [-]
$α_p$	0.4514	η≤17	0.224	L≤190	0	B≤29.1	0
$β_p$	589.4	η≤20	0.776	190<L≤261	0.014L–2.68	29.1<B≤50	0.048B-1.4
$α_t$	0.4378	η>20	1.000	L>261	1	L>50	1
$β_t$	709.1						

Using these predictor variables, a polynomial regression model is made for respectively the expected damage length y_l and penetration depth y_t:

$$y_l = exp\left(h_l(x|\hat{\beta}^l)\right) \quad \text{(Eq. 4)}$$

$$y_t = exp\left(h_t(x|\hat{\beta}^t)\right) \quad \text{(Eq. 5)}$$

with:

$$h_l(x|\hat{\beta}^l) = \sum_{i=1}^{5} \hat{\beta}_0^l + \sum_{j=1}^{5} \hat{\beta}_{i,j}^l x_j^i \quad \text{(Eq. 6)}$$

$$h_t(x|\hat{\beta}^t) = \sum_{i=1}^{5} \hat{\beta}_0^t + \sum_{j=1}^{5} \hat{\beta}_{i,j}^t x_j^i \quad \text{(Eq. 7)}$$

The regression coefficients for the expressions h_l and h_t are given in Table 2.

Table 2: Regression coefficients of polynomial expressions for h_l and h_t, from [12]

	$\hat{\beta}_{i,j}^l$						$\hat{\beta}_{i,j}^t$					
	i=0	i=1	i=2	i=3	i=4	i=5	i=0	i=1	i=2	i=3	i=4	i=5
j=0	-2.63						-3.68					
j=1		-0.12	4.67	-1.97	1.16	0.05		6.65	3.99	0.427	0.051	0.044
j=2		5.79	/	16.82	-0.57	/		-3.76	-4.33	/	/	/
j=3		/	-5.76	-53.7	/	/		/	/	-9.29	/	/
j=4		-10.9	0	69.4	/	/		/	/	20.69	/	/
j=5		7.798	4.031	-31.2	/	/		1.83	1.87	-12.4	-0.35	/

Using the above expressions, the damage length y_l and penetration depth y_t can be evaluated based on the impact scenario parameters. Note that m_1 and m_2 have as units tonnes, v_1 and v_2 are in knots, φ in degrees with φ=0 bow-bow collision and η in degrees. Index 1 and 2 denote striking and struck vessel, respectively.

The determination of the maximum and minimum location of the longitudinal damage extent, respectively y_{l1} and y_{l2}, depends on the damage length y_l, but also on the relative damage location l, the ship length L and the damage direction θ:

$$y_{l1} = (1 - \theta)y_l + (1 - l)L \quad \text{(Eq. 8)}$$

$$y_{l2} = -\theta y_l + (1 - l)L \quad \text{(Eq. 9)}$$

Naturally, y_{l1} and y_{l2} cannot exceed the position of the fore or aft perpendicular. The damage direction θ accounts for the phenomenon that the longitudinal damage extent will not necessarily be symmetrical around the impact location. In van de Wiel and van Dorp [12], it is assumed that θ depends on the impact angle φ and the relative tangential velocity v_t as follows:

$$\theta = \begin{cases} 0 & \text{if } \varphi = 0 \\ \left(\frac{1}{2}\left(\frac{\varphi}{90}\right)^n\right)^{\exp(mv_t)} & \text{if } 0 < \varphi < 90 \\ \left(1 - \frac{1}{2}\left(\frac{180-\varphi}{90}\right)^n\right)^{\exp(mv_t)} & \text{if } 90 \leq \varphi < 180 \\ 1 & \text{if } \varphi = 0 \end{cases} \qquad \text{(Eq. 10)}$$

where $v_t = -v_1\cos\varphi - v_2$, m=0.091 and n=5.62.

The penetration depth y_t is applied to evaluate which longitudinal bulkheads are breached and hence from which tank compartments in the transverse direction oil can spill. Likewise, the longitudinal limits of the collision damage, y_{l1} and y_{l2}, are applied to evaluate which transverse bulkheads are breached and hence from which tank compartments in the longitudinal direction oil can spill.

3. A MODEL FOR ESTIMATING BULKHEAD LOCATION AND TANK VOLUMES

3.1. Aim and tanker arrangement data

The overall aim of the model for tanker tank arrangement is to determine, based on limited data of a given ship, a reasonable estimate for the location of transversal and longitudinal bulkheads and the corresponding tank volumes. As mentioned in the introduction, AIS data typically only contains very crude ship related data such as vessel type, length and width. The model presented below aims to allow estimates of tank arrangements if only these variables are known. The approach is based on a data set containing tank arrangement parameters for 219 product tanker designs which operate in the Baltic Sea. The data is obtained from IHS Maritime [23]. Some abridged tank arrangement data used in the analysis is shown in Table 3. L, B and D represent respectively the vessel length, width and depth. DISPL is the displacement (weight of the ship) and DWT the deadweight (carrying capacity). TT is the tank type, where TT1 signifies a DH tanker with no longitudinal bulkhead, TT2 a DH tanker with one longitudinal bulkhead and TT3 a DH tanker with two longitudinal bulkheads. PST, CT and SBT are the number of port side, center and starboard side tanks. Of the 219 tanker designs, 93% has is tank type 2, 5% tank type 3 and 2% tank type 1, showing that for the product tanker class, the most common configuration is with one longitudinal bulkhead on the center line.

Table 3: Basic information concerning tanker layout, excerpt from [23]

	L [m]	B [m]	D [m]	DISPL [tonnes]	DWT [tonnes]	TT	PST	CT	SBT
Ship 1	127.6	20.8	11	18179	13781	2	6	/	6
Ship 2	134.8	22	12.8	23838	18008	2	7	/	7
Ship 3	147.2	24.5	13.4	27502	19831	1	/	8	/
Ship 4	160	24.6	13.5	29842	23400	3	5	5	5

3.2. Methodology for finding bulkhead locations and tank volumes

The methodology applied in this paper is to a large extent based on the procedure proposed by Smailys and Česnauskis [11], but is for the analysis in Section 4 calculated on the tanker database set as outlined in Section 3.1. The main parameters relevant for the determination of the tank volumes and the location of the transverse and longitudinal bulkheads are shown in Figure 2. L_A and L_F are the horizontal distance from the aft perpendicular to the aft cargo tank compartment and the horizontal distance from the fore perpendicular to the frontmost cargo tank compartment. L_T, B_T and D_T are the cargo tank compartment length, width and depth and V_i the volume of tank i. The double hull width is denoted w and the double bottom height has notation h.

Figure 2: Definition of tank dimensions and ship parameters

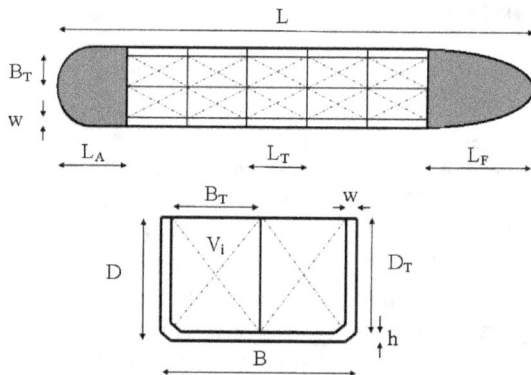

The volume V_i of a given tank is determined as:

$$V_i = C_i B_T L_T D_T \tag{Eq. 11}$$

where C_i is a volumetric coefficient, accounting for the actual shape of the tank in comparison with a rectangular prism. Values for this factor are given in Table 4, taken as averages of an analysis by Smailys and Česnauskis [11]. The tank length, width and depth L_T, B_T and D_T are determined as:

$$L_T = \frac{(L - L_A - L_F)}{n} \tag{Eq. 12}$$

$$B_T = \frac{(B - 2w)}{m} \tag{Eq. 13}$$

$$D_T = D - h \tag{Eq. 14}$$

where n is the number of tanks in the longitudinal direction and m the number of tanks in the transversal direction. It is thus assumed that all tanks have the same width B_T and length L_T. Values for L_A and L_F are given in Table 4, taken as average values reported by Smailys and Česnauskis [11]. The double bottom height h and double hull width w are determined based on the relevant rules for classification of ships [24].

Table 4: Basic information concerning tanker layout, based on [11]

Layout	Cargo tank	10k-35k DWT	35k-50k DWT	50k-60k DWT
TT1	Front	0.7	0.74	0.74
	Middle	1	1	1
	Aft	0.91	0.92	0.92
TT2	Front	0.72	0.75	0.75
	Middle	1	1	1
	Aft	0.91	0.92	0.92
TT3	Front outer	0.68	0.7	0.7
	Front internal	0.84	0.85	0.85
	Middle	1	1	1
	Aft internal	0.93	0.94	0.94
	Aft outer	0.84	0.85	0.85
	L_A	0.24 L	0.22 L	0.21 L
	L_F	0.06 L	0.055 L	0.055 L

The above information can be used to determine the set of positions of the longitudinal and transversal bulkheads, respectively noted LBH and TBH, as follows:

$$TBH = \{L_A + kL_T \mid k = 0 \ldots n\} \tag{Eq. 15}$$

$$LBH = \{w + kB_T \mid k = 0 \dots m\} \qquad \text{(Eq. 16)}$$

3.3. Validation

As the procedure to determine tank arrangement is based on a series of simplifying assumptions, the methodology presented in Section 3.2 is validated by comparing the total cargo tank volume as calculated with the DWT as available from the data of the 219 tankers, see Table 3. Figure 3 shows a comparison between the DWT as available in the tanker database (DWT_D) with the DWT as calculated from the cargo tank volume (DWT_C), assuming an oil density of 0.9 tonne/m³. It is seen that the calculation procedure generally overestimates the cargo tonnage. The histogram shows that the cargo tonnage is overestimated by ca. 15% on average, ranging from an underestimate of ca. 20% to a maximum overestimate of ca. 35%. Overall, the procedure thus leads to a conservative estimate for the possible oil outflow, especially for the larger vessels.

Figure 3: Comparison of DWT_C and DWT_D

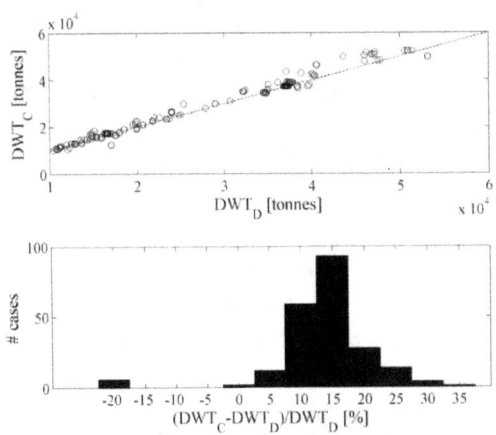

While important for the evaluation of the oil outflow, it is not possible to validate the methodology in terms of bulkhead locations as the detailed tanker layouts are not available. A limited study by Smailys and Česnauskis [11] indicates reasonable agreement for this aspect as well. Nonetheless, some uncertainty is inevitable in this model aspect.

3.4. Determining the oil outflow volume

In actual collision cases, the damage location can be at a range of vertical positions above or below the waterline. Calculations show that the spilled volume can significantly vary depending on the vertical position of the damage [25]. However, there is considerable uncertainty regarding the impact location in accident scenarios. None of the available impact scenario models [5] account for this factor and the vertical damage location will amongst other depend on the striking vessel's depth, bow shape, loading condition (draft and trim) and on the presence of a bulbous bow. None of these parameters can be derived with a reasonable degree of accuracy based on information available in AIS data and uncertainty related to such factors is high in risk assessment of maritime transportation.

Other factors affecting the oil outflow are e.g. the damage opening size, the ship stability and wave conditions. Various combinations of these can affect the spilled volume of oil, but in maritime traffic risk assessment, there typically is high uncertainty concerning these conditions.

To minimize uncertainty, the assumption is made that all cargo of the breached cargo compartments is spilled, see Fig. 4. This is a conservative estimate which is also applied by e.g. van Dorp and Merrick [3]. The determination of which cargo components are breached is based on a comparison of the penetration depth y_t with the position(s) of the longitudinal bulkhead(s) LBH, respectively the maximum and minimum location of the longitudinal damage extent (y_{l1} and y_{l2}) with the positions of the transversal bulkheads TBH.

Figure 4: Definition of oil outflow given a damage extent

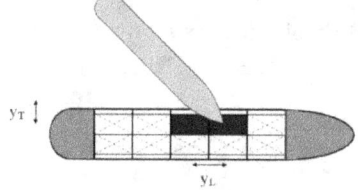

4. BAYESIAN BELIEF NETWORK MODEL FOR OIL OUTFLOW

4.1. Model construction rationale

The definition of the BBN is based on an integration of the model for the collision damage extent conditional to impact scenarios with the oil outflow model conditional to tanker layout.

For each ship, the tank layout is generated according to the procedure in Section 3.1 and 3.2. Subsequently, the impact scenario variables, discretized as shown in Table 4 and Fig. 5, are probabilistically sampled per combination, based on which the CPTs for the regression variables x_1, x_2, x_3 x_4 and θ, the damage extent variables y_L and y_T and the oil outflow is computed as explained in Section 2.2 and 3.4. For each combination of parent node variable classes, 100 samples are generated, from which the probability distribution over the child node variable classes are computed.

As for the ship specific parameters as shown in Table 3, the deadweight is selected as parent node for the oil outflow volume. The other ship parameters are linked to the deadweight through the Greedy Thick Thinning Bayesian learning algorithm. The CPT results for the oil outflow for each ship design are aggregated in the respective DWT classes. The complete BN aggregates the probability distributions for each individual ship, accounting for the DWT classes.

Table 4: Discretization of the impact scenario and oil outflow nodes in the BN

Variable	Discretization	Variable	Discretization
v_1	0, 3, 6, 9, 12, 15, 18, 21, 24	x_1	0, 0.1, 0.2, 0.3, 0.4, 0.5, 0.6, 0.7, 0.8, 0.9, 1
v_2	0, 3, 6, 9, 12, 15, 18	x_2	0, 0.1, 0.2, 0.3, 0.4, 0.5, 0.6, 0.7, 0.8, 0.9, 1
φ	0, 36, 72, 108, 144, 180	x_3	0, 0.1, 0.2, 0.3, 0.4, 0.5, 0.6, 0.7, 0.8, 0.9, 1
η	< 17, 17-20, > 20	x_4	0.224, 0.776, 1
l	0, 0.1, 0.2, 0.3, 0.4, 0.5, 0.6, 0.7, 0.8, 0.9, 1	y_L	0, 2, 4, 6, 8, 10, 12
m_2	10k, 20k, 30k, 40k, 50k, 60k	y_T	0, 5, 10, 15, 20, 25, 30, 35
m_1	0, 10k, 20k, 30k, 40k, 50k, 60k, 70k, 80k, 90k, 100k, 110k, 120k, 130k, 140k, 150k, 160k, 170k, 180k, 190k, 200k	OILOUT	0, 2k, 4k, 6k, 8k, 10k, 12k, 14k, 16k, > 16k
θ	0, 0.2, 0.4, 0.6, 0.8, 1		
DWT	< 15k, 15k, 22k5, 30k, 37k5, >37k5		

4.2. Results and example application

The resulting BBN is shown in Fig. 5. The variable OILOUT is the output of the model, providing the estimate of the oil outflow in ship-ship collisions. Note that the variable x_5 of Eq. 1 is not implemented as for the considered ship sizes, this predictor variable always has value 0, see also Table 1.

Figure 5: Resulting BBN oil outflow model for product tanker collisions

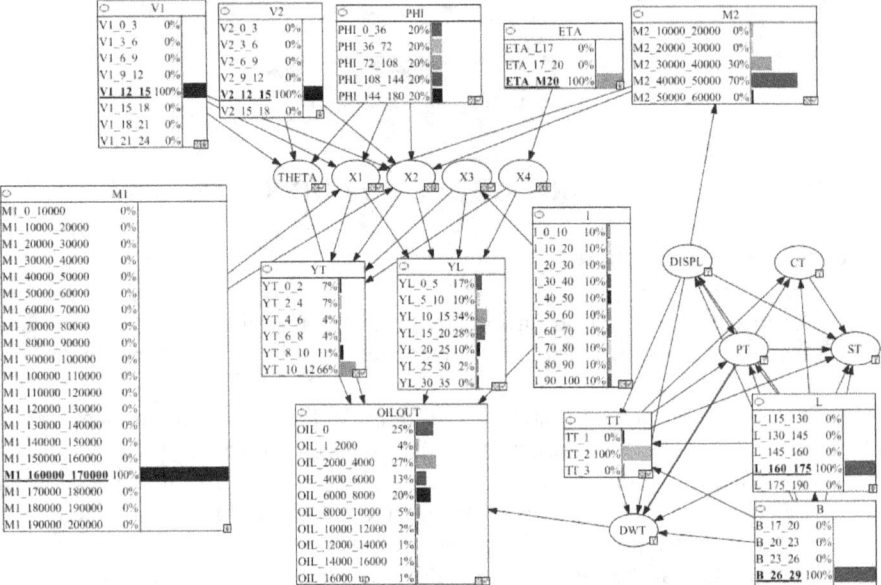

The example application shows a scenario where a relatively large product tanker is hit by a large vessel (e.g. a suezmax bulk carrier or tanker), both at cruising speed. The resulting oil outflow ranges from 0 to more than 10000 tonnes. The relatively large range of probable oil outflows is due to the significant sensitivity of the oil outflow results to the values of the relative damage location and the impact angle, as illustrated in a sensitivity analysis in Fig. 6. In risk assessment of maritime transportation, vessel masses (M1 and M2) and impact speeds (V1 and V2) can be estimated rather well, whereas there is high uncertainty regarding damage location l and impact angle PHI as the process from encounter to impact is poorly understood. The BBN approach shows its value in making such uncertainty explicit rather than providing a potentially uninformative expected value estimate.

Figure 6: Sensitivity analysis of BBN, in relation to OILOUT
(darker shades: higher sensitivity)

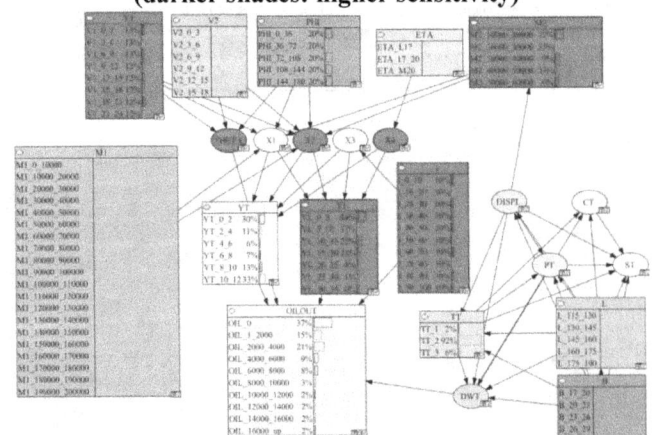

The presented model provides an extensive insight in possible oil outflows in ship-ship collision accidents involving a product tanker. However, several limitations should be noted. First, the oil outflow is only considered as a direct result of the ship-ship collision. Potentially occurring explosions, subsequent progressive structural failures and ship capsizing or sinking could lead to bigger oil outflows than calculated using the above outlined procedure. The results from the model should be evaluated in light of this additional outcome uncertainty. Second, no distinction is made between the type of spilled oil, which however is important for calculating costs [16] and environmental effects [19].

5. CONCLUSION

This paper has presented a BN model for the evaluation of accidental oil outflow in double hull product tanker collisions. The main intended application area for this model is risk assessment of maritime transportation. For such applications, uncertainty related to the specific tanker layouts is high as only very generic data on the vessels operating in the area is available. Furthermore, the uncertainty related to the exact conditions at impact, conditional to encounter scenarios is high. The application of BBNs provides a platform for consistent reasoning under such uncertainty.

The model integrates results from two models found in the literature. A first model provides an estimate of tanker layouts when only limited data is available for a given ship. This model is applied to a set of tanker layouts which are representative for the Baltic Sea area. A second model provides a link between impact conditions in ship-ship collisions and the resulting damage extent.

Based on a large set of simulated damage cases and oil outflows, a Bayesian Belief Network is constructed, to probabilistically evaluate the possible oil outflows conditional to impact scenarios. The resulting network provides a reasonable, relatively conservative estimate of spill sizes.

Acknowledgements

We used GeNie modeling environment developed at the DSL, University of Pittsburgh: http://genie.sis.pitt.edu. The authors appreciate the financial contribution of the following entities: the city of Kotka, the Republic of Finland, the European Union and the Russian Federation, as this work was co-funded by the RescOp project (2011-2014) and the FAROS project (2012-2015). Merenkulun Säätiö is acknowledged for the travel grant.

References

[1] J. Montewka, P. Krata, F. Goerlandt, A. Mazaheri, and P. Kujala, "Marine traffic risk modelling - an innovative approach and a case study," *Proc. Inst. Mech. Eng. Part O J. Risk Reliab.*, vol. 225, no. 3, pp. 307–322, Sep. 2011.
[2] P. Friis-Hansen and B. C. Simonsen, "GRACAT: software for grounding and collision risk analysis," *Mar. Struct.*, vol. 15, no. 4–5, pp. 383–401, Jul. 2002.
[3] J. R. van Dorp and J. R. . Merrick, "On a risk management analysis of oil spill risk using maritime transportation system simulation," *Ann. Oper. Res.*, no. 187, pp. 249–277, 2011.
[4] J. Montewka, F. Goerlandt, and P. Kujala, "Determination of collision criteria and causation factors appropriate to a model for estimating the probability of maritime accidents," *Ocean Eng.*, vol. 40, pp. 50–61, Feb. 2012.
[5] F. Goerlandt, K. Ståhlberg, and P. Kujala, "Influence of impact scenario models on collision risk analysis," *Ocean Eng.*, vol. 47, pp. 74–87, Jun. 2012.
[6] K. Ståhlberg, F. Goerlandt, S. Ehlers, and P. Kujala, "Impact scenario models for probabilistic risk-based design for ship-ship collision," *Mar. Struct.*, vol. 33, pp. 238–264, 2013.

[7] K. Ståhlberg, F. Goerlandt, J. Montewka, and P. Kujala, "Uncertainty in analytical collision dynamics model due to assumptions in dynamic parameters," *TransNav Int. J. Mar. Navig. Saf. Sea Transp.*, vol. 6, no. 1, pp. 47–54, 2012.

[8] M. Przywarty, "Probabilistic model of ships navigational safety assessment on large sea areas," in *Proceedings of the 16th International Symposium on Electronics in Transport*, Ljubljana, Slovenia, 2008.

[9] J. Montewka, K. Ståhlberg, T. Seppala, and P. Kujala, "Elements of risk analysis for collision of oil tankers," in *Risk, Reliability and Safety*, London: Taylor & Francis Group, 2010, pp. 1005–1013.

[10] IMO, "Revised interim guidelines for the approval of alternative methods of design and construction of oil tankers. Regulation 13F(5) of Annex I of MARPOL 73/78 Resolution MEPC.110(49)." International Maritime Organization, 2003.

[11] V. Smailys and M. Česnauskis, "Estimation of expected cargo oil outflow from tanker involved in casualty," *Transport*, vol. 21, no. 4, pp. 293–300, 2006.

[12] G. van de Wiel and J. R. van Dorp, "An oil outflow model for tanker collisions and groundings," *Ann. Oper. Res.*, vol. 187, no. 1, pp. 279–304, 2011.

[13] NRC, "Environmental performance of tanker designs in collision and grounding, Special report 259," National Research Council, 2001.

[14] O.-V. E. Sormunen, S. Ehlers, and P. Kujala, "Collision consequence estimation model for chemical tankers," *Proc. Inst. Mech. Eng. Part M J. Eng. Marit. Environ.*, vol. 227, no. 2, pp. 98–106, 2013.

[15] J. Evangelista, Ed., "Scaling the tanker market." American Bureau of Shipping, 2002.

[16] J. Montewka, M. Weckström, and P. Kujala, "A probabilistic model estimating oil spill clean-up costs - A case study for the Gulf of Finland," *Mar. Pollut. Bull.*, vol. 76, no. 1–2, pp. 61–71, Nov. 2013.

[17] I. Helle, T. Lecklin, A. Jolma, and S. Kuikka, "Modeling the effectiveness of oil combating from an ecological perspective - A Bayesian network for the Gulf of Finland; the Baltic Sea," *J. Hazard. Mater.*, vol. 185, pp. 182–192, 2011.

[18] A. Lehikoinen, E. Luoma, S. Mäntyniemi, and S. Kuikka, "Optimizing the recovery efficiency of Finnish oil combating vessels in the Gulf of Finland using Bayesian Networks," *Environ. Sci. Technol.*, vol. 47, no. 4, pp. 1792–1799, Feb. 2013.

[19] T. Lecklin, R. Ryömä, and S. Kuikka, "A Bayesian network for analyzing biological acute and long-term impacts of an oil spill in the Gulf of Finland," *Mar. Pollut. Bull.*, vol. 62, pp. 2822–2835, 2011.

[20] P. Terndrup Pedersen and S. Zhang, "On Impact mechanics in ship collisions," *Mar. Struct.*, vol. 11, no. 10, pp. 429–449, Dec. 1998.

[21] P. T. Pedersen, "Review and application of ship collision and grounding analysis procedures," *Mar. Struct.*, vol. 23, no. 3, pp. 241–262, Jul. 2010.

[22] A. J. Brown and D. Chen, "Probabilistic method for predicting ship collision damage," *Ocean. Eng. Int.*, vol. 6, pp. 54–65, 2002.

[23] IHS Maritime, "Maritime Insight & Information." IHS Maritime, 2013.

[24] Det Norske Veritas, *Rules for classification of ships*. Høvik: Det Norkse Veritas, 2007.

[25] M. T. Tavakoli, J. Amdahl, and B. Leira, "Analytical and numerical modelling of oil spill from a side damaged tank," in *Proceedings of the Fifth International Conference on Collision and Grounding, ICCGS*, Espoo, Finland: Aalto University, 2010.

Ship Grounding Damage Estimation Using Statistical Models

Sormunen, Otto-Ville[*][a]
[a] Aalto University, Espoo, Finland

Abstract: This paper presents a generalizable and computationally fast method of estimating maximum grounding damage extent in case of grounding based on damage statistics of groundings in Finnish waters. The damage is measured in relative maximum damage depth into the bottom structure, total damage length as well as the damage two-dimensional area.

Keywords: Ship groundings, damage estimation, statistical models, Bayesian Belief Networks

1. INTRODUCTION

Groundings are among the most frequent of maritime accidents, sometimes with catastrophic consequences for human life and the maritime environment such as the Exxon Valdez and Costa Concordia accidents. In order to mitigate risk, procedures such as IMO's [1] Formal Safety Assessment (FSA) have been presented where the cost-effectiveness of various risk control options is evaluated. In order to do this, first general and reliable risk analysis tools are required.

Various models have been presented for modeling ship structural damage caused by groundings, see e.g. Wang et al. [2] for a comprehensive overview. The use of these models in a comprehensive risk analysis is challenging: The detailed models that are usually based on the Finite Element Method (FEM), which not only is very computationally intensive [3, 4] but usually also models only a limited number of grounding scenarios for a very limited number of ship and rock types, see e.g. van de Wiel and van Dorp [5]. This limits applicability in more comprehensive risk analysis, where one would be interested in e.g. modeling the grounding damage on all expected groundings in a given sea area over a certain time period.

More general models based on accident statistics such as IMO [6], Zhu, James and Zhang [7] and Papanikolaou et al. [8] on the other hand can be used quite generally, but the usual limitation in these models is not linking the ship particulars (velocity, mass, etc.) with the resulting damage. The damage is usually modeled with an (empirical) statistical distribution that can be used directly to estimate damage as a percentage of total ship length, beam or draft. However, if the ship particulars are not linked with the resulting damage, one ends up with models where a tanker of 100 000 DWT sailing at 5 knots has the same probability of having a grounding damage extending 1/10 of ship draft as a 5000 DWT tanker sailing at 15 knots in case of a grounding [9]. Montewka, et al. [10] developed a function for expected grounding oil spill volume as a function of ship size, however other ship particulars such as velocity and other local conditions are not taken into account.

Besides these, there are simplified analytical grounding damage models (e.g. Cerup-Simonsen, Törnqvist and Lützen [11], Zhang, [12], Zhu, James and Zhang [7]) which usually estimate either the damaged volume of steel or damage extent along one dimension when damage extent along the other dimension(s) are known. In order to use these models in more comprehensive risk analysis one must combine the analytical models with models that estimate the damage along the other dimensions first. For this reason these models alone are not enough to estimate grounding damage. Some of the models presented in this paper can be used to estimate the damage along the other dimension(s) for the simplified analytical models.

---

[*] E-mail: otto.sormunen@aalto.fi

When it comes to making a grounding consequence analysis for a large number of ships sailing in a given sea area, a further challenge arises: The availability of information regarding grounding depths and/or the bottom shape. Models that are not just based on non-dimensional damage extent such as IMO [6] require detailed information regarding the bottom. This information is incomplete for most cases and therefore usually grounding damage models use relatively simplified assumptions or consider limited cases, see e.g. Zhu, James and Zhang [7] and van de Wiel and van Dorp [5]. This makes the outcome of the models reliant on the assumptions.

Table 1: Overview of different grounding damage model types

Model type	Computation	Info needed on bottom	Link ship size and speed with damage	Generally applicable	Other requirements
FEM	Slow	Detailed	Yes	Each case needs own simulation	
Statistical	Fast	No	No/limited	Yes, many or all ships	
Analytical	Fast	Detailed	Yes	Yes	Damage extent**
This model	Fast	No	Yes	Yes*	

*Model is limited by the available data, which e.g. did not include ships over 57 000 tonnes, see table 2.
**In order to model damage depth, width or length, one or two of the three aforementioned must usually be known in advance.

1.1. Aim

This paper aims at presenting different generalizable and computationally fast grounding damage assessment models based on grounding statistics– without requiring knowledge of the grounding depth, bottom type or damage extent along other dimensions first. In this way the models presented here overcome some of the limitations presented in table 1.

Models presented here link the ship particulars such as velocity, mass and double bottom height with the resulting maximum height of damage bottom structure in groundings as well as the resulting bottom damage length and area.

1.2. Methodology

Detailed grounding damage from accidents in Finnish waters is analyzed using Bayesian Belief Networks as well as different regression models. The models describe grounding damage extent based on grounding velocity, ship size and other factors.

2. DATA

The data used for the analysis consists of 18 ship grounding damage cases from Luukkonen [13], who measured ship grounding damage in Finnish territorial waters, see table 2. These cases were classified as dry cargo, tankers, roro or passenger vessels. In this paper, one case with missing double bottom height as well as one case where the grounded vessel had a draft of less than 2 meters were left out from the data. The following variables from the report are included in the analysis:

v_1 = initial velocity right before grounding [kn]
v_2 = velocity after grounding (0 if ship stuck)
Δv = velocity reduction during grounding (v_1- v_2)
E = grounding energy [MJ], calculated as

$$E = \frac{1}{2}(1 + C_a)m(v_1^2 - v_2^2),\qquad(1)$$

where
$C_a = 0.1$ is the added surge mass and
m = displacement [tonnes]
d = maximum vertical depth of damaged material [m]
l = total damage length [m] (can exceed ship length if multiple damage tracks on bottom, see figure 1)
s = total damage length [m] from aft to fore (cannot exceed ship length L, see figure 1)
L = ship length [m]
T = ship draft [m]
A = bottom 2-dimensional damage area [m^2]
DB_H = ship double bottom height [m]

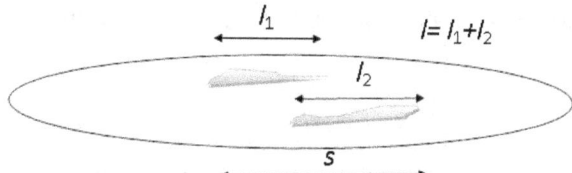

Figure 1: Example of grounding damage seen from below, indicating how *l* and s are calculated

The relevant data is presented in the following table:

Table 2: Grounding damage data, adapted from Luukkonen [13]

E [MJ]	DB_H [m]	max d [m]	l [m]	s [m]	A [m^2]	L [m]	T[m]	m [tonnes]	v1 [kn]	v2 [kn]	Δv
690.9	1.35	1.35	156.6	93.6	638	159.2	9.1	21100	15	0	15
208.1	1.8	1.5	125	108.9	475	150	9.5	22700	8	1	7
380.3	1.8	1.6	93.8	101.5	549	150	9.5	22100	13.5	8	5.5
2.9	1.2	0.75	7.2	7.2	20	130	6.1	8980	1.5	0	1.5
86.9	1.2	1.2	14.4	50.4	45	130	6.1	8780	18	16	2
3.2	1.2	0.65	13.6	13.6	15	130	6.1	9830	1.5	0	1.5
342.3	1.9	1.9	172.4	92.2	1124	146	7.3	12000	14	0	14
43.6	1.99	0.98	33.7	37.2	152	180.5	11.7	57000	2.5	1	1.5
57.8	1.18	0.8	22.5	22.5	100	142.4	5.8	11025	6	0	6
487.3	1.2	1.2	118.4	107.2	929	139.8	5.9	12029	19	10	9
115.5	1.6	0.95	25.3	25.3	79	118.5	6.4	9800	9	0	9
33.6	2.1	0.8	8.8	8.8	20	171.6	6.8	25673	3	0	3
144.1	1.9	1	28.8	28.8	120	146	8.5	14200	10	5.5	4.5
174.2	1.95	1.95	67	50.2	327	126.5	6.6	9050	11.5	0	11.5
53.9	1.5	1.1	16.1	32.9	35	134.3	7.3	16100	12	11	1
300.1	1.8	2	98.3	46.8	479	128.8	8.2	12200	13	0	13
244.8	1.2	1.5	60.8	60.8	442	130	6.1	8582	14	0	14
24.5	1.8	0.6	23.4	40.2	61	129	8.2	18700	3	0	3

The ship displacements were relatively small ranging from ~8500 to 57 000 tonnes. Interestingly, in half of the cases the velocity at the start of the grounding accident was above 10 knots. In 7 of the cases the maximum damage extent d was equal to or exceeded the double bottom height. The problem in the data is the that d does not tell us whether the damaged bottom structure material ruptured or just deformed, thus is difficult to know when exactly the double bottom structure was penetrated.

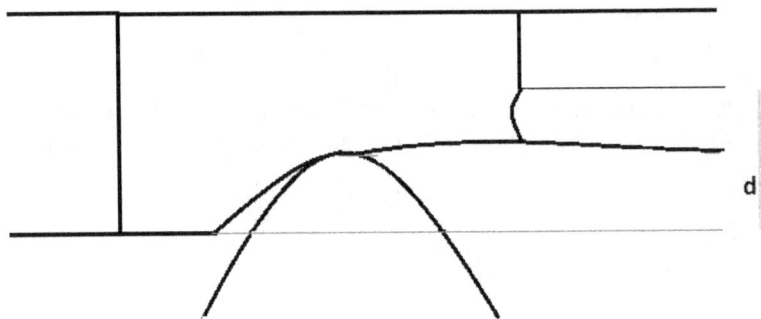

Figure 2: Example of maximum damage depth d in a grounding scenario

2.1. Data Analysis

To find out which variables influenced grounding damage the most, first qualitative visual analysis was used where the different variables were plotted against each other in a two-dimensional co-ordinate system. This was done to see which kind of (if any) dependency existed between variables. Furthermore, quantitative analysis was carried out; the correlation between the different variables was calculated.

Most notably, collision energy E has a positive, statistically significant correlation (2-way test, 5 % significance level) with: d (r = 0.577), $d \geq DB_H$ (0.622), l (0.867), s (0.805), v_1 (0.709), A (0.819) and Δv (0.745).

The damage area A seems to follow a linear dependency with the collision energy especially when the collision energy is no more than 300 MJ as illustrated in the following Figure 1 (with the linear equation as well as the coefficient of determination R^2):

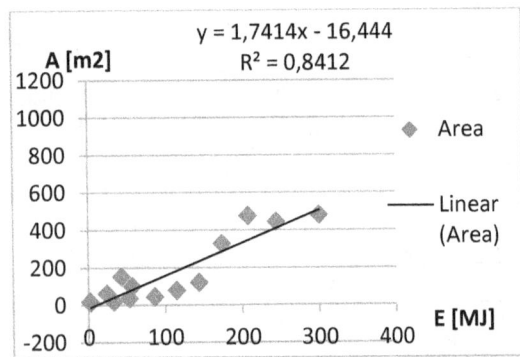 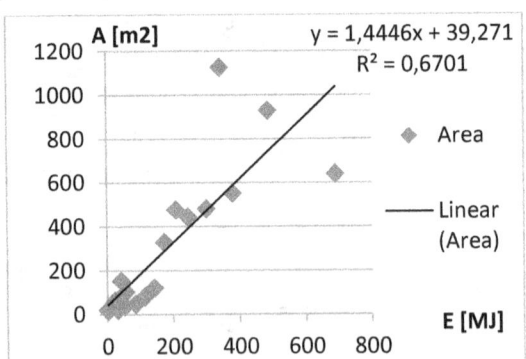

Figure 3: Damage area A [m^2] as a function of grounding energy E [MJ]. Left: cases with E < 300 MJ, right: all cases.

The total damage length l also depends quite linearly on the energy:

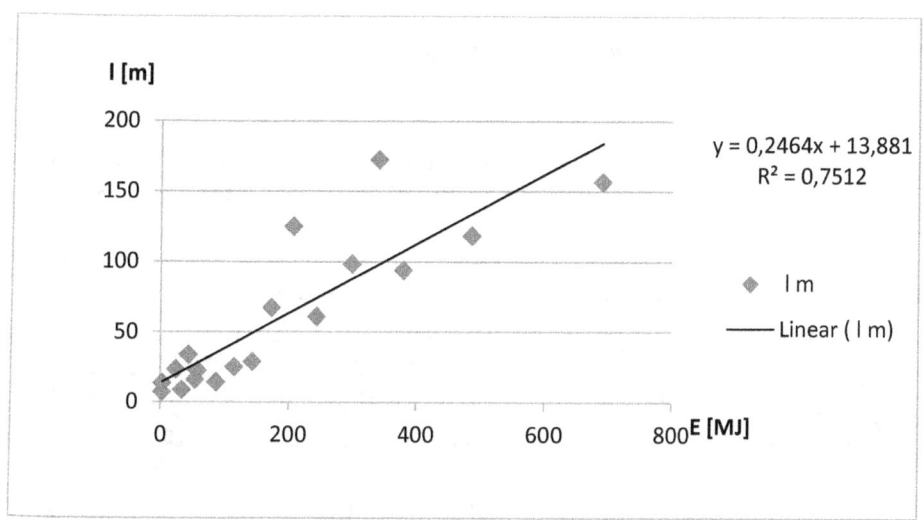

Figure 4: Damage total length *l* [m] as a function of grounding energy *E* [MJ].

The maximum damage depth d also depends somehow linearly on the grounding energy but only in the cases where $E < 400$ MJ.

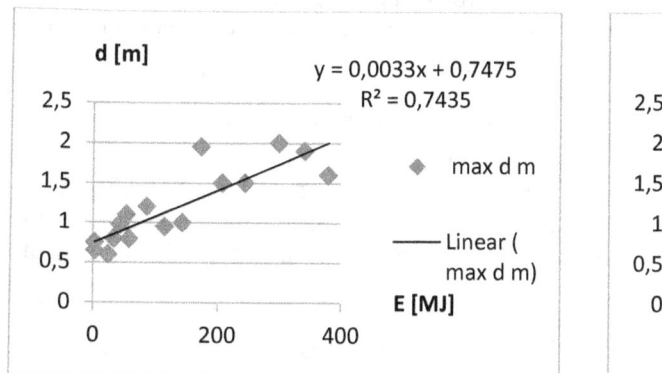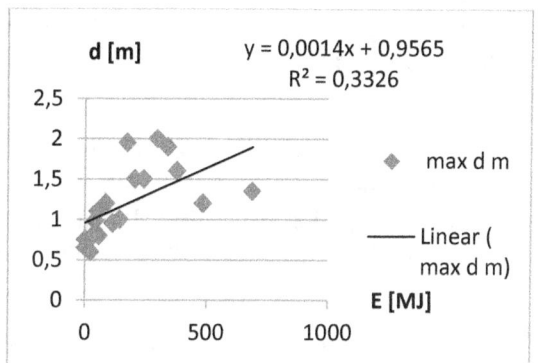

Figure 5: Damage maximum depth *d* [m] as a function of grounding energy *E* [MJ].
Left: cases with $E < 400$ MJ, right: all cases.

In order to estimate whether d equals or exceeds DB_H, a logistic regression model is used which has the following form [14]:

$$\Pr(d \geq DB_H) = \frac{1}{1 + e^{-z}}, \qquad (2)$$

where $z = b_0 + b_1 x_1 + b_2 x_2 + \ldots + b_n x_n$, x_i being the predictor variables and b_i their coefficients.

If $\Pr(d \geq DB_H) > 0.5$, it is assumed that that the grounding damage in the case equaled or exceeded the double bottom height and vice versa. Trying with different predictor variables x, the best prediction was achieved with just the grounding energy E:

$$\Pr(d \geq DB_H) = \frac{1}{1 + e^{2.475 - 0.011\, E}} \tag{3}$$

This model correctly classifies 83.3 % of the cases from Luukkonen [13], performing significantly better for the $d < DB_H$ cases (prediction accuracy: 90.9 %) compared to the cases where $d \geq DB_H$ (accuracy: 71.4 %).

Table 3: Correct Classification Table

Observed in Luukkonen's data		Logistic regression model prediction*		Percentage Correct
		$d \geq DB_H$		
		0	1	
$d \geq DB_H =$	0	10	1	90.9
$d \geq DB_H =$	1	2	5	71.4
Overall Percentage				83.3

*The cut value is 0.5

Adding other variables such as initial velocity or double bottom height causes one or more of the x-variables to have a p-value of more than 0.05; i.e. that the variables are no longer statistically significant. The same problem is present in the other regression models presented earlier.
A logistic regression model for estimating whether $d > DB_H$ or not in case of grounding could not be constructed: The correct classification % was 0 (0/2 correct).

From equation 2 the cut-off point between $d \geq DB_H$ or not is at 225 MJ, which according to the model means that if the collision energy is 225 MJ or greater, then $d \geq DB_H$. This, however, is not entirely true: As can be seen in Table 1 there were 2 cases where MJ < 225 but still $d \geq DB_H$ and 1 case where $E = 380.3$ MJ but still $d < DB_H$. Looking at the cases there seems to be no obvious reason for the difference, this might be connected to bottom shape and type; information which is not included in the data.

Simplified this means that ships sailing in high-risk grounding areas can keep the probability of damage depth exceeding double bottom height low by keeping their velocity below a certain threshold, namely energy < 225 MJ (assuming $v_2 = 0$). As the grounding energy also depends on the mass, heavier vessels must sail slower to avoid $d \geq DB_H$ in case of grounding as illustrated in Figure 4:

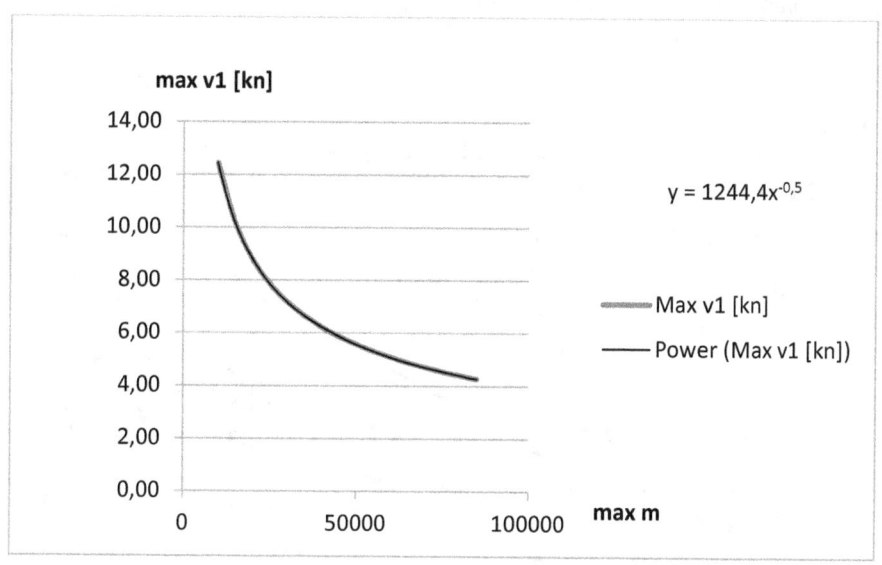

Figure 6: Maximum allowed velocity to avoid $d \geq DB_H$ in case of grounding

Though it must be noted that the vessels in this study had a displacement of 8500 – 57 000 tonnes, meaning extrapolating the results beyond this range might not be accurate, especially towards the lighter vessels as the allowed velocity grows drastically as the mass decreases; different (size) vessels have different hull strength. Note that one case with a displacement (~mass) of 9050 tonnes had a energy below 225 MJ (174.2) still had $d \geq DB_H$ with $v_1 = 11.5$ and $v_2 = 0$ meaning that any results obtained here should be interpreted with care. Luukkonen (1999) noted that groundings with energy of less than 144 MJ resulted in only damage to the ship shoulder or bow.

2.2. Bayesian Belief Networks

Even though the logistic regression model described above is fairly accurate, it cannot describe $d > DB_H$ nor take several x-variables into account with the given data. To make a more sophisticated model where more ship particulars can be used in estimation, two Bayesian Belief Networks (BBN) were constructed to estimate whether $d \geq DB_H$ and $d > DB_H$.

The structure of the BBN is based on the previously calculated correlations and a qualitative understanding of the causal effect dependencies. The network was built and learned using GeNIe software. See DSL [15] for further information on GeNIe and BBNs. For other applications of BBNs in maritime risk analysis see e.g. Goerlandt and Montewka [16], Hänninen et al. [17] or Montewka et al. [18]. The data for the BBNs shown in figures 7 and 8 was discretized as follows.

Table 4: Discretized data for BBN

GeNIe category tag	m (disp) [tonnes]	$d \geq DB_H$	$d > DB_H$	E [MJ]	DB_H [m]	v1 [kn]	Δv
s1	<10 000	0	0	< 50	<1.3	<4	<3
s2	10 -15 000	1	1	50-150	1.3-1.6	4-12	3-7
s3	15- 22 000			150-300	1.6-1.85	12-15	7-12
s4	>22 000			>300	>1.85	>15	>12

The so-called naïve Bayesian networks had the best prediction accuracies. Multi-level networks with many nodes and complex interdependencies could not achieve the same correct classification rates in terms of $d \geq DB_H$ and $d > DB_H$.

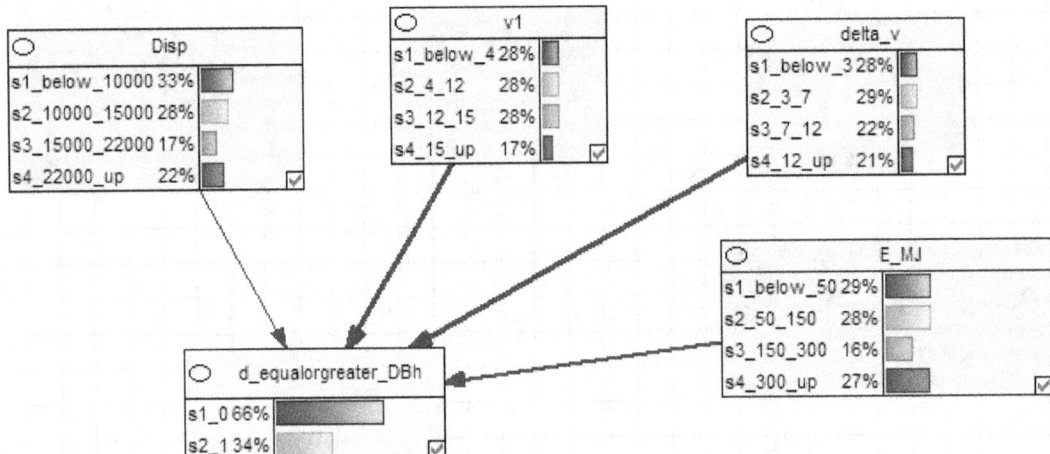

Figure 7: Discrete BBN for estimating $d \geq DB_H$ in grounding

This BBN can predict $d \geq DB_H$ with 100 % accuracy regardless of validation method (k-fold, leave one out). The thickness of the arrows in the figure indicates how significantly the different variables affect the results with v_1 and Δv being the most important variables.

The BBN for $d > DB_H$ looks as follows:

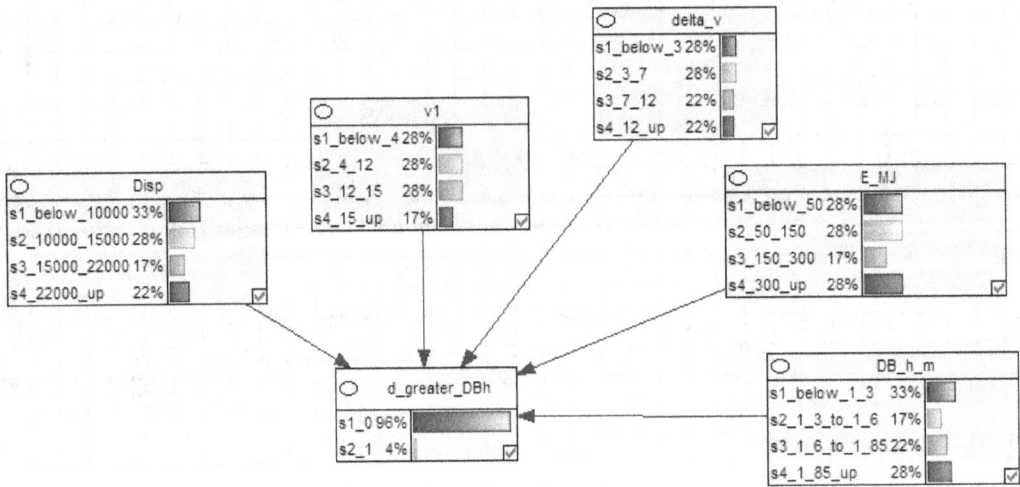

Figure 8: Discrete BBN for estimating $d > DB_H$ in grounding

Also in this case a 100 % prediction accuracy is achieved with k-fold validation (k=2). However, there are only 2 cases with $d > DB_H$ which means that in this model there is a noticeable deal of uncertainty.

3. CONCLUSIONS AND DISCUSSION

The models presented here could relatively accurately predict damage length and area in grounding situations. Surprisingly, this is possible even though information regarding bottom and grounding depth is missing. The most important equations are as follows:

Ship bottom damage area based on grounding energy:

$$A = 39.27 + 1.44\,E \tag{4}$$

Total damage length:

$$l = 13.88 + 0.25\,E \tag{5}$$

Damage depth (if $E < 400$ MJ):

$$d = 0.75 + 0.0033\,E \tag{6}$$

Probability of $d \geq DB_H$:

$$\Pr(d \geq DB_H) = \frac{1}{1 + e^{2.475 - 0.011\,E}} \tag{3}$$

Furthermore, this paper presents a logistic regression model and two BBNs to predict whether the maximum damage extent is equal to or greater than the double bottom height. Knowing damage extent in one dimension (together with grounding energy, bottom shape and depth) can be used in models such as the ones presented by e.g. Cerup-Simonsen, Törnqvist and Lützen [11], Zhang [12] or Zhu, James and Zhang [7] to estimate the damage extent along the other dimensions as well.

Knowledge of damage extent along all dimensions (width, depth and length) can be used comprehensive risk analysis; e.g. to assess oil spill size in case of groundings. However, as stated in Luukkonen [13], d describes the maximum depth of damaged bottom structure – this means that having $d \geq DB_H$ does not equal that the inner bottom is actually penetrated. Also, the data is relatively limited with only 18 observations from Finnish territorial waters, based on ships with displacement of ~8500 to 57 000 tonnes. For these reasons, further research is proposed: Similar analysis should be run for a larger data set, the exact point where the inner bottom is ruptured should be recorded along with the grounding water depth and the bottom shape.

3.1. Acknowledgements

This paper was written for the CHEMBALTIC project, which is funded by TEKES and the EU in co-operation with Merikotka, NesteOil Ltd., Vopak Ltd., the Port of HaminaKotka, TraFi, the Finnish Port Association and the Finnish Shipowners' Association. The author would like to thank Arsham Mazaheri as well as the reviewers for valuable feedback.

3.2. References

[1] International Maritime Organization (IMO). Guidelines for Formal Safety Assessment (FSA) for use in the IMO rule-making process, *MSC/Circ.1023/MEPC/Circ.392*, IMO, 2002, London.

[2] G. Wang, J. Spencer and Y. Chen. *Assessment of a ship's performance in accidents*, Marine Structures, 15 (4-5), pp. 313-333, (2002).

[3] M. Heinvee, K. Tabri and M. Kõrgesaar. *A simplified approach to predict the bottom damage in tanker grounding*, In: Proceedings of the 6th International Conference on Collision and Grounding of Ship and Offshore Structures, Trondheim 17-19. June (2013). Available online at: http://www.ntnu.no/iccgs. Accessed: 23.1.2014.

[4] O. Kitamura. *FEM approach to the simulation of collision and grounding damage*, Marine Structures, 15, pp. 403–428, (2002).

[5] G. van de Wiel and J. van Dorp. *An oil outflow model for tanker collisions and groundings*. Annals of Operations Research, 26 (1), pp.1-26, (2009).

[6] International Maritime Organization (IMO). Interim guidelines for approval of alternative methods of design and construction of oil tankers under regulation 13F(5) of Annex I of MARPOL 73/78. *Technical Report, Resolution MEPC, 66(37)*, pp. 1-40, IMO, 1995, London.

[7] L. Zhu, P. James and S.Zhang. *Statistics and damage assessment of ship grounding*, Marine Structures, 15 (4–5), pp. 515–530, (2002).

[8] A. Papanikolaou, R. Hamann, B. Lee, C. Mains, O. Olufsen, D. Vassalos and G. Zaraphonitis. *GOALDS—Goal Based Damage Ship Stability and safety standards*, Accident Analysis and Prevention, 60, November, pp. 353-365, (2013).

[9] P.T. Pedersen and S. Zhang, S. *Effect of ship structure and size on grounding and collision damage distributions*, Ocean Engineering, 27, pp. 1161–1179, (2000).

[10] J. Montewka, K. Ståhlberg, T. Seppälä and P. Kujala. Elements of risk analysis for collision of oil tankers. In: *Risk, Reliability and Safety*, pp. 1005–1013, Taylor and Francis Group, 2010, London.

[11] B. Cerup-Simonsen, R. Törnqvist and M. Lützen. *A simplified grounding damage prediction method and its application in modern damage stability requirements*, Marine Structures, 22(1), pp. 62–83, (2009).

[12] S. Zhang. *Plate tearing and bottom damage in ship grounding*, Marine Structures, 15 (2), pp. 101–117, (2002).

[13] J. Luukkonen.. *Vauriotilastot Suomen aluevesien karilleajoista ja pohjakosketuksista*, M-240 report, Espoo: Helsinki University of Technology, (1999).

[14] M. Norušis. *IBM SPSS Statistics 19 Advanced Statistical Procedures Companion*. NJ: Prentice Hall, 2011, Englewood Cliffs.

[15] Decision System Laboratory (DSL). *GeNIe Documentation*. University of Pittsburgh, 2010, Pittsburgh. Available online at: http://genie.sis.pitt.edu/wiki/GeNIe_Documentation. Accessed: 23.1.2014

[16] F. Goerlandt and J. Montewka: *A probabilistic model for accidental cargo oil outflow from product tankers in a ship-ship collision*, Marine Pollution Bulletin, DOI: 10.1016/ j.marpolbul.2013.12.026, (2014).

[17] M. Hänninen, A. Mazaheri, P. Kujala, J. Montewka, P. Laaksonen, M. Salmiovirta, and M. Klang. *Expert elicitation of a navigation service implementation effects on ship groundings and collision in the Gulf of Finland,* Proceedings of the Institution of Mechanical Engineers, Part O, Journal of Risk and Reliability, doi: 10.1177/1748006X13494533, (2014).

[18] J. Montewka, M. Weckström and P. Kujala. *A probabilistic model estimating oil spill clean-up costs – A case study for the Gulf of Finland,* Marine Pollution Bulletin, 76(1-2), pp. 61-71, (2013).

Effects of the Background and Experience on the Experts' Judgments through Knowledge Extraction from Accident Reports

Noora Hyttinen[a], Arsham Mazaheri[b*], and Pentti Kujala[c]

[a] Aalto University, Department of Applied Mathematics, School of Science, Espoo, Finland
[b] Aalto University, Department of Applied Mechanics, School of Engineering, Espoo, Finland
and Kotka Maritime Research Center (Merikotka), Kotka, Finland
[c] Aalto University, Department of Applied Mechanics, School of Engineering, Espoo, Finland

Abstract: Available risk models for maritime risk analysis are not proper enough for risk management purposes as they are not evidence-based. One of the sources of evidence that can be used for accident modeling is the accident reports. The reports need to be reviewed to extract the presented knowledge. This study investigates how the differences in the background and expertise of the reviewers can affect the extracted knowledge from the accident reports. The study is conducted by utilizing three-round Delphi method and using two test groups as researchers and mariners to review four grounding accident reports prepared by Finnish Accident Investigation Board. The results of the study show that although neither of the groups have superiority over the other with regard to the extracted knowledge, there are some categories that are chosen more frequently by specific group. Mariners chose more often the causes related to navigation and the actions of the crew, while the researchers tend to see more organizational and environmental related causes. Thus, the background of the reviewer should be considered in evidence-based modeling, as it affects the resulting models and thus the implementing risk control options suggested by the constructed models.

Keywords: Grounding, Accident Reports, Knowledge Extraction, Belief Networks, Hamming-Distance

1. INTRODUCTION

Marine accidents are not only seen as the threat to the maritime transportation itself because of the posing risks to the human lives and the environment, they are also threats to the international trade and industry by interrupting the supply networks [1]. Thus, it is of great importance to keep the likelihood and consequence of an accident as low as reasonably possible. In order to do so, risk managers usually use risk models to understand the system and to mitigate the involved risk by implementing proper risk control options. In this regard, a suitable model for risk management purposes should reflect the true available knowledge on the system with satisfying accuracy [2]. However, currently available risk models for maritime risk analysis are not proper enough for risk management purposes as they are not based on the evidences rather on the intuition of the developers, thus not presenting the true available knowledge of the system [3]. To mitigate this flaw, evidence-based modeling that deal with real accident scenarios, as opposed to imaginary scenarios, is encouraged [3,4]. One of the rich sources of the evidence useful for modeling is the accident reports prepared by the accident investigation boards. When building a database required for modeling, the required causal networks of events need to be extracted from the accident reports. Often researchers, as system experts, read the accident reports or use an algorithm [5-7] to identify the causal networks of the accidents. The problem with researchers reading the reports is that they may not necessarily have the appropriate expertise to get the needed knowledge from the reports. There is also a possibility of misunderstanding because the reports are written in natural language [8]. When reading the accident reports it is likely that different people perceive the events of the accident differently [9].

---

* Corresponding author: arsham.mazaheri@aalto.fi

To reduce the uncertainty in mathematical models that are built based on the reports, it is necessary to study if certain individuals are better equipped to extract the knowledge presented in the accident reports. Thus, since the expertise most likely gives the expert better understanding of the events leading to an accident, it seems necessary to know how the background and expertise of the reader affects the way she sees the causal network of events and thus affects the extracted knowledge. The general opinion may be that the people with insight in the field of the researched subject are more qualified to review accident reports to extract the knowledge about the accident causes. The mariners may be less aware of accident theories and accident causality whereas the researchers may have less understanding of the ship's operational factors and navigation. Researchers that study accident theories and the causes of accidents may see connections between the causes better and are better equipped to review large amounts of data; while mariners may find the individual causes from the accident reports better. Therefore, the objective of this study is to find out how a person's background and knowledge affect the causal network that is extracted from accident reports and the mathematical model that is built based on that– especially whether researchers can identify the same causal networks as mariners with years of nautical experience.

The outline of this paper is as follow: the research procedure, the methods, and the reports used for the study together with the composition of the participants are presented in the next section; the results of the study based on the collected responses from the participants are summarized in Section 3; the results are then analyzed and discussed in Section 4; and finally the paper is concluded in Section 5.

2. METHODOLOGY

The study focuses on grounding accidents. Delphi method [10] with three rounds was implemented to perform the study and to collect the responds of the participants (**Figure 1**). The first round consisted of a questionnaire with open-ended questions. Open-ended questions were used in order not to influence the participants' judgments. The aim of the first part was to get the participants to extract proximate causal events from accident reports. Thus, four grounding accident reports prepared by the Finnish Accident Investigation Board (**Table 1**) were chosen from among the reports that have English translations and sent to the participants for the analysis. The participants were asked to review the accident reports and collect the causes of the accident and list them in a timeline manner that ends to a grounding accident.

Figure 1: Three-round Delphi method is used in the study

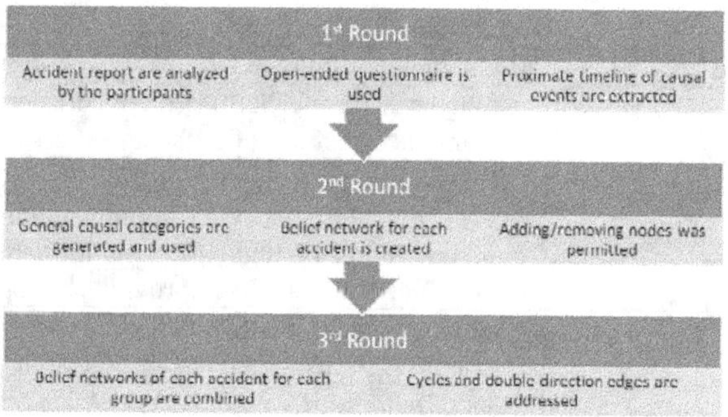

1st Round		
Accident report are analyzed by the participants	Open-ended questionnaire is used	Proximate timeline of causal events are extracted

2nd Round		
General causal categories are generated and used	Belief network for each accident is created	Adding/removing nodes was permitted

3rd Round	
Belief networks of each accident for each group are combined	Cycles and double direction edges are addressed

The participants were two groups of experts. The first group consisted of six marine specialists with nautical education and work experience in either navigation or marine accident investigation with average of 16 years of nautical experience. The second group consisted of six researchers in the maritime transportation domain with the average of 3.5 years of research experience but no formal nautical education or experience. The questionnaire of the first round was pilot tested by a person who had both nautical education and research work experience.

After all answers of the first round were collected, the extracted causes are grouped into 22 general categories (**Table 2**). Categories from Ref. [11] and Ref. [12] are used as guidelines in building the categories related to human factors. The three largest human factors in marine accidents are fatigue, inadequate communications and cooperation between pilot and bridge crew, and inadequate technical knowledge [11]. All of these three causes were found in the studied accident reports. The above causes can be classified as human factors and they are mostly errors that are made by the bridge crew. In addition there were some organizational factors in the studied accident reports. These are mostly errors made by the shipping companies and other outside operators. In addition to human and organizational factors, environmental factors were thought to have caused groundings. Environmental factors are considered to be factors that cannot be controlled by the crew or the shipping company.

Table 1: Grounding accident reports used in the study

Name of the vessel	Time of the accident	Location
MV EMSRunner	11.12.2009	Kalajoki
MS Pauline Russ	20.01.2005	Hanko
MS Claudia	23.10.2007	Tornio
MS Superfast VII	12.11.2004	Hanko

Table 2: List of the categories that were found in the reviewed accident reports

Categories	Short Description of the causes in the category
Human Factors	
Fatigue	Fatigue or other personal reasons, workload
Faulty practices	Faulty standards, policies or practices
Hazardous natural environment	Not adjusting operations based on hazardous natural environment (sea clutter, wind)
Inadequate communications	Lacking communications and cooperation, route plan not discussed
Inadequate route plan	Route plan missing, not on the radar, inadequate, bypassing distances missing
Inappropriate use of navigational equipment	Lack of or incorrect inappropriate use of navigational equipment
Incompetence	Lack of competence, knowledge and ability, other violation of good seamanship practices
Lack of redundancy	Lack of redundancy
Lack of situational awareness	Lack of situational awareness or monitoring
Manning	Too few people on bridge
Mistake	Complacency, mistake made by crew, decisions based on inadequate information
Route plan not followed	Route plan not followed
Wrong steering decision	Wrong steering decision, corrective manoeuvres were made too late, the incorrect use of tugboats
Organizational Factors	
Lack of guidelines	Lack of guidelines from the shipping company, including guidelines on using tugboats, wind limitations etc. (SMS), lack of risk analysis, systems improving ship operations, feedback
Lack of training	Lack of simulator practice/other education/inadequate instructions for using the navigation equipment
Lack of VTS	Lack of help from VTS or other outside operator, VTS not involved
Poor design of automation	Poor design of automation, rudimentary or inadequate navigation or other equipment
Schedule	Schedule, time pressure from shipping company, productivity
Environmental Factors	
Bad visibility	Darkness, fog, dazzling lights of another vessel, sun glare
Fairway	Wrongly marked shallow, fairway or narrow fairway, pilot station placed unfavourably
Outside distractions	Outside factors, dredging or other disturbing factor on fairway
Wind	Wind

At the second round, the created categories of causal events were sent back to the participants and they have been asked to structure a belief network (BN) for each accident using the general categories as nodes, in the way that reflects their beliefs on the causal network that ends to the accident. A brief instruction of BN was attached to the questionnaire to help those who were not familiar with the concept. To make this round easier, each participant has her own questionnaire with categories (nodes) that were related to the causal events that she has reported in the first round. A list that includes all categories from all participants and all four studied accident reports was also attached to the questionnaire. The list gives some directions on how other participants interpreted the causes in the accident reports. All of the categories were in the same list to prevent the participants of seeing which causes were thought to have been present in which accident. Adding and removing nodes from the list was permitted in the process of constructing BN, though only few participants used this possibility. In general, there were no limitations in the instructions on how many nodes there should be in one network; therefore the number of nodes varied from one (excluding the node "grounding") to eleven nodes per BN. As the result, none of the structured BNs were exactly the same for same accident and there were some BNs of the same accident that did not have even a single similar node in them.

After collecting BNs from the two groups for each accident, the networks for each accident were combined for each group to make two networks for each accident, one for mariners and one for researchers. The network of a group was built from the majority opinion; meaning if a node or an edge (arc) is in at least half of the participants' networks it is selected to the group's network. In case of a tie, the participants who have more experience in accident investigation or accident research are weighted 50% higher than the other participants. The resulted grouped networks were then used for comparing the two groups.

Since combining networks may create cycles and double direction edges, the third round of the questionnaire is used to eliminate those problems in the combined networks. Thus, the participants were asked about the contradicting edges in their own groups' combined networks. The third questionnaire consisted of multiple choice questions about the directions of the edges between certain nodes. The assumption in the third part was that the nodes in question were causes of the accident. The participants were asked to either choose the direction of the edge or not to use an edge.

3. RESULTS

The structured BNs are compared in three ways: 1) the distances between different individual BNs are calculated using a redefined version of the Hamming-Distance. The distances are compared in order to find the general existing differences in the views of each group; 2) the nodes of the individual and combined BNs are compared by calculating the appearance frequency of the nodes in the BNs of the two test groups. This helps us to find the most common causes that are detected in the reports by each group, and to understand the possible existing differences between their views; 3) the edges of the individual and combined BNs are compared by assessing the nodes that are connected with each edge. This helps us to find believes of each group on the dependencies of different causes, and to understand the possible existing differences between their views.

3.1. Distances

The networks of the two groups were compared with each other by calculating an alternative distance to the Hamming-Distance. The Hamming-Distance describes the number of changes that have to be made to a network for it to turn into the one that is being compared with [13]. In calculating the Hamming-Distance, both networks have to have the same nodes. Moreover, a reference network, which would be the network that other networks are compared to, is needed. The problem in this study was that the reference networks did not exist. Therefore, instead of using the original Hamming-Distances (***Eq.1***), a different method of calculating differences of two networks, as Alternative-Distance, was introduced (***Eq.4***) and used. Hamming-Distance is calculated as:

$$H = A + D + I \tag{1}$$

where A is the added, D deleted, and I incorrectly oriented edges in the constructed network in comparison to the reference network [14]. The added edges in the Hamming-Distance represent the edges that need to be removed from the BN that is being compared with the reference network. The deleted edges represent the edges that the reference network has but the compared BN does not.

The research setting of this study was such that the created networks did not include all of the possible nodes. The nodes had as much significance in this study as the edges. Neither of the BNs is considered to be the reference network. Thus, if we rename the reference network as BN1 and the BN that it is compared with as BN2, we can say that the added edges are extra edges in BN2 and the deleted edges are extra edges in BN1. The changes in the names of the variables are shown in **Table 3**.

Table 3: Changes in the names of the variables from the Hamming-Distance, when using the introduced Alternative-Distance

Hamming-Distance (H)	Alternative-Distance ($Dist$)
Reference network	BN_1
Compared network	BN_2
Added edges (A)	Extra edges in BN_2 (E_{BN2})
Deleted edges (D)	Extra edges in BN_1 (E_{BN1})
Incorrectly oriented edges (I)	Differently oriented edges (I)

Because the BNs in the study are of different sizes, the extra edges are divided so that they are proportional to the number of all possible edges in the BN. The edges that are left out of the BNs also tell us about the dependencies of the variables, and they should also be taken into account in calculating the distance between two BNs.

The Hamming-Distance is divided by all possible edges in a BN (***Eq.2***) that includes all nodes from the BNs that are compared with each other (see ***Eq.3***). The node *"grounding"* is not counted to the total number of nodes in the network as it is included in every network. Because of that, the number of all possible places for edges in the combined BN can be calculated as:

$$Edg = \sum_{i=1}^{n} i \qquad (2)$$

where n is the number of different nodes in the two BNs. *Edg* tells us how many edges are possible in the combined BN at the same time. It is possible to have only one edge between two nodes, otherwise a cycle would be created. It is also possible to have edges between all nodes without having any cycles in the network. Therefore, the Hamming-Distance for our study should be calculated as:

$$Dist = \frac{E_{BN1} + E_{BN2} + I}{Edg} \qquad (3)$$

In this study we were also interested in how many nodes the BNs have in common, because the BNs that have fewer different nodes can be considered to be more similar than those that have more different nodes. Therefore, to account the differences in the nodes of the BNs, we have added an extra part to ***Eq. 3***, and thus have introduced the *Alternative-Distance* ***(Eq.4)***.

$$Dist = \frac{E_{BN1} + E_{BN2} + I}{Edg} + \frac{N_{BN_1} + N_{BN_2}}{Nod} \qquad (4)$$

where N_k is extra nodes in network k and *Nod* is the number of nodes in the combined network.

The difference of Alternative-Distance to the definition of the Hamming-Distance is that ***Eq.4*** calculates the part of nodes and edges that have to be removed (extra edges and nodes) or changed (differently oriented edges) to form identical networks, rather than only the total amount of edges that need to be removed (added edges), added (deleted edges) or changed (incorrectly oriented edges) as in the Hamming-Distance.

The minimum value given by **Eq.4** when the BNs are identical is zero. When none of the edges are same, that is $E_{BN1} + E_{BN2} = Edg$ and $I = 0$, the maximum of the first part of **Eq.4** is one. For all edges to be different all nodes have to be the same. The maximum value of the second part of **Eq.4** is also one when none of the nodes are same, that is $N_{BN1} + N_{BN2} = Nod$. Thus the Alternative-Distance between two BNs cannot have the value of more than two, though it can never be exactly two either.

Figure 2 shows the average Alternative-Distances between individual BNs of the participants. The mariner-mariner and researcher-researcher distances were calculated by comparing each participant's BN to one another inside each group and calculating the average of all those distances. The mariner-researcher distance was calculated similarly by comparing every mariner's BN with each researcher's BN. The average distances show the variation inside each group and the variation between the groups. The lower average distance indicates that when the individual BNs in the group are compared with each other they have more similarities than in the groups that have higher average distances. **Figure 2** shows that the researchers' BNs are more uniform than the mariners'; meaning less variation between the individual researchers is seen than between individual mariners. This may be the sign of more consistency in the common knowledge of the researchers than the mariners, which can be the result of the smaller differences in the years of experience of the individuals in the researchers' group that is five years in compare with the mariners' that is 17 years. Nevertheless, in general, no specific trends or significant differences between mariners and researchers can be recognized from **Figure 2**, which indicates that neither of the groups have superiority over the other.

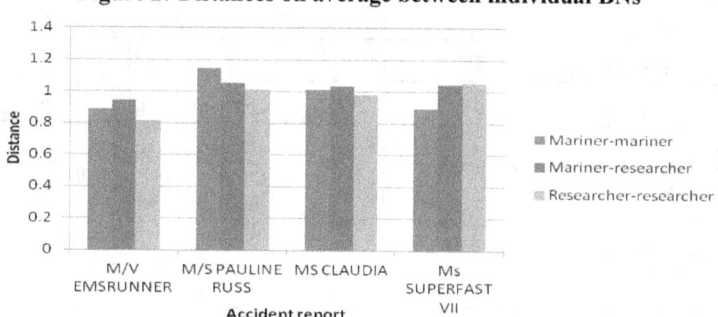

Figure 2: Distances on average between individual BNs

3.2. Nodes

The nodes of the BNs were collected in **Table 4**. The table shows the number of participants from each test group who detected the cause as an affecting factor in the reviewed accident reports. The maximum of each group is six. The differences between the groups can be better compared with the combined results of all reports and their differences in the last three columns on the right (**Table 4**).

In order to find which categories more frequently are chosen by each group as well as in general, the frequencies of each category are calculated using **Eq.5** and are shown in **Table 5**.

$$Frequency = \frac{Times\ that\ the\ cause\ occured\ in\ all\ four\ reports'BNs}{4 * Number\ of\ participants\ in\ the\ group} \quad (5)$$

The rank order of the frequencies of the categories for the mariners and the researchers in **Table 5** shows Spearman rank correlation of 0.71. This shows that the order of the causes almost matches between the groups. There are some clear differences though, as is seen from **Table 4**. For instance, *inadequate route plan* was thought to cause groundings three times more often by mariners than researchers. This is the biggest difference between the groups. Other categories that the mariners had chosen more frequently in their BNs are *wrong steering decision*, *poor design of automation*, and *lack of training*. **Table 4** also shows that, for instance, *incompetence* was chosen only by the mariners;

while *mistake, lack of guidelines, wind,* and *bad visibility* are chosen more frequently by the researchers than the mariners as a cause of accident. In general, it can be seen that the causes that are more related to navigation and ship handling as well as actions of the crew are detected and chosen more frequently by the mariners, while the researchers tend to detect and choose organizational and environmental causes more frequently.

Table 4: The frequency of each category in all accident reports, recognized by M = mariners and R = researchers

	Category	MV EMSRUNNER		MS PAULINE RUSS		MS CLAUDIA		MS SUPERFAST VII		Total		Difference (R – M)
		M	R	M	R	M	R	M	R	M	R	
More frequently chosen by the MARINERS	Inadequate route plan	6	0	3	0	4	4	0	0	13	4	-9
	Wrong steering decision	2	0	3	4	0	0	5	0	10	4	-6
	Lack of situational awareness	4	2	1	0	2	0	2	2	9	4	-5
	Poor design of automation	4	1	0	0	3	1	0	0	7	2	-5
	Incompetence	2	0	1	0	1	0	1	0	5	0	-5
	Lack of training	0	0	0	0	0	0	4	1	4	1	-3
	Manning	0	0	3	1	0	0	1	0	4	1	-3
Almost INDIFFERENCE between mariners and researchers	Lack of VTS	6	4	0	0	0	2	1	0	7	6	-1
	Outside distractions	0	0	0	0	3	2	0	0	3	2	-1
	Route plan not followed	1	0	0	0	0	0	0	0	1	0	-1
	Faulty practices	3	5	4	3	0	1	5	3	12	12	0
	Inappropriate use of navigational equipment	1	2	0	0	5	3	4	5	10	10	0
	Hazardous natural environment	0	0	3	4	2	1	3	3	8	8	0
	Fairway	4	5	0	0	0	1	1	0	5	6	1
	Lack of redundancy	0	2	0	0	1	0	0	0	1	2	1
More frequently chosen by the RESEARCHERS	Schedule	2	2	0	0	0	2	0	0	2	4	2
	Fatigue	1	3	0	0	0	0	0	0	1	3	2
	Inadequate communications	3	6	4	5	5	3	2	3	14	17	3
	Wind	0	0	1	2	0	0	1	3	2	5	3
	Bad visibility	1	2	0	0	3	6	0	0	4	8	4
	Lack of guidelines	0	1	3	3	2	5	5	6	10	15	5
	Mistake	3	6	2	3	0	1	0	0	5	10	5

There are some drawbacks of such generalization over the whole groups though, which one may want to consider. For example, none of the researchers thought that incompetence was a cause in the studied accidents, while only two of the six mariners thought that incompetence was a cause of the groundings; one of them chose incompetence to all of their BNs and the other to one BN. This may represent the opinion of only one mariner rather than the opinion of the whole group. For some other categories that are chosen too seldom, like *route plan not followed* or *lack of redundancy*, the drawn conclusions might be even weaker especially if they are divided between more than one accident. The differences are as likely caused by randomness than the differences in the knowledge of the test groups.

Table 5: Frequencies [0-1] of each category from all four accident reports

Category	Mariners	Researchers	Total
Inadequate communications	0,583	0,708	0,646
Lack of guidelines	0,417	0,625	0,521
Faulty practices	0,500	0,500	0,500
Inappropriate use of navigational equipment	0,417	0,417	0,417
Inadequate route plan	0,542	0,167	0,354
Hazardous natural environment	0,333	0,333	0,333
Mistake	0,208	0,417	0,313
Wrong steering decision	0,417	0,167	0,292
Lack of situational awareness	0,375	0,167	0,271
Lack of VTS	0,292	0,250	0,271
Bad visibility	0,167	0,333	0,250
Fairway	0,208	0,250	0,229
Poor design of automation	0,292	0,083	0,188
Wind	0,083	0,208	0,146
Schedule	0,083	0,167	0,125
Incompetence	0,208	0,000	0,104
Manning	0,167	0,042	0,104
Outside distractions	0,125	0,083	0,104
Lack of training	0,167	0,042	0,104
Fatigue	0,042	0,125	0,083
Lack of redundancy	0,042	0,083	0,063
Route plan not followed	0,042	0,000	0,021

3.3. Edges

The comparison of the edges is more complicated than the nodes as there are much more combinations. The edges in the BNs also vary generally more than the nodes because all of the participants do not have all same nodes in their BNs. Assuming that identifying the causes from accident reports does not require much knowledge on the interdependencies of the causes, the driven results from the nodes are considered to be more reliable than what can be driven from the edges. This is because constructing a BN by adding directed edges to the causes may need prior knowledge on causality, BN, and accident theories.

With that being said, looking at the most frequent edges from each node we can compare the structures of the BNs. We concentrated on the top 12 most frequent nodes in **Table 5**, where both test groups have in at least four of their individual BNs. The less frequent nodes have mostly single edges connected to them. **Figure 3** shows how many participants in each test group on average chose same edges to their BN of the same accident. It can be seen that researchers are more unanimous about the edges in their networks than the mariners, which again may be the result of the differences in the years of experience of the individuals in each test group.

Figure 3: Number of same edges in one report in relation to all edges

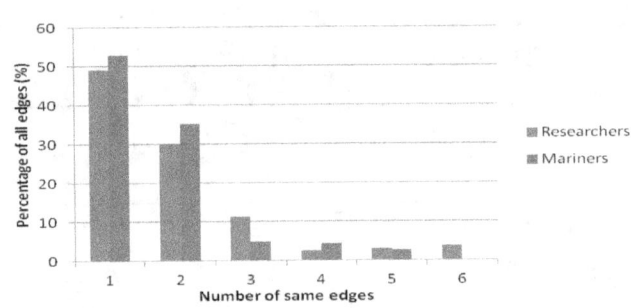

The dependencies in the BNs can also be viewed in terms of how often there is an edge between certain pair of nodes (in both directions) when both nodes are in the same BN. **Table 6** shows the dependencies between the most frequent nodes from **Table 5**. The numbers in **Table 6** show the fractions of the number of the BNs in a group that have specific pair of nodes and have indicated the dependency between that pair of nodes. For instance, there is dependency between fairway and lack of situational awareness in all of the researchers' BNs (100%), but in only 80 % of the mariners' BNs that have both of the two nodes (see **Table 6**). In pairs that are missing a number in the table, the dependencies are not known because none of the participants of the test groups had the pair in their individual BNs. **Table 6**, for instance, shows that although mariners believe that there is dependency between lack of VTS and lack of situational awareness, none of the researchers believe in that connection. What **Table 6** does not show is the mariners and researchers' beliefs regarding the causality links between the nodes (i.e. direction of the edges). Interested readers are referred to the complete report of the study [15], in where the direction of the edges by each group is discussed more comprehensively.

Table 6: The dependencies between the most frequent nodes in all accident reports. The number of connections has been divided with the number of BNs that have both nodes in them. The empty spaces in the table indicate that there are no BNs that include both nodes

	Bad visibility	Inadequate communications	Faulty practices	Hazardous natural environment	Mistake	Inadequate route plan	Inappropriate use of navigational equipment	Wrong steering decision	Lack of situational awareness	Fairway	Lack of VTS	Lack of guidelines	Grounding
Bad visibility		0,0	-	-	-	0,3	0,0	-	0,5	-	0,0	0,0	0,3
Inadequate communications	0,6		1,0	0,0	1,0	0,3	0,2	0,6	1,0	0,0	0,7	0,3	0,2
Faulty practices	0,0	0,7		0,5	0,5	0,4	0,6	0,2	0,3	0,7	0,5	1,0	0,2
Hazardous natural environment	0,0	0,2	0,5		1,0	0,7	0,0	0,5	0,0	-	-	0,0	0,4
Mistake	0,7	0,4	0,2	0,5		0,0	-	1,0	1,0	0,3	0,0	0,0	0,8
Inadequate route plan	0,3	0,0	1,0	0,0	1,0		0,4	0,5	0,6	0,0	0,2	0,7	0,2
Inappropriate use of navigational equipment	0,3	0,4	0,6	0,3	0,3	0,5		0,7	1,0	0,0	0,0	0,7	0,2
Wrong steering decision	-	0,7	0,5	0,0	1,0	-	-		0,8	0,0	0,0	0,2	0,7
Lack of situational awareness	-	0,3	0,5	0,0	1,0	-	0,3	-		0,8	1,0	0,0	0,7
Fairway	0,0	0,2	0,0	-	0,4	-	0,0	-	1,0		0,3	0,0	0,0
Lack of VTS	0,0	0,8	0,3	0,0	0,2	0,0	-	-	0,0	0,0		0,0	0,1
Lack of guidelines	0,0	0,7	0,2	0,4	0,0	0,7	0,4	0,5	0,0	0,0	0,0		0,0
Grounding	0,6	0,5	0,1	0,8	0,9	0,8	0,4	0,8	0,5	0,3	0,2	0,3	

Researchers / Mariners

4. DISCUSSION

The results of the study show that although in general neither of the groups have superiority over the other with regard to the extracted knowledge, there are some categories that are more frequently chosen by specific group. Mariners chose more often the causes related to navigation and the actions of the crew, while the researchers tend to see more organizational and environmental related causes. This was somehow expected as for instance the researchers may not know the importance of missing sea maps or inadequate route plan whereas mariners have personal experience on the matter. Therefore, causes like *mistake*, *wind*, and *fatigue* that are not related to navigation and thus are easier to find in the reports by people with no nautical background are more frequent in researchers' answers. Additionally, the researchers often connected darkness to bad visibility, while none of the mariners had darkness in their answers. The mariners may think that darkness should not affect the safety because it is unavoidable and vessels normally have enough equipment like radar to sail safely in darkness. There are also some categories that were chosen by the majority of the mariners but by none of the researchers in certain reports. For instance in the grounding of MV EMSRUNNER *inadequate route plan* was thought to have been a cause of accident by all mariners and in the grounding of MS PAULINE RUSS by half of the mariners but by none of the researchers. Also in the grounding of MS SUPERFAST VII the majority of mariners thought that *wrong steering decision* was one of the causes that led to the grounding. It was said in the answers that the use of tugboats was inadequate or the tugboats were not used correctly. None of the researchers thought that wrong steering decision caused the accident. These differences may be caused by the lack of navigational knowledge among the researchers as, for instance, the effect of tugboats may not have been directly pointed out in the report by the investigators thus making it harder for the researchers to see it as a contributing factor. This suggests that there are some causes that are mentioned in the storyline of the event but have not been investigated further in the accident reports; thus non-mariners are less likely to recognize them as contributing factors in the accident.

From another perspective, by looking at the results from an accident theory point of view like Reason's Swiss-Cheese [16], it is observable that the mariners have more "active failures" in their BNs whereas the categories that the researchers have more frequently in their BNs are "latent conditions". This means that the mariners are more concentrated in the person approach of the human factors than the researchers. On the other hand the mariners did also have latent conditions in their BNs but the way that they structured their BNs shows that they know latent conditions do not cause the accidents by themselves and often the crew is also responsible for the groundings. Of course the researchers also had categories from active failures in their BNs but not as much as the mariners. The mariners may know better from personal experience the types of active failures that cause accidents and how accidents can be avoided with different actions. The researchers may not be familiar with the rules and practices and if they are not mentioned in the accident reports clearly as accident causes the researchers may not know if all good-seamanship practices were followed.

On the other hand, the comparison of the whole BNs by calculating distances between the BNs shows that the consistency within the groups are not on average always more than between the groups. There were some instances where the distance in one group was higher than the distance between the groups. There are also many instances where there is a BN in one group that has smaller distances with the BNs of the other groups than with its own groups. Since the distances do not show us the differences in individual nodes and edges and merely show the differences in the networks as a whole, results of the distance measurement suggest that the two groups are not necessarily as different as what the comparison of the nodes showed when the whole networks are compared instead of individual nodes.

Despite of the above discussion and drawn results, there are uncertainties involved in the study that need to be mentioned and taken into account. Due to the lack of marine experts that were willing to participate in the study, the sample size in the study is limited. The sample size of each group should be at least ten to get test statistics that are good enough for statistical testing [17]. With only six observations in total, the statistical reliability of the study is not high enough. This is unfortunate though unavoidable because the study requires plenty of the participants' time. With such small sample size more accurate assumptions about the differences in the nodes should not be made. With bigger sample sizes the results could be analyzed statistically to get a better view on the differences between the groups. Now the variance of the results is too big to get reliable results on whether there is a significant difference between the two groups. Only the causes that were clearly picked by only one group can be said to have differential effects between the groups. Moreover, the inconsistency in the years of experience between the test groups may have introduced unknown uncertainties to the results. Therefore, the years of experience is certainly a factor that has potential for further studies in this regard.

In addition, different interpretation of the categories (nodes), like *inappropriate communication* that might be outside communication or bridge communication, might result in different ways of connecting the nodes with edges. Therefore, it is not easy to accurately say if the frequency of the edges varied between the groups' BNs. One reason for the differences may be that the causalities can be interpreted differently in building BNs. The most significant differences in the edges are the differently oriented edges. For instance in the accident report of the grounding of MS SUPERFAST VII, although there were no connection mentioned in the report between two causes as *wind* and *lack of guidelines*, there were two differently directed edges between the two nodes in the collected answers. It was said in the accident report that "*The instructions did not contain harbor maneuvering in a storm, which resulted in a defective estimation of the wind effect.*" and "*The wind limit had been exceeded*" [18]. The first sentence may imply that there should be an edge from lack of guidelines to hazardous natural environment as wind, which may come from an interpretation as: "Because of the lack of guidelines about wind limits the wind becomes a problem, whereas if there had been adequate guidelines the wind would not have been a problem." Two of the researchers and none of the mariners chose this combination to their BNs. Two other researchers though chose the opposite direction for this edge. These participants may have interpreted the interdependency differently as: "The high winds in a harbor area create a need for guidelines and the lack of guidelines creates a dangerous situation." Therefore, future studies needs to cover causality interpretation aspect more closely, either by providing a unique interpretation for each possible link, or by in-deep interviews regarding the implemented interpretation for each link.

5. CONCLUSION

In general, although there is no superiority between the groups for reviewing the accident reports with regard to the extracted knowledge, it seems that the researchers cannot identify the same BNs as the mariners and vice versa. There are clear differences between the two test groups of the study. Mariners chose more often the causes related to navigation and the actions of the crew, while the researchers tend to see more organizational and environmental related causes. Therefore, as was expected, the causal network of the same accidents varied both within and between the groups. Even with a finite set of possible categories none of the participants identified identical BNs from the accident reports of the study. From the study we can conclude that there are differences in the mathematical models that different individuals build from existing accident reports.

Acknowledgements

This study was conducted as a part of "Minimizing risks of maritime oil transport by holistic safety strategies" (MIMIC) project. The MIMIC project is funded by the European Union and the financing comes from the European Regional Development Fund, The Central Baltic INTERREG IV A Programme 2007-2013; the City of Kotka; Kotka-Hamina Regional Development Company (Cursor Oy); Centre for Economic Development, and Transport and the Environment of Southwest Finland (VARELY).

Merenkulun säätiö is also appreciated for providing the financial support for attending the PSAM12 conference to present this paper.

References

[1] Mazaheri A., and Ekwall D. (2009) *"Impacts of the ISPS code on port activities: a case study on Swedish ports"*, World Review of Intermodal Transportation Research, Vol.2, No.4, pp.326-342, doi: 10.1504/WRITR.2009.026211

[2] Aven T. (2013) *"A conceptual framework for linking risk and the elements of the data-information-knowledge-wisdom (DIKW) hierarchy"*, Reliability Engineering System Safety Vol.111, pp.30–36

[3] Mazaheri A., Montewka J., Kujala P. (2013) *"Modeling the risk of ship grounding - A literature review from a risk management perspective"*, Published online in WMU-Journal of Maritime Affairs, doi: 10.1007/s13437-013-0056-3

[4] Kristiansen S. (2010) *"A BBN approach for analysis of maritime accident scenarios"* In: Ale BJM, Papazpglou IA, Zio E (eds) ESREL. Taylor & Francis, Rhodes

[5] Tirunagari S., Hänninen M., Ståhlberg K., Kujala P. (2012) *"Mining Causal Relations and Concepts in Maritime Accidents Investigation Reports"*, International Conference cum Exhibition on Technology of the Sea. Visakhapatnam: IMU

[6] Grech M. R., Horberry T., Smith A. (2002) *"Human error in maritime operations: analyses of accident reports using the Leximancer tool"*, Proceedings of the Human Factors and Ergonomics Society 46th Annual Meeting. 46, pp. 1718-1722. Baltimore: Human Factors and Ergonomics Society. doi:10.1177/154193120204601906

[7] Mazaheri A., Sormunen O. V. E., Hyttinen N., Montewka J., Kujala P. (2013) *"Comparison of the learning algorithms for evidence-based BBN modeling – A case study on ship grounding accidents"*, Proceedings of the Annual European Safety and Reliability Conference (ESREL), pp. 193-200, September 30th - October 2nd, Amsterdam, the Netherlands; ISBN 978-1-138-00123-7

[8] Johnson C. W., Botting R. M. (1999) *"Using Reason's Model of Organisational Accidents in Formalising Accident Reports"* Cognition, Technology & Work, Vol.1, pp.107-118

[9] Lekberg A. (1997) *"Different approaches to incident investigation-how the analyst makes a difference"*, System Safety Society Conference, Hazard Prevention 33:4

[10] Dalkey N. C. (1969) *"The Delphi Method: An experimental study of group opinion"* Santa Monica, California: The Rand Corporation.

[11] Rothbaum A. R. (2000) *"Human Error and Marine Safety"*, National Safety Council Congress and Expo. Orlando

[12] McCafferty D., Baker C. (2006) *"Trending the causes of marine incidents"* ABS Technical Papers 2006, 1-9.

[13] de Jongh M., Druzdzel M. J. (2009) *"A Comparison of Structural Distance Measures for Causal Bayesian Network Models"*, In: M. Klopotek, A. Przepiorkowski, S. Wierzchon, & K. Trojanowski (Eds.), Recent Advances in Intelligent Information Systems, Challenging Problems of Science, Computer Science series (Vol. 1, pp. 443-456). Warsaw, Poland: Academic Publishing House EXIT.

[14] Acid S., de Campos L. M., Fernández-Luna J. M., Rodríguez S., Rodríguez J. M., Salcedo J. L. (2004) *"A comparison of learning algorithms for Bayesian networks: a case study based on data from an emergency medical service"*, Artificial Inteligence in Medicine, 30, 215-232.

[15] Hyttinen N. (2013) *"The effect of experience on knowledge extraction from accident reports"*, Master's thesis, School of Science, Aalto University, Espoo, Finland, p.80

[16] Reason J. (1990) *"Human Error"*, Cambridge University Press, Cambridge, UK.

[17] Mellin I. (2006) *"Testejä laatuasteikollisille muuttujille"*, Retrieved 07 09, 2013, from math.aalto.fi/opetus/sovtoda/oppikirja/Testit.pdf

[18] Safety Investigation Authority (2004) *"MS Superfast VII, grounding off Hanko on 12.11.2004"*, Report No. B7/2004M

A Study for Adapting a Human Reliability Analysis Technique to Marine Accidents

Kenji Yoshimura[a], Takahiro Takemoto[b], Shin Murata[c], and Nobuo Mitomo[d]

[a] National Maritime Research Institute, Mitaka, Japan
[b] Tokyo University of Marine Science and Technology, Tokyo, Japan
[c] National Institute for Sea Training, Yokohama, Japan
[d] Nihon University, Funabashi, Japan

Abstract: The deck officer who has the duty of navigation and keeping watch on a ship's bridge is known as the officer of the watch (OOW). The OOW is a qualified and capable person with knowledge of ship navigation. According to the Japan Marine Accidents Inquiry Agency, however, "inadequate lookout" is the cause of 84% of collision accidents. In 41% of accidents, "the OOW couldn't find the target until collision," and in 32% of collisions, "even though the OOW had found the target, they didn't maintain a proper lookout." Many of the causes behind accidents pertain to not only the OOW's knowledge and capability, but background factors.

The Japan Transport Safety Board (JTSB) has been established in order to prevent recurrences and to mitigate damages caused by accidents. The JTSB considers introducing analysis method with objective/scientific processes.

The Cognitive Reliability and Error Analysis Method (CREAM) is a technique for analysing human reliability. CREAM organizes interactions between humans and the environment using the human-technology-organization triad. CREAM defines common performance conditions (CPC), the dependencies between them, and the links between antecedents and consequences to clarify the background factors that affect human performance.

This method has mainly been used in the nuclear industry. When analysing the causes of accidents, it is necessary to clarify how much influence conditions have on human performance and the dependencies between CPCs. Since these conditions change across domains, the CPCs will apply differently to domains other than the nuclear industry. For example, in comparing the nuclear industry and the maritime industry, there are significant differences in the influence the work environment has on behaviour and human performance. Therefore, the dependencies between CPCs and priority are now evaluated according to the expert judgment of each domain. To facilitate simple and objective analysis, the CPCs and the dependencies, and the links need to be fitted to each domain.

From a point of view described above, we first proposed CPCs adapted to maritime collision accidents. Secondly, we administered a questionnaire to OOWs for the purpose of quantifying the priority of CPCs.

Though our research is ongoing, we have reached certain conclusions. We herein provide an outline of our findings and the results of the questionnaire survey. We also specifically discuss the priority of the CPCs that were adapted to maritime collision accidents.

Keywords: CREAM, Common Performance Conditions, Pairwise Comparison Method

1. INTRODUCTION

The deck officer who has the duty of navigation and keeping watch on a ship's bridge is known as the *officer of the watch* (OOW). The OOW is a qualified and capable person with knowledge of ship navigation. According to the Japan Marine Accidents Inquiry Agency [1], however, "inadequate lookout" is the cause of 84% of collision accidents. In 41% of accidents, "the OOW couldn't find the target until collision," and in 32% of collisions, "even though the OOW had found the target, they didn't maintain a proper lookout." Many of the causes behind accidents pertain to not only the OOW's knowledge and capability, but background factors.

The *cognitive reliability and error analysis method* (CREAM) is a technique for analyzing human reliability [2]. CREAM organizes interactions between humans and the environment using the *human-technology-organization* triad. CREAM defines *common performance conditions* (CPC) and the dependencies between them to arrange the background factors that affect performance. Fig.1 shows the dependencies. This method has mainly been used in the nuclear industry. When analyzing the causes of accidents, it is necessary to clarify how much influence conditions have on performance and the dependencies between CPCs. Since these conditions change across domains, the CPCs will apply differently to domains other than the nuclear industry. For example, in comparing the nuclear industry and the maritime industry, there are significant differences in the influence the work environment has on behavior and performance [3]. Therefore, the dependencies between CPCs are now evaluated according to the expert judgment of each domain.

To facilitate simple and objective analysis, the dependencies between CPCs and the priority of CPCs need to be fitted to each domain. In principle, all background factors should be investigated. Given our limited resources, however, it is difficult to investigate all background factors. Therefore, we propose that adopt method of pairwise comparisons to define the priority and the weight of CPCs.

2. FACTOR ANALYSIS METHOD FOR MARINE ACCIDENTS

2.1. Overview of CREAM

Applying a human reliability analysis method to analyze marine accidents requires an assessment of decision-making errors made during emergencies. Second-generation human reliability analysis methods are appropriate since they can assess cognitive processes during emergencies. Methods such as CREAM and ATHEANA [6] are examples of second-generation human reliability analysis. CREAM in particular has been utilized for analyzing some marine accidents. There are two techniques associated with CREAM: one involves analysis by screening and the other involves detailed analysis. Analysis by screening is used in this study. The concept of CPCs is introduced to qualitatively analyze human behavioral environments. Nine categories of CPCs are defined. In addition, dependencies are defined between each of the CPCs. Figure 1 shows CPCs and dependencies. Since CREAM is utilized for general working tasks, it cannot apply directly to marine accidents. For the purposes of this study, we have adapted CPCs for marine accidents as in Table 1.

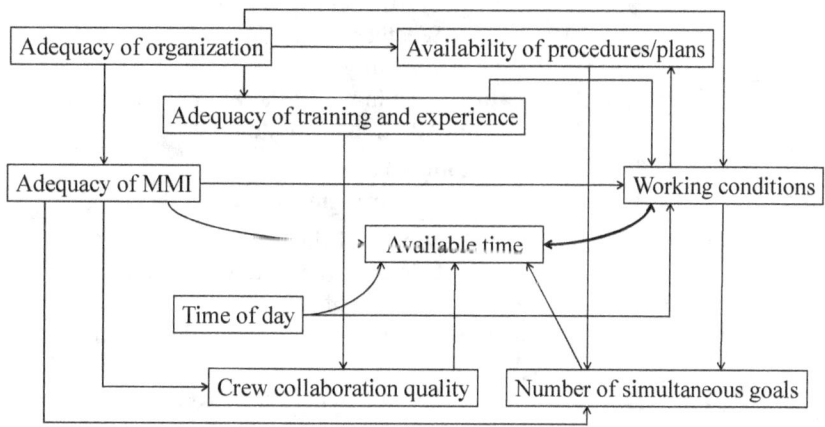

Figure 1: The dependencies between CPCs defined by CREAM

Table 1: Adapted CPC for marine accidents

CPC	Adapted CPC for marine accidents

Adequacy of organization	**Adequacy of safety management**
	    Contents of educational training
	    Systems of educational training
	    Management of navigational watch
	    Outside support and communication

Working conditions	**Navigation conditions**
	    Traffic density
	    Weather conditions
	    Visibility
	    Area in which the vessel is navigating
	    Ship maneuvering characteristics
	    Watch condition on the bridge

Adequacy of MMI and operational support	**Adequacy of human-machine interface**
	    Installed navigational equipment
	    Accessibility of navigational aids

Availability of procedures / plans	**Adequacy of navigation manuals**
	    Chain of command
	    Criteria for avoiding action
	    Navigational conditions
	    Standards, procedures, and guidance
	    Unification of terminology
	    Readability of manual

Number of simultaneous goals	**Number of simultaneous goals**
	    Traffic density
	    Weather conditions
	    Visibility
	    Area in which the vessel is navigating
	    Additional workload

Available time	**Available time**
	    Traffic density
	    Weather conditions
	    Visibility
	    Area in which the vessel is navigating
	    Additional workload

Time of day (circadian rhythm)	**Time of day** Day or night

Adequacy of training and experience	**Resource of the officer**
	    Contents of educational training
	    Systems of educational training
	    Knowledge of and confidence in the professional watch
	    Experience of the officer of the watch

Crew collaboration quality	**Communication and information sharing**
	    Communication and information sharing with the bridge team
	    Communication and information sharing with other ships
	    External assistant communication and information sharing

2.2. Human Error in Navigational Watch

Human error in navigational watch is considered the same mechanism of human malfunction proposed by Rasmussen [7]. "Mechanisms of human malfunction" and "internal human malfunction" are affected by external conditions such as "performance shaping factors," "situation factors," and "personnel task."

These categories may be adapt to marine accidents as follows: Causes of human malfunction; bad weather conditions (effect of tide, wind, and visibility), traffic density, and illness. External mode of malfunction; not taking evasive action (asleep, invisible), take wrong evasive action (inadequate knowledge of regulation, own ship handling character, and weather conditions), take evasive action appropriately but couldn't avoid collision (other ship movement). Situation factors; excessive demand, conflict problem, inadequate education and training, lack of manual.

Rothblum [8] summarized that human factors issues in the maritime industry as follows, Fatigue, Inadequate Communications, Inadequate General Technical Knowledge, Inadequate Knowledge of Own Ship Systems, Poor design of Automation, Decision based on Inadequate Information, Faulty standards, policies, or practices, Poor maintenance, and Hazardous natural environment.

We reported the dependencies between CPCs that were adapted to marine collision accidents based on the results of a questionnaire survey [9]. As in Figure 2, these results confirm the essence and character of the maritime industry, the details of which are described below. The dependencies that were common in the definition of CREAM—such as the influence of "Adequacy of human-machine interface"—were extracted. "Available time" affects many other conditions. The influence of "Navigation conditions" is restrictive. The OOW's knowledge and capabilities are influenced by many conditions that differ from those of other industrial domains.

As mentioned above, CREAM organizes interactions between humans and the environment using the human-technology-organization triad. These studies pointed out that various factors affect the human error process. Especially, it supposed that the working conditions of navigational watch is affected by great fluctuation of natural environment than other working conditions.

3. METHOD

The questionnaire survey for the OOWs, including OOWs on training ships, was administered in cooperation with the National Institute for Sea Training, Japan. We created the pairwise comparison

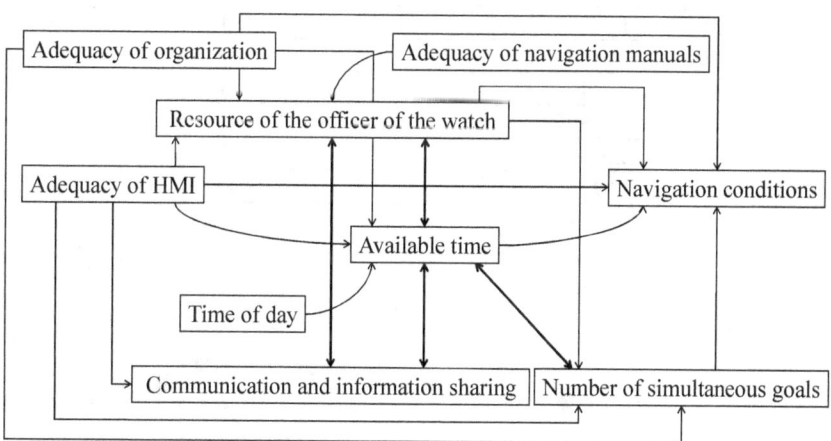

Figure 2: The dependencies between CPCs

questionnaire to clarify the priority of CPCs that were adapted to marine collision accidents. There were 36 combinations. The participants compared and answered that which is more important for prevention of collision accident. Twenty-eight copies were distributed; all copies were collected. The time period for tallying the results of the questionnaires was July 2013 through September 2013.

4. RESULTS AND DISCUSSION

Table 2 shows the profile of the participants. The results which were obtained in the way described above are shown in Table 3. Table 3 shows a pairwise comparison matrix that is based on the Thurstone model. The numbers in the table represent the inverse of the cumulative distribution function of the standard normal distribution. Figure 3 shows the ranking and the weights calculated from pairwise comparison matrix. The weights represent along X-axis. This paper will consider that the ranking and the weights. The highest ranking was "Resource of the officer", and the weight of CPC was exceptionally high. "Available time", "Communication and information sharing", "Number of simultaneous goals" becomes lower in the order of descending priorities. These high weight CPC are presumed to be relating to human or organization. These results shows that CPCs related human or organization are considered to be important factor to prevent collision accident.

On the contrary, the lowest evaluated CPC was "Time of day". This results is similar to the results of preceding survey. For example, "Time of day" was not affected by any CPCs as shown in Figure 2. The number of affected CPC corresponds to the ranking of the CPC. On the other hand, "Resource of the officer" was affected by 5 CPCs, and "Available time" was affected by 6 CPCs. These results showed that the number of affected CPC and the ranking of the CPC having high correlation.

As a result, we may conclude that it may be started investigate background factor at CPCs related human and organization, such as "Knowledge of and confidence in the professional watch", or "Experience of the officer". Considering that the ranking and weight of CPCs, the background factor can investigate more efficiency.

Table 2: Characteristics of survey participants

	No (%) of respondents
Ranks	
Captain	2 (7)
Instructor	3 (11)
Chief Officer	11 (39)
2nd Officer	4 (14)
3rd Officer	8 (29)
Age	
60-	0 (0)
50-59	2 (7)
40-49	9 (32)
30-39	8 (29)
20-29	9 (32)
-19	0 (0)

Table 3: Pairwise comparison matrix for criteria and weights

	Adequacy of safety management	Navigation conditions	Adequacy of Human Machine Interface	Adequacy of navigation manuals	Number of simultaneous goals	Available time	Time of day	Resource of the officer	Communication and Information sharing
Adequacy of safety management	0.00	0.57	0.46	0.37	0.79	1.07	-0.37	1.07	0.79
Navigation conditions	-0.57	0.00	-0.37	-0.57	-0.09	0.18	-0.57	1.24	0.37
Adequacy of Human Machine Interface	-0.46	0.37	0.00	-0.09	0.67	0.57	-0.92	1.07	0.37
Adequacy of navigation manuals	-0.37	0.57	0.09	0.00	0.67	0.79	-0.46	0.92	0.79
Number of simultaneous goals	-0.79	0.09	-0.67	-0.67	0.00	-0.09	-0.92	0.67	0.00
Available time	-1.07	-0.18	-0.57	-0.79	0.09	0.00	-1.80	0.67	0.09
Time of day	0.37	0.57	0.92	0.46	0.92	1.80	0.00	1.24	1.07
Resource of the officer	-1.07	-1.24	-1.07	-0.92	-0.67	-0.67	-1.24	0.00	-0.57
Communication and Information sharing	-0.79	-0.37	-0.37	-0.79	0.00	-0.09	-1.07	0.57	0.00
Total	-4.75	0.37	-1.57	-3.00	2.39	3.55	-7.35	7.45	2.91
Average	-0.95	0.07	-0.31	-0.60	0.48	0.71	-1.47	1.49	0.58

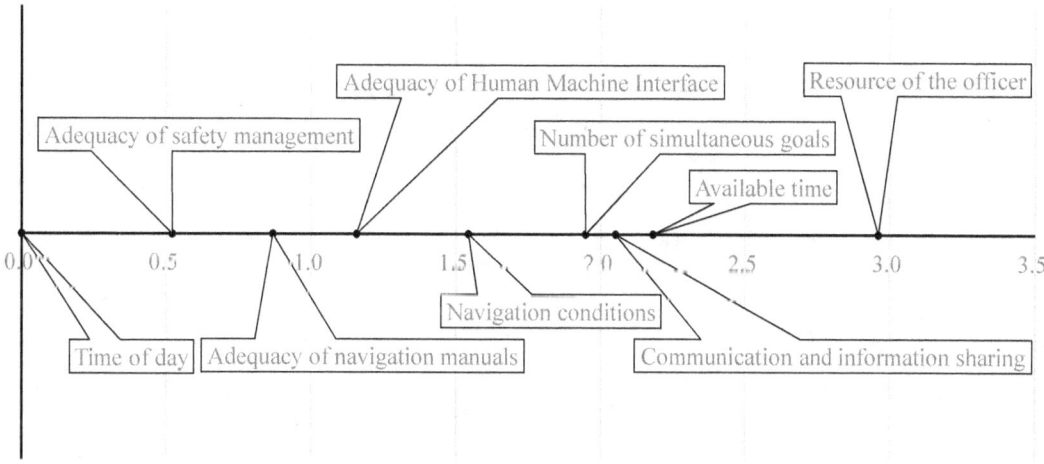

Figure 3: The ranking and weights of CPCs

5. CONCLUSION

To improve reliability assessment, we propose CPCs that are adapted to collision accidents. In addition, we report the results of the questionnaire for clarifying the priority of CPCs for marine collision accidents. These results indicate possibilities that will improve the reliability of CREAM when it is fitted to each domain.

References

[1] Japan Marine Accident Inquiry Agency. "*Kainan Report 2008*", pp. 38-45, (2008).
[2] E. Hollnagel. "*Cognitive Reliability and Error Analysis Method*", pp. 107-117, Elsevier, London, (1998).
[3] A.P.-J. Thunem, J. Ferkingstad, and V. Frette. "*A Comparison Between the Nuclear and the Maritime Domains on Challenges Related to Technological Advances*", ESREL 2012 / International Probabilistic Safety Assessment and Management Conference, 11 (PSAM), Helsinki, pp. 6460-6469, (2012).
[4] T. Takemoto, N. Mitomo, K. Hikida, and K. Yoshimura. "*A Study on Human Factors Analysis for Possible Factors of Marine Accident-Modifying CPC for Marine Accident Analysis*", The Journal of Japan Institute of Navigation, Vol. 127, pp. 95-101, (2012).
[5] N. Mitomo, K. Hikida, K. Yoshimura, C. Nishizaki, and T. Takemoto. "*Common Performance Condition for Marine Accident-Experimental Approach*", 2012 Fifth International Conference on Emerging Trends in Engineering and Technology, pp. 100-104, (2012).
[6] S.E. Cooper, A.M. Ramey-Smith, J. Wreathall, G.W. Parry, D.C. Bley, W.J. Luckas, J.H. Taylor, and M.T. Barriere. "*A Technique for Human Error Analysis (AHTEANA)*", NUREG/CR-6350, US-NRC, (1996).
[7] J. Rasmussen, O.M. Pedersen, G. Mancini, A. Carnino, M. Griffon, and P. Gagnolet. "*Classification System for Reporting Events Involving Human Malfunction*", Riso-M-2240, (1981).
[8] Rothblum, A.M.. "*Human Error and Marine Safety*", Proceedings of the Maritime Human Factors Conference, Maryland, USA, pp. 1-10, (2000).
[9] K. Yoshimura, C. Nishizaki, A. Kimura, S. Murata, N. Mitomo, and T. Takemoto. "*Questionnaire Survey for Adapting Common Performance Conditions to Marine Accidents*", Proceedings of 2013 IEEE International Conference on Systems, Man, and Cybernetics, (2013).

Quantifying the effect of noise, vibration and motion on human performance in ship collision and grounding risk assessment

Jakub Montewka[a*], Floris Goerlandt[a], Gemma Innes-Jones[b], Douglas Owen[b], Yasmine Hifi[c], Markus Porthin[d]

[a] Aalto University, Department of Applied Mechanics, Marine Technology, Research Group on Maritime Risk and Safety, Espoo, Finland
b – Lloyd's Register, EMEA, Bristol, UK
c – Brookes Bell R&D, Glasgow, UK
d - VTT Technical Research Centre of Finland, Espoo, Finland

Abstract: Risk-based design (RBD) methodology for ships is a relatively new and a fast developing discipline. However, quantification of human error contribution to the risk of collision or grounding within RBD has not been considered before. This paper introduces probabilistic models linking the effect of ship motion, vibration and noise with risk through the mediating agent of a crewmember. The models utilize the concept of Attention Management, which combines the theories described by Dynamic Adaptability Model, Cognitive Control Model and Malleable Attentional Resources Theory. To model the risk, an uncertainty-based approach is taken, under which the available background knowledge is systematically translated into a coherent network and the evidential uncertainty is qualitatively assessed.

The obtained results are promising as the models are responsive to changes in the GDF nodes as expected. The models may be used as intended by naval architects and vessel designers, to facilitate risk-based ship design.

Keywords: Risk-Based Ship Design, Bayesian Belief Networks, Risk Assessment, Collision Probability, Grounding Probability.

1. INTRODUCTION

In the Risk-Based Ship Design (RBSD) methodology the assessment of the risk level of a new ship is conducted in the early design stage, where a design modification is easy and cost-effective [1]. In this approach, risk is evaluated alongside conventional design performance measures like sufficient strength and stability, low resistance, cargo carrying capacity, propulsion and maneuvering capability. Risk is thus treated as a design objective rather than a constraint imposed by prescriptive safety rules.

In line with an increased focus on the human element in the maritime domain, it is possible to focus on human performance in the early design phase. This paper presents advances to the RBSD methodology, focusing on the effect of specific performance-shaping factors (PSFs) known as global design factors (GDF) on human performance and ultimately on ship collision and grounding risk. These GDFs comprise ship motion, noise and whole-body vibrations (WBV). Considerable scientific attention has been paid to developing risk models for ship collision and a variety of approaches have been presented quantifying the specific effect of GDFs on human error and collision risk has not been considered before, for the review of the existing methods see for example [2]–[5]. Quantifying this effect is however challenging as common human error quantification frameworks do not readily account for the specific effect of the GDFs, see for example [6]–[10].

A workable approach for human performance risk modelling has emerged from the literature on the effects of exposure to ship motion, noise and vibration GDFs focussing on attention management, [11]. It is based on three theories: the Dynamic Adaptability Model, - [12] - Cognitive Control Model - [13] - and Malleable Attentional Resources Theory - [14].

These foundations are used as a guide for constructing the risk model, which is developed using Bayesian Belief Networks (BBNs). The methodology applied here for the risk assessment stems from an uncertainty-based risk perspective, where the assessment is used to transform the available background knowledge in a coherent framework to assess uncertainty about the occurrence of a collision or grounding event. Special attention is given to the evidential uncertainties underlying the BBNs structure and probabilities, which are considered alongside sensitivities. BBNs, as probabilistic tools are capable of representing background knowledge about the collision and grounding phenomena, the quantification of associated uncertainties, efficient reasoning and updating in light of new evidence, see for example [2], [15], [16]. Two models are presented in this paper. They are stand-alone models that behave in response to GDF inputs as intended. The results are comparable with other existing models, and the data from the marine industry. The models can enable comparative assessment of ship designs, which is their primary intention.

The remainder of the paper is organized as follows: Section 2 introduces the adopted risk perspective, upon which general structure of the model is developed, as presented in Section 3. Section 4 elaborates on the human performance model. Section 5 introduces and discusses two risk models developed here, whereas Section 6 concludes.

2. ADOPTED RISK PERSPECTIVE

In this paper we adopted an uncertainty perspective of risk, where risk is seen as follows, [17]:
$$R \sim C \& U \qquad (1)$$
This means that risk assessment is an expression of an assessor's uncertainty (U) about the occurrence of events and the associated consequences (C). Following this perspective, risk assessment can always be performed, as the risk model is seen as a tool to describe and convey uncertainties rather than a tool to uncover the truth. For this purpose, the risk model encompasses the events and the consequences of the events when they become true.

The models presented here focuses on two events namely ship-ship encounter and ship-ground encounter. The consequences of the above events are ship-ship collision and ship running aground respectively. It is assumed, that these accidents stem from the inability of a navigator to perform evasive action, when exposed to an encounter with another ship or shallow waters. This inability results from the reduced performance, which has been affected by the GDFs. Thus, the model links the effect of GDFs on human performance with the probability of an accident.

Finally, the risk model presented in this paper delivers the probability of an accident given encounter, accounting for the available background knowledge about the analysed phenomena. The background knowledge in terms of available data, models, theories and expert judgement, is assessed qualitatively, which makes it possible to elaborate on the evidential uncertainty of the risk model. Such information coupled together with the results of the sensitivity analysis of the model provides information about the outcome uncertainty of the model. This way of analysing the uncertainty allows an analyst determining the areas of the domain under study, which needs further research in order to improve the overall performance of the risk model.

The adopted risk perspective allows defining the plethora of consequences and associated uncertainties. This makes it possible to expand the model with desired consequences, which can be very specific, depending on the ship type under analysis. For instance, in case of a RoPax ship, the societal impact may be of interest, thus the risk in this case will be expressed through the probability of a number of fatalities, see for example [2]. Whereas, in case of a tanker, the environmental impact of an accident may be a key issue, thus the risk is expressed as the probability of oil spill of certain size [18]. However, detailed quantification of the specific consequences of an accident is out of the scope of this paper.

3. GENERAL STRUCTURE OF THE RISK MODEL

To describe the process through which exposure to GDFs causally affects the probability of the specified unwanted outcomes, a causal pathway was developed through the mediating agent of the crewmember. Importantly, the causal chain represents the effects of GDFs exposure on human performance in a way that could be developed and elaborated in the risk model. It served to do three things:
1. Represent the mechanism by which GDFs exposure impacts collision and grounding risk.
2. Describe the overall topography of the final model.
3. Facilitate the identification of nodes.

GDFs can be considered a type of PSF, where PSFs are an aspect of the human's individual characteristics, environment, organisation, or task that specifically decrements or improves human performance, thus increasing or decreasing the likelihood of human error respectively, [19]. While there are many other PSFs that can affect human behaviour – for instance training, experience, competence, time available, workload, job design, manning, ergonomics of the equipment and procedures - these are excluded from the collision and grounding risk models as they are not affected by exposure to GDFs. All the excluded PSFs are implicitly assumed to remain constant within the model.

Other potentially relevant factors, which are not considered, are the long terms effects of GDFs on the crew performance. For example, we do not consider the hearing loss due to long-term noise exposure either individually or in combination with other GDF effects. In practical terms only the effect of GDF-affected human performance on the possible occurrence of collision and grounding in combination with the safety critical task being performed are considered.

In the presented model the inputs and outputs of the model are predetermined. The GDFs form the three inputs: ship motion, noise and vibration. The effect of the latter is considered through WBV. The unwanted outcomes form the two outputs: collision, grounding. In reality, crew exposure to GDFs is likely to result in a plethora of effects on human performance and subsequent outcomes. Likewise, the unwanted outcomes are likely to have numerous causal inputs, which may include GDFs. However, to remain within the scope of our study the causal representation is limited to describing only those mechanisms that can describe the relationships between the predetermined inputs and outputs, as depicted in Figure 1.

Two main paths linking GDF exposure to human behaviour, and subsequently to collision and grounding, have been identified:
- Path 1: Stressor effects. Exposure to a GDF acts as a stressor and can affect the perceptual, cognitive and physical capabilities of an individual (e.g. attention management), which can subsequently impair the performance of the individual (i.e. the actual behaviour produced).
- Path 2: Physical effects. Exposure to a GDF can have specific and direct effects on the behaviour produced. For example, Ship motion can result in Motion-Induced Interruptions (MII). MII does not affect the underlying human capabilities of balance or fine motor control, but it exceeds the ability of the human to compensate and produce the intended behaviour. Similarly, WBV can directly impact the actual behaviour produced.

These two paths show how GDF exposure affects human behaviour, which in turn influences the performance of safety critical tasks. It is the outcomes of an individual's actions and behaviour that determine the success or failure of a safety critical task. Insufficient performance of the safety critical tasks associated with maintaining safe vessel navigation and avoiding collision or grounding create an antecedent for the unwanted outcome. However, insufficient task performance alone does not determine whether or not a collision or grounding occurs; the vessel must also be exposed to the collision or grounding hazard, as follows:
- For a collision to occur, another vessel must be on a collision course.
- For a grounding to occur, the ship must be in shallow water.

This causal mechanism makes the following assumptions:
- While we recognise that individuals have differing cognitive and physical abilities, it is assumed that all individuals have the same basic set of capabilities (i.e. all individuals can manage their attention, irrespective of the extent of this capability).
- Human behaviour is influenced by diffuse and acute effects of GDF exposure as represented by the paths in Figure 1.
- The crew perform safety critical tasks related to collision and grounding.
- Tasks are appropriate, processes and procedures are optimised, and are undertaken by a competent operator.
- Safety critical tasks must be performed correctly to maintain safe vessel operation.
- Safety critical tasks manage the exposure of the vessel to the collision and grounding hazard.
- While it is recognised that interaction effects between GDFs within each pathway are likely to exist, these are excluded from the model, as the literature does not provide any information describing this interaction [20]. While no representation is included of any potential interaction effects from the exposure to multiple GDFs acting through a single pathway (i.e. stressor or physical effects), a cumulative effect is represented for the presence of GDFs acting both through stressor and physical effects of GDFs simultaneously.

Figure 1: Causal chain describing the relationship between crew GDF exposure and unwanted outcomes

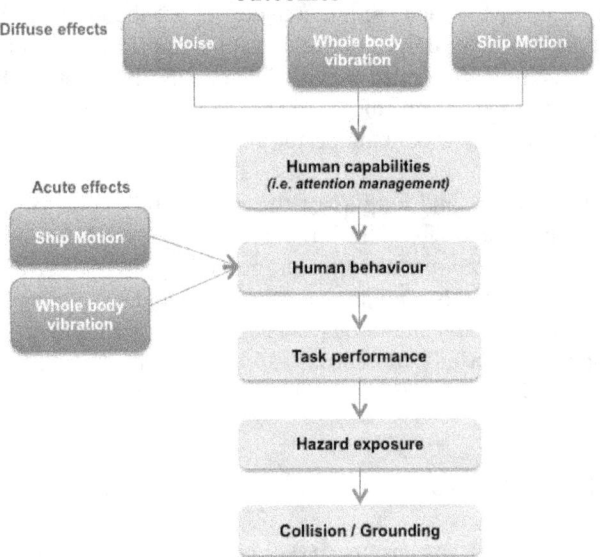

4. HUMAN PERFORMANCE MODEL

It was found that the data on the specific GDF effects of ship motion (with the exception of MII), noise, WBV on human performance are sparse and in many, but not all, cases generated under very specific, often non-marine, conditions. Data shows that there is certainly evidence for GDFs having some effect on human performance. However, the direct effects of GDF exposure on human performance tend to be weak, whereas secondary effects acting through another mechanism (e.g. fatigue, Motion Induced Sickness (MIS)) tend to be stronger and more pervasive. Specifically there are some data that describe the:
- Impact of GDFs on specific human capabilities, [21].
- Impact of GDFs on specific human behaviours, [22].
- Impact of errors on task performance, [23].

However, there is very little data about the link between the following components:
- Degraded human capabilities and collision or grounding related performance
- Degraded task performance and exposure to the collision / grounding hazard

Figure 2 demonstrates the links in the causal chain for which some quantitative data are available (in green) and the links for which there is no data (in red). In addition to this gap, a given level of exposure to GDFs of certain intensity or duration may not affect all individuals equally. For example, while a given frequency and amplitude of ship motion may be generally MIS-inducing, individual experiences may range from significant nausea to no negative affects whatsoever, depending on their underlying susceptibility to MIS and the degree to which they have acclimatized. Moreover, with the possible exception of secondary effects on human performance caused by fatigue, attributable to sleep disruption, a holistic view could not readily be derived directly from the individual findings. As such, our approach was guided by the relevant theoretical models available in the scientific literature.

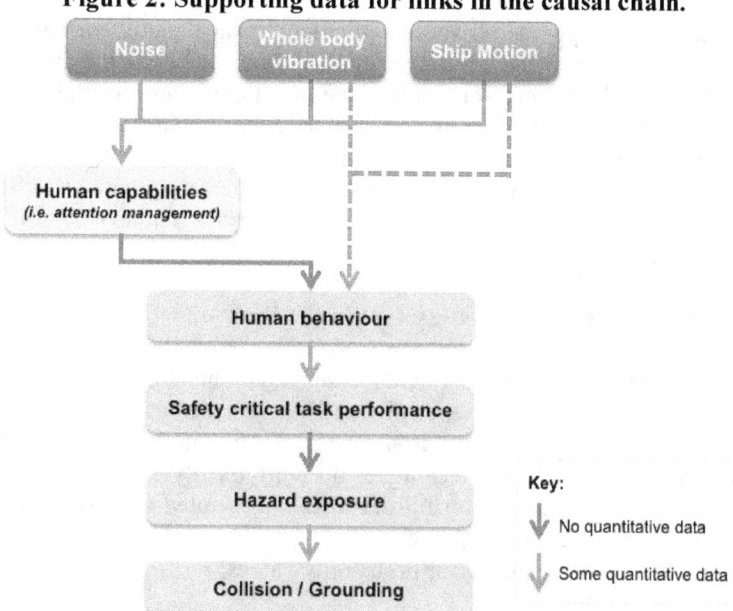

Figure 2: Supporting data for links in the causal chain.

4.1. Theoretical models of human performance

The approach taken here to describe a mechanism that accounts for the impact of stressors on human performance, has been based on the principles of attention management, [24]. It combines the principles from three theoretical models:
1. Dynamic Adaptability Model (DAM), [12].
2. Cognitive Control Model (CMM), [13].
3. Malleable Attentional Resources Theory (MART), [14].

Under the DAM paradigm, GDFs are seen as types of physical stressor that affect human capabilities associated with maintaining a desired level of task performance either directly or indirectly (e.g. via fatigue). When exposed to GDFs, CCM describes humans compensating through the effortful direction of more cognitive resources at the task, typically at the cost of performance in other areas. Despite the sophisticated, and potentially subconscious, strategies humans have at their disposal, there is a limit to how much an individual can compensate without experiencing degradation in primary or secondary task performance. In addition, the extent to which human can compensate for task demands is not fixed. MART describes this compensatory capability changing as a function of task demands and associated arousal an individual experiences – attentional resources available vary as a function of load. When humans are in a state of under-load (i.e. bored) their pool of attentional resources is relatively small and will increase proportionately with the demands placed on them. However, there is a limit to how much the pool of attentional resources can grow. When task demands exceed the pool of attentional resources available (either transiently or when the upper attentional resource limit is exceeded), performance can breakdown and errors may be made.

Generally, task performance is only expected to degrade and become insufficient when compensatory mechanisms have failed. However, the literature does not allow prediction of how and when (chronologically) an operator would fail, under what conditions of GDF exposure, and what the specific effect on behaviour (i.e. type of error) would be.

In the risk models presented here, the main task around which the models revolve is to perform the accident evasive action. This task is complex and distributed in time, but it can be decomposed into three major phases:

1. detection (D),
2. assessment (A),
3. action (Act).

These three phases (DAAct) reflect the basic cognitive functions of observation, interpretation and planning, and execution, see for example [25], [26].

In terms of risk modelling, an approach based on attention management theory allows representation of the effect of GDF exposure as a stressor that sits either above or below the threshold of attentional capacity for any given task. If the stressor exceeds the attentional capacity then a negative effect is expected, whereas no negative effect on human performance would result if the stressor can be managed within the available attentional capacity.

Representing ship motion, noise and WBV GDFs as stressors interacting with an individual's attention management capabilities provides an evidence-based mechanism for human performance that has been used to develop the risk models presented here.

4.2. Integration of Human Reliability Assessment in the Risk model

Due to the limitations in data on the effects of GDF exposure on human performance, one cannot find precise values in the scientific literature. Hence probabilistic representation of the human performance component in the risk model was potentially problematic. A solution was found in Human Reliability Analysis (HRA) techniques. While HRA techniques do not typically cover the specific GDFs or the maritime environment, the human error probabilities (HEPs) generated by HRA allow sensible bounds to be determined. While imperfect, this approach at least allows calibration of probabilities in the risk model against established generic human error probability values.

The HRA method Nuclear Action Reliability Assessment (NARA) was selected to provide the HEPs associated within collision and grounding model, [27]. NARA is a third generation HRA method that, while nuclear industry focussed, uses a broad range of industries in the CORE-DATA dataset underlying the HEP calculation, arguably making it more suitable for navigation tasks performed on the bridge. NARA was adopted to enhance the accuracy of the risk model through the generation of validated (albeit non-marine specific) HEPs associated with task characteristics that are compatible with tasks performed by the Officer of the Watch (OOW) and helmsman. NARA also provided baseline error rates for a given Generic Task Type (GTT) unaffected by GDFs. This allowed probabilistic estimation of the effect of GDF exposure on HEPs.

However, NARA was not used to represent the direct physical effect exposure some GDFs may have as physical aspects of task performance. This is out the scope the intended application of the NARA method. The probability of insufficient human performance resulting from physical effects of GDF exposure was estimated based on judgement alone.

NARA categorises the factors that negatively influence human performance as one of eighteen Error Producing Conditions (EPCs). The EPC that best represented the causal mechanism from GDF exposure to human performance was EPC No. 15: *Poor Environment*. This EPC represents the stressor effect of GDF exposure on attention management capability. The potential strength of effect of this EPC was set using the Assessed Proportion of Affect (APOA) variable. The APOA level was set based on the application of the NARA methodology to subjectively determine an appropriate value, nominally between 0 (no effect) and 1 (maximum effect). However, based on the guidance available for NARA, it was decided to cap the maximum APOA associated with the EPC to 0.1.

Table 1: Maximum APOA Value Caps for the EPC associated with Attention Management

GDF Effect Path	NARA EPC Reference	NARA EPC Description	Maximum APOA Cap	Justification
Attention Management	EPC No. 15	Poor environment	0.1 (low)	The task environment is generally benign even in the presence of GDFs. An APOA of 1.0 would represent an extreme physical environment in which crew are required to wear PPE.

NARA was also used to determine a baseline human error rate (Nominal HEP) to set the task performance HEP unaffected by GDF exposure. One of the limitations of the application of NARA in this context was highlighted by the selection of the task from the predefined list within NARA from which the HEP was established to represent task performance associated with the hazard exposure. To limit the complexity of the model, a single GTT was sought to represent all relevant navigational tasks performed by the OOW that are important in managing collision or grounding risk. The GTT that is most analogous is:

Task C1 – Simple response to alarms/indications providing clear indication of situation (Simple diagnosis required) Response might be direct execution of simple actions or initiating other actions separately assessed. (Nominal HEP = 0.0005)

A second GTT was identified to account for possibility that a helmsman may also be present. In this case the helmsman is steering the ship based on verbal instructions communicated by the OOW. The GTT that is most analogous is:

Task D1 - Verbal communication of safety critical data

While having a helmsman present may introduce the possibility of a miscommunication error with the OOW, NARA also recognises a mitigating effect of a team. The NARA Human Performance Limiting Value for 'Actions taken by a team of operators' was used to cap the potential error rate at 1E-4 for the condition where a helmsman is present. The same value is taken for the probability of potential error of not performing evasive action by another ship involved in the encounter.

The NARA calculation allows inclusion of multiple EPCs and an Extended Time Factor (ETF). In this risk model for collision and grounding, GDFs are represented using only one EPC and there is little justification to include the ETF. Thus, the HEP was calculated based on the following formula:

$$HEP = GTT \times [(EPC-1) \times (APOA + 1)] \qquad (2)$$

5. RISK MODEL DEVELOPMENT

In this section we briefly present the process of risk model development, which is seen as translation of background knowledge into a coherent structure, which allows inference and decision-making under uncertainty. Also the results of the quantitative uncertainty assessment are presented in this section. Two risk models are developed, as depicted in figures 3 and 4, for collision and grounding respectively.

5.1. Risk model quantification

The model starts with the evaluation whether or not the levels of GDFs exceed the threshold. The GDFs may have either an acute or diffuse effect. In case of an acute effect, the threshold refers to the GDFs level at which an individual may be unable to physically compensate for GDF exposure and perform actions as intended. In case of diffuse effect, the threshold refers to the amount of motion, vibration or noise an individual can endure before it acts as a stressor (with a corresponding stress response). The exact value of the threshold will vary between individuals and it is dependent upon previous experience, exposure duration and sensitivity. If the thresholds are not exceeded, then the attention of a navigator is not affected, otherwise the attention management capability is degraded. Attention management is the supervisory human capability that directs, allocates and regulates the attentional resources required to perform various tasks. This high-level supervisory capability manages lower-level tasks such as perception, cognition, decision-making, memory, fine motor control and locomotion.

Representing GDF exposure effects on safety behaviour via the attention management path provides a structure compatible with the introduction of an Error Producing Conditions (EPC) using NARA.
In the presented model a variable called *C1 - Detection, Assessment and execution of simple actions* (NARA Generic Task Type – GTT No. C1) determines whether or not the performance of detection, assessment and actions required to avoid collision or grounding is sufficient or insufficient. This is an all-encompassing definition including whatever tasks are required to maintain situational awareness and to respond appropriately to avoid collision or grounding.
The probability of insufficient 'Detection, assessment and execution of simple actions' is calculated based on the integration of GDF Physical Effect and Attention Management Capability nodes. The calculation of Attention Management Capability of insufficient performance is performed using the NARA calculation.

The GDF Physical Effect on the probability of insufficient safety behaviour is assumed to be very weak (p=0.001) when Attention Management Capability (AMC) is *normal*. However, this effect is represented as being more significant in combination with *a degraded* AMC (p=0.0011), thus representing an additional drain on cognitive resources compensating for physical task disruption. These values were estimated using judgement.
To set the probability of *insufficient* performance unaffected by GDF exposure to reflect a baseline error rate the NARA GTT value for Task C1 of 0.0005 is used. The HEP of 0.0006 was calculated using NARA given exposure to EPC No. 15 representing the effect of GDF exposure via the AMC node.

If evasive manoeuvres are performed with the presence of helmsman, which requires appropriate communication of information between the OOW and helmsman, the node *D1 - verbal communication of safety critical data* is evaluated. This node (NARA GTT No. D1) determines whether or not the verbal communication of safety critical data to the helmsman required to avoid collisions is sufficient or insufficient. This node is only active if the *Helmsman present* node is set to *Yes* to reflect the fact that the OOW is not controlling the vessel directly. The NARA GTT value for Task D1 of 0.006 is used to set the probability of *insufficient* performance unaffected by GDF exposure to reflect a baseline error rate. The HEP of 0.0072 was calculated using NARA given exposure to EPC No. 15 representing the effect of GDF exposure via the *Attention Management Capability* node. It is assumed that *Verbal communication of safety critical data* is unaffected by the *GDF Physical Effect*.

The probability that OOW successfully executes actions required to manoeuvre the vessel to take evasive action is assessed in the node *Evasive Action*. It is assumed that if *Detection, assessment and execution of simple actions* and *Verbal communication of safety critical data*, where applicable, are performed sufficiently then *Evasive Action* will be executed. If *Helmsman present* is *Yes* then the HEP for *Evasive Action* being in a state *Not executed* is capped at 1E-4 in line with the NARA Human Performance Limiting Value for 'Actions taken by a team of operators'.

Technical failure node quantifies the probability of the relevant systems not functioning as a result of lack of maintenance or poor maintenance caused by the GDFs. This node was included in recognition of the importance of maintenance in sustaining the functionality of vessel equipment such that it performs as it is designed to. Errors during maintenance on systems that provide the manoeuvring capability of the vessel can limit the vessel's response to control inputs associated with evasive action, hence affecting the probability of an unwanted outcome. *Technical failure* node determines whether or not maintenance actions performed on equipment that provides the vessel's manoeuvring capability has been completed successfully or not. The probability of insufficient *Maintenance Task Performance* is calculated based on the integration of *GDF Stressor Technical* and *Attention*

Management Capability-2 nodes. To set the probability of *insufficient* performance unaffected by GDF exposure to reflect a baseline error rate the NARA GTT value for Task C1 of 0.0005 is used. The HEP of 0.0006 was calculated using NARA given exposure to EPC No. 15 representing the effect of GDF exposure via the *Attention Management Capability-2* node.

The GDF Physical Effect on the probability of insufficient safety behaviour is not anticipated. This comes from an assumption about lack of preventive maintenance carried out if the levels of GDFs are above thresholds. This is in line with most of the operations guidelines of ships, where crews abstain from certain tasks if the weather conditions do not allow for safe performance of these tasks.

Evasive action of another ship is a node which accounts for the behaviour of OOW on a ship that is encountered by the own vessel. The probability for this node being in a state *Not executed* is capped at 1E-4 in line with the NARA Human Performance Limiting Value for 'Actions taken by a team of operators'. The last node of the model is *Collision,* and it exists in the state *Yes*, if and only if:
- *Evasive Action of Another Ship=No, and Evasive Action=No, Technical Failure=Yes or No,*
- *Evasive Action of Another Ship=No, Evasive Action=Yes and Technical Failure=Yes.*

Otherwise, the node *Collision* is set to its state *No*.

5.2. Sensitivity, uncertainty and importance assessment of the risk model

The sensitivity analysis is performed to identify the essential variables that have the highest impact on the outcome of the model. For this purpose, every conditional and prior probability in the BBNs is systematically varied in turn while keeping the others unchanged. This allows the effects on the output probabilities computed from the network to be examined. To determine the sensitivity of an output variable to a given parameter of the model a sensitivity function is estimated for each single variable. This function describes outcome of the model as a function of the parameter z, which takes the following form:

$$z = p(Y = y_i|\pi) \qquad (3)$$

Where y_i is one state of a network variable Y, and π is a combination of states for *Y*'s parent nodes. The sensitivity values were obtained using dedicated tools, which are implemented in the software package GeNIe, used for the development of the models presented here. Based on the findings from the sensitivity analysis, the following can be concluded:
- Both models are highly sensitive to the following parameters: *Maintenance Task Performance, C1 - Detection, Assessment and execution of simple actions and D1 - verbal communication of safety critical data.*
- The model assessing the probability of collisions is also sensitive to *Evasive action of another ship and Helmsman present.* However, the effect that these parameters have on the output is significantly lower than the effects of *C1* and *D1*, as specified above.
- The model assessing the probability of grounding is sensitive to *Helmsman present.*
- The remaining nodes have very low sensitivity values, meaning that their effects on the models outputs are rather minor.

Secondly, the evidential uncertainty assessment has been carried out on the mot sensitive model parameters. To rank the uncertainty, the following qualitative scoring system is applied, see [28]:
Significant uncertainty
One or more of the following conditions are met:
- The phenomena involved are not well understood; models are non-existent or known/believed to give poor predictions.
- The assumptions made represent strong simplifications.
- Data are not available, or are unreliable.
- There is lack of agreement/consensus among experts.

Moderate uncertainty
Conditions between those characterising significant and minor uncertainty, e.g.:
- The phenomena involved are well understood, but the models used are considered simple/crude.
- Some reliable data are available.

Minor uncertainty
All of the following conditions are met:
- The phenomena involved are well understood; the models used are known to give predictions with the required accuracy.
- The assumptions made are seen as very reasonable.
- Much reliable data are available.
- There is broad agreement among experts.

Table 2: The qualitative assessment of evidential uncertainty for models assessing the probability of an accident

Model parameter	*Justification for the evidential uncertainty score*	*Evidential uncertainty score*
Maintenance Task Performance C1 - Detection, Assessment and execution of simple actions	This node represents the performance of navigation tasks critical in collision or grounding avoidance and provides a structure compatible with the introduction of a NARA GTT, potentially affected by EPC No. 15 via 'Attention Management Capability' and GDF Physical effects	Moderate
D1 - verbal communication of safety critical data	This node represents the communication of vessel manoeuvring instructions critical in collision or grounding avoidance with a helmsman present and introduction of a NARA GTT, potentially affected by EPC No. 15 via 'Attention Management Capability'.	Moderate
Evasive action of another ship	The node represents the performance of navigation tasks critical in accident avoidance on board other ship. It is quantified based on NARA.	Moderate
Helmsman present	At the moment this node is quantified fully based on judgement. However, more detailed assessment is possible, by performing survey among shipping companies.	Moderate

Finally, by combining the results of sensitive and uncertainty assessment, the parameter importance ranking is carried out. It allows for screening the model for both uncertain and sensitive parameters. For the models presented here, three parameters have high importance score, namely *Maintenance Task Performance, C1-Detection, Assessment and Execution of Simple Action, D1 – verbal communication on safety critical data.* This means, that states of these parameters should be carefully selected for any analysis. Moreover, when making the comparison between two designs of a ship, these two parameters shall not be changed, otherwise, their effect may outshine the effect of GDFs, which is much weaker.

Table 3: The qualitative assessment of model parameters importance for models assessing the probability of an accident.

Model parameter	*Evidential uncertainty score*	*Sensitivity score*	*Importance score*
Maintenance Task Performance	Moderate	High	High
C1 - Detection, Assessment and execution of simple actions	Moderate	High	High
D1 - verbal communication of safety critical data	Moderate	High	High
Evasive action of another ship	Moderate	Moderate	Moderate
Helmsman present	Moderate	Moderate	Moderate

6. CONCLUSION

The models presented in this paper offer a novel, evidence-based approach to modelling risk of ship-ship collision and grounding. They provide a flexible framework that could readily be extended to encompass the actions of third parties and mechanical failures in the future. The flexibility to extend the model's application is provided by the causal mechanism represented within the model that describes occurrence of an accident as the result of insufficient performance of an individual when exposed to hazardous situation.

The models focus on modelling improper performance in critical situations. This is also compatible with the general conceptualisation of human error within the Human Factors (HF) domain and its relationship to task performance.

As expected, the paucity of data on GDF effects presented a particular challenge. However, attention management theory successfully provided a means to represent the mechanism by which ship motion, noise and WBV affect cognitive performance. Due to the supervisory role attention management has in human cognition, this approach may also be readily generalizable. However, caution is required to ensure that this framework can be applied to model other performance shaping factors outside of the GDFs. It should be noted that DAM, CCM and MART are theories that best fit the current scientific data on GDFs, but more research is required to understand whether the integration of these three models into a combined model of 'attention management theory' is a robust and valid representation of human performance within different complex environments.

The integration of HRA (specifically NARA) to support the calculation of HEPs within the risk models has clear positive and negatives. On the one hand, it provided a facility to generate 'reasonable' HEPs using a well-known method, which would not have been possible otherwise. On the other hand, the application of NARA to physical tasks associated with physical effect of the GDFs on *Detection, Assessment and Execution of Simple Actions*, is stretching its application to, and perhaps beyond, its limit.

Despite the limitations and the paucity in data supporting certain hypotheses, the application of BBNs as a modelling tools, allows for clear representation of the modelled problem and comprehensive distribution of all the recognised uncertainties. By adopting BBNs and performing the importance analysis, we learned that the crucial elements of the models are the nodes, where the human error probabilities are quantified. Whereas the detailed quantification of the levels of GDFs associated with a given ship design or their effect on the attention management capability is less important.

The inherent feature of BBNs, two-ways reasoning, allows not only forward propagation of the evidences resulting in an outcome, but also the back propagation of the evidences, and estimation of the input variables, given a selected state of the output is possible.

Finally, comparative assessment of vessel designs based on manipulation of the GDF input nodes is possible in principle. The models are responsive to changes in the GDF nodes as expected. The models may be used by naval architects, vessel designers, and vessel system designers as intended, provided access to HF expertise is available to assist with application and interpretation. It is important to recognise the relevance of human factors input during its eventual application. HF provides the understanding of the complexities of human behaviour in operational settings, its interdependencies and interactions.

Figure 3: Risk model for ship-ship collision.

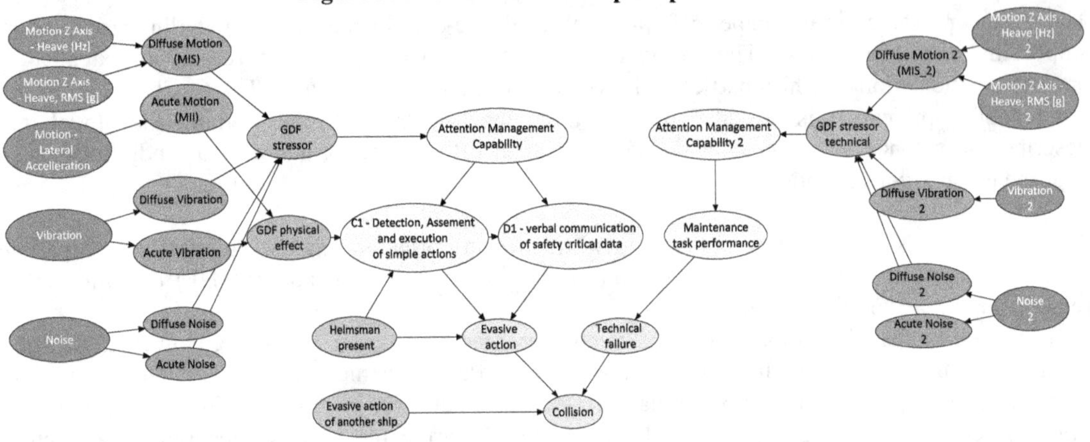

Figure 4: Risk model for ship-ship grounding.

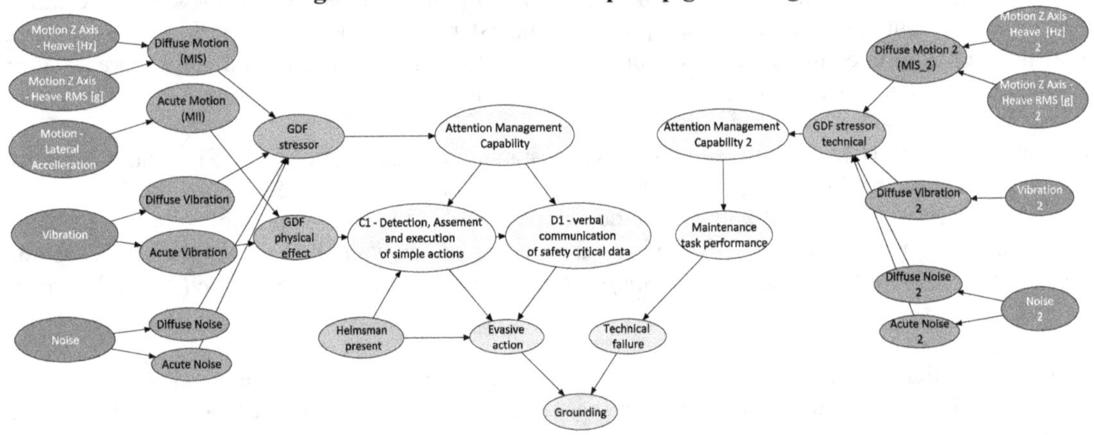

Acknowledgements

The authors appreciate the financial contributions of the EU, as this research was co-funded by the FAROS project (2012–2015). The probabilistic models introduced in this paper were created using the GeNie modeling environment developed at the Decision Systems Laboratory, University of Pittsburgh, available from http:// genie.sis.pitt.edu/.

References

[1] A. Papanikolaou, Ed., *Risk-Based Ship Design: Methods, Tools and Applications*. Springer, 2009, p. 376.
[2] J. Montewka, S. Ehlers, F. Goerlandt, T. Hinz, K. Tabri, and P. Kujala, "A framework for risk assessment for maritime transportation systems—A case study for open sea collisions involving RoPax vessels," *Reliab. Eng. Syst. Saf.*, vol. 124, pp. 142–157, 2014.
[3] P. T. Pedersen, "Review and application of ship collision and grounding analysis procedures," *Mar. Struct.*, vol. 23, no. 3, pp. 241–262, Jul. 2010.
[4] A. Mazaheri, J. Montewka, and P. Kujala, "Modeling the risk of ship grounding—a literature review from a risk management perspective," *WMU J. Marit. Aff.*, vol. 12, Dec. 2013.
[5] S. Li, Q. Meng, and X. Qu, "An Overview of Maritime Waterway Quantitative Risk Assessment Models," *Risk Anal.*, vol. 32, no. 3, pp. 496–512, 2012.
[6] DNV, "Formal Safety Assessment - Large Passenger Ships, ANNEX II: Risk Assesment - Large Passenger Ships - Navigation," 2003.
[7] M. Hänninen and P. Kujala, "Influences of variables on ship collision probability in a Bayesian belief network model," *Reliab. Eng. Syst. Saf.*, vol. 102, no. 0, pp. 27–40, Jun. 2012.
[8] M. C. Leva, P. Friis-Hansen, E. S. Ravn, and A. Lepsoe, "SAFEDOR: A practical approach to model the action of an officer of the watch in collision scenarios," in *European safety and reliability conference; Safety and reliability for managing risk; ESREL 2006*, 2006, pp. 2795–2804.

[9] M. R. Martins and M. C. Maturana, "Application of Bayesian Belief networks to the human reliability analysis of an oil tanker operation focusing on collision accidents," *Reliab. Eng. Syst. Saf.*, vol. 110, pp. 89–109, Feb. 2013.

[10] M. C. Kim, P. H. Seong, and E. Hollnagel, "A probabilistic approach for determining the control mode in CREAM," *Reliab. Eng. Syst. Saf.*, vol. 91, no. 2, pp. 191–199, Feb. 2006.

[11] S. Kivimaa, A. Rantanen, T. Nyman, D. Owen, T. Garner, and B. Davies, "Ship motions, vibration and noise influence on crew performance and well-being studies in FAROS project," in *Transport Research Arena Conference TRA*, 2014.

[12] P. A. Hancock, "A dynamic model of stress and sustained attention," *Hum. Factors*, vol. 31, no. 5, pp. 519–537, 1989.

[13] G. Robert and J. Hockey, "Compensatory control in the regulation of human performance under stress and high workload: A cognitive-energetical framework," *Biol. Psychol.*, vol. 45, no. 1, pp. 73–93, 1997.

[14] M. S. Young and N. A. Stanton, "Malleable attentional resources theory: a new explanation for the effects of mental underload on performance.," *Hum. Factors*, vol. 44, no. 3, pp. 365–75, Jan. 2002.

[15] A. Darwiche, *Modeling and Reasoning with Bayesian Networks*, 1st ed. Cambridge University Press, 2009.

[16] J. Montewka, F. Goerlandt, and P. Kujala, "On a systematic perspective on risk for Formal Safety Assessment (FSA)," *Reliab. Eng. Syst. Saf.*, vol. 127, no. July, pp. 77–85, Apr. 2014.

[17] T. Aven and O. Renn, "On risk defined as an event where the outcome is uncertain," *J. Risk Res.*, vol. 12, no. 1, pp. 1–11, Jan. 2009.

[18] F. Goerlandt and J. Montewka, "A probabilistic model for accidental cargo oil outflow from product tankers in a ship-ship collision," *Mar. Pollut. Bull.*, vol. 79, no. 1–2, pp. 130–144, 2014.

[19] H. S. Blackman, D. I. Gertman, and R. L. Boring, "Human error quantification using performance shaping factors in the SPAR-H method," in *Proceedings of the Human Factors and Ergonomics Society Annual Meeting*, 2008, vol. 52, no. 21, pp. 1733–1737.

[20] J. Montewka, "Summarizing literature review (Human Factors in Risk-Based Ship Design Methodology, Project no 314817)," 2014.

[21] D. M. Jones, W. J. Macken, and N. A. Mosdell, "The role of habituation in the disruption of recall performance by irrelevant sound," *Br. J. Psychol.*, vol. 88, no. 4, pp. 549–564, Nov. 1997.

[22] P. Crossland and K. J. N. . Rich, "A method for deriving MII criteria," in *Human factors in ship design and operation; RINA international conference*, 2000.

[23] G. E. Conway, J. L. Szalma, and P. A. Hancock, "A quantitative meta-analytic examination of whole-body vibration effects on human performance.," *Ergonomics*, vol. 50, no. 2, pp. 228–45, Feb. 2007.

[24] D. Owen, S. Pozzi, and J. Montewka, "Validation results and amendments made." p. 18, 2014.

[25] E. Hollnagel, *Cognitive Reliability and Error Analysis Method (CREAM)*. Elsevier Science, 1998, p. 302.

[26] X. He, Y. Wang, Z. Shen, and X. Huang, "A simplified CREAM prospective quantification process and its application," *Reliab. Eng. Syst. Saf.*, vol. 93, no. 2, pp. 298–306, Feb. 2008.

[27] C. Spitzer, U. Schmocker, and V. N. Dang, Eds., *Probabilistic Safety Assessment and Management*. London: Springer London, 2004.

[28] R. Flage and T. Aven, "Expressing and communicating uncertainty in relation to quantitative risk analysis," *Reliab. Risk Anal. Theory Appl.*, vol. 2, no. 13, pp. 9–18, 2009.

Further Development of the GRS Common Cause Failure Quantification Method

Jan Stiller*, Albert Kreuser, Claus Verstegen
Gesellschaft für Anlagen- und Reaktorsicherheit mbH (GRS), Cologne, Germany

Abstract: For the quantification of common cause failures (CCF), GRS has developed the coupling model. This model has two important features: Firstly, estimation uncertainties which arise from different sources, e.g. statistical uncertainties, uncertainties of expert judgments or uncertainties due to inhomogeneities of statistical populations, are taken into account in a consistent way. Secondly, it automatically allows for the extrapolation of CCF events two groups of different sizes ("mapping"). This feature has been very important since for most component types groups of several different sizes can be found in German NPP. The model assumptions necessary to allow for this feature, however, also lead to undesirable convergence properties when a large amount of operating experience is available. Therefore GRS has started a project to research possible improvements of CCF modeling with respect to this aspect including the development of models that avoid making use of the restrictive modeling assumptions, which allow a comprehensive treatment of uncertainties, and which are applicable to data available in the German CCF data pool which does not contain information on single failures. Two different models have been developed, including a conservative mapping procedure. Comparisons of the results of the new estimation procedures and the coupling model show that the results are compatible.

Keywords: PRA, CCF, Quantification, Uncertainty Analysis

1. INTRODUCTION

Common cause failures (CCF) contribute to a large extent to the unavailability of redundant systems, especially for highly redundant systems. Probabilistic safety assessments (PSA) have shown that these unavailabilities may make a significant or even dominant contribution to the estimate of the core damage frequency of a nuclear power plant. Therefore an appropriate estimation of CCF probabilities including an adequate uncertainty analysis is of great importance. Since in most cases CCFs are very rare events statistical uncertainties have to be considered. Uncertainties arising from other sources like uncertainties of expert judgments on the impairments of components in CCF events or the possible inhomogeneity of populations have to be considered as well. To accomplish this, GRS has developed the coupling model and associated estimation procedure [1]. In the coupling model, uncertainties are treated in a consistent way by applying Bayesian statistical methods. The coupling model and the procedures for event assessment and parameter estimation have been continuously advanced in recent years [2,3] including a new procedure to consistently represent the remaining uncertainties related e.g. to a possible inhomogeneity of populations. One main feature of the coupling model is that it automatically allows for the extrapolation of CCF events to groups of different sizes. Model assumptions associated with this feature also cause undesirable convergence properties which may become more important since an increasing part of German operating experience has been evaluated recently with regard to CCF [4,5] and hence the number of CCF events available has increased. Therefore GRS has started a project to research possible improvements of CCF modeling with respect to this aspect. In the present paper, the first results of these efforts are discussed.

The paper is organized as follows: In chapter 2 the present coupling model and estimation procedure are described, including a recent development for the modeling of sources of additional uncertainties not included before (section 2.4). The convergence properties are also discussed (section 2.5). In chapter 3 two alternative models are developed which avoid the restrictive modelling assumptions leading to the undesirable convergence properties of the coupling model. In chapter 4 the estimation results of the different models are compared and discussed. In chapter 5 the use of operating experience from component groups of different size (so-called mapping) is briefly discussed. A simple mapping procedure is introduced and compared to estimation results of the coupling model. In chapter 6 conclusions are made.

For the sake of simplicity and compactness, no complete mathematical treatments are presented in this paper. These will be given in [6].

2. PRESENT COUPLING MODEL AND ESTIMATION PROCEDURE

2.1. Basic Equations of the Coupling Model

Like in the binomial failure rate (BFR) model [7], it is assumed in the coupling model that if a CCF phenomenon occurs in a component group, the individual components fail independently of each other with a probability η or remain unaffected with a probability $1-\eta$. The parameter η is denoted "coupling parameter". Unlike in the BFR model, it is not assumed that the coupling parameter is identical for all CCF phenomena. Therefore CCF failure probabilities are estimated separately for all observed CCF phenomena. The total common cause probability of a (k out of r) failure is calculated as the sum over all phenomena.

The probability $q_{k\backslash r;j}$ of a common cause (k out of r)-failure due to the CCF phenomenon j is given by

$$q_{k\backslash r;j} = \varphi_j \binom{r}{k} \eta_j^k (1-\eta_j)^{r-k} \qquad (1)$$

with r denoting the size of the target component group, η_j denoting the coupling parameter of CCF phenomenon η and φ_j denoting the probability that a CCF due to phenomenon j occurs in the target component group.

Equation (1) has the form of a product of the probability that a CCF due to phenomenon j occurs in the target component group and the conditional probability that k out of r components fail, given that a CCF due to phenomenon j has occurred. Under the assumptions mentioned above, the number of failed components follows a binomial distribution with parameter η_j, when CCF phenomenon j occurred in the target component group.

The probability φ_j that a CCF due to phenomenon j occurs in the target component group is calculated as

$$\varphi_j = f_j\, t_j\, \lambda_j \qquad (2)$$

with f_j denoting the applicability factor, t_j denoting the expected failure detection time for the CCF phenomenon j and λ_j denoting the rate of CCF phenomenon j in the observed population. The applicability factor is defined as the relative rate of an occurrence of the CCF phenomenon in the target component group with respect to the component group in which the CCF event occurred. The estimation of f_j is based on possible technical and operational differences between the observed and the target component group. If no substantial technical and operational differences exist, the observed CCF event j is fully applicable to the target group and therefore $f_j = 1$ holds, which applies in most cases.

Equation (2) is only valid if λ_j is small, i.e. $f_j\, t_j\, \lambda_j \ll 1$ holds. This generally is true in normal applications.

As noted before, the total common cause probability of a (k out of r) failure $q_{k\backslash r}$ is calculated as the sum of the probabilities $q_{k\backslash r;j}$ over all observed CCF phenomena:

$$q_{k\backslash r} = \sum_{j=1}^{N} q_{k\backslash r;j} \quad (3)$$

with N denoting the number of CCF events that occurred in the observed population of component groups.

2.2. Estimation of Model Parameters

To estimate the CCF failure probabilities with the coupling model, first the coupling parameters η_j have to be estimated (see equation (1)). Since the number of components m in the component group where a CCF took place is usually quite small, there is a large uncertainty associated with this estimation. This uncertainty is treated by using a Bayesian approach to estimate the coupling parameters. Using a non-informative prior [8] $\pi(\eta_j) \propto 1/\sqrt{\eta_j(1-\eta_j)}$ for the coupling parameter the a posteriori probability distribution is given by the Beta distribution

$$p(\eta_j) = \frac{\Gamma(r+1)}{\Gamma(k+1/2)\,\Gamma(r-k+1/2)} \eta_j^{k-1/2} (1-\eta_j)^{r-k-1/2} \quad (4)$$

if k out of r components failed during the CCF event j.

Operating experience shows that in many CCF events components are found which are more or less severely degraded but have not failed yet. In the coupling model, however, it is assumed that if a CCF phenomenon occurs in a component group, the components either fail or remain unaffected. Hence, degraded components are not directly modeled. This is done for two reasons: Firstly, in a PSA component states are also only modeled as failed or unaffected. Secondly, a CCF model would require additional parameters to model degraded states. Since the number of CCF events found in operating experience is very limited, estimation of such parameters would be difficult and would lead to large estimation uncertainties. In the coupling model this is resolved by interpreting degradations as probabilities of failure. Therefore, the number of failed components on which the estimation of the coupling parameter η_j (equation (1)) is based has to be treated as an uncertain quantity. This uncertainty has been termed "interpretation uncertainty" in [2]. It is treated in the following way: The probability $w_{k\backslash r}$ that k out of r components would fail during an additional demand is estimated for all $k = 0 \dots m$ using engineering judgment. These probabilities are represented by a so-called interpretation vector $W = \{w_{0\backslash r}, w_{1\backslash r}, \dots, w_{r\backslash r}\}$ where the condition $\sum_{k=0}^{r} w_{k\backslash r} = 1$ must hold.

Generally, it is not feasible for technical experts to directly assess such kinds of subjective probabilities. Therefore, a method was developed which automatically generates an interpretation vector W. This method consists of assessing degradation levels for each component of the component group where the CCF event took place. The degradation level is interpreted as the probability that the component would fail during the next demand due to the CCF phenomenon observed (e.g. 1 for failed components and 0 for completely unaffected components). Given the component impairments the interpretation vector W is determined using probability calculus [2]. This approach is well proven and is also used for other CCF models [9].

After determination of the interpretation vector W, the coupling parameter η_j for each CCF event j can be estimated. For each single interpretation alternative, the a posteriori distribution is determined according to equation (4). This results in an a posteriori distribution of the coupling parameter η_j which has the form of a weighted mixture of Beta distributions:

$$p(\eta_j) = \sum_{i=0}^{r} w_{i \backslash r} \frac{\Gamma(r+1)}{\Gamma(i+1/2)\,\Gamma(r-i+1/2)} \, \eta_j^{i-1/2} (1-\eta_j)^{r-i-1/2} \qquad (5)$$

This distribution expresses the uncertainty about the coupling parameter η_j taking into account both the statistical and the interpretation uncertainty.

To include the statistical uncertainty of the rates of the CCF events, the Bayes a posteriori distribution $p(\lambda_j)$ is calculated using the non-informative prior [8] $\pi(\lambda_j) \propto 1/\sqrt{\lambda_j}$. This results in a Gamma distribution as a posteriori distribution of λ_j:

$$p(\lambda_j) = \frac{T^{3/2}}{\Gamma(3/2)} \, \lambda_j^{1/2} \, e^{-\lambda_j T} = \frac{2}{\sqrt{\pi}} \sqrt{T^3 \lambda_j} \, e^{-\lambda_j T} \qquad (6)$$

Using equation (2), the probability distribution of $q_{k \backslash r;j}$ can be calculated. This is done with Monte Carlo methods.

2.3. Uncertainty Related to Expert Judgments

Since expert judgments are afflicted with uncertainties, a considerable number (usually 4 or more) of expert judgments on the impairment of the components and on the applicability factor is collected for each event observed. The procedure for clarifying the technical facts and performing the expert judgments is described in detail in [10]. Below, the entirety of expert assessments is denoted by $\mathfrak{E}$.

The calculation of the distributions of $q_{k \backslash r;j}$ as described in the last chapter is carried out individually for all experts, resulting in an expert-specific subjectivist probability distribution of $q_{k \backslash r;j}$. To combine the judgments of the different experts, the mixture distribution is calculated from these individual expert-specific subjectivist distributions. According to equation (3) the total probability of a CCF event with k failures is the sum of the individual phenomenon-specific probabilities $q_{k \backslash r;j}$. These calculations are carried out using Monte Carlo methods.

2.4. Consideration of Additional Uncertainties

Finally, the remaining uncertainties not considered before have to be included. Most prominent is the uncertainty related to a possible inhomogeneity of populations. This possible inhomogeneity implies that if $q_{k \backslash r}$ is the probability of a (k out of r)-CCF estimated from operating experience from a specific population, the probability of a (k out of r) CCF $\hat{q}_{k \backslash r}$ in a specific CCF group modelled in a PRA in general deviates from that value. This is quantified by a conditional distribution $p(\hat{q}_{k \backslash r}|q_{k \backslash r})$. The probability distribution of $\hat{q}_{k \backslash r}$ can be expressed as

$$p(\hat{q}_{k \backslash r}) = \int_0^1 p(\hat{q}_{k \backslash r}|q_{k \backslash r}) p(q_{k \backslash r}) dq_{k \backslash r} \qquad (7)$$

In general there is no information that the different sources of uncertainty take effect differently in different component groups or component types. Hence a universal distribution $p(\hat{q}_{k \backslash r}|q_{k \backslash r})$ is assumed for all component groups. Due to the limited number of CCF events it is not possible to estimate the functional form of $p(\hat{q}_{k \backslash r}|q_{k \backslash r})$ or its characteristics from operating experience. Therefore, the following assumptions are made:

1. **Preservation of expectation value:** It can neither be expected that $\hat{q}_{k\backslash r}$ is larger nor that it is smaller than $q_{k\backslash r}$ Therefore it is assumed that $p(\hat{q}_{k\backslash r}|q_{k\backslash r})$ preserves the expectation value, which implies $\int_0^1 \hat{q}_{k\backslash r} p(\hat{q}_{k\backslash r}|q_{k\backslash r}) d\hat{q}_{k\backslash r} = q_{k\backslash r}$.
2. **Scale-independence:** The shape and relative width of p shall be independent of $q_{k\backslash r}$. This implies that the standard deviation of $p(\hat{q}_{k\backslash r}|q_{k\backslash r})$ is proportional to $q_{k\backslash r}$ (with the proportionality factor denoted by ϱ).
3. **Minimal width:** The 95%-quantile should at least be 4 times as large as the mean. This criterion has been determined by expert judgments and is in agreement with the previous "broadening" described in German PRA guidelines [11].

These assumptions are applicable for $q_{k\backslash r} \ll 1$. This is generally valid in normal applications.

$p(\hat{q}_{k\backslash r}|q_{k\backslash r})$ is assumed to be a beta distribution, i.e.:

$$p(\hat{q}_{k\backslash r}|q_{k\backslash r}) = \frac{(1-\hat{q}_{k\backslash r})^{\beta-1} \hat{q}_{k\backslash r}^{\alpha-1}}{f_\beta(\alpha,\beta)} \qquad (8)$$

with $f_\beta(\alpha,\beta)$ denoting the beta function.

The assumptions described above allow to determine the two ($q_{k\backslash r}$-dependent) parameters as

$$\alpha = \frac{1 - q_{k\backslash r} - \varrho^2 q_{k\backslash r}}{\varrho^2}$$
$$\beta = \frac{1 - 2 q_{k\backslash r} - \varrho^2 q_{k\backslash r} + (q_{k\backslash r})^2 + \varrho^2 (q_{k\backslash r})^2}{\varrho^2} \qquad (9)$$

Assumption 3 implies $\varrho = 0.9463$. Equation (8) can seamlessly be implemented in the Monte Carlo simulation described above.

2.5. Convergence Properties

As mentioned above, the coupling model has been developed to estimate CCF probabilities from operating experience data comprising usually only a small number of events for each component type. However, the model assumptions introduced to facilitate this also imply undesired convergence properties. Firstly, if the data is not compatible with the model assumption that the number of failed component obeys a Binomial distribution for each CCF phenomenon (eq. 1), a convergence to the true values $q_{k\backslash r}$ is generally not possible. More importantly, the assumption that different CCF phenomena may be characterized by different coupling parameters implies a separate estimation of the coupling factor for each event (eq. 4). This implies that distributions $p(q_{k\backslash r})$ are not getting narrower when the number of events grows. As an example, the resulting distributions are identical if one (2 out of 4)-event has been observed during an observation time T to a case where ten (2 out of 4)-events have been observed during observation time $10\,T$. This does not reflect the fact that the evidence on the probabilities of the various failure combinations has grown significantly, which should lead to a smaller estimation uncertainty and thus narrower width of the distributions.

As an increasing part of German operating experience has been evaluated with regard to CCF and hence the number of CCF events available has increased considerably [4,5] this issue is growing in importance. Therefore, GRS has started a research project to evaluate methods to improve CCF modelling with respect to this aspect, which will be described in the following chapter.

3. ALTERNATIVE MODELING OF CCF

3.1. Boundary Conditions of CCF Probability Estimation

Any further development of the model and the procedures for CCF probability estimation has to take into account the boundary conditions prevailing in Germany. According to German PRA guidelines, plant specific operating experience has to be used for quantification whenever possible. Therefore usually plant specific operating experience is used to quantify unavailabilities due to independent failures, while common cause failures are quantified using generic operating experience. As a consequence, only CCF-related information is available in the German CCF data pool. Events with non-systematic (single) failures are not included. Therefore, models which relate the CCF rate to single component failure rates – like the alpha-factor model – cannot directly be applied. The exact number of demands of stand-by components is not available. Instead, observation times have been determined. For the failures during demands operating times have also been calculated. Therefore CCF failure rates have to be estimated (see eq. 1). Not only true CCF, but also "potential CCF" events where multiple components were impaired due to a systematic cause while only one or even no component actually failed are included in the data pool. To improve the accuracy of estimations, this information should also be utilized for CCF quantification. For all events, component impairments and applicability factors have been quantitatively assessed by several experts to allow the consideration of the uncertainty of expert assessments. These uncertainties should be adequately represented in the improved model as well. Generally, the treatment of uncertainties should be as comprehensive and consistent as in the coupling model. Also, the results of CCF quantification (a posteriori distributions) should be representable in a form suitable for practically carrying out PRA calculations including data handling. Ideally they would be independent parametric distributions of the CCF probabilities or well approximable by such distributions.

Modelling approaches to fulfil these requirements are presented in the following section.

3.2. Alternative CCF models

To avoid the undesirable convergence properties associated with the model assumptions discussed above, they would need to be replaced with less restrictive assumptions. An obvious way to achieve this is to replace the phenomenon-dependent binomial distribution with a categorical distribution. Such distributions however do not have group-size independent parameters; hence the "auto-mapping" feature of the coupling model is lost. Therefore, separate mapping algorithms are needed. Alternatively, (k out of r)-CCF with different k could be considered independent elementary events. Here, separate mapping algorithms are needed as well. In the following, two model structures are discussed:

Model A

In model A, CCF events with k out of r failed components occur with a rate $\lambda_{k\backslash r}$. The probability of a (k out of r)-CCF is

$$q_{k\backslash r} = t\, \lambda_{k\backslash r} \qquad (10)$$

Model B

In model B, CCF events occur with a rate λ. With conditional probability $\omega_{k\backslash r}$ k of r components fail if a CCF occurs (with $k = 2 \ldots r$). Hence $\sum_{k=2}^{r} \omega_{k\backslash r} = 1$ is valid. The probability of a (k out of r)-CCF is

$$q_{k\backslash r} = t\, \lambda\, \omega_{k\backslash r} \qquad (11)$$

The model structures are shown in Figure 1. OK denotes a state where no CCF has occurred.

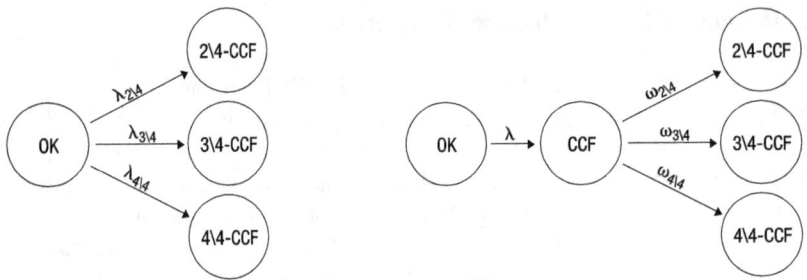

Figure 1: Model structures for model A (left) and B (right) for a CCF group of size 4.

It should be noted that these models are equivalent since $\lambda_{k\backslash r} = \lambda\, \omega_{k\backslash r}$ is valid. Therefore, estimations from operating experience should lead to the very same results, if equivalent a priori distributions are used. However, the different model structures suggest using different a priori distributions, as will be discussed later.

For these models the total observation time T and the numbers $m_{k\backslash r}$ of CCF with k failed components ($k = 2 \dots r$) form a sufficient statistic. It can be written as $(r-1)$-tuple $M := \{m_{2\backslash r}, \dots, m_{r\backslash r}\}$. In general M is not precisely known as mentioned in chapter 2.2. A probability distribution $p(M|\mathfrak{E})$ has to be calculated from the expert assessments $\mathfrak{E}$ of component impairments and applicability factors. This can be done using probability calculus.

It is possible, however, to implement $p(M|\mathfrak{E})$ in a Monte Carlo procedure without calculating the $(r-2)$-dimensional distribution $p(M|\mathfrak{E})$ by noting that for each event the interpretation vector W (see section 2.2) is an expert-specific distribution of the number of failed components. The Monte Carlo procedure consists of first randomly choosing one of the available experts for each event. This reflects the assumption that all experts are equally competent in assessing the events. Then, a number of failed components is drawn from W as estimated by that expert. With probability f (applicability factor) this number is accepted, with $1 - f$ the number of failed components is set to 0. This is repeated for every event observed. The total number of cases where $2, 3, \dots, r$ component failures occurred is summed up over all events to calculate the sample of M.

Given M, the model parameters can be calculated. For model A, the model parameter $\lambda_{k\backslash r}$ is only dependent on $m_{k\backslash r}$. This suggests choosing independent a priori distributions for all model parameters, i.e. $\pi(\lambda_{2\backslash r}, \dots, \lambda_{r\backslash r}) = \prod_{k=2}^{r} \pi_{k\backslash r}(\lambda_{k\backslash r})$. If a non-informative approach is followed, all $\pi_{k\backslash r}$ are chosen identically. Using Jeffreys' rule the a priori distribution becomes

$$\pi(\lambda_{2\backslash r}, \dots, \lambda_{r\backslash r}) \propto \prod_{k=2}^{r} (\lambda_{k\backslash r})^{-1/2} \qquad (11)$$

The a posteriori distribution then also factorizes:

$$P(\lambda_{2\backslash r}, \dots, \lambda_{r\backslash r}|M) = \prod_{k=2}^{r} \frac{T^{m_{k\backslash r}+1/2}}{\Gamma(m_{k\backslash r} + 1/2)} (\lambda_{k\backslash r})^{m_{k\backslash r}-1/2} e^{-\lambda_{k\backslash r} T} \qquad (12)$$

with Γ denoting the Gamma function. Hence the model parameters are independent and distributed according to Gamma distributions with parameters $m_{k\backslash r} + 1/2$ and T.

For model B the model parameter λ is only dependent on the total number of CCF events $m_T := \sum_{k=2}^{r} m_{k\backslash r}$, not the individual $m_{j\backslash r}$. This suggests choosing the a priori distribution $\pi(\lambda, \Omega) = \pi(\lambda)\pi(\Omega)$ with $\Omega = \{\omega_{2\backslash r}, \ldots, \omega_{r\backslash r}\}$ denoting the parameter set of the categorical distribution. If a non-informative approach is followed and Jeffreys' rule is applied the a priori is

$$\pi(\lambda, \omega_{2\backslash r}, \ldots, \omega_{r\backslash r}) \propto \lambda^{-1/2} \prod_{k=2}^{r} (\omega_{k\backslash r})^{-1/2} \qquad (13)$$

This implies for the a posteriori distribution

$$P(\lambda, \Omega|M) = \frac{T^{m_T+1/2}}{\Gamma(m_T + 1/2)} (\lambda_{k\backslash r})^{m_{k\backslash r}-1/2} e^{-\lambda T} \frac{\prod_{i=2}^{r}(\omega_{i\backslash r})^{m_{k\backslash r}-1/2}}{B(m_{2\backslash r} + 1/2, \ldots, m_{r\backslash r} + 1/2)} \qquad (14)$$

with $B(a_{2\backslash r}, \ldots a_r) = \prod_{i=2}^{r}\Gamma(a_{i\backslash r})/\Gamma(\sum_{i=2}^{r} a_{i\backslash r})$. λ is distributed according to Gamma distribution with parameter $m_T + 1/2$ and T. Ω is distributed according to a Dirichlet distribution with parameter set $\{m_{2\backslash r} + 1/2, \ldots, m_{r\backslash r} + 1/2\}$. This mathematically is similar to the alpha factor model; it should be noted though that the meaning of the model parameters is different (see above).

It is worth noting that $P(\lambda_{2\backslash r}, \ldots, \lambda_{r\backslash r}|M)$ factorizes into $r-1$ one-dimensional Gamma distributions and $P(\lambda, \omega_{2\backslash r}, \ldots, \omega_{r\backslash r}|M)$ factorizes into a Gamma and a Dirichlet distribution, while $p(M|\mathfrak{E})$ does in general not factorize. Hence the resulting distributions of model parameters, given the expert assessments, also do not factorize. The same is true of the CCF probabilities $q_{k\backslash r}$. Therefore, in principle, joint probabilities should be used when carrying out uncertainty analyses. The effect of properly including the statistical dependencies in PRA uncertainty calculations is currently under investigation in a GRS research project.

Well-known properties of Gamma and Dirichlet distributions imply that the estimates (12) and (14) in the limit of an infinite number of events converge to their true values. If the number of relevant events is relatively small, however, the different a priori assumptions of model A and B may have a considerable effect on the estimation results.

If, for example, no failures occurred, in model A all $\lambda_{k\backslash r}$ are (independently) distributed with a Gamma distribution with parameters $1/2$ and T. Hence the expectation values are $\langle q_{k\backslash r}\rangle = t/(2T)$. In model B λ is distributed with a Gamma distribution with Parameters $1/2$ and T. $\{\omega_{2\backslash r}, \ldots, \omega_{r\backslash r}\}$ are distributed according to a Dirichlet distribution with parameters $\{1/2, \ldots, 1/2\}$. Hence, the expectation values are $\langle q_{k\backslash r}\rangle = t/(2T(r-1))$, which is smaller by a factor of $(r-1)$. This may be significant for large CCF groups.

Generally, due to the different a priori assumptions for a finite number of events the expectation values $\langle q_{k\backslash r}\rangle$ are always larger for model A than for model B. Also the width of the uncertainty distributions is larger (see also figures 2-4).

4. COMPARISON OF ESTIMATION RESULTS

To compare the different models, they were used for CCF quantification in various different CCF data sets of German operating experience. The information contained in the German CCF database is partly proprietary. Therefore, representative modified data sets have been prepared which allow to compare the quantification results while protecting the proprietary information. Two of these datasets will be discussed below. Data set one represents populations with a large number of observed events, while data set two represents populations with a very small number of observed events. Data set one has been created by first dropping from an original dataset all events that occurred in a CCF group of a size different from 4 and then dropping a small number of random additional events and multiplying the observation time T by a random number. This data set comprises 15 events. In some events all components were impaired. No more than two components completely failed. Data set two has been created by picking a typical event from component datasets where very few events occurred. In this particular event, one component failed and the remaining three were considered impaired by three experts and unaffected by one expert. The observation time T was selected arbitrarily. The intermediate results before the additional uncertainties (see section 2.4) are included are shown in figure 2.

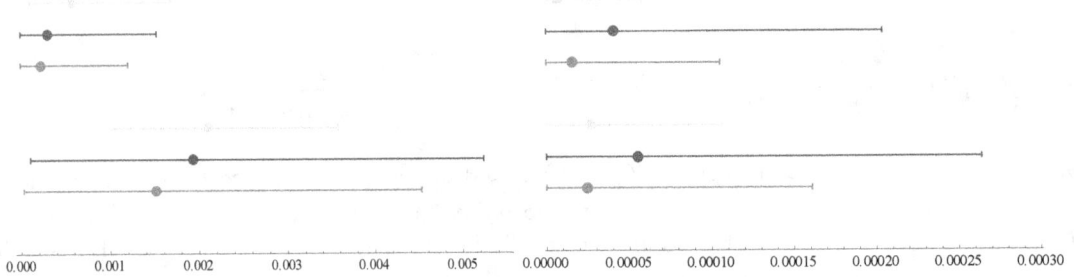

Figure 2: Intermediate estimation results for dataset 1 (left) and dataset 2 (right). Shown are the expectation values (circles) and 95%-confidence intervals (error bars) of estimates of $q_{4\backslash 4}$ (top) and $q_{2\backslash 4}$ (bottom). Estimates of model A are shown in blue, estimates of model B in red, estimates of the coupling model in green.

For the parameter describing failures where substantial empirical evidence is present ($q_{2\backslash 4}$ for data set 1) the results are quite similar (deviations of less than a factor 1.4 in the mean and 1.5 in the 97.5%-quantile). For parameters describing failures that are "extrapolated" from the events observed the deviations between all models are somewhat larger (deviations of up to a factor 5.67 in the mean and 5.52 in the 97.5%-quantile). This can be attributed to the different modelling and a priori assumptions.

The 95%-confidence intervals $[Q_{2.5\%}, Q_{97.5\%}]$ show a very large overlap. In all cases the expectation values of all models lie within the confidence intervals of all other models.

In figure 3 the final results after considering the additional uncertainties are shown.

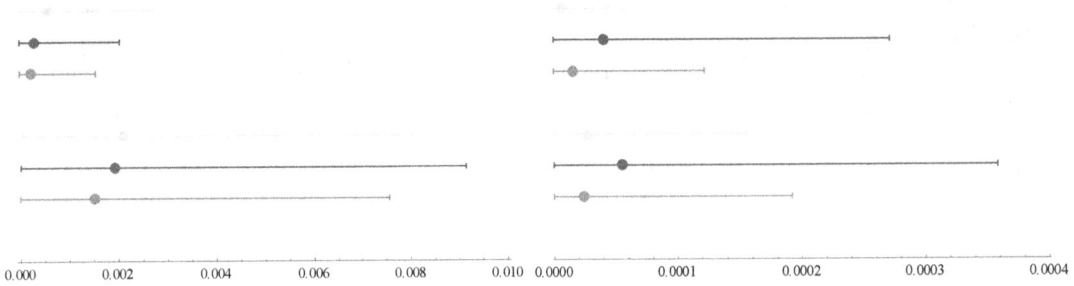

Figure 3: Final estimation results for dataset 1 (left) and dataset 2 (right) after "broadening". Shown are the expectation values (circles) and 95%-confidence interval (error bars) of estimates of $q_{4\backslash 4}$ (top) and $q_{2\backslash 4}$ (bottom). Estimates of model A are shown in blue, estimates of model B in red, estimates of the coupling model in green.

The final results after inclusion of the additional uncertainties are qualitatively the same. While the confidence intervals have grown the mean has not changed (see chapter 2.4).

5. MAPPING

As noted before, for models A and B – in contrast to the coupling model – only those operating experience events can be used directly that occurred in groups with size identical to the component groups CCF probabilities are estimated for. Therefore, if not enough such operating experience is available, separate mapping algorithms have to be used to determine how many components would have failed in groups of different size. This would be especially important in Germany where in many cases group sizes are different in different plants. For example, emergency diesel generators groups of size 2, 3, 4, 5 and 6 exist or existed. Several approaches have been discussed but since CCF genesis and detection are complex and multifarious processes it appears difficult to rate the different approaches or justify a specific approach (see e.g. [12,13] and references therein). Different aspects like the assessment of the modelling uncertainty or the compatibility with a priori beliefs need to be considered. E.g. the assumption that a CCF group is statistically equivalent to an arbitrary subgroup of a larger group is generally not consistent with choosing a non-informative a priori distribution by simply applying Jeffreys' rule like in eq. (11) and (13), an approach that is usually also used for the alpha factor model [12]. This can be easily seen by a simple example: If a group of size three is considered a subgroup of a group of size four $\omega_{2\backslash 3} = 1/2\, \omega_{2\backslash 4} + 3/4\, \omega_{3\backslash 4}$ holds which implies $\omega_{2\backslash 3} \leq 3/4$. This is not consistent with the non-informative prior chosen according to Jeffreys' rule $\pi(\lambda, \omega_{2\backslash 3}, \omega_{3\backslash 3}) \propto 1/\sqrt{\lambda \omega_{2\backslash 3} \omega_{3\backslash 3}}$ where no bounds on $\omega_{2\backslash 3}$ are present. This is due to the fact that the information mentioned above is not accounted for. Similarly, the assumption of a Binomial distribution often used for mapping is conflicting with the a priori belief of eq. (11) and (13). Estimation procedures utilizing such assumptions and conflicting non-informative priors would suffer from inconsistency.

Therefore, for the present studies a simple procedure was applied. When mapping down, it consists of "deleting" the least affected components. When mapping up, it consists of "duplicating" the most affected components. This means an event with 2 failed and 2 impaired components would be mapped to an event with 2 failed and one impaired component for group size three or to an event with 3 failed and 2 impaired components for group size five. It is evident that this approach is conservative under any circumstances. While for large differences in group sizes this approach appears to be overly conservative, for small differences, e.g. group sizes differing by only one it may lead to reasonable results. One example is shown in figure 4. Here a data set comprising CCF groups of size 3 and 4 was considered. The observation time for CCF groups of size 3 was 5 % of the total observation time. There is one event in a component group of size 3 and 15 events in groups of size 4. Estimates using model A have been calculated both from data from groups of size 3 and from data from groups of sizes 3 and 4, applying the mapping algorithm described above. The coupling model has been applied to data from both groups of sizes 3 and 4, making use of the "auto mapping" feature.

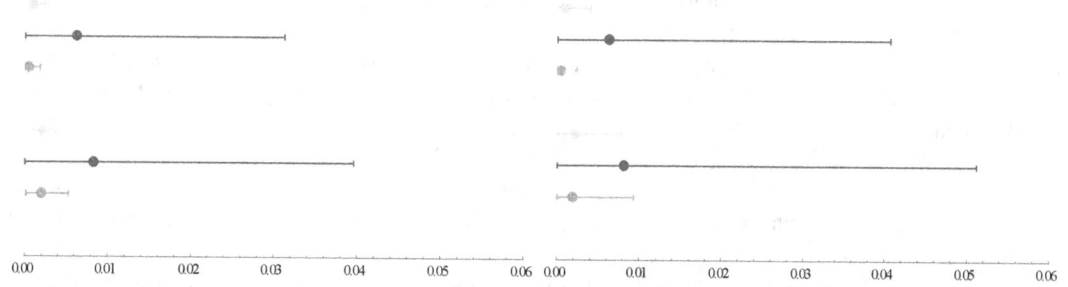

Figure 4: Intermediate (left) and final estimation results (right). Shown are the expectation values (circles) and 95%-confidence interval (error bars) of estimates of $q_{3\backslash 3}$ (top) and $q_{2\backslash 3}$ (bottom). Estimates of model A using the data set of events and observation time in component groups of size 3 are shown in blue, estimates of model A using the data set of events in component groups of size 3 and mapped events of group size 4 and appropriate observation time are shown in magenta. Estimates with the coupling model are shown in green.

The results for model B using the data set including mapped data and the coupling model result in similar estimates. In all four cases the expectation values lie within the confidence intervals of the other model. The estimates using operating experience only of groups of size 3 are significantly different. Both uncertainty and expectation values are larger. This can be attributed to the considerably smaller amount of operating experience both in terms of observation time and number of events.

6. CONCLUSION

The methods for quantification of CCF applied by GRS have been continuously improved. A new procedure for the consideration of additional uncertainties not treated before which is consistent with the Bayesian framework used in CCF quantification has been established. The convergence properties of the coupling model have been researched. Theoretical considerations show that for very large data sets the coupling model has undesirable properties preventing an adequate representation of the reduction of statistical uncertainty. Therefore comparative studies have been carried out to assess the relevance for the actual operating experience with regard to CCF in Germany. Two different models have been developed which are suitable for estimating CCF probabilities from the information available in the German CCF database. They do not use the restrictive modelling assumption on which the coupling model is based and therefore no undesirable convergence properties are present. Comparisons show that for present German operating experience no significant deviations are found. The deviations of the numerical results between all the three different methods are comparable. Therefore, modeling decisions need to be based on theoretical properties and practical considerations. While the coupling model has the convenient "auto-mapping" feature, Model A is free from any, possibly non-conservative, modeling assumptions. In the further course of the research project these properties will be evaluated in more detail. The development and evaluation of suitable mapping algorithms will be an additional focus of future research.

Acknowledgement

This research has been funded by the German Federal Ministry of Economic Affairs and Energy.

References

[1] A. Kreuser and J. Peschke: "*Coupling Model: A Common-Cause-Failure-Model with Consideration of Interpretation Uncertainties*", Nuclear Technology, Vol. 136, (2001).

[2] A. Kreuser, J. Peschke and J. C. Stiller: "*Further Development of the Coupling Model*", Kerntechnik, 71, (2006).

[3] J. Stiller, A. Kreuser, and C. Verstegen: "*Consideration of Additional Uncertainties in the Coupling Model for the Estimation of Unavailabilities due to Common Cause Failures*", Proceedings of the 9th International Conference on Probabilistic Safety Assessment & Management, Hong Kong, (2008).

[4] A. Kreuser, and C. Verstegen: "*Common-Cause Failure Analysis – Recent Developments in Germany*", Proceedings of the 10th International Conference on Probabilistic Safety Assessment & Management, Seattle, (2008).

[5] J. C. Stiller, L. Gallner, H. Holtschmidt, A. Kreuser, M. Leberecht, C. Verstegen: "*Development of an integrated program and database system for the estimation of CCF probabilities*", the ANS PSA 2011 International Topical Meeting on Probabilistic Safety Assessment and Analysis, Wilmington, (2011).

[6] J. C. Stiller: „Weiterentwicklung des Quantifizierungsverfahrens für GVA zur Vermeidung von Schätzfehlern aufgrund vereinfachender Modellannahmen", GRS-322, (to appear in 2014).

[7] C. L. Atwood: "*Estimators for the Binomial Failure Rate Common Cause Model*", NUREG/CR-1401, US-NRC, (1980).

[8] G. E. P. Box and E. G. Tiao: "*Bayesian Inference in Statistical Analysis*", Addison-Wesley, (1973).

[9] H. M. Stromberg, F. M. Marshall, A. Mosleh, and D. M. Rasmuson: *"Common Cause Failure Data Collection and Analysis System, Vol. 2, Definition and Classification of Common-Cause Failure Events"*, INEL-94/0064, Idaho National Engineering Laboratory, Idaho Falls, Idaho, (1995).

[10] A. Kreuser, J. Peschke and C. Verstegen: *"Assessing Operation Experience for Generating Reliability Data"*, Proceedings of the 6th International Conference on Probabilistic Safety Assessment & Management, Puerto Rico, (2002).

[11] Facharbeitskreis (FAK) Probabilistische Sicherheitsanalyse für Kernkraftwerke: *"Daten zur Quantifizierung von Ereignisablaufdiagrammen und Fehlerbäumen"*, Bundesamt für Strahlenschutz, BfS-SCHR-38/05, (2005).

[12] A. Mosleh, D. M. Rasmuson, and F. M. Marshall: *"Guidelines on Modeling Common-Cause Failures in Probabilistic Risk Assessment"*, NUREG/CR-5485, INEEL/EX-97-01327, Idaho National Engineering Laboratory, Idaho Falls, Idaho, (1998).

[13] J. Vaurio: *"Consistent mapping of common cause failure rates and alpha factors"*, Reliability Engineering and System Safety, 92, (2007).

Plant-Specific Uncertainty Analysis for a Severe Accident Pressure Load Leading to a Late Containment Failure

S.Y.Park[a*] and K.I.Ahn[a]

[a]Korea Atomic Energy Research Institute, Daeduk-Daero 989-111, Yusong, Daejeon, KOREA, 305-353

*E-mail of corresponding author: sypark@kaeri.re.kr

Abstract: Typical containment performance analyses for a level 2 probabilistic safety analysis (PSA) have made use of a containment event tree (CET) modeling approach, to model the containment responses by depicting the various phenomenological processes, containment conditions, and containment failure modes that can occur during severe accidents. A general approach in the quantification of the containment event tree is to use a decomposition event tree (DET) to allow a more detailed treatment of the top event. A quantification of the physical phenomena in the decomposition event tree is achieved based on the results obtained through validated code calculations or expert judgments. The phenomenological modeling in the event tree still entails a high level of uncertainty because of our incomplete understanding of reactor systems and severe accident phenomena. This paper includes an uncertainty analysis of a containment pressure behavior during severe accidents for the optimum assessment of a late containment failure model of a decomposition event tree.

Keywords: Uncertainty Analysis, PSA, MAAP Code, Containment Pressure Load.

1. INTRODUCTION

A level 2 probabilistic safety analysis (PSA) is used to assess the performance of the containment in mitigating severe accidents. The analysis includes an evaluation of the accident progression in the containment; an estimation of the timing, location, and mode of containment failure; and an estimation of the source term characteristics. Typical containment performance analyses have made use of a containment event tree (CET) modeling approach, to model the containment responses by depicting the various phenomenological processes, containment conditions, and containment failure modes that can occur during severe accidents. A level 2 PSA of an OPR-1000, which is the reference plant of this analysis, has made use of a CET modeling approach, where a general approach in the quantification of a small event tree is to use a decomposition event tree (DET) to allow a more detailed treatment of the top event. A quantification of the physical phenomena in the DET is achieved based on the results obtained by validated the code calculations or expert judgments. The phenomenological modeling in the event tree still entails a high level of uncertainty. Such uncertainty exists because of our incomplete understanding of reactor systems and severe accident phenomena.

This paper illustrates the application of a severe accident analysis code, MAAP [1], to the uncertainty evaluation of a late containment failure DET, which is one of the CET top events in the reference plant of this study. An uncertainty analysis of a containment pressure behavior during severe accidents has been performed for the optimum assessment of a late containment failure model. The MAAP code is a system level computer code capable of performing integral analyses of potential severe accident progressions in nuclear power plants, whose main purpose is to support a level 2 probabilistic safety assessment or severe accident management strategy developments. The code employs lots of user-options for supporting a sensitivity and uncertainty analysis. The present application is mainly focused on determining an estimate of the containment building pressure load caused by severe accident sequences. Key modeling parameters and phenomenological models employed for the present uncertainty analysis are closely related to in-vessel hydrogen generation, gas combustion in the containment, corium distribution in the containment after a reactor vessel failure, corium coolability in the reactor cavity, and molten-corium interaction with concrete.

2. ANALYSIS METHODOLOGY

The basic approach of this methodology is to 1) develop severe accident scenarios for which the containment pressure loads should be performed based on a level 2 PSA, 2) identify severe accident phenomena relevant to a late containment failure, 3) identify the MAAP input parameters, sensitivity coefficients, and modeling options that describe or influence the late containment failure phenomena, 4) prescribe likelihood descriptions of the potential range of these parameters, and 5) evaluate the code predictions using a number of random combinations of parameter inputs sampled from the likelihood distributions; in addition 6) the results have been summarized and displayed for the important output variables.

To quantify the uncertainties addressed in the MAAP code, a computer program, MOSAIQUE [2], has been applied, which was recently developed by the Korea Atomic Energy Research Institute. The program consists of fully-automated software to quantify the uncertainties addressed in the thermal hydraulic analysis models or codes. MOSAIQUE employs a methodology of sampling-based uncertainty analysis using thermal hydraulic or severe accident analysis codes [3][4][5]. The Korean standardized nuclear power plant, the OPR-1000, has been selected as a reference plant for this analysis.

2.1. Development of DET Scenarios for the late containment failure

A late containment failure is defined as a failure of the containment long after a reactor vessel failure. The time frame for a late containment failure begins many hours after the vessel has failed and continues to three days after accident initiation. Three days are considered enough to control the containment pressurization. The primary cause of a failure of the containment is the steam over-pressurization resulting from the loss of the containment heat removal. Containment heat removal can be achieved by the operation of recirculation sprays or fan coolers. The steam over-pressurization process is slow and it takes times to reach the containment failure pressure. The possibility of late containment failure due to a late hydrogen burn can also be considered. To evaluate the early containment failure, the total pressure inside the containment should be calculated. In addition to the base pressure and reactor coolant system (RCS) blow-down pressure, the pressurization owing to steam generation in the cavity, gas generation by molten corium-concrete interaction (MCCI), late hydrogen burn in the containment have been considered. Eventually, the probability of a containment failure and its failure mode will be calculated using the containment fragility curve, which is out of scope of this paper.

The first and second top headings of a late containment failure DET are the RCS pressure at the reactor vessel failure and the amount of corium remained in the cavity, respectively. The driving force of corium ejection out of the reactor cavity is the RCS pressure. It is known that corium can escape the reactor cavity when the RCS pressure exceeds a certain value. The higher the RCS pressure, the more corium that can be ejected out of the cavity. The third top event is the availability of secondary heat removal using a motor-driven auxiliary feedwater pump and main steam safety valves. The fourth concern of the DET is a reactor cavity condition. Three discretized regimes, which are 'flooded', 'wet', or 'dry', have been selected to represent the cavity condition. There is a water flow path from the containment sump level to the reactor cavity in the reference plant. This path allows the cavity to be flooded if the inventory of the refueling water tank is injected into the containment through the high pressure safety injection (HPSI) or the low pressure safety injection (LPSI) system. In the case of the HPSI or LPSI operation, the cavity condition is assigned as a 'flood'. On the other hand, the inventory of a safety injection tank (SIT) is only injected into the cavity, and the cavity condition is allocated as 'wet'. The fifth event is a containment pressure load due to a late hydrogen burn. The late hydrogen burn has a dependency on the cavity condition. If the cavity condition is 'flooded' or 'wet', the containment pressure load increase by the late hydrogen burn will be limited owing to the higher steam generation in the reactor cavity.

Nine scenarios were developed as DET scenarios of a late containment failure. The developed DET scenarios are shown in Fig. 1 and Table 1: three large loss of coolant (LOCA) initiated scenarios for the sequences of low reactor vessel pressure at vessel failure, three loss of offsite power (LOOP) initiated scenarios for the sequences of high reactor vessel pressure at vessel failure, and three LOOP initiated scenarios for cases of high reactor vessel pressure at vessel failure with secondary heat removal available. Three cavity flooded cases by LPSI or HPSI operation, three cavity wet cases by SIT operation, and three cavity dry cases are included.

Fig. 1: DET scenarios for uncertainty analysis of the pressure load for late containment failure

Events	RCS Pressure at RV Failure	Corium Mass in Cavity	Secondary Heat Removal	CAVITY CONDITION	Early Hydrogen Burn	SEQ #	STC #
	Low	High	Unavailable	FLOODED	NOT BURNABLE	1	LLFLD
				WET	NOT BURNABLE	2	LLWET
				DRY	BURNABLE	3	LLDRY
	High	Not High	Unavailable	FLOODED	NOT BURNABLE	4	LPFLD
				WET	NOT BURNABLE	5	LPWET
				DRY	BURNABLE	6	LPDRY
			Available	FLOODED	NOT BURNABLE	7	LPFLDSG
				WET	NOT BURNABLE	8	
				DRY	BURNABLE	9	LPDRYSG

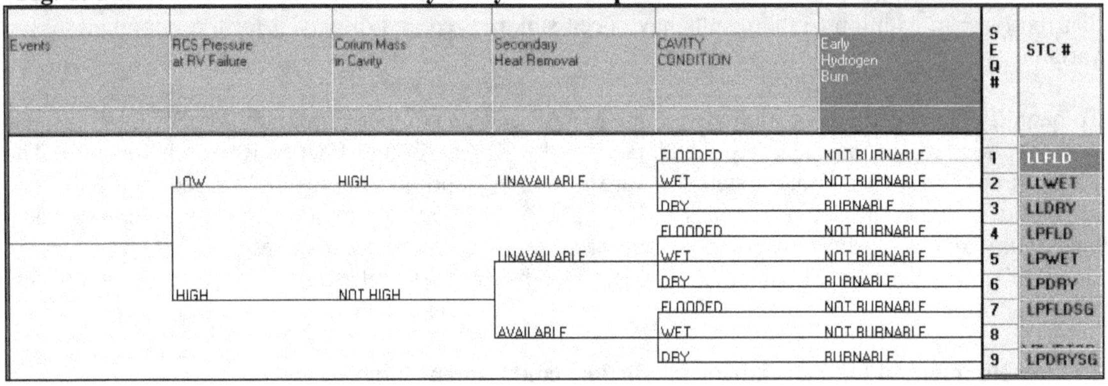

Table 1: Tabularized DET scenarios for an uncertainty analysis of the pressure load for late containment failure

RCS Pressure at RV Failure	Corium Mass in Cavity	Secondary Heat Removal	Cavity Condition	Late Hydrogen Burn	Sequence ID
Low	High	Unavailable	Flooded	Not Burnable	LLFLD
			Wet	Not Burnable	LLWET
			Dry	Burnable	LLDRY
High	Not High	Unavailable	Flooded	Not Burnable	LPFLD
			Wet	Not Burnable	LPWET
			Dry	Burnable	LPDRY
		Available	Flooded	Not Burnable	LPFLDSG
			Wet	Not Burnable	LPWETSG
			Dry	Burnable	LPDRYSG

2.2. Selection of MAAP Modeling Parameter and Sampling

In the severe accident analysis, there were uncertainties in the physical phenomena. There were also uncertainties in the MAAP phenomenological models. Users had control over the uncertainties through the so-called 'model parameters' of the MAAP program. They were either used as an input to a given physical model or to select between different physical models. This feature of the code architecture was included specifically to facilitate sensitivity or uncertainty in the analysis. In this study, input variables assigned as the model parameters to affect the pressure load of containment building during the late state of a severe accident were identified, and their uncertainty was characterized using a user specified distribution. These parameters were selected based on MAAP input parameter files.

For the present uncertainty analysis, 20 input variables were selected, which include six variables of steam and non-condensable gas generation, eight variables of in-vessel hydrogen generation, three variables of high pressure melt ejection, and three variables of hydrogen combustion in the containment. The list of variables and descriptions of the listed parameters were defined as shown in Table 2. The corresponding default values and uncertainty distributions of the parameters were defined as shown in Table 3. User assumption was given for the assigned range of modeling parameters and uncertainty distributions based on engineering judgments. To propagate these uncertain inputs through the MAAP code, they were sampled using a random sampling technique. The Monte Carlo Sampling method with a size of 200 for each scenario was used to sample the input parameter distributions, and 200 MAAP calculations were then performed.

Table 2: The list of parameters considered in the uncertainty analysis of pressure load for the late containment failure

Phenomena	MAAP Parameter	Parameter Description
Steam and Non-condensable gas generation in Cavity	HTCMCR	Downward heat transfer coefficient for convective heat transfer from molten corium to the lower crust in MCCI
	HTCMCS	Sideward heat transfer coefficient for convective heat transfer from molten corium to the side crust in MCCI
	TCNNP	Melting temperature of concrete
	FCHF	Flat plate critical heat flux Kutateladze number
	HTFB	Coefficient for film boiling heat transfer from corium to an overlying pool
	ACMPLB(1)	Floor surface area which the corium debris pool may occupy in cavity
High Pressure Melt Ejection	FKUTA	Kutateladze coefficient in the debris entrainment criterion
	FWEBER	Weber number used in the calculation of the diameter of the debris particles during the entrainment process
	ENT0C	Jet entrainment coefficient for the Ricou-Spalding correlation
In-vessel Hydrogen Generation	FAOX	Multiplier for the cladding outside surface area 새 calculate oxidation
	TCLMAX	Temperature to lead to rupture if the cladding is at this temperature for 0.01 hr. Larson-Miller parameter is calculated from TCLMAX
	LMCOL0	Collapse criteria parameters for a Larson-Miller-like functional dependence
	LMCOL1	
	LMCOL2	
	LMCOL3	
	EPSCUT	Cutoff porosity below which the flow area and the hydraulic diameter of core node are zero, i.e., the node is fully blocked
	EPSCU2	
Hydrogen Burn	TAUTO	Auto-ignition temperature for H_2 and CO burns
	XSTIA	Steam mole fraction required to inert an H_2-Air-H_2O mixture at incipient auto-ignition
	TJBRN	Temperature of H_2 jet entering a non-inerted compartment which is sufficient to cause a local burn

Table 3: The default values and uncertainty distributions of MAAP modeling parameters considered in the uncertainty analysis

MAAP Parameter	Default Value	Assigned Range [min-max]	Distribution
HTCMCR	3,500 W/m^{2}C	[1000, 5,000]	Triangle
HTCMCS	3,000 W/m^{2}C	[1000, 5,000]	Triangle
TCNNP	1,450 K	[1,450-1,750]	Triangle
FCHF	0.1	[0.02, 0.25]	log uniform
HTFB	300 W/m^{2}C	[100, 400]	Triangle
ACMPLB(1)	62.54 m^2	[43.78-62.54]	uniform
FKUTA	2.46	[2.46-3.7]	uniform
FWEBER	10.0	[1.0-100]	log uniform
ENT0C	0.045	[0.025-0.06]	uniform
FAOX	1.0	[1.0-2.0]	uniform
TCLMAX	2500 K	[2000-3000]	uniform
LMCOL0	50.0	[48-54]	uniform
LMCOL1	50.0	[48-54]	uniform
LMCOL2	50.0	[48-54]	uniform
LMCOL3	50.0	[48-54]	uniform
EPSCUT	0.1	[0-0.25]	Triangle
EPSCU2	0.2	[0-0.35]	Triangle
TAUTO	983 K	[750-1200]	Triangle
XSTIA	0.75	[0.55-0.75]	Triangle
TJBRN	1060 K	[900-1900]	Triangle

3. ANALYSIS RESULTS

3.1. Accident Progression Analyses of Representative Sequences

In advance of uncertainty analyses, accident progression analyses have been performed for the representative DET scenarios. The selected accident sequences are large loss of coolant (LOCA), loss of offsite power (LOOP) with auxiliary feedwater system (AFWS), and LOOP without AFWS sequences which are shown in Table 1. Each sequence has three cavity conditions: 'flood'. 'wet', and 'dry' cases. For the LLFLD scenario, none of engineered safety features (ESF) such as high pressure safety injection system (HPSIS), containment spray, or reactor containment fan cooler is available. The only available system is a low pressure safety injection system (LPSIS). For the LPFLD scenario, the only water available to cool the core is from the HPSIS. For the LOOP sequences, the injected water from four safety injection tanks (SITs) or HPSIS is available only after the system pressure decreases after the reactor vessel failure.

Complete coverage of corium behavior both in-vessel and ex-vessel, and the corresponding containment responses, can be predicted in the MAAP code analyses. The in-vessel progressions include the thermal hydraulics in the primary system, core heat up, hydrogen generation, and melt progression up to the reactor vessel breach. The ex-vessel progressions include high pressure melt ejection, direct containment heating, gas combustion phenomena, molten-corium concrete interaction and the pressure behavior in the containment atmosphere. The values of the MAAP uncertain input parameters for these scenarios are taken from the default values in Table 3. The calculation results for the timing of key events and the pressure loads in the containment are summarized in Table 4.

Table 4: Timing of key events occurrence and containment pressure load for the representative DET scenarios in OPR-1000

Sequence ID	Simulated Scenario		Timing of Key Event Occurrence (seconds)			Containment Pressure (MPa)
	Initiating Event	Safety System Availability	Steam Generator Dryout	Core Uncovery	Reactor Vessel Failure	Peak Pressure at 72 hours
LLFLD	Large Loss of Coolant Accident	LPSIS	No Dryout	< 10.0	21,170	1.384
LLWET		SIT	No Dryout	< 10.0	9,796	1.055
LLDRY		N/A	No Dryout	< 10.0	5,043	0.774
LPFLD	Loss of Offsite Power Accident	HPSIS	5,340	6,994	14,914	1.296
LPWET		SIT	5,340	6,994	14,914	1.244
LPDRY		N/A	5,340	6,994	14,914	0.812
LPFLDSG		AFWS, HPSIS	55,869	60779	75,089	0.734
LPWETSG		AFWS, SIT	54,516	59,392	73,360	1.031
LPDRYSG		AFWS	54,516	59,392	73,360	0.649

3.2. Uncertainty Analysis

The results of the 200 MAAP analyses constitute samples of the distribution of the containment pressure load related variables given the uncertainties expressed in Table 3. In this study, any dependency between parameters was not considered in the sampling process, and thus all parameters were treated as independent. The results of all 200 MAAP analyses of the uncertain code parameters for the 9 scenarios are shown in Table 5. Since this application was focused on determining an estimate of the pressure load in the containment building, the calculation results of the relevant variables are shown from Fig. 2 through Fig. 7 for the case of the LPWER scenario as examples. The figures show the calculation results of the pressure behavior in the containment building, axial concrete erosion depth in the cavity, hydrogen combustion mass in the containment, and their distributions.

Table 5: Calculation results for the pressure load related variables of the late containment failure in the uncertainty analysis

Calculation Scenario ID	Peak Pressure (MPa)		Axial Concrete Erosion (m)		H_2 Burn Mass (kg)	
	Mean	Deviation	Mean	Deviation	Mean	Deviation
LLFLD	1.428	0.037	0.463	0.396	11.2	77.0
LLWET	1.044	0.023	3.10	0.84	6.56	0.78
LLDRY	0.776	0.053	3.74	1.19	1695	183
LPFLD	1.292	0.041	0.065	0.157	108.2	129.9
LPWET	1.228	0.050	2.21	0.67	127.2	295.5
LPDRY	0.767	0.026	3.02	0.94	1500	445
LPFLDSG	0.753	0.023	0.005	0.007	86.8	115.1
LPWETSG	0.990	0.069	0.99	0.56	75.6	112.9
LPDRYSG	0.592	0.056	2.13	0.85	127.2	295.5

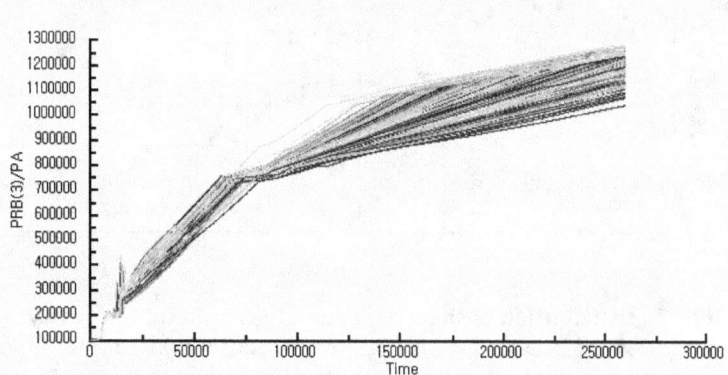

Fig. 2: Pressure behavior in the containment building
(LPWET case) (time: seconds)

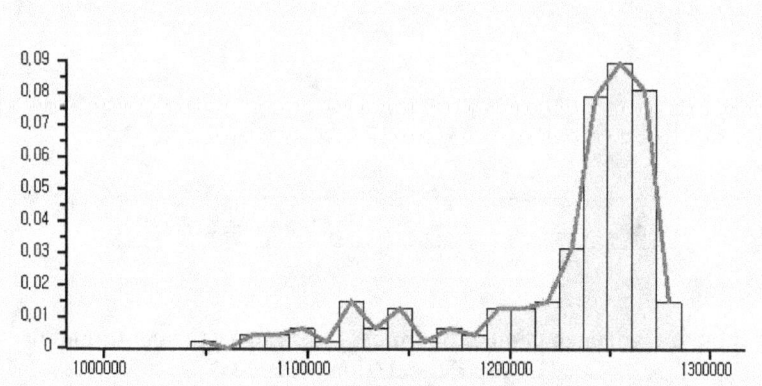

Fig. 3: Distribution of peak pressure in the containment building
(LPWET case) (Mean: 1.228 MPa, Deviation: 0.050 MPa)

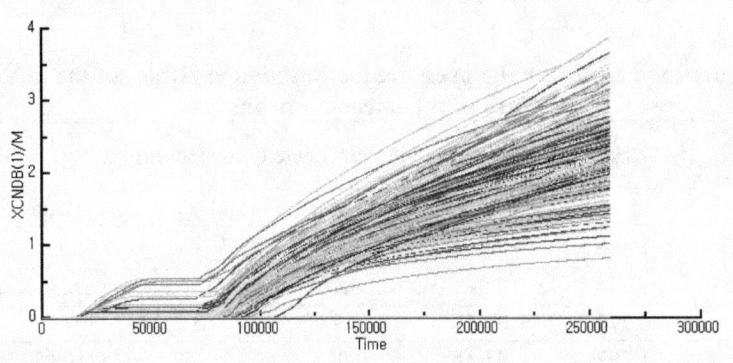

Fig. 4: Axial concrete erosion behavior in the cavity
(LPWET case) (time: seconds)

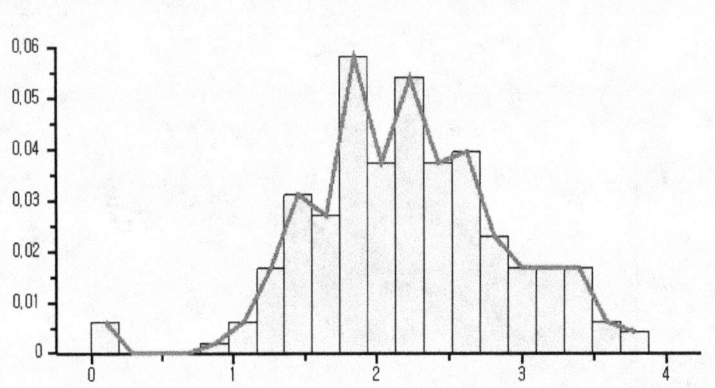

Fig. 5: Distribution of the axial concrete erosion in the cavity
(LPWET case) (Mean: 2.21 m, Deviation: 0.67 m)

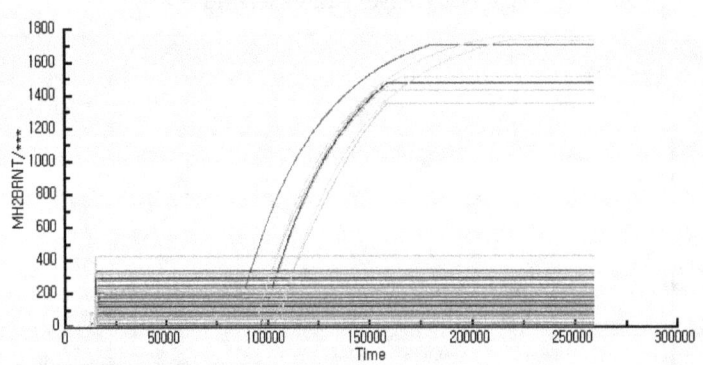

Fig. 6: Hydrogen combustion mass behavior in the containment
(LPWET case) (time: seconds)

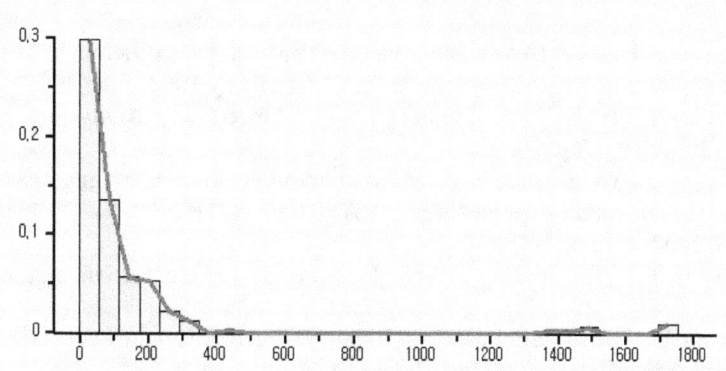

Fig. 7: Distribution of the hydrogen combustion mass in the containment
(LPWET case) (Mean: 127.2 kg, Deviation: 295.5 kg)

4. SUMMARY AND CONCLUSION

The phenomenology of severe accidents is extremely complex. Severe accident evaluation methodologies are associated with large uncertainties. Thus, a quantitative evaluation of the uncertainties associated with the results of a level 2 PSA requires knowledge of the uncertainties in the severe accident phenomenology. Such epistemic uncertainties are the major source of uncertainty in the results of a level 2 PSA [6].

In this paper, a sampling-based phenomenological uncertainty analysis was performed to statistically quantify uncertainties associated with the pressure load of a containment building for a late containment failure evaluation, based on the key modeling parameters employed in the MAAP code and random samples for those parameters. Phenomenological issues surrounding the late containment failure mode are highly complex. Included are the pressurization owing to steam generation in the cavity, molten corium-concrete interaction, late hydrogen burn in the containment, and the secondary heat removal availability. The methodology and calculation results can be applied for the optimum assessment of a late containment failure model. The accident sequences considered were a loss of coolant accidents and loss of offsite accidents expected in the OPR-1000 plant. As a result, uncertainties addressed in the pressure load of the containment building were quantified as a function of time.

A realistic evaluation of the mean and variance estimates provides a more complete characterization of the risks than conservative point value estimates. Therefore, the analysis methodologies demonstrated by these phenomenological uncertainty studies can be much preferable over deterministic methods employing a conservative selection of the code parameters. This methodology provides an alternative to simple deterministic analyses and sensitivity studies for use in the containment performance analysis of a level 2 PSA, and provides insight into identify the additional research area to reduce the uncertainties associated with severe accident phenomena by an investigation of the responsible uncertain parameters, and provides a useful tool in establishing risk-informed or severe accident related regulation to the nuclear industry.

Acknowledgements

This work was supported by Nuclear Research & Development Program of the National Research Foundation of Korea (NRF) grant, funded by the Korean government, Ministry of Science, Ict & future Planning (MSIP).

References

[1] Fauske & Associates, LLC, "MAAP4 Modular Accident Analysis Program for LWR Power Plants User's Manual", Project RP3131-02 (prepared for EPRI), (May 1994–June 2005)

[2] Ho G. LIM, Sang H. HAN, "Development of T/H Uncertainty Analysis S/W MOSAIQUE", Proceeding of KJPSA 10, May, 2009.

[3] Kwang I. Ahn, Joon E. Yang, Dong H. Kim,"Methodologies for uncertainty analysis in the level 2 PSA and their implementation procedures", KAERI/TR-2151/2002, KAERI Technical Report, Daejeon, Korea, 2002.

[4] Helton, J.C., Davis, F.J., "Illustration of Sampling-Based Methods for Uncertainty and Sensitivity Analysis", Risk Analysis, vol.22(3), p.591-622, 2002.

[5] Crécy, A., Bazin, P., "BEMUSE Phase III Report: Uncertainty and Sensitivity Analysis of the LOFT L2-5 Test", NEA/CSNI/R(2007)4, OECD, 2007.

[6] Hossein P. Nourbakhsh and Thomas S. Kress, "Assessment of Phenomenological Uncertainties in Level 2 PRAs", USNRC, OECD Workshop Proceeding, Aix-en-Provence, November, 2005.

Comparison of Uncertainty and Sensitivity Analyses Methods Under Different Noise Levels

David Esh[a*] and Christopher Grossman[a]
[a] US Nuclear Regulatory Commission, Washington, DC, USA

Abstract: Uncertainty and sensitivity analyses are an integral part of probabilistic assessment methods used to evaluate the safety of a variety of different systems. In many cases the systems are complex, information is sparse, and resources are limited. Models are used to represent and analyze the systems. To incorporate uncertainty, the developed models are commonly probabilistic. Uncertainty and sensitivity analyses are used to focus iterative model development activities, facilitate regulatory review of the model, and enhance interpretation of the model results. A large variety of uncertainty and sensitivity analyses techniques have been developed as modeling has advanced and become more prevalent. This paper compares the practical performance of six different uncertainty and sensitivity analyses techniques over ten different test functions under different noise levels. In addition, insights from two real-world examples are developed.

Keywords: Uncertainty, Sensitivity, Probabilistic, Model, Radioactive.

1. INTRODUCTION

Decisions regarding the capability of radioactive waste disposal facilities to contain and isolate radioactive waste from the environment are typically informed by the results of performance assessments. Performance assessments of radioactive waste disposal facilities use modeling to estimate potential radiological risk in the distant future, because the facilities cannot be monitored indefinitely. Therefore, modeling is used and uncertainties must be accounted for to support reliable decision making. The performance assessments commonly have a large number of uncertain inputs, and complex, non-linear, and sometimes non-monotonic responses of the outputs. The performance assessment can be thought of as a complex, non-linear, multi-dimensional numerical experiment. A performance assessment can involve many models to represent a variety of different features, events, and processes. Commonly the performance and degradation of engineered barriers, geochemistry and release of radionuclides from a wasteform, hydrogeologic transport through environmental media, and exposure to humans in the environment are represented by models within a performance assessment. However there are no defined limits on the number of models, the types of models, or how the models are integrated and interact. These models can interact in complex and non-intuitive ways. Decision-makers using the results of performance assessments are often interested in discerning which features, events, or processes are expected to significantly affect waste isolation and the level of uncertainty in the understanding of those features, events, and processes in order to improve confidence in the decision.

Sensitivity and uncertainty analysis is important to be able to develop understanding of the behavior of the numerical models, such as in performance assessments, focus research on the real-world system to reduce uncertainties, as well as to efficiently focus review effort on the most significant aspects. Different researchers use different terminology to refer to the process of estimating the important input parameters in a computational model incorporating uncertainty. This paper examines the sensitivity of outputs to uncertain inputs for a variety of models.

A number of issues arise when sensitivity analyses are applied to performance assessments. First, the number of uncertain variables may be very large while the truly important variables from the standpoint of driving the variance in the output may be quite small. For many performance

---
[*] david.esh@nrc.gov

assessments of radioactive waste disposal facilities uncertain parameters often number in the hundreds or greater whereas only on the order of tens of them are typically believed to be "important" to waste isolation. The large numbers of inputs often result in significant noise when discerning the sensitivity of the complex performance assessments to input parameters. 'Noise' is defined here as anything that can confound the identification of the truly important uncertain variables.

There are various types of noise that can impact sensitivity analysis techniques. The first type of noise considered in this evaluation is unimportant input distributions. As mentioned previously, a performance assessment may have hundreds or in a few cases thousands of uncertain input distributions to represent variability and uncertainty in physical processes and phenomena. This is primarily driven by the radiological source term that has a large number of different isotopes (e.g. 50 to 100). Processes that represent the potential release and transport of these isotopes through the environment are uncertain, and the parameters used to represent the processes are different for different elements. Generally only a handful of the parameters will be driving the variance in the results at a particular time or for a particular scenario. A good sensitivity analyses technique should be able to avoid identifying unimportant inputs (confounding influences) as important while identifying the important inputs. The second type of noise (e.g. Gaussian noise) represents irreducible uncertainty in the output, such as a result of measurement uncertainty. Many different types of features and processes can contribute to this second type of noise. These sources of noise could lead to failure to identify an important input as well as improperly classifying an important input as non-important.

A second issue that arises when sensitivity analyses are applied to performance assessments results because performance assessments can be quite complex and the computational burden to execute them can be significant. A single probabilistic realization can take anywhere from seconds to days. At the higher end of the range of execution times, the input and output data produced would generally be considered to be sparse or very sparse. Finally, the output response is generally not normally distributed, may span many orders of magnitude, may contain many zero (null) results, and is dynamic with time. The sensitivity analyses techniques need to be able to handle these challenges. This paper compares the performance of six different uncertainty and sensitivity analyses techniques over ten different test functions under different noise levels. In addition to the ten analytical test functions, most of the methods are also evaluated against some real-world datasets from more complex computational models.

2. METHOD

Analytical test functions described in the literature were implemented in a probabilistic computational model using the GoldSim® dynamic simulation platform. Table 1 provides the test functions used in the analyses. The test functions were selected to provide different amounts of non-linearity and non-monotonicity in functions of different dimension. The dimensionality of the functions range from 2 to 10 as implemented in this analysis.

To represent the various types of noise that can impact sensitivity analysis techniques, seven additional uncertain input parameters that have no impact on the output of each test function were included in the determination of the sensitivity of the output to the input. The seven additional uncertain parameters were varied uniformly from zero to one. Further, to all of these test functions, a noise term, to represent irreducible uncertainty, was added to the output of the function prior to performing the sensitivity analysis. The noise term was developed using a discrete probability distribution that could be adjusted from 0 (no noise) to 5 (high noise) and a unit normal distribution with the standard deviation adjusted as follows. The range of each function without the noise was first determined, and then the standard deviation of the unit normal distribution was adjusted so that the noise would be of the same magnitude as the product of the range of the function without noise and the value of the discrete probability distribution mentioned previously (i.e. 0 to 5). Figure 1 shows the simple function output as a function of the two inputs (x_4 and x_5) for a noise level of 0.1 and a noise level of 1.

Table 1: Analytical Test Functions

Test Function Name	Description	Input Description[a]	Ref #
Simple	$2x_1 + 5x_2$	$x_1 = U[-1,1]$ $x_2 = U[-1,1]$	NA
Moon	$x_1 + x_2 + 3x_1 x_3$	$x_1, x_2, x_3 = U[0,1]$	[1]
Webster	$x_1{}^2 + x_2{}^3$	$x_1 = U[1,10]$ $x_2 = N[\mu=2, \sigma=1]$	[2]
Eldred	$\dfrac{x_1}{x_2}$	$x_1 = LogN[\mu=1, \sigma=0.5]$ $x_2 = LogN[\mu=1, \sigma=0.5]$ $r = 0.3$ (x_1 and x_2)	[3]
Park	$\dfrac{x_1}{2}\left[\sqrt{1 + (x_2 + x_3{}^2)\dfrac{x_4}{x_1{}^2}} - 1\right] + exp[1 + sin(x_3)]$	$x_1, x_2, x_3, x_4 = U[0,1]$	[4]
Currin	$\left[1 - exp\left(\dfrac{-1}{2x_2}\right)\right]\dfrac{2300x_1{}^3 + 1900x_1{}^2 + 2092x_1 + 60}{100x_1{}^3 + 500x_1{}^2 + 4x_1 + 20}$	$x_1, x_2 = U[0,1]$	[5]
Ishigami	$sin(x_1) + a\, sin^2(x_2) + b\, x_3{}^4 sin(x_1)$	$x_1, x_2, x_3 = U[-\pi, \pi]$	[6]
G-Function	$\displaystyle\prod_{i=1}^{d}\dfrac{\lvert 4x_i - 2 \rvert + a_i}{1 + a_i}$ where $a_i = \dfrac{i-2}{2}$ for all $i = 1, \ldots, d$	$x_i = U[0,1]$; d=3, 5, or 10	[7]

[a] U = uniform, N = normal, LogN = lognormal, r = correlation

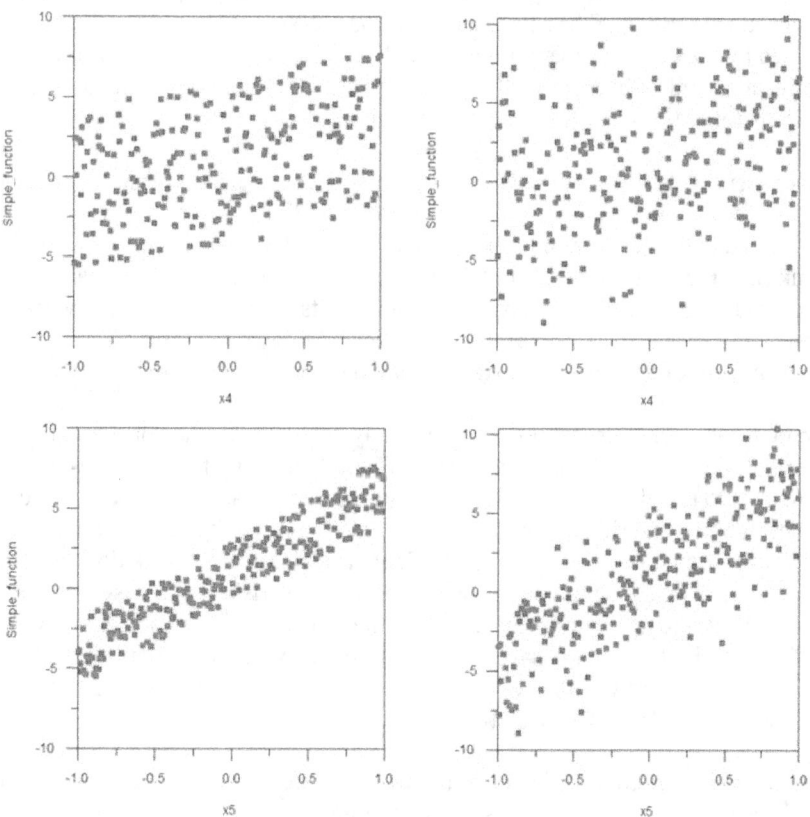

Figure 1: Comparison of the Simple Function Output for Two Different Noise Levels (a = 0.1, b = 1)

The six sensitivity analysis techniques evaluated comprised regression-based and other methods. Regression-based methods are most useful for monotonic relationships (i.e., the dependent variable y only increases or decreases with increasing value of the independent variable x). When this is not the case, regression based methods are at a disadvantage because a non-monotonic relationship can be highly correlated but have a low correlation coefficient. The six uncertainty and sensitivity techniques evaluated were the importance measure, correlation coefficient, standardized regression coefficient (SRC), partial correlation coefficient, a genetic algorithms based measure (GA-m), and a method based on the Extended Fourier Amplitude Sensitivity Test (extended FAST) [8,9,10,11,12]. Outside of the GA-m measure, all of these methods are well-documented in the literature and therefore are not described in further detail here.

The genetic algorithm based measure was developed during the analysis of a performance assessment describing the degradation of a cement wasteform [12]. It was then evaluated on the data from performance assessment calculations associated with the proposed high-level waste repository at Yucca Mountain, Nevada [13]. In each of those cases the technique was found to perform well at identifying important inputs.

The GA-m approach used neural network software developed by Neuralware [14]. Neuralworks Predict® is an add-in to Microsoft Excel that can be used to build neural networks. The approach used was to export sampled stochastic input variables along with the pertinent output variables from the test functions or real-world datasets to Excel and then to build a neural network using Neuralworks Predict. The variable selection algorithms were used to select the most important input variables needed to develop a neural network to predict the output. The neural network itself was not used, just the variable selection algorithm. The variable select algorithm uses a genetic algorithm to search for synergistic sets of input variables that are good predictors of the output. The software also can perform a pre-selection of variables using a cascaded genetic algorithm approach. This method gives more consistent variable sets by eliminating variables that are consistently rejected during different invocations of the genetic algorithm. The sensitivity analysis measure (GA-m) was the frequency that a variable was retained by the GA over many iterations. The GA-m technique may work well on performance assessment-type data because of the relatively large (compared to other techniques) amount of variable transformations that are used by the technique.

The analyses in this paper for the first four techniques (importance measure, correlation coefficient, standardized regression coefficient (SRC), partial correlation coefficient) were performed on the raw data and rank transformed data. Rank transformation, a dimensionless transform, replaces the value of a variable by its rank (i.e., the position in a list that has been sorted from largest to smallest values). Analyses with ranks tend to show a greater sensitivity than results with untransformed variables. For performance assessments, if the distribution of results is skewed toward the low end, which is usually the case, rank transformation of the dependent variable can over-weight the lower results.

Variable transformation can be an important step in the analysis process. The correlation between input and output variables in statistical methods can often be enhanced by transforming the variables. Transformations are used to (i) eliminate dimensionality of the variables, (ii) reduce the role of points at the tails of the distributions, and (iii) properly scale the resulting sensitivities to the variability of the input variables. However, transformation of the dependent variable (usually peak dose in performance assessments) can skew the results in an undesirable way because the risk, being based on the average dose, is weighted heavily by the largest doses. Therefore, the sensitivities should reflect what matters most to the risk and give weights proportional to the doses. Although transformations of the dependent variable often will improve the goodness of fit in the sensitivity analysis, in general they should only be applied to the independent variables.

The ten test functions varied in complexity in terms of non-linearity and the dimensionality of the function. The coefficient of determination was calculated and used to determine the fraction of the variance in the output that can be explained by a linear relationship for each test function. Results were generated for all of the test functions, and were further categorized into the five functions that were more linear (i.e. a high coefficient of determination) and were more non-linear (i.e. a low coefficient of determination). In order to evaluate the performance of the techniques, the numerical value of the uncertainty measure was ranked and the statistics across all functions were compiled. For example, if a function had two uncertain inputs (A and B) and seven uncertain dummy variables (C through I), then a rank of 1 and 2 for the uncertainty measure of inputs A and B would mean the technique successfully identified the two important inputs and avoided spurious correlation with a noise input. The GA-m technique was different from the other techniques in terms of its output because inputs are screened to determine those that are suitable for building the neural network. The results are a confusion matrix whereas the other techniques only produce a ranking of the numerical values. For metrics that can have both positive and negative values as the output (e.g. correlation coefficient) the absolute value was used to determine the ranks for scoring purposes.

3.0 RESULTS

3.1. Test Function Analyses

In addition to the magnitude of the noise added to the output of the functions, the analysis also considered the number of probabilistic realizations. As previously discussed, performance assessments can be complex and computationally expensive. In some cases generating more than a few hundred realizations may be impractical. Figure 2 provides the results of the sensitivity analyses techniques for 250 realizations with no noise added to the output function. The metric plotted is the average classification rate over the ten test functions. The extended FAST and importance factor consistently rated highest, followed by the GA-m. However, the results for the total sensitivity index for the extended FAST method (STi) may not be directly comparable to the other methods. To perform the analyses to produce Figure 2, all of the other methods were given 250 realizations of results. For the extended FAST method shown in the chart, 65 outputs were generated for each input and 4 replicates were performed. For a function with 10 inputs and 7 "dummy" inputs, a total of 4420 calculations of the function would be performed. Calculations were performed with 10,000 realizations for the other methods and the classification percentages increased 10-15%, closing the gap on the STi of the extended FAST method though it was still superior.

Figure 3 provides the results of the sensitivity analyses techniques as a function of the number of realizations with no noise function added to the output of the functions. The more common (and simpler) measures show limited sensitivity to the number of realizations over the range investigated. However, the importance factor showed a strong decrease in performance as the number of realizations decreased. Even at the lowest end of the range shown in Figure 3 the ratio of the number of realizations to inputs (~ 25) is much higher than many performance assessment applications. In some cases the ratio of the number of realizations (calculations of model output) to uncertain inputs can be less than unity.

The results were further broken down by separating the results into those associated with the five simpler test functions and those associated with the more complex test functions (the last five in Table 1). The distinction between simpler and complex was made on the basis of the resultant coefficient of determination and the dimension. The mean coefficient of determination of the simpler functions was 0.87 compared to 0.19 for the more complex test functions.

The mean dimension of the simpler functions was 2.6 compared to 4.6 for the more complex functions. Table 2 provides the difference in overall classification percentages for the sensitivity analyses methods between the simpler and more complex test functions as a function of noise level (250 realizations). In general, the performance on the complex test functions is lower, as would be expected. For all the methods at moderate to low noise levels the performance is comparable. However at high noise levels the importance method and GA-m show deteriorating performance for the simple test functions. For the complex test functions, the performance of the importance measure and GA-m is better than the simpler methods, with the importance measure really standing out. The simpler methods struggle with the complex functions.

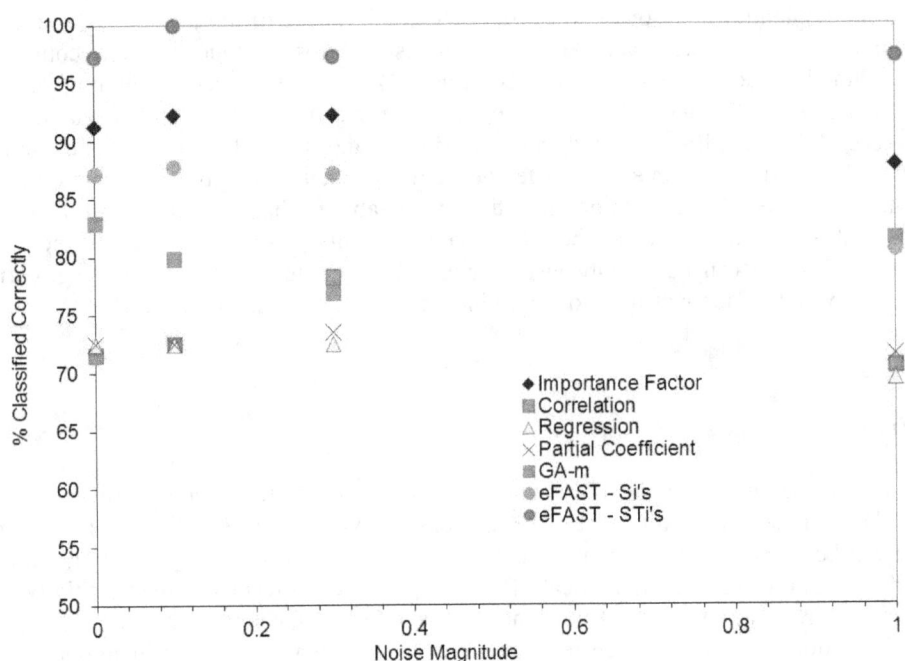

Figure 2: Impact of the Magnitude of Noise on Average Classification Rate (Number of Realizations = 250)[†]

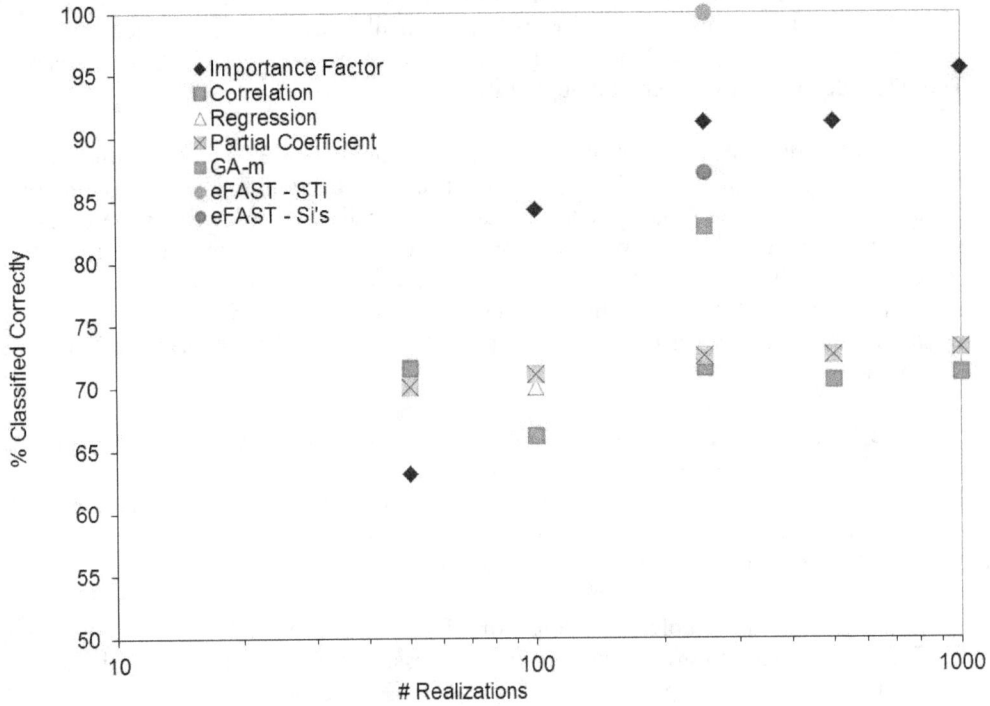

Figure 3: Impact of the Number of Realizations on Average Classification Rate (Noise level = 0)

[†] As discussed in the text, the results for the extended FAST method may not be directly comparable. Not all methods were executed for each comparison provided in this paper.

Table 2: Performance for Simple vs. Complex Functions (% Classified Correctly)

	Noise	Importance	Correlation	Regression	Partial Coefficient	GA-m
Simple	0	92	100	92	92	92
	0.1	92	100	92	92	92
	0.3	92	100	100	100	92
	1	85	92	92	100	92
	3	62	85	85	85	54
	5	54	69	85	77	46
	Noise	Importance	Correlation	Regression	Partial Coefficient	GA-m
Complex	0	83	48	52	52	65
	0.1	87	57	52	52	70
	0.3	87	57	52	52	65
	1	83	43	43	43	65
	3	78	52	48	48	57
	5	65	43	48	48	48

For performance assessment, it is very important that the sensitivity analysis method can determine the important inputs when there are a very large number of non-contributing inputs and a low ratio of results to uncertain inputs. To evaluate the sensitivity of the results to the number of non-contributing inputs, the number of "dummy" inputs was increased from 7 to 20 for each test function. For 250 realizations and a noise level of 1.0, the performance of the importance measure, correlation coefficient, standardized regression coefficient, and partial correlation coefficients did not change significantly for the simple test functions (compared to the results in Table 2). However, for the complex test functions the importance measure held up well (65% vs. 83%), but the correlation coefficient, standardized regression coefficient, and partial correlation coefficients performance decreased substantially. In fact the classification percentages are no better than would be expected by chance (22%). Resources did not allow the authors to complete similar analyses with the GA-m and extended FAST methods though they may be evaluated for sensitivity to increasing the number of inputs in future research.

3.2. Performance Assessment Analyses

The final analysis that was performed was comparison of some of the methods on real world datasets from performance assessments. The performance assessments are very complex, can take an extremely long time to execute, and have a very large number of uncertain inputs. The output can have many zero results and dynamic behavior that change over the many thousands or hundreds of thousands of years of simulated performance. Many methods struggle with this type of problem due to the sparseness of data to analyze or inherent limitations of the methods with complex datasets. In comparison to a performance assessment, the test functions evaluated in the analysis are extremely computationally efficient. Producing 10,000 realizations of output required only a second of computer time.

The Nuclear Regulatory Commission (NRC) staff has a long history of performing sensitivity analysis on performance assessment and other complex models. Much of that work has recently been summarized in Mohanty et al. [13]. Many methods were applied to the same dataset. The dataset was produced by an assessment describing the performance of a proposed high-level nuclear waste repository, the Total-system Performance Assessment code (TPA) [15]. The TPA code has on the order of 1,000 parameters, of which 200-400 are typically sampled in a given calculation and 40-50 are cross-correlated to other sampled parameters. A large variety of sensitivity analyses techniques were applied to TPA code output. NRC staff experience has found that the best way to identify important parameters in these large complex models is to run a variety of techniques and rank the parameters based on how many of the techniques identify the parameters as important. The GA-m method was then run on the same dataset produced by the TPA code that was evaluated with the other methods. The selection algorithm identified six parameters, all with a high frequency. The parameters pertained to the degradation of the waste form, the quantity of radionuclides directly consumed in drinking water, or the dilution of contaminants in the saturated zone during transport and transport time. From a physical standpoint, the parameters selected were in strong agreement with the analysts' conceptual understanding of the performance of the disposal system. Using modified raw data, the GA-m method successfully identified 6 of the 10 parameters ranked as most significant by compiling the results across all methods. Screening out realizations that had zero doses (about 200 of the 512) identified 7 of the 10 parameters. Using both sets of results identified 8 of the 10 parameters, which was a very strong result for an individual method.

All of the methods except the extended FAST were applied to a second performance assessment model. This model describes the degradation of a cement wasteform and potential release of radioactivity into the environment for 41 species (isotopes) including decay chains [12]. Figure 4a shows the time histories of projected dose (mean values) for 1,000 realizations. Figure 4b is the distribution of peak doses, and Figure 4c is the distribution of the time of peak doses. The peak dose and the time of the peak dose were moderately correlated (-0.26). The sensitivity analysis techniques

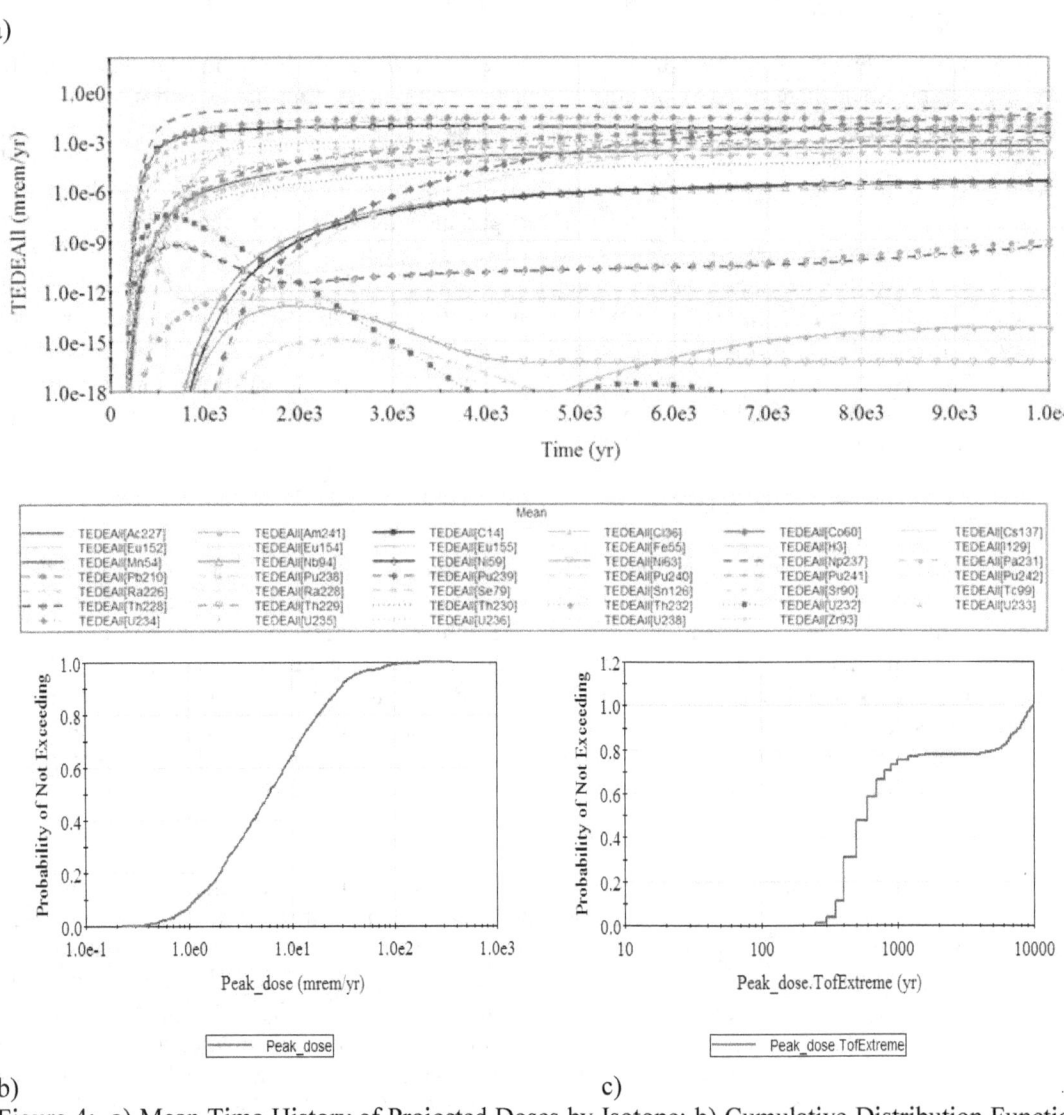

Figure 4: a) Mean Time History of Projected Doses by Isotope; b) Cumulative Distribution Function of Peak Doses; c) Distribution of the Time of Peak Doses.

were applied to this model. The model had 374 uncertain input parameters, some of which were correlated. Latin Hypercube Sampling (LHS) was used to sample the parameters for probabilistic simulations.

One of the problems analyzing performance assessments is that the "true" answer is not known. The models provide projections into the future but validation cannot be achieved in the strict sense. Rather, indirect methods must be used to provide confidence in the model results. To evaluate if the sensitivity analysis methods results are reasonable, the analyst has to use the information such as that shown in Figure 4 combined with interaction with subject matter experts (if the analyst is not familiar with the features, events, and processes represented in the model). Analysts can use their experience to determine if the results of the sensitivity analysis methods make sense. For instance, if the sensitivity analysis method identifies parameters associated with an isotope that contributes a trivial amount to peak dose, it is likely that sensitivity analysis results at and below that level are indeterminate at best.

Table 3 provides the list of the top ten parameters identified by the sensitivity analysis techniques. The parameters highlighted in green are, based on experience, believed to be accurate determinations of important parameters. The parameters highlighted in blue are parameters that can confidently be determined as being insignificant. The parameters highlighted in yellow are plausible that can't be confirmed or eliminated. The standardized regression coefficient method provided the poorest results, whereas the GA-m was the best performer. There was a rather substantial improvement in the results for the importance measure when the ranks were used as compared to the raw data.

To test the reasonableness of the results, a multivariate plot of the top two parameters from the GA-m method was developed. For large, complex datasets visualization can be useful especially to confirm or refute the results of analysis methods. Figure 4a and 4b are the multivariate plots of the raw data and ranks of the total dose at 10,000 years as a function of the bound waste degradation rate parameter and the fracture spacing parameter. The plots confirm the importance of these parameters to the results.

By cautiously eliminating most of the inputs that contribute very little to the output variance, the ability of the sensitivity analysis methods to identify the important parameters can be significantly enhanced. This is in agreement with the results presented earlier that showed as the number of "dummy" inputs was increased the classification percentages decreased. In other words, and experience-based screening step of the initial results can greatly improve the identification of important parameters. As discussed in [12], the number of important parameters that were identified with confidence was doubled when the same technique was applied on a shortened variable list.

Table 3: Performance Assessment Model Top Ten Parameters Identified

Importance	Importance - Ranks	Correlation	Regression	PCC	GA-m
Frac Spacing	Frac Spacing	Tc	SZ pipe length	Tc	Bnd W deg rate
H Sol	GW flow	Frac Spacing	Kd UZ Np	Nm	Frac Spacing
Kd W Pu Ox	Water Intake	GW flow	Cap life	Kd SZ Cs	GW flow
Kd Soil Cl	Nm	Kd UZ Se	U Sol	Erosion rate	Infiltration rate
Kd W Am deg	Soil T lateral	Water Intake	Kd SZ I	SO4 conc	Water Intake
Pb Sol	Kd Sz Ac	Kd W U Ox	Sr Sol	Np Sol	Kd W Tc
Kd Soil Cs	Kd UZ Sn	C	Frac Diet Animal	Kd UZ Tc	Nm
Kd Garden C	Kd W Pb Ox	Th Sol	Tc	Kd SZ Pa	SO4 conc
Rn	Soil int farm	Cl Sol	Th Sol	Fe	Kd W Np
Kd UZ Cs	Kd W Mn deg	Kd SZ Mn	Porous length	Co Sol	Kd W Pu

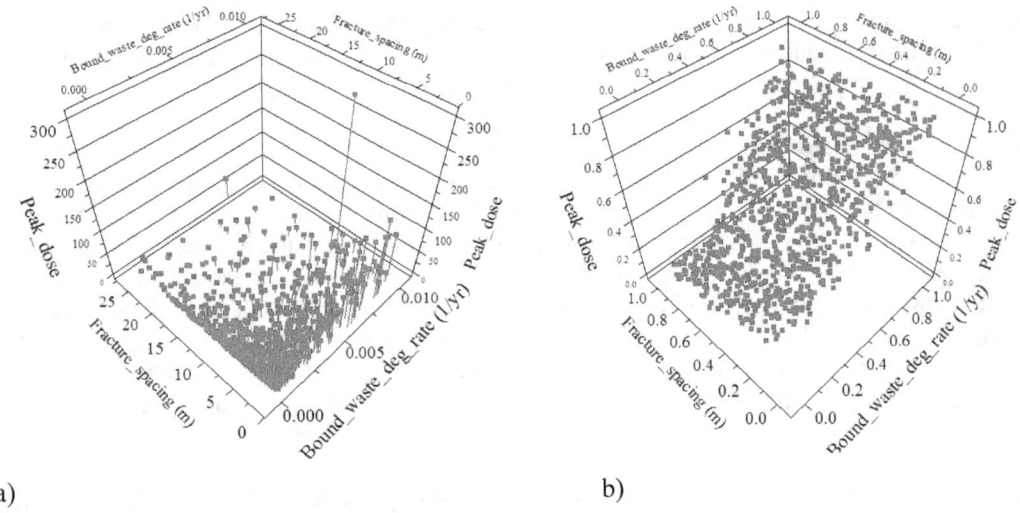

Figure 4: a) Multivariate Plot of the Top Two Parameters Identified with the GA-m Method for a) the Raw Data and b) the Ranks of the Data

4. CONCLUSIONS

Developing sensitivity analysis methods for application to performance assessments is a challenging problem. Computational limitations may limit the number of realizations (calculations) that can be performed to provide data to the methods. In addition, the number of uncertain inputs can be many hundred or more. The results of the analyses identified methods that had more computationally efficient performance over others (i.e. the extended FAST). Some methods were more sensitive to noise, and some showed greater sensitivity to the amount of data available. Some techniques were much more favorable to use when non-linearity was present. Overall conclusions were:

- Simpler methods are reliable for simpler functions.
- The total sensitivity index (STi) of the extended FAST method produced very strong results for the cases where it was evaluated. Future research may involve more extended evaluation of this method on performance assessment models if computational resources allow.
- Applying numerous methods and averaging or ranking the results appears to lead to the most reliable identification of important parameters.
- The GA-m method produced adequate results on the test functions with moderate to low noise levels, but much stronger results on performance assessment data. At this time it is not clear why and may be a subject of future research.
- The importance measure using ranks performed much better on performance assessment data compared to other methods and compared to using unranked data.
- A screening step to remove clearly unimportant uncertainties can greatly improve the results for many techniques, especially when the number of inputs is large relative to the number of probabilistic realizations. Using expert judgment as a screening step to remove unimportant variables may improve results as much or more than advances in different sensitivity analyses techniques.

References

[1] H. Moon, "*Design and Analysis of Computer Experiments for Screening Input Variables*", Doctoral dissertation, Ohio State University, Columbus, OH, (2010).
[2] M. D. Webster, M. A. Tatang, and G. J. McRae, "*Application of Probabilistic Collocation Method for an Uncertainty Analysis of a Simple Ocean Model*", Technical report, MIT joint program on the science and policy of global change, report series no. 4, MIT, (1996).

[3] M. S. Eldred, H. Agrawal, V. M. Perez, S. F. Wojtkiewicz, and J. E. Renaud, "*Investigation of Reliability Method Formulations in DAKOTA/UQ*", Structure and Infrastructure Engineering, 3(3), pp. 199-213, (2007).
[4] D. D. Cox, J. S. Park, C. E. Singer, "*A Statistical Method for Tuning a Computer Code to a Data Base*", Computational Statistics and Data Analysis, 37(1), pp. 77-92, (2001).
[5] C. Currin, T. Mitchell, M. Morris, and D. Ylvisaker, "*Bayesian Prediction of Deterministic Functions, with Applications to the Design and Analysis of Computer Experiments*", Journal of the American Statistical Association, 86(416), pp. 953-963, (1991).
[6] T. Ishigami and T. Homma, "*An importance quantification technique in uncertainty analysis for computer models*", Proceedings of the First International Symposium on Uncertainty Modeling and Analysis, IEEE, pp. 398-403, (1990).
[7] A. Marrel, B. Iooss, B. Laurent, and O. Roustant, "*Calculation of Sobol Indices for the Gaussian Process Metamodel*", Reliability Engineering and System Safety, 94(3), pp. 742-751, (2009).
[8] J. R. Benjamin and C. A. Cornell, "*Probability, Statistics, and Decision for Civil Engineers*", McGraw-Hill, 1970, New York.
[9] A. Saltelli and S. Tarantola, "*On the Relative Importance of Input Factors in Mathematical Models: Safety Assessment for Nuclear Waste Disposal*", J. Am. Stat. Ass., Vol. 97, No. 459, (2002).
[10] R. L. Iman, et al., "*A FORTRAN Program and User's Guide for the Calculation of Partial Correlation and Standardized Regression Coefficients*", NUREG/CR-4122, SAND85-0044, US Nuclear Regulatory Commission, Sandia National Laboratory, (1985).
[11] A. Saltelli, S. Tarantola, and K. P.-S. Chan, "*A Quantitative Model-Independent Method for Global Sensitivity Analysis of Model Output*", Technometrics, 41:1, pp. 39-56, (1999).
[12] D. W. Esh, A. C. Ridge, and M. Thaggard, "*Development of Risk Insights for Regulatory Review of a Near-Surface Disposal Facility for Radioactive Waste*", Proceedings of the WM'06 Conference, February 26-March 2, 2006, Tuscon, AZ, (2006).
[13] S. Mohanty, et al., "*History and Value of Uncertainty and Sensitivity Analyses at the Nuclear Regulatory Commission and Center for Nuclear Waste Regulatory Analyses*", ML112720123, US Nuclear Regulatory Commission, (2011).
[14] Neuralware, NeuralWorks Predict® Product Version 2.40, Carnegie, PA, (2001).
[15] B. W. Leslie, C. Grossman, and J. Durham, "*Total-System Performance Assessment (TPA) Version 5.1 Code Module Descriptions and User Guide*", ML080510329, The Center for Nuclear Waste Regulatory Analyses, San Antonio, TX, (2007).

Understanding Relative Risk: An Analysis of Uncertainty and Time at Risk

A. El-Shanawany[a,b]

[a] Imperial College London, London, United Kingdom
[b] Corporate Risk Associates, London, United Kingdom
ashanawany@c-risk-a.co.uk

Abstract: Risk at nuclear facilities in the UK is managed through a combination of the ALARP principle (As Low As Reasonably Practicable), and numerical targets. The baseline risk of a plant is calculated through the use of Probabilistic Safety Assessment (PSA) models, which are also used to estimate the risk in various plant states, including maintenance states. Taking safety equipment out of service for maintenance yields a temporary increase in risk. Software tools such as RiskWatcher can be used to monitor the real time level of risk at plant. In combination with software tools to estimate the instantaneous risk, time at risk arguments are frequently employed to justify safety during plant modifications or maintenance activities. In this paper we consider the effect of using conservative estimates for the probability of failure on demand of safety critical components compared to using a full uncertainty distribution. It is found that conservatism in the base case model translates to a hidden optimism when used in time at risk arguments. While it is known and accepted that quantified risks are necessarily approximate, useful insights can be gained through risk modelling by considering relative risks. Anything that distorts relative risks impacts on the usefulness of the risk modelling. The important point of the effect discussed here is that it has the potential to distort relative risks. The mapping between the base case conservatism and the time at risk optimism is characterised, and the effect is illustrated using simple hypothetical examples. These simple examples show that the shape of the full uncertainty distributions of model parameters have important and direct consequences for time at risk arguments, and must be considered in order to avoid distorting the risk profile.

Keywords: PRA, Uncertainty, Bayesian, Risk, Modelling.

1. INTRODUCTION

Conservative estimates are a mainstay feature of probabilistic risk models used for safety analysis. Conservative arguments are frequently invoked, often in cases when it would be hard to confidently provide an accurate estimate, and are strongly linked with the assessment of uncertainty. The use of conservative arguments implicitly restricts the value of quantitative risk analysis (QRA) to statements such as "the frequency of core melt is lower than x per year". This fails to do justice to the potential uses of QRA. The value of QRA can extend beyond the identification of "negative insights" and high risk areas, to informing plant operators about "positive insights" such as where the safety margin is very high and could potentially be relaxed [1]. The issue of uncertainty in risk analysis has been discussed extensively in the literature, and the importance of an adequate representation of uncertainty has also been presented [2, 3]. This paper stresses the point that conservative arguments are not an adequate approach to uncertainty by demonstrating that conservative estimates distort the risk profile of the plant, sometimes in non-obvious ways. This lends extra weight to the viewpoint that conservative estimates should always be replaced with best estimates coupled with uncertainty estimates. The conservative distortion is illustrated using the concepts of time at risk and maintenance, in which case conservatisms can hide the true risk.

Time at risk is a fundamental concept when considering risk. In most quantitative risk models, time at risk is used to represent the effect of maintenance, the degradation of plant components, and their susceptibility to various hazards. This paper will explore the concept of time at risk and uncertainty in parameter estimates using the example of maintenance states and considering how plant unavailability due to maintenance affects the prediction of risk.

Maintenance is known to have both potentially positive and negative effects on the risk at a plant. Maintenance is used to identify and fix defects that occur due to anticipated wear and tear on plant equipment due to normal operation, and is vital to ensure the continued operation of the plant. However, incorrectly performed maintenance can also leave a plant in a worse state than before; the classic example is that of misaligned valves. In addition, during a maintenance outage safety equipment is unavailable to perform its duty, hence the plant incurs an increase in risk during the maintenance period. Given that maintenance can have a significant impact on the risk at a plant, it is important to be able to estimate all aspects of maintenance risk as precisely as possible in order to design maintenance schedules that are as close to risk optimal as possible. The significant effects of maintenance on risk at industrial facilities have been extensively discussed in the literature [4, 5, 6]. Although there are numerous dimensions to the interaction of maintenance with plant risk, this paper will start from the reasonable assumption that maintenance is essential and then consider only the impact due to unavailability of plant systems. It is demonstrated that the method used to estimate the failure parameters in the risk model has a significant effect on the estimation of the risk significance of maintenance outages.

Before proceeding, it is worthwhile to make more precise the key concepts used in this paper. It is noted that there are numerous formulations of the definition of risk, and that the word is often used differently in various contexts. Frequently, risk is defined as probability "multiplied" by consequence, and it is commonly expressed in terms of a frequency of an undesirable event per unit of time. This is an excellent intuitive description of risk, although it is noted that there are certain deficiencies with the definition, such as a precise definition of the multiplication operation. In this document Kaplan and Garrick's [7] definition of risk as a triplet consisting of scenario, probability and consequence is used. However, the consequence part of the triplet will be a constant throughout this document, and will simply be considered to be "failure to perform a prescribed safety function". The scenario will switch only between a nominal "normal state" and a "maintenance state". Hence, of the triplet defined above, the main concern in this paper is the probability component of risk. When considering time at risk, the concept of cumulative risk is key; at its most general, cumulative risk can be defined as the time integral of instantaneous risk. The instantaneous risk will refer to the risk at some specified time point, while the cumulative risk is the total risk experienced over a period of time. In this document "risk" will be used as a synonym for "instantaneous risk". For example, the risk due to a specified hazard depends on the length of time over which the hazard could potentially occur. Uncertainty itself is a complex topic, but in this document the phrase uncertainty is restricted only to statistical uncertainty. The expansion to the full range of uncertainties is considered in the further work section.

The cumulative risk and instantaneous risk in a maintenance case are shown in Figure 1 below.

Figure 1: Maintenance Outage and Instantaneous Risk

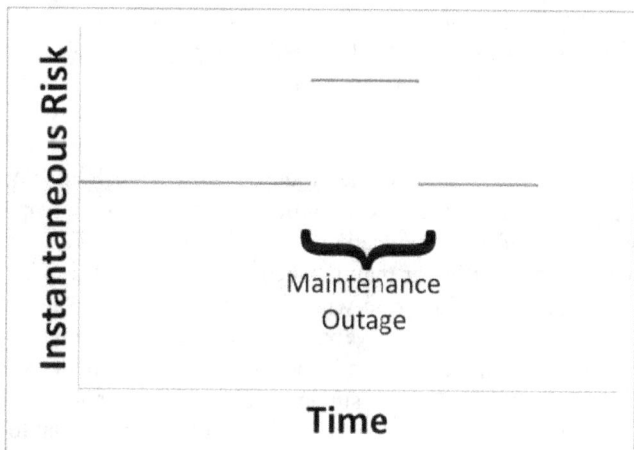

It should be noted that in Figure 1, above, a constant failure rate is assumed and the cumulative risk has been calculated using a Poisson distribution and is not actually proportional to the time; in particular the cumulative probability does not rise linearly to one. For realistic values of the Poisson parameter and short time periods the relationship is very close to being proportional to the time near the origin. In general the failure rate of a component is not constant, for example a bathtub shaped curve could be modelled using a Weibull distribution, but over the short time period that maintenance occurs, a constant failure rate is a reasonable assumption.

To consider maintenance cases in a time varying model, the variation with time of the probability of an event must be considered. A saw tooth curve is produced by the basic assumption that a component's failure rate is constant with time, and that maintenance perfectly restores a component to working order.

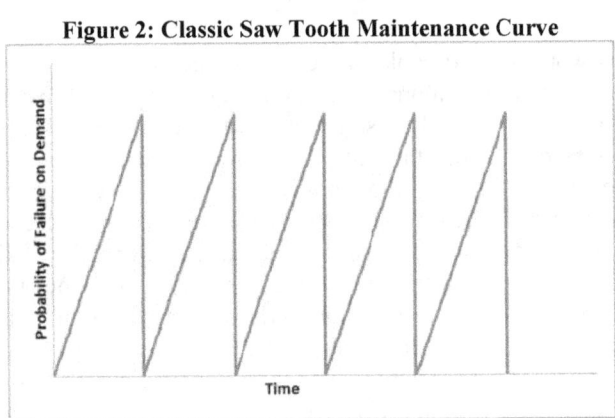

Figure 2: Classic Saw Tooth Maintenance Curve

The contribution of maintenance outages can be usefully considered as a proportion of "baseline risk". It is shown that the uncertain nature of the probability component can distort the relative importance of time effects depending on the way in which the uncertainty is handled. Historically, uncertainty in probability estimates has been handled by using conservative estimates for the probability. Despite a shift in thinking towards best estimate values, in practice conservative judgement is frequently invoked in difficult situations. The purpose of making the argument presented in this paper is to reinforce the viewpoint that even where a best estimate is difficult, it should not be replaced by a conservative estimate; in the opinion of the author a large uncertainty distribution (if necessary) is preferable to recourse to a conservative estimate. It is shown that the *combination* of conservative estimates, in the sense of multiple lines of protection, yields an *optimistic* viewpoint of the relative risk during time periods in which a line of protection is removed. This is first demonstrated by comparing conservative estimates and best estimates, and then the case of using best estimates plus uncertainty for the failure parameters is considered. The best estimate plus uncertainty method yields similar results as the best estimate method, but retains the possibility to interpret the results in a conservative way.

An appropriately designed maintenance schedule is an important contributor to the safe operation of an nuclear power plant (NPP). The issue of estimating the effect of maintenance is a multi-faceted problem, and numerous models of maintenance have been developed [4, 5], but the argument made in this paper is not dependent on the particular maintenance model used. Hence, for clarity, only the most simple model assuming a constant failure rate with time, as shown in Figure 2 above is considered here. Two cases are considered, using a very basic model in which safety systems have a fixed probability of failure on demand, that is time independent. Using this model we seek to estimate the increase in the risk due to unavailability of a single system. Time at risk is the key concept in this formulation since the risk due to a particular plant state is directly proportional to the time spent in that plant state using the simplifying assumptions. The theoretical consequences of conservative estimates of the probabilities of failure on demand are considered in Section 2. This is illustrated using a simple example in Section 3. Section 4 then further develops the argument to include uncertainty distributions for the failure parameters, rather than point estimates. Section 5 discusses the implications of the results and Section 6 presents areas for further work. Section 7 presents the conclusions of the paper.

2. THEORETICAL JUSTIFICATION

This section provides a theoretical justification for why conservative estimates of failure parameters lead to the underestimation of the contribution to the total risk of maintenance outages. This is done by building up the algebra for a simple system with 'n' diverse lines of protection.

Let X be a system with n protective systems, each of which provides an independent protective barrier to the failure of system X. Note, common cause failures are not considered in this setup, as each line of protection is considered to be a diverse system and hence genuinely independent of each other. Logically, this can be represented by the fault tree shown in Figure 3 below.

Figure 3: Simple Example – A System with n Lines of Protection

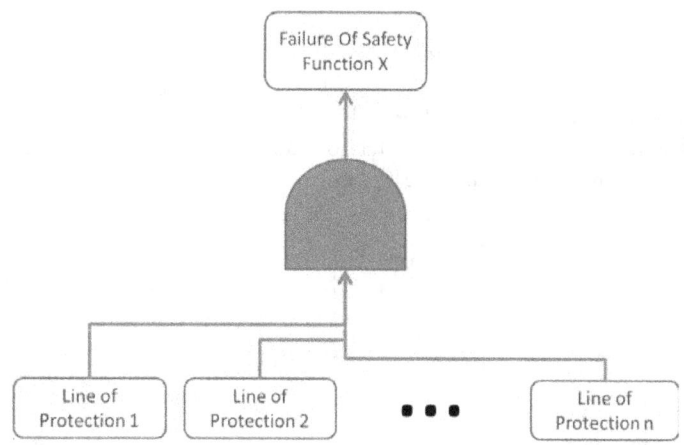

Let $P(B_i)$ be the probability of failure of the ith protective barrier. In general the probability of failure of a barrier can be described by some unknown distribution D_i; i.e. $P(B_i) \sim D(\alpha, \beta)$. The estimation of the distribution D is, in general, a difficult task with numerous associated difficulties for which there is an existing and extensive literature; References 8, 9, 10 and 11 provide a good background on some of the estimation methods used in modern risk analysis and discussion of the associated difficulties. In practice the distribution D is never known, although some of the sources of variability may have been partially estimated. Historically conservative values have been used, usually attempting to estimate the 95th percentile of the distribution D. The conservative estimate will be represented by $\hat{P}(B_i)_{95}$. The "best estimate" of $P(B_i)$ is some measure of central tendency. Usually in risk analysis models the mean is used. However, for purity of the results, and strictness of inequalities the median is used in this document; it is noted that the extension to the use of a mean estimate is straightforward, except for the existence of certain caveats relating to heavily skewed distributions. The best estimate median value is represented by $\hat{P}(B_i)_{50}$. It is noted that for any probability distribution that is not a single point the following strict inequality holds:

$$P(B_i)_{95} > P(B_i)_{50} \qquad (1)$$

It is hence reasonable to assume that for any "good" estimate, the following strict inequality will also hold:

$$\hat{P}(B_i)_{95} > \hat{P}(B_i)_{50} \qquad (2)$$

It is noted that for most distributions:

$$P(B_i)_{95} > P(B_i)_\mu \tag{3}$$

Hence the results presented in this document will almost always also hold if the mean of the distribution D is used in place of the median.

Let P(X) be shorthand for the probability that system X fails. Then given independence of the lines of protection we see that:

$$P(X) = \prod P(B_i) \tag{4}$$

Using the notation above we can find a best estimate of the probability of failure of X using:

$$\hat{P}(X)_{50} = \prod \hat{P}(B_i)_{50} \tag{5}$$

And a conservative estimate of the probability of failure of X using:

$$\hat{P}(X)_{95} = \prod \hat{P}(B_i)_{95} \tag{6}$$

Now, consider the effect of removing one train of protection, for example for maintenance. Without loss of generality assume that the jth line of protection is removed. Using the subscript 'mj' to denote maintenance of barrier 'j', the system failure estimates now become:

$$\hat{P}(X)_{50,mj} = \prod_{i \neq j} \hat{P}(B_i)_{50} \tag{7}$$

And:

$$\hat{P}(X)_{95,mj} = \prod_{i \neq j} \hat{P}(B_i)_{95} \tag{8}$$

Risk models for complex engineering systems are acknowledged to be approximate tools. Most analysts agree that the absolute value of probabilities calculated using the risk model are very approximate. However, ranking of risks and estimating relative magnitudes is still a useful output, even in absence of good absolute measures. The estimate of the relative risk of different plant components and configurations is a valuable output from risk models. For this reason the risks above can be usefully considered in the context of the baseline risk when all lines of protection are available. The relative risks are:

$$\frac{\hat{P}(X)_{50,mj}}{\hat{P}(X)_{50}} = \frac{1}{\hat{P}(B_j)_{50}} \tag{9}$$

And:

$$\frac{\hat{P}(X)_{95,mj}}{\hat{P}(X)_{95}} = \frac{1}{\hat{P}(B_j)_{95}} \tag{10}$$

Now, noting that the 95th percentile estimate is larger than the 50th percentile estimate we see that:

$$\frac{1}{\hat{P}(B_j)_{95}} < \frac{1}{\hat{P}(B_j)_{50}} \tag{11}$$

This equation tells us that using a 95th percentile estimate of every line of protection gives a lower estimate of the relative increase in risk during maintenance compared to the relative increase in risk that occurs if a median estimate of the probability of failure of each line of protection is used. In general, this means that conservatively estimating failure probabilities results in optimistic estimates of the relative risk increases during maintenance. Furthermore, the level of optimism is proportional to the level of conservatism, if we define "conservatism" to mean a multiplication factor from the best estimate.

It is noted that the above demonstration did not require the introduction of time at all into the argument. The time argument remains a linear argument that affects only the magnitude of the above effect. The extension to consider a time variant model is trivial but provides the same message with more complicated algebra. The next section considers the effect of the conservatism described above on a simple example model, and shows that the using conservative values gives an under estimate of the proportion of risk that is incurred during maintenance compared to during normal operation with all plant available.

3. SIMPLE EXAMPLE

This section works through an example fault tree representation of a simple system, to demonstrate the effect described in section 2. A system with two lines of protection, instead of n lines of protection, is used for clarity. The failure logic of this simple system is shown as a fault tree in Figure 4 below:

Figure 4: Simple Example – Two Protective Barriers

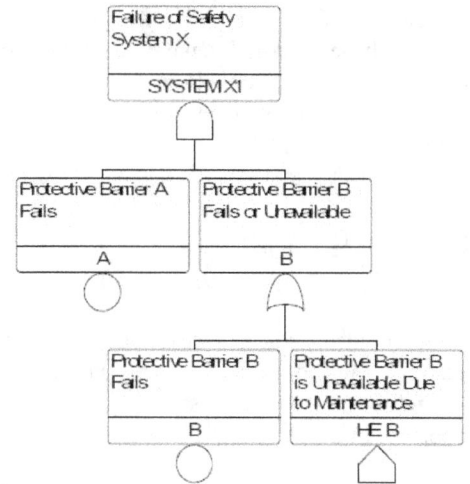

Figure 4 shows the base case for the system, in which the house event is set to false. It is assumed that each line of protection provides has a probability of failure on demand, as per the equations developed in section 2 above. A maintenance case can be considered using the same fault tree by setting the house event to true. This has the effect of taking one line of protection out of service.

A hypothetical conservative and best estimate cases are considered. The probability of failure on demand of each protective barrier in each case is shown in Table 1 below.

Table 1: Probability of Failure on Demand of Each Protective Barrier vs Estimation Technique

	P(A Fails on Demand)	P(B Fails on Demand)
Conservative	1E-02	1E-02
Best Estimate	1E-03	1E-03

Table 2 below shows the probability of failure on demand of safety system X, for each system state and for each failure probability assumption.

Table 2: Probability of Failure on Demand of Safety System X

	Base Case	Maintenance Case
Conservative	1E-04	1E-02
Best Estimate	1E-06	1E-03

The key point is demonstrated in Table 3 below, which shows the increase in the probability of failure on demand of safety system X during maintenance under conservative and best estimate assumptions.

Table 3: Ratio of the probability of failure on demand of system X to failure probability on demand in the base case.

	Ratio of Maintenance: Base case probability of failure on demand of System X
Conservative	100
Best Estimate	1,000

Let the base case failure on demand be $P(X)_{C,BC}$ and $P(X)_{B,BC}$ under conservative and best estimate assumptions respectively, and similarly let the maintenance case failure on demand be $P(X)_{C,M}$ and $P(X)_{B,M}$ under conservative and best estimate assumptions respectively. Further assume that a fixed proportion p_M of the time is spent in the maintenance state. The proportion of time spent in the base case is then $1- p_M$. This proportion is a constant across both cases. The proportioned probabilities of failure on demand of the system are $P(TX)_{C,P}$ and $P(X)_{B,P}$ respectively. Then we have:

$$P(TX)_{C,P} = (1 - p_m)P(X)_{C,BC} + p_m P(X)_{C,M} \qquad (12)$$
$$P(X)_{B,P} = (1 - p_m)P(X)_{B,BC} + p_m P(X)_{B,M} \qquad (13)$$

A sensible question to ask is "what contribution of the weighted probability of failure on demand is does the maintenance state make?" This contribution is $p_M P(X)_{C,M} / P(TX)_{C,P}$ and $p_M P(X)_{B,M} / P(TX)_{B,P}$ for the conservative and best estimate case respectively. Table 4 considers how this contribution changes as the proportion of time spent in the maintenance state changes.

Table 4: The percentage contribution of maintenance

	Percentage Contribution of Maintenance	
Proportion of time in the Maintenance State	Conservative	Best Estimate
p_M = 12 hours per 365 days	12%	58%
p_M = 4 days per 365 days	53%	92%

The first row of Table 4 represents a case in which the time that spent in the maintenance case is very low, only 12 hours per year. The conservative analysis predicts that only one eighth of the total risk is due to the maintenance state, while the best estimate shows that in fact the majority of the annual risk (58%) is incurred during the twelve hour maintenance period.

Although this observation is very simple, it has important implications for how risk models are interpreted. Most quantitative analysts acknowledge that the absolute numerical prediction of the risk is not the most important contribution of risk models. As this example demonstrates, any conservative bias can result in the distortion of the risk profile, which may affect decisions and the allocation of resources. At present a culture of erring towards conservatism in safety risk models still exists, and this example provides a cogent reason to avoid conservatism.

Section 4 discusses the case when uncertainty in the probability of failure on demand is also taken into account.

4. BEST ESTIMATE PLUS UNCERTAINTY

The preceding sections have established that the use of conservative estimates can significantly alter the risk profile of a plant. However, there is a useful concept implicitly encoded in the use of conservative estimates; namely that of uncertainty in the estimate. Although it is not usually explicitly described, by vague allusions to the 95th percentile confidence interval, conservative values do in fact implicitly take account of uncertainty in estimated values. However, this is not true of a best estimate point value, and represents a significant shortcoming of using best estimates in isolation. Where a conservative estimate overly penalises the risk and in so doing distorts the risk profile, best estimates alone ignore the issue of uncertainty making the results vulnerable to estimates which are not precise. The next logical stage in the process is to include uncertainty explicitly, based on a probability distribution around the best estimate. This has the effect of preserving the uncertainty information that is only encoded implicitly in conservative estimates.

It is useful to note that there are several common operations on distributions such as convolution, product distribution and multiplication of functions; more information on these is available from any standard text. Convolution of distributions is the required operation to combine events using an OR gate, however, this document exclusively considers AND gates so only the product distribution is required. The general form of this for 2 inputs to an AND gate is given below, where Z is the product distribution XY formed from the random variables X and Y.

$$P(Z) = \iint_{x,z} P\left(x, \frac{z}{x}\right) |\frac{1}{x}| dx dz \qquad (14)$$

The general form for n inputs can be found by iteratively re-naming the product distribution. The resultant product distribution can look very different depending on the input distributions. However, a typical form is shown in Figure 5, below, for the product distribution of two lognormal distributions.

Figure 5: Product Distribution of Two Lognormal Distributions

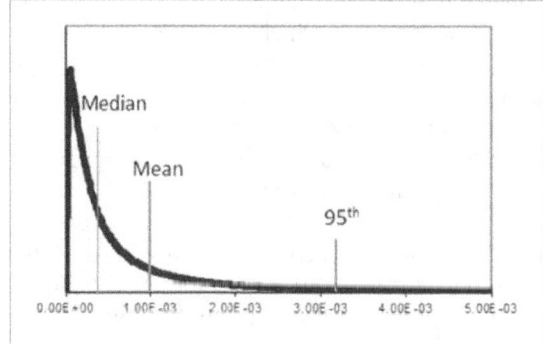
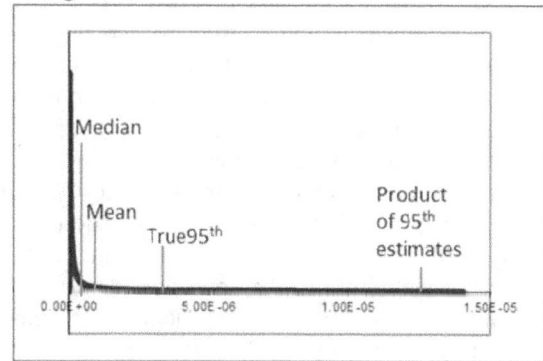

Lognormal Distribution　　　　　　　　　　Product Distribution

The mean, median and 95th percentile are marked on the distribution in Figure 5. The product of two 95th estimates is also marked on the product distribution to illustrate that it does not coincide with the true 95th percentile of the product distribution. The true 95th percentile of the product distribution (in the given example) is 3.7E-06, while the 95th percentile calculated by just multiplying two 95th percentiles together is 1.4E-05, which is actually the 99th percentile of the true product distribution. This is a further distortive effect of conservative estimates that occurs after the use of just a single AND gate, and provides additional justification for not using them. If the mean is used as the best estimate then this estimate at least is not distorted by the process of using an AND gate; i.e. the value of the product of mean best estimates equals the mean of the product distribution of the two "true" distributions. However, using the mean alone fails to recognize the uncertainty information. The purpose of including a full uncertainty distribution is to add back in the information which is "lost" when moving from a conservative estimate to a best estimate paradigm. This "loss" of information is easily seen by considering two distributions with the same mean, but different levels of uncertainty. One distribution may be very narrow, while the other may be broad, but using the mean estimate for both would appear to equate the two. Hence, in most scenarios, simply using a best estimate will result in a naïve understanding of the risk. To complete the picture it is essential to have an explicit estimate of the uncertainty as well as using a best estimate of the central tendency.

5. DISCUSSION

The results presented here provide additional confirmation that the use of conservative judgements in risk analysis does not provide results which properly reflect the risk profile. In particular it has been shown that in the case of maintenance arguments, it leads to undervaluing the contribution to risk of removing lines of protection which have been conservatively assessed. This extent of this distortion increases the greater the conservatism. This can have a real impact on decision making; for example a particular barrier may afford (in reality) excellent protection but where analysis for it is very conservative, this would lead to a significant under-valuing of its protective capability. This conservatism may not be apparent in the cutset results and risk importance results. However, during maintenance it means that the risk model would fail to inform the analyst about the risk spike which would occur when that excellent barrier was unavailable due to maintenance.

There are parallels here with assessment of software. Software testing is an exceptionally hard problem which continues to challenge the software community [12]. A major contributor to the difficulty is the high dimension of the parameter space which needs to be checked, meaning that only a small volume can practically be checked. For this reason the current approach in assessing software in risk analyses is to use an ultra-conservative approach. To the author this is an outdated viewpoint, which needs to be addressed. While it is acknowledged that predicting software reliability is hard, it should still be subject to the best estimate philosophy; uncertainty estimates then provide a way to qualify that best estimate and to, rightfully, acknowledge that the software reliability is currently approximate.

An explanation for this attraction to conservative estimates is found in what appears to be the basic psychological wiring of humans showing an aversion to uncertainty, which has been dubbed the "uncertainty effect" [13]. Indeed, the uncertainty effect goes further than merely devaluing a package compared to the mean due to uncertainty; a package including uncertainty is often valued, subjectively, as worth less than the worst possible outcome. For example an uncertain lottery in which payouts are gift certificates with a face value between $50 and $100 is valued, by a significant proportion of people, as being worth less than a certain payout of a gift certificate with a face value of $50 [13]. This is a surprising result indeed, but the only point drawn from this result here is that this type of observation is indicative of the level of human aversion to uncertainty, even if the precise characterisation of that aversion requires further investigation in the psychological literature. However, this type of psychological bias must not be allowed to creep into the way in which risk analysis is performed. Even if psychological biases are unavoidable in the eventual decision making, psychological effects should be deferred as far down the process as possible; i.e. it should not feature until the "end" of the quantitative risk analysis estimation problem, after the risk profile has been estimated as faithfully as possible, including estimates of uncertainties as far as possible. This helps to avoid a compounding of these effects throughout the process.

Human perception of risk and reward is known to be a complex topic and subject to numerous psychological effects [14, 15]. Humans are bad at internalising small probabilities, and are known to distort the value of small probabilities, with a strong tendency over-value them. The value gradients are observed to be steep near certainty and near impossibility [14]. Since the absolute magnitude of small probabilities is not readily processed by humans, this presents a strong argument for the use of relative probabilities in assessing different scenarios, as far as possible. Relative probabilities are, in general, closer to the range of 50:50, which is a probability region in which humans tend to respond more rationally [14].

The aim of this type of work is to attempt to erase the prevailing mindset that it is better to be conservative than optimistic in risk estimates. This type of thought process certainly makes sense when it comes to design but is absolutely flawed when it comes to quantitative risk assessment, since it is like trying to push down a lump in a carpet. If you are conservative in one area (push down the lump) then you inadvertently neglect another area by distorting the risk profile (the lump pops up somewhere else). Not only does this type of behavior reduce how informative the analysis we have performed on areas of the system we understand well, but it also has the less well defined and pernicious effect of permitting the belief that, since we have been conservative in all our assessments, the overall values we are calculating are themselves conservative. The ill-stated implication is that, by being conservative for known sequences, we have implicitly allowed for model completeness uncertainty. Unfortunately, this is demonstrably untrue by a comparison between predicted values from PSA studies and the observed figures of reactor core melts and total reactor operating years accumulated worldwide, which is (at least) 3 severe accidents (Fukushima, Chernobyl, and Three Mile Island) in approximately 15,000 operating reactor years. There are strong mitigating arguments against this type of simplistic frequency observation, including the location dependence of hazards and the evolution of reactor design compared to reactors that suffered severe accidents. Nonetheless it provides a strong indication that current risk models may be missing significant risk contributors. The use of conservative assessments in the development of risk models could be acting to mask this conclusion by appearing to imply that risk models in their totality are also conservative; this in turn provides a loose rationale for the neglect of model completeness issues.

6. FURTHER WORK

This paper has demonstrated the distortion of the risk profile due to maintenance outages for a simplified model of a system in which there are 'n' lines of diverse protection, resulting in a particularly simple class of fault tree, but this analysis could be extended to more complex models. The analysis was greatly simplified by assuming that only AND gates were necessary. Events under OR gates could be replaced by a single new basic event with a different failure parameter; it is noted that by doing this the uncertainty distribution associated with the new basic event would be more complex, but this does not affect the overall argument above since no assumptions were made about the form of the uncertainty distribution $D(P_i)$. For this reason the results presented here are applicable to a general fault tree model, although further work could be done to definitively prove this claim using more complex fault trees. In addition to the level of complexity of the model, other aspects of risk models typically found in PSA models could also be included in the analysis. For example the use of time varying models, event trees and the use of boundary conditions to define scenarios of interest. This would lend even greater weight to the need for the use of best estimates, especially when the results of risk modelling are being used to inform decision making. Beyond strengthening the motivation for quantitatively assessing uncertainty, there are numerous maintenance analyses that could be usefully re-evaluated including uncertainty. For example the design of 'optimal' maintenance schedules could be strongly affected by the inclusion of uncertainty in failure parameter estimates.

The type of uncertainty considered in this document has only been statistical uncertainty. There are numerous other forms of uncertainty in PSA models, for example scenario uncertainty, success criteria uncertainty, accident progression and operator reliability uncertainty. Incorporating, explicitly, uncertainty from these sources would greatly benefit the predictions and insights that can be gained from PSA models. It is acknowledged that this represents a significant body of work and many uncertainties will require a bespoke method to incorporate. An example of the assessment of a "hidden" conservatism resulting from supporting neutronic analysis is presented in Reference 16.

7. CONCLUSION

The distortive effect of conservative estimates has been examined. This paper has demonstrated that the risk due to maintenance outages is underestimated if conservative values are used for failure parameters instead of best estimate values. It was then acknowledged that the conservative estimates, relied upon historically in the risk community, actually have intrinsic estimates of uncertainty bound up in them, and this partially justifies their use. It was shown that, in order not to distort the risk profile of a system, but while also retaining the uncertainty information implicit in conservative estimates, that best estimates alone are not sufficient and that best estimates plus uncertainty distributions are required. While there are clearly challenges in quantitatively finding a best estimate, and an estimate of the uncertainty, the author maintains that this is not a fundamentally different or more difficult task than producing a conservative estimate. A major difference is in fact exposure; whereas it is almost always possible to find a conservative number that few people would challenge, a best estimate is intrinsically more vulnerable to criticism. This is not necessarily a trite consideration; in some legal settings this could be of significance. However, from a pure risk quantification perspective it should always be desirable to develop the most accurate risk profile possible.

Acknowledgements

The author is very grateful to the EPSRC for funding and to Corporate Risk Associates for additional funding and technical support. Thanks in particular to Garth Rowlands and Rebecca Brewer for reviewing this paper and providing useful feedback. The author would also like to thank Dr. Simon Walker of Imperial College London and Jasbir Sidhu of Corporate Risk Associates for their continued support of this work.

REFERENCES

[1] G. E. Apostolakis, *"How Useful Is Quantitative Risk Assessment?"*, Risk Analysis, Vol.24, No.3, 2004.

[2] E. Zio and T Aven, *"Industrial disasters: Extreme events, extremely rare. Some reflections on the treatment of uncertainties in the assessment of the associated risks"*, Process Safety and Environmental Protection 91, 2013, pp 31–45.

[3] J. M. Reinert and George E. Apostolakis, *"Including model uncertainty in risk-informed decision making"*, Annals of Nuclear Energy 33, 2006, pp 354–369.

[4] N.S. Arunraj and J. Maiti, *"Risk-based maintenance—Techniques and applications"*, Journal of Hazardous Materials 142, 653–661, (2007).

[5] S. Turner, *"The Representation of Unavailability in Fault Trees and the Optimisation of Maintenance Actions"*, PhD Thesis, University of Birmingham, (1997).

[6] M. Bertolini, M. Bevilacqua, F.E. Ciarapica, G. Giacchetta, *"Development of Risk-Based Inspection and Maintenance procedures for an oil refinery"*, Journal of Loss Prevention in the Process Industries 22, pp. 244–253, (2009).

[7] S. Kaplan and B. J. Garrick, *"On The Quantitative Definition of Risk"*, Risk Analysis, Volume 1, No.1, pp.11-27, (1981).

[8] C. Bunea and T. Charitos & R.M. Cooke, *"Two-stage Bayesian models – application to ZEDB project"*, Reliability Engineering and System Safety, pp 321-329, (2005).

[9] E.L. Droguett, F. Groen & A. Mosleh, *"The combined use of data and expert estimates in population variability analysis"*, Reliability Engineering & System Safety, pp311-321, (2003).

[10] A. El-Shanawany, *"A Comparison of Bayesian Analysis Methods for Reliability Parameter Estimation in PSA"*, IET International System Safety Conference, 2010.

[11] K. Pörn, *"On Empirical Bayesian Inference Applied to Poisson Probability Models"*, Linköping Studies in Science and Technology, Dissertation No. 234, (1990).

[12] V. R. Basili and R. W. Selby, *"Comparing the Effectiveness of Software Testing Strategies"*, IEEE Transactions on Software Engineering, vol. se-13, no. 12, December 1987.

[13] Gneezy, Uri, John A. List, and George Wu. "The uncertainty effect: When a risky prospect is valued less than its worst possible outcome." The Quarterly Journal of Economics 121, no. 4 (2006): 1283-1309.

[14] Y. Rottenstreich, and K. H. Christopher, *"Money, kisses, and electric shocks: On the affective psychology of risk"*, Psychological Science 12, no. 3 (2001): 185-190.

[15] D. Kahneman and D. Lovallo, *"Timid Choices and Bold Forecasts: A Cognitive Perspective on Risk Taking"*, Management Science, Vol. 39, No. 1 (1993), pp. 17-31.

[16] A. El-Shanawany et al, *"Propagating Uncertainty in Phenomenological Analysis into Probabilistic Safety Analysis"*, Submitted to PSAM12, 2014.

Understanding the Long-term Behavior of Sealing Systems and Neutron Shielding Material for Extended Dry Cask Storage

Dietmar Wolff*[*,a], Matthias Jaunich[a], Ulrich Probst[a], and Sven Nagelschmidt[a]

[a] Federal Institute for Materials Research and Testing (BAM), 12200 Berlin, Germany

Abstract: In Germany, the concept of dry interim storage in dual purpose metal casks before disposal is being pursued for spent nuclear fuel (SF) and high active waste (HAW) management. However, since there is no repository available today, the initially planned and established dry interim storage license duration of up to 40 years will be too short and its extension will become necessary. For such a storage license extension it is required to assess the long-term performance of SF and all safety related storage system components in order to confirm the viability of extended storage.

The main safety relevant components are the thick-walled dual purpose metal casks. These casks consist of a monolithic cask body with integrated neutron shielding components (polymers, e.g. polyethylene) and a monitored double lid barrier system with metal and elastomeric seals. The metal seals of this bolted closure system guarantee the required leak-tightness whereas the elastomeric seals allow for leakage rate measurement of the metal seals.

This paper presents an update on running long-term tests on metal seals at different temperatures under static conditions over longer periods of time. In addition, first results of our approach to understand the aging behavior of different elastomeric seals and neutron radiation shielding material polyethylene are discussed.

Keywords: Extended Storage, Dual Purpose Casks, Metal seals, Elastomeric Seals, Neutron Shielding Material

1. INTRODUCTION

In Germany, the decision was made to phase out of energy production by nuclear power stations. But since there is no repository available today this does not imply that questions of spent fuel storage are no longer relevant. Without an available repository it is expected that the initially planned and established dry interim storage license duration of 40 years in Germany will be too short and its extension will become necessary [1]. But since all data used for the safety case so far are also limited to this time span, for a storage license extension it is necessary to assess the long-term performance of spent fuel and all safety related storage system components in order to confirm the viability of extended storage. A topical overview on the German aging management approach for dry SF and HLW storage in dual purpose casks and the influence of aging mechanisms on transport safety and reliability of such casks is given in [1, 2].

The main safety relevant components for spent fuel storage are the thick-walled dual purpose metal casks which are approved for transportation and storage. These casks consist of a monolithic cask body with integrated neutron shielding components (polymers, e.g. polyethylene) and a monitored double lid barrier system with metal and elastomeric seals. The metal seals of this bolted closure system guarantee the required leak-tightness whereas the elastomeric seals allow leakage rate measurement of the metal seals. Irrespective of this application of elastomeric seals in SF and HAW storage casks only as auxiliary seals, they are of major interest for cask designs for low and intermediate level radioactive wastes.

---

* dietmar.wolff@bam.de

In order to ensure their full safety relevant functional capability during the whole extended storage period and for transportation after storage, the change of material properties of metal and elastomeric seals as well as neutron shielding polymers has to be investigated for long-term storage conditions.

For sealing systems, the sealing performance during long-term storage is crucial. This is mainly influenced by the reduction of the compression load which means also a reduction of the usable resilience. Therefore, for both types of seals the time-dependent sealing performance will be determined for different thermal and mechanical loads. For elastomeric materials also radiological aging has to be considered.

In case of neutron shielding polymers general changes of material parameters are investigated with special interest in changes induced by gamma radiation and aging, e.g. hydrogen release, crosslinking and mechanical behavior.

This paper presents an update on running long-term tests on metal seals at different temperatures under static conditions over longer periods of time. Due to creeping effects a reduction of the pressure force over deflection during loading and unloading does appear depending on prior holding times. Analytical approaches on basis of test data are developed and discussed to extrapolate seal performance (pressure force, elastic recovery, leak-tightness) to longer periods of cask operation.

In addition, first results of our approach to understand the aging behavior of different elastomer seals and neutron radiation shielding material polyethylene are discussed.

2. METAL AND ELASTOMERIC SEALS

Figure 1 shows a schematic representation of a German casks common lid and sealing system. The metal seals are responsible for the main sealing function and have to meet the high requirements of the specified leakage rate of 10^{-8} $Pa \cdot m^3 \cdot s^{-1}$. The elastomeric seals are auxiliary seals to create a cavity which is necessary to measure and validate the leakage rate of the metal seals after installation.

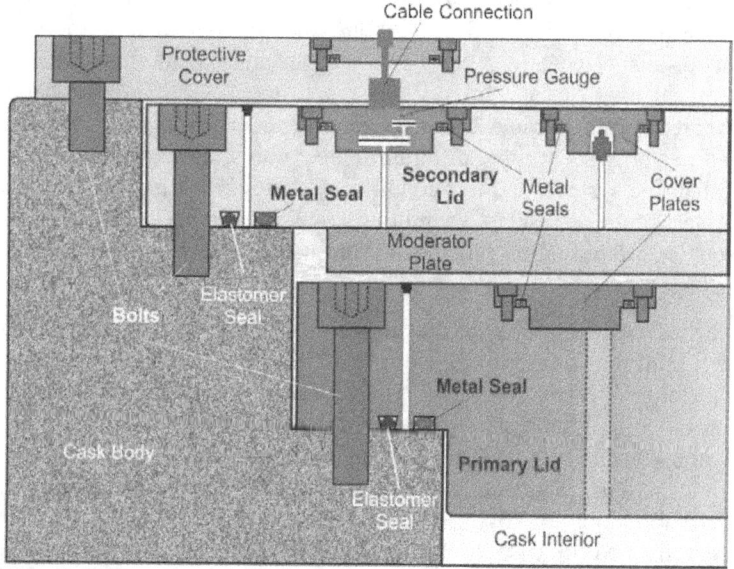

Figure 1: Lid area and sealing system of a transport and storage cask

The bolts fixing the lids onto the cask body have to be tightened with a suitable pre-tension for compressing the metal seals to their proper assembly situation. The lids are equipped with grooves to carry the seals. In case of metal seals there has to be a specific groove depth, which corresponds with the given operation point and optimal pressure force of the seals. By screwing the primary or the secondary lid to the contact position onto the cask body, the correct compression of metal and elastomeric seals is given.

For the safe long-term operation of such cask closure systems it is important to understand time dependent degradation mechanisms like loss of seal pressure forces and screw pre-stresses due to creeping and relaxation effects. For that reason systematic investigations of seal function under consideration of installation conditions, material properties, operation temperatures and periods of time gain necessary information for further long-term safety assessments.

2.1. Metal seal investigation

In Germany, metal seals used in dual purpose casks for SF dry interim storage are usually of the HELICOFLEX® HN200 type as illustrated in Figure 2. Such seals consist of an inner helical metal spring and two thin metal jackets. The outer metal jacket is made of aluminium (Al-seals) or silver (Ag-seals) to achieve and maintain tight contact between seal and lid or cask body surfaces.

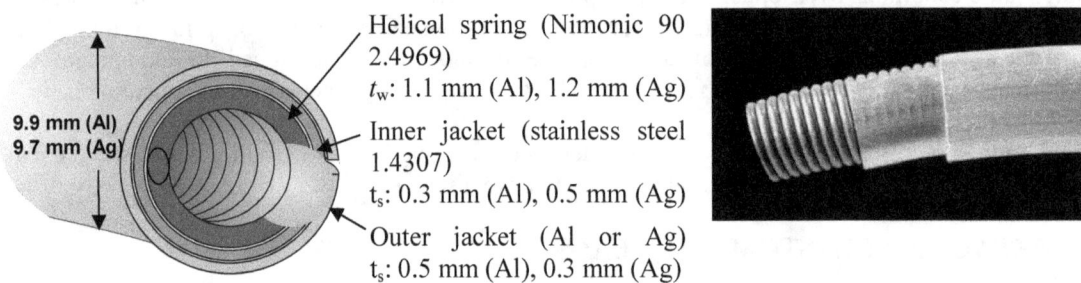

Figure 2: HELICOFLEX® HN200 seal type applied in test series and in dry storage casks

BAM is already performing investigation programs with such type of metal seals. To allow appropriable dimensions of our test setup, we use seals with a smaller overall diameter compared to seals that are installed in casks. However, the seal cross section diameter of about 10 mm as well as materials and dimensions of spring and jackets are identical to dual purpose cask seals to get representative test results. Depending on cask design and spent fuel decay heat, maximum temperatures at the seal area of cask lid systems reach about 110°C at the beginning of storage. Knowing that higher temperatures can accelerate aging mechanisms, in February 2009 and November 2010 BAM decided to start tests at three different temperatures: 20°C (ambient temperature), 100°C, and 150°C. The test flange systems and test conditions are described in detail in previous papers, e.g. in [3] and [6]. The test setup represents assembling conditions in casks and allows for simultaneous measurement of seal load-deformation relationship during compression and relieving procedures and standard helium leakage rate.

Major outcomes of these test series are decreasing pressure forces F_r and decreasing elastic recovery (useable resilience r_u until the specified leakage rate of 10^{-8} $Pa \cdot m^3 \cdot s^{-1}$ is exceeded) with holding times at the different temperature levels for both seal types (aluminum, silver). For example, the reduction of F_r and r_u depending on holding time and temperature is plotted over a logarithmic time scale for Al-seals in Figures 3 and 4, which illustrates a proper linear correlation and allows extrapolating very easily to longer time periods. This seal relaxation correlates to the reduction of the outer jacket thickness by plastic deformation improving the contact of the metal surfaces. As a result a proper seal function was observed even in case of nearly complete loss of pressure force. More detailed information can be found in [4, 5, 6]. In case of real casks, an exceeding of the specified leakage rate may be caused either by mechanical loads under accident scenarios or by reduction of the restoring seal force F_r due to time depending creeping processes of seals and/or bolts.

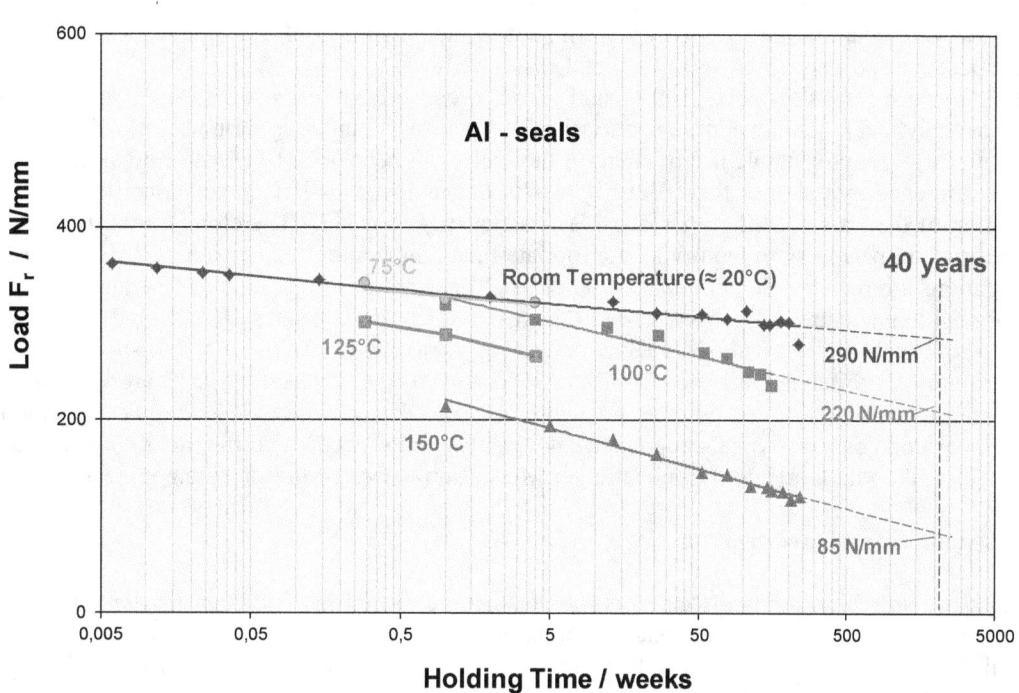

Figure 3: Reduction of restoring seal force F_r depending on holding time and temperature for current test periods and extrapolation up to 40 years (dashed lines)

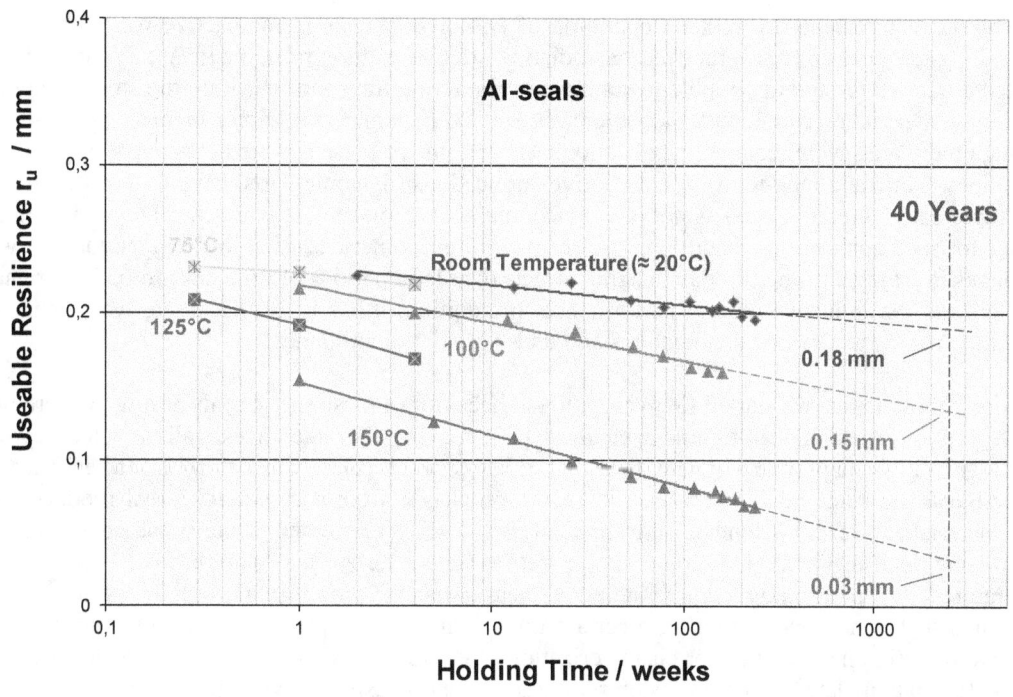

Figure 4: Reduction of useable resilience r_u depending on holding time and temperature for current test periods and extrapolation up to 40 years (dashed lines)

Recently performed investigations take account of the Larson-Miller relationship [7], which was developed with regard to the long-term performance of metallic materials under consideration of time and temperature and also widely used for metal seals, see [8, 9, 10].

By using our test results for 20°C, 100°C, and 150°C, we also have discussed different approaches to derive an analytical time-temperature relationship between short- and long-term tests at these different temperatures to get predictable information on the long-term behavior of the restoring seal force and useable resilience in a shorter time. But the results of our investigations have shown that, e.g., the application of the Larson-Miller approach is not adequate so far [11]. Therefore it was decided that some efforts are necessary to modify an appropriate time-temperature relationship and that additional investigations are required, in particular for additional temperatures to gain more test data. Such tests at additional temperatures have been started this year for Al- and Ag-seals at 75°C and 125°C. First results for Al-seals at these additional temperatures are presented in Figures 3 and 4, showing that (i) there is almost no difference between 20°C and 75°C and (ii) that values for 125°C are well between those for 100°C and 150°C. As soon as we will have more data for the two additional temperatures, we will extend our analytical approach taking also into account different time-temperature relationships like Manson-Haferd formulation or Mendelson-Roberts-Manson parameterization [12].

2.2. Elastomeric seal investigation

It is known, that typical aging effects of elastomers can be caused by oxidation, irradiation and high temperature [13]. The consequence can be additional crosslinking and/or chain scission. Material properties of elastomers at low temperatures are determined by the rubber-glass transition (abbr. glass transition). During continuous cooling, the material changes from rubber-like entropy-elastic to stiff energy-elastic behavior, that allows nearly no strain or retraction due to the glass transition. Hence elastomers are normally used above their glass transition but the minimum working temperature limit is not defined precisely. Moreover, the influence of the above mentioned aging effects on changes of the minimum working temperature has to be discussed carefully.

Challenging requirements for reliable operation of elastomeric seals in radioactive waste containers are e.g. long-term use up to several decades, radiation effects resulting from the inventory, operation at elevated temperatures and at possible low temperatures during transport or within the storage facility, operation under static conditions and potentially under dynamic conditions in case of accidents. Although there are several common applications where some of these requirements apply as well, e.g. water pipes (static and long-term) or automotive applications (dynamic parts, temperature variations), the complete set of requirements is not often encountered as replacement is much easier and common practice for such applications. Therefore, the behavior of elastomeric seals at low temperatures as well as changes in material properties due to aging effects over long periods of time have to be investigated to understand the influence on seal performance. In addition, it has to be discussed, which changes occur due to aging at elevated temperatures and under irradiation influence.

To address these topics we started to investigate the behavior of elastomeric seals at low temperatures with regard to potential leak-tightness changes [14, 15, 16]. For the investigations, fluorocarbon (FKM) and ethylene-propylene-diene (EPDM) rubbers were selected as they are often used in radioactive waste containers. Some materials were purchased from a commercial seal producer and some materials were compounded and cured at BAM. The elastomers where studied by several thermo-analytical methods and compression set to characterize the material behavior at low temperatures. By performing compression set measurements, the degree of deformation can be determined that is not recovered after a certain time span after sample release. In the course of this investigation also a new method for characterization of elastomeric seal materials has been developed [17, 18, 19] that emulates the compression set measurement by using a Dynamic Mechanical Analysis device. By this technique time-dependent and temperature-dependent measurements can be performed semiautomatic and much faster than by standardized procedures. For instance, a clear temperature dependency of the compression set values was observed as the compression set increases with decreasing temperature. For example, this behavior is shown in Figure 5 for an FKM material. As the glass transition of this sample is in the range of -10°C to -23°C [19], the compression set increase on cooling is connected with the material change from rubber-like entropy-elastic to stiff energy-elastic behavior, allowing nearly no retraction at even lower temperatures.

In addition to the described investigations, component tests were performed to determine the breakdown temperature of the sealing function of complete elastomeric O-rings and to correlate the temperature dependence of compression set with leakage rate, see e.g. [14, 15, 19, 20].

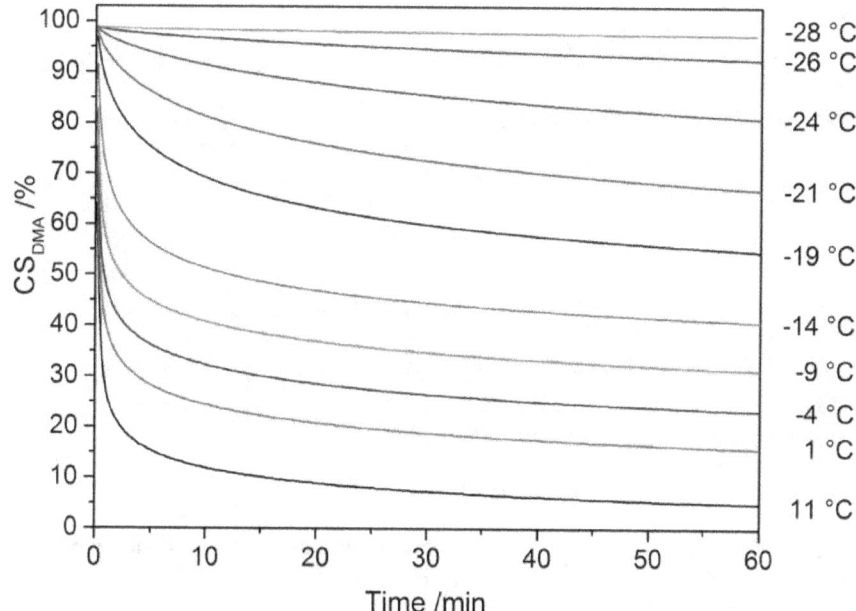

Figure 5: Temperature dependency of compression set values for an FKM rubber over time

To understand long-term performance of elastomeric seals accelerated experiments have to be performed. There are several standards that describe how to determine aging effects and how to interpret the results (see e.g. [21, 22]). These standards generally use a lifetime criterion and assume an Arrhenius-like behavior of aging processes. The question whether these assumptions are correct especially for the extrapolation over long periods of time is under discussion and several studies imply non-Arrhenius behavior, e.g. [23]. This is the reason for us to actually start an extensive experimental aging program with several materials to investigate the occurring changes of material properties of elastomeric seals by aging performed at different temperatures for up to two years. One goal is to determine the influence of compression during aging periods and therefore, compressed samples and O-rings under assembly conditions will be stored and later on analysed simultaneously. Aging temperatures are selected in the range from 75°C up to 150°C (i) to ensure significant acceleration of aging processes, (ii) to obtain sufficient amount of data, (iii) to prove the applicability of typical time-temperature-superposition approaches as e.g. Arrhenius, and (iv) to verify that no additional sample degradation occurs as a result of high temperatures selected for appropriate accelerated aging. The seal dimensions were chosen with a rather large torus diameter of 10 mm and an inner diameter of 190 mm to have sufficient amount of sample material for different analytical methods to be performed.

To allow a correlation of measured changes in material properties with seal function, a setup with seals mounted in flanges allows for leakage rate measurement by pressure rise method at chosen aging times.

As these tests will not consider irradiated samples, in addition we investigate elastomeric seals that were subjected to gamma radiation of up to 600 kGy to simulate a rather high gamma dose compared to seal application in casks. First results obtained for FKM samples indicate an increase of the lower application temperature limit. For example, as can be seen from Figure 6, there is a significant dose dependent increase of compression set comparable to the material behavior on cooling as displayed in

Figure 5. Detailed investigations are on the way to understand this observation in more detail, including known chemical changes of FKM samples induced by gamma irradiation.

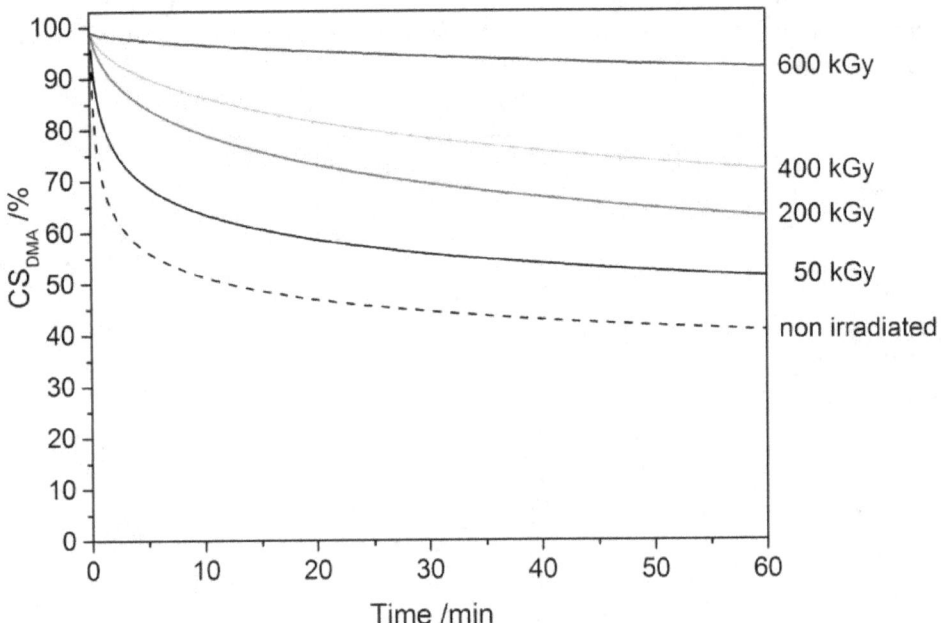

Figure 6: Gamma dose dependent increase of compression set values for an FKM rubber
(measurements at -15°C)

3. INVESTIGATION OF (U)HMW-PE FOR NEUTRON RADIATION SHIELDING

Due to their extreme high hydrogen contents ultra-high molecular weight polyethylene UHMW-PE and high molecular weight polyethylene HMW-PE are a comprehensible choice as neutron radiation shielding material in casks for storage and transport of radioactive materials. As HMW-PE has a higher density compared to UHMW-PE, it is more efficient for neutron radiation shielding. However, in the broad operational temperature range from below room temperature up to, e.g., 160°C, which has to be considered depending on the heat generation of the cask radioactive inventory, HMW-PE shows substantial flow at higher temperatures above its melting region. Therefore, rods of HMW-PE are used only in cask positions at which the maximal operational temperature does not reach the main melting region; it is below 120°C. In contrast to this, UHMW-PE shows a better dimensional stability due to its higher molecular weight, even above the melting temperature of the crystalline phase, i.e., the polymer remains in a solid-like condition as the viscosity is still very high.

As a consequence of the radioactive inventory, for this application (U)HMW-PE have to withstand any type of gamma radiation induced degradation affecting safety relevant aspects. Several effects of ionizing radiation affecting structure and properties of the PE materials may be expected, such as chain scission, crosslinking, oxidation, provided that oxygen is present, and release of hydrogen. Especially the balance of chain scission and crosslinking determines the resulting changes in mechanical properties as it may affect the thermal expansion directly, and indirectly the recrystallization after (partial) melting of the semi-crystalline material. With regard to the application as neutron radiation shielding material, several safety relevant aspects have to be considered:
    (i)   neutron radiation shielding efficiency, which is influenced by thermal expansion and density changes,
    (ii)   influence on overall cask mechanical stability, which can be influenced by pressure build-up resulting from thermal expansion or hydrogen released as a result of decomposition of polyethylene chain molecules,

(iii) behavior under accident conditions (e.g. fire scenario, mechanical impact).

In order to determine (U)HMW-PE material changes we applied a set of characterization methods in order to identify changes in the neutron radiation shielding material that was gamma irradiated with a dose of up to 600 kGy. Such a high dose represents a typical value of gamma irradiation that may occur within 40 years of interim storage (currently approved duration of interim storage license in Germany) leading to major possible changes of the material. Detailed results were published in [24, 25]. It was confirmed that with the applied methods it is possible to detect and characterize structural changes of (U)HMW-PE induced by gamma radiation. In order to assure a sufficient neutron radiation shielding caused by the radioactive inventory, the obtained variation in density has to be considered to determine the (U)HMW-PE rods diameter. Until now, the obtained results indicate that the observed degree of changes of the irradiated material does not affect safety relevant issues for long-term neutron radiation shielding purposes under normal operation conditions. As the thermal expansion coefficient decreases due to the formation of crosslinks this aspect is not important under normal operational conditions. But for unusual high temperatures (e.g. under accident conditions), it might be of interest especially in combination with reduced flow properties of the irradiated HMW-PE.

To understand the influence of combined gamma radiation and thermal impact on (U)HMW-PE material property changes, in the future we intend to subject irradiated and not irradiated samples to accelerated aging at elevated temperatures as described for elastomeric seals aging experiments.

4. CONCLUSION

Due to major delays of siting and establishment of a final repository in Germany extended interim storage periods are inevitable in the future. To be prepared for lifetime extensions of interim storage facilities, systematic collection and evaluation of data on long-term behavior of relevant cask material and components is inevitable. One main criterion for long-term performance is the leak tightness of bolted lid closure systems.

Even if today two decades remain until most storage licenses expire in Germany, investigation programs for reliable long-term predictions are time consuming that necessitates starting such programs early. In particular, analytic approaches describing time and temperature dependent material behavior are of interest to gain reliable predictions and to establish accelerating test configurations. BAM has already commenced such test series for metal and elastomeric seals as well as neutron shielding polymers to understand material aging mechanisms in more detail for extended storage usage.

References

[1] H. Völzke „*The German Aging Management Approach for Dry Spent Fuel Storage in Dual Purpose Casks*", paper #183, PSAM2014, Honolulu, Hawaii, USA.

[2] B. Droste, S. Komann, F. Wille, A. Rolle, U. Probst, and S. Schubert „*Consideration of Aging Mechanisms Influence on Transport Safety and Reliability of Dual Purpose Casks for Spent Nuclear Fuel or HLW*", paper #180, PSAM2014, Honolulu, Hawaii, USA.

[3] H. Völzke, U. Probst, D. Wolff, S. Nagelschmidt, and S. Schulz „*Investigations on the Long Term Behavior of Metal Seals for Spent Fuel Storage Casks*", Proceedings of the 52nd INMM Annual Meeting, Palm Desert, CA, USA, July 17-21, (2011).

[4] H. Völzke and D. Wolff „*Safety Aspects of Long Dry Interim Cask Storage of Spent Fuel in Germany*", Proceedings of the 13th International High-Level Radioactive Waste Management Conference (IHLRWMC), Albuquerque, NM, USA, April 10-14, (2011), American Nuclear Society ANS (2011).

[5] H. Völzke, U. Probst, D. Wolff, S. Nagelschmidt, and S. Schulz „*Seal and Closure Performance in Long Term Storage*", Proceedings of the PSAM11 & ESREL 2012 Conference, Helsinki, Finland, June 25-29, (2012).

[6]   H. Völzke, D. Wolff, U. Probst, S. Nagelschmidt, and S. Schulz „Long-term Performance of Metal Seals for Transport and Storage Casks", 17th International Symposium on the Packaging and Transport of Radioactive Materials (PATRAM 2013), San Francisco, CA, USA, August 18-23, (2013).

[7]   F.R. Larson and J. Miller „A time-temperature relationship for rupture and creep stresses", Trans. ASME, vol. 74, pp. 765-775, (1952).

[8]   H. Sassoulas, L. Morice, P. Caplain, C. Rounaud, L. Mirabel, and F. Beal „Ageing of metallic gaskets for spent fuel casks: Century-long life forecast from 25,000-h-long experiments", Nuclear Engineering and Design 236, pp. 2411-2417, (2006).

[9]   A. Erhard, H. Völzke, U. Probst, and D. Wolff „Ageing Management for Long Term Interim Storage Casks", Paper #61, 16th International Symposium on the Packaging and Transport of Radioactive Materials (PATRAM 2010), London, UK, October 03-08, (2010).

[10] K. Shirai, M. Wataru, T. Saegusa, and C. Ito „Long-Term Containment Performance Test of Metal Cask", Paper #3332, 13th International High-Level Radioactive Waste Management (IHLRWM) Conference, Albuquerque, NM, USA, April 10-14, (2011).

[11] S. Nagelschmidt, U. Probst, U. Herbrich, D. Wolff, and H. Völzke „Test Results and Analyses in Terms of Aging Mechanisms of Metal Seals in Casks for Dry Storage of SNF – 14255", WM2014 Conference, Phoenix, Arizona, USA, March 2 – 6, (2014).

[12] E. and G. A. Young, T.-L. (Sam) Sham „A Unified View of Engineering Creep Parameters", PVP2008-61129, 2008 ASME Pressure Vessels and Piping Division Conference, Chicago, IL, USA, July (2008).

[13] G.W. Ehrenstein and S. Pongratz „Beständigkeit von Kunststoffen Band 1", Carl Hanser Verlag, 2007, München.

[14] M. Jaunich „Tieftemperaturverhalten von Elastomeren im Dichtungseinsatz", BAM Bundesanstalt für Materialforschung und –prüfung, pp. 1-143, 2012, Berlin.

[15] M. Jaunich, K. von der Ehe, D. Wolff, H. Völzke, and W. Stark „Understanding low temperature properties of elastomer seals", Packaging, transport, storage & security of radioactive materials (RAMTRANS), 22(2), pp. 83-88, (2011).

[16] M. Jaunich, W. Stark, and D. Wolff „Low Temperature Properties of Rubber Seals", KGK-Kautschuk Gummi Kunststoffe, 64(3), pp. 52-55, (2011).

[17] M. Jaunich, W. Stark, and D. Wolff „A new method to evaluate the low temperature function of rubber sealing materials", Polymer Testing, 29(7), pp. 815-823, (2010).

[18] M. Jaunich, W. Stark, and D. Wolff „Comparison of low temperature properties of different elastomer materials investigated by a new method for compression set measurement", Polymer Testing, 31(8), pp. 987-992, (2012).

[19] M. Jaunich, W. Stark, D. Wolff, and H. Völzke „Investigation of Elastomer Seal Behavior for Transport and Storage Packages", 17th International Symposium on the Packaging and Transport of Radioactive Materials (PATRAM 2013), San Francisco, CA, USA, August 18-23, (2013).

[20] M. Jaunich, W. Stark, and D. Wolff „Low Temperature Properties of Rubber Seals - Results of component Tests", KGK-Kautschuk Gummi Kunststoffe, 66(7-8), pp. 26-30, (2013).

[21] DIN 53505, Prüfung von Kautschuk und Elastomeren - Künstliche Alterung, 2000, DIN.

[22] EN ISO 2578: Plastics - Determination of time temperature limits after prolonged exposure to heat, 2000.

[23] M. Celina, K.T. Gillen, and R.A. Assink „Accelerated aging and lifetime prediction: Review of non-Arrhenius behaviour due to two competing processes", Polymer Degradation and Stability, 90(3), pp. 395-404, (2005).

[24] D. Wolff, K. von der Ehe, M. Jaunich, and M. Böhning „Performance of Neutron Radiation Shielding Material (U)HMW-PE Influenced by Gamma Radiation", Proceedings of the PSAM11 & ESREL 2012 Conference, Helsinki, Finland, June 25-29, (2012).

[25] D. Wolff, K. von der Ehe, M. Jaunich, M. Böhning, and H. Goering „(U)HMWPE as Neutron Radiation Shielding Materials: Impact of Gamma Radiation on Structure and Properties", in „Effects of Radiation on Nuclear Materials" 25th Volume, STP 1547, ASTM, (2012).

Gap Analysis Examples from Periodical Reviews of Transport Package Design Safety Reports of SNF/HLW Dual Purpose Casks

Steffen Komann[a*], Frank Wille[a], Bernhard Droste[a]
[a] Federal Institute for Materials Research and Testing (BAM), Berlin, Germany

Abstract: Storage of spent nuclear fuel and high-level waste in dual purpose casks (DPC) is related with the challenge of maintaining safety for transportation over several decades of storage. Beside consideration of aging mechanisms by appropriate design, material selection and operational controls to assure technical reliability by aging management measures, an essential issue is the continuous control and update of the DPC safety case. Not only the technical objects are subject of aging but also the safety demonstration basis is subject of "aging" due to possible changes of regulations, standards and scientific/technical knowledge. The basic document, defining the transport safety conditions, is the package design safety report (PDSR) for the transport version of the DPC. To ensure a safe transport in future to a destination which is not known yet (because of not yet existing repository sites) periodical reviews of the PDSR, in connection with periodic renewals of package design approval certificates, have to be carried out. The main reviewing tool is a gap analysis. A gap analysis for a PDSR is the assessment of the state of technical knowledge, standards and regulations regarding safety functions of structures, systems and components.

Keywords: Dual Purpose Casks, Aging, Transportation, Periodical Review.

1. INTRODUCTION

For interim storage of spent fuel or HLW in many countries transport casks are used. The design of these "dual purpose casks" (DPC) has to be assessed and approved according to transport regulations (based on IAEA SSR-6), and to be assessed within the storage facility licensing procedure. Although the transport cask design differs from the storage cask design, e.g. by use of impact limiters, the majority of cask components is identical for both.

Differences occur also in the acceptance criteria; these are for the transport case defined in IAEA SSR-6 [1], and have to be developed for the storage case based on the storage facility requirements [11]. Considering transport after several decades of storage requires the implementation of ageing behavior into the transport safety case. Additionally the transport package design safety case has to be maintained in an up-to-date state, considering potential regulatory changes and development of scientific and technical knowledge. The review of a transport package design safety case has to be done periodically, implemented in periodic re-assessment for renewal of the package design approval certificate. The review process (as a kind of "intellectual periodic inspection") should be part of the approved applicant's management system and part of the administrative competent authority system.

From experience we have seen that stability of regulatory requirements for Type B(U) packages was not a major problem, but consideration of ageing and developments regarding the state-of-the-art technology can cause necessary adjustments of specific technical evaluation, with the result of confirmation of package safety, or with the development of appropriate compensatory measures to reach the required level of safety.

2. BACKGROUND OF DPC STORAGE IN GERMANY

Casks for interim storage are dual purpose casks in Germany, it means the casks are used not only for transportation but also for interim storage of spent nuclear fuel or radioactive waste inside an interim storage facility. A DPCs in an interim storage facility are shown in figure 1.

Figure 1: DPCs in an Interim Storage Facility Gorleben (Photo: GNS)

Decommissioning of spent nuclear fuel (SNF) requires several decades of storage before direct disposal. In case of reprocessing the fission products are transferred to vitrified high level waste (HLW) which has to be stored also over several decades before it can be disposed in a repository.

Most of the spent fuel produced in Germany until 2005 went to reprocessing in La Hague, France and Sellafield, UK. Since July 2005 German utilities are forced by law to store spent fuel in storage facilities located at the NPP sites. Before that decision, the old decommissioning policy was based on two central storage facilities in Ahaus and Gorleben. Since 1979 it was decided in Germany to store SNF and HLW under dry conditions in transport casks.

This concept of Dual Purpose Casks (DPC) was at the first time developed with the types of CASTOR® casks by GNS (Gesellschaft fuer Nuklear-Service GmbH, Essen, Germany). The central transport cask storage facility Ahaus was mainly used for the storage of 305 CASTOR® THTR/AVR casks with the complete inventory of the Thorium-High-Temperature-Reactor (THTR) after its decommissioning in 1994. The central transport cask storage facility Gorleben was mainly used to store 108 casks (CASTOR® HAW 20/28CG, TS 28 V, TN85, CASTOR® HAW 28M) with vitrified HLW received back from France. Besides 12 transport cask storage sites at NPPs (there were stored 332 CASTOR® V/19 and V/52 SNF casks at the end of December 2013), there are additionally two storage sites at Research Centre Jülich (for 152 CASTOR® THTR/AVR casks with the complete inventory of the research reactor AVR) and in Lubmin the ZLN storage site at the decommissioned former GDR NPPs with 65 CASTOR® 440/84 casks.

All these several hundred SNF and HLW transport and storage casks have to be transported in future after a storage period which is currently per license limited to 40 years, but which is expected to be some decades longer due to outstanding evaluation, selection and licensing of a high active waste repository. Responsibility for future generations requires from the beginning that there will be a safe transfer of the existing DPC to their currently unknown destination.

All stakeholders (vendors, transport and storage operators, authorities, regulators, technical experts) involved in that process need to follow a strict course of keeping the foundation for transport safety, the transport package safety report (safety case) effective through the entire lifetime of these objects.

3. DPC TRANSPORT PACKAGE DESIGN APPROVAL

The casks being used have an approved package design in accordance with the international transport regulations. The license for dry storage is granted on the German Atomic Energy Act with respect to the recently revised "Guidelines for dry interim storage of irradiated fuel assemblies and heat-generating radioactive waste in casks" by the German Waste Management Commission (ESK) [11].

Every storage cask has to have a transport approval certificate at time of storage placement. Usually a Type B(U) approval (accident resistance package according to IAEA transport regulations [1]) is necessary. In accordance to German guideline R003 [2] the package design assessment and the approval procedure are conducted by the Federal Office for Radiation Protection (BfS) and the Federal Institute for Materials Research and Testing (BAM). The assessment has to base on a Safety Case, provided by the applicant.

Figure 2: Structure of a PDSR; here for a SNF Dual Purpose Cask [3]

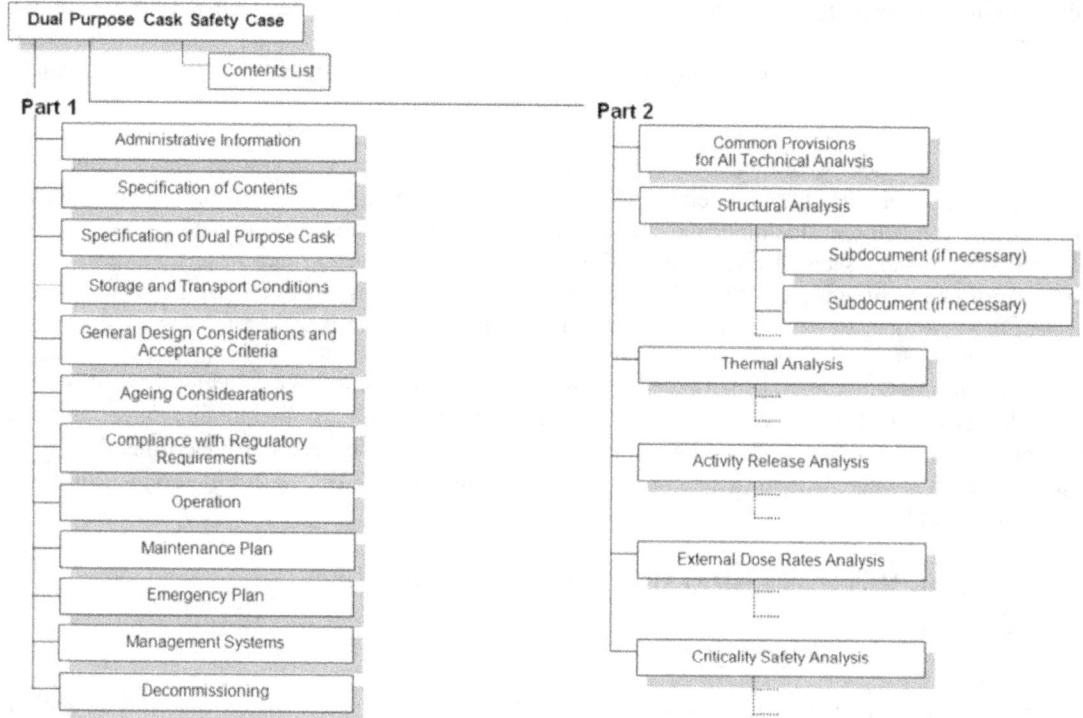

The Package Design Safety Report (PDSR) is a collection of scientific and technical arguments, including safety assessment and package design, manufacturing and operation specifications required to demonstrate compliance with the applicable transport regulations.

The "European Association of Competent Authorities for the Safe Transport of Radioactive Materials" issued the "Technical Guide – Package Design Safety Report for the Transport of Radioactive Materials" [3]. This "European PDSR Guide" is a useful guidance for structure and safety assessment details of a package design safety report. Based upon the same structure an IAEA working group developed the document "Guidance for preparation of a safety case for a dual purpose cask containing spent fuel" [4].

Figure 2 shows the structure of a Dual Purpose Cask Safety Case (DPCSC). Important for the long-term safety preservation of dual purpose casks are the requirements for ageing considerations in the safety case, ageing management during storage and inspection programs before transport after storage.

The DPCSC document also addresses problems of adjusting the differences between licensing types of storage and transport package design approval. A storage license is issued for a storage period of several decades. A transport package design approval is normally issued for a period of a few to several (between 3 and 10) years in Germany. Before the end of the approval period the certificate needs to be extended for the next period by a demonstration of compliance with the current transport regulations.

4. LONG TERM BEHAVIOR OF COMPONENTS AND MATERIALS

During long term interim storage the main driving forces of aging effects are gamma radiation, neutron radiation, decay heat, outer corrosion effects (e. g. moisture, and air pollution), relaxation, creep, corrosion of bolted and sealed lid systems, basket, and fuel rods.

Degradation effects strongly depend on the type of material. All main cask components responsible for the safe enclosure are usually made of metal like cask body, lids, main seals, and bolts. Additionally, polymer components are used for supplementary neutron shielding components, auxiliary seals and decontamination coatings. In general, damaging effects of radiation depend on dose rates, type of radiation and material structure. Metals are generally more resistant than polymers. Degradation effects may result in quantitative changes of specific material properties or modifications in material structure which may decrease the effectiveness of cask components.

Current investigations performed by BAM focus on the long term behavior of metal seals as the essential component for the safe enclosure, on the long term behavior of polymer materials as components for neutron shielding and on the aging mechanisms and low temperature behavior of elastomeric auxiliary seals. These investigations shall generate a better data base for understanding and quantification of aging effects and have been summarized by Wolff et al. [5]. More details with respect to aging management fundamentals and the investigation programs performed by BAM were published by Erhard et al. [6], and Jaunich et al. [8]. Latest results from the BAM metal seal investigation program are published by Völzke et al. [7]. These include extrapolation of seal pressure force decrease and decrease of elastic seal recovery depending on the temperature level and with respect to seal performance evaluation under normal operation and accident conditions during and after long term storage. The influence of aging mechanisms on the DPC transport safety is discussed in Droste et al. [12].

5. PERIODIC REVIEW AND GAP ANALYSES EXAMPLES

5.1 General

Not only DPC components and materials are subject to aging, this is also the case for regulations, standards, technical and scientific knowledge. Their aging mechanism is the change. Therefore, it is essential, for keeping a PDSR up-to-date for periodic package design approval renewal, to evaluate in a periodic review the impacts of these changes onto package safety. The method for that is a gap analysis. In [4] gap analysis is defined: "A gap analysis for a DPCSC is an assessment of the state of technical knowledge, standards, and regulations regarding safety functions of structures, systems and components. Gap analysis consists of

i) listing of characteristic factors, such as the state of technical knowledge, regulations, and standards of the safety case,
ii) evaluation of the effect of changes of technical knowledge, and standards on the safety of the DPC package, and then
iii) high-lighting the gaps that exist and need to be filled."

Periodic safety reviews and gap analyses are to be performed to keep a DPCSC up-to-date. Those periodical reviews are an important part of knowledge management, and force DPC designers, storage operators and regulators to keep knowledge on DPC safety present to all relevant stakeholders during the several decades lasting operation period. Periodic reviews are the only method which allows the tracking of safety knowledge, independently from institutional and personnel changes too.

Nitsche et al. [9] and Wille et al. [10] describe the German regulatory concept of transport package design approval for DPCs during interim storage period in detail. If manufacturing and loading of the casks are not done anymore and no transports are planned, BfS and BAM allow application of package design certificate with a validity period of 10 years.

A step wise procedure of evaluation of documents of the PDSR over the validity period is defined. This procedure includes the evaluation of consequences at enactment of new regulations over the entire validity period of the certificate. Beyond that after 5 years the certificate holder has to provide an evaluation that all safety related technical standards and codes, and safety demonstrations of the PDSR are valid.

After 10 years an extension of the package design certificate is necessary. The complete evaluation regarding state-of-the art technology of all parts of the PDSR has to be done. The consequence could be that e. g. new analysis methods for safety demonstration have to be applied. The advantage of this substantial work is the reflection of the knowledge about the package design and the concept of safety demonstration. We understand this procedure as the aging management concerning knowledge of the PDSR and the safety concept behind.

5.2 Examples of Safety Case Gap Analyses

Important for the periodical review of the package design safety report are both the experience and feedback out of manufacturing and the state-of-the art of safety analyses. This includes not only the analyses methods but also the validity of the applicable standards.

This periodically report has to contain an assessment of all components of the packaging including radioactive inventory regarding their condition and ageing influences and the comparison with the package design safety report based on the valid approval certificate.

Figure 3: Static and Dynamic Numerical Model of Cask Body

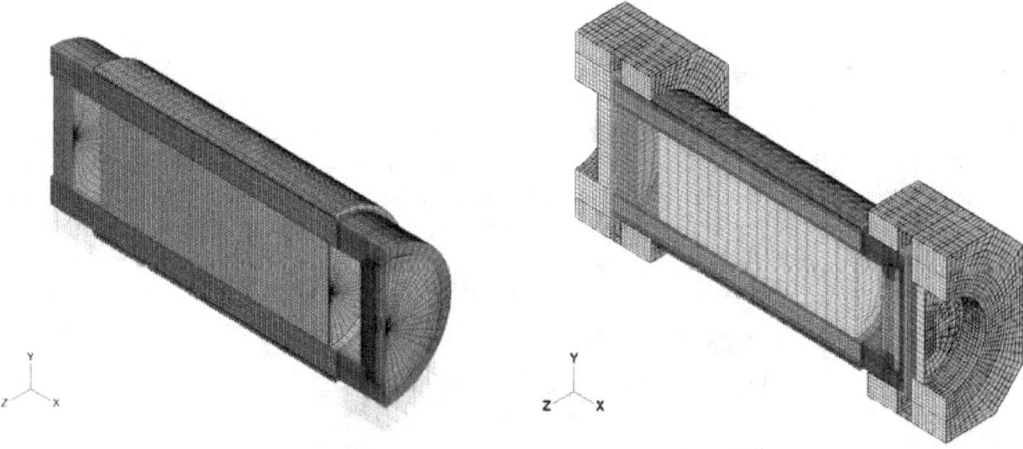

Cask Body Structural Analysis

It is to check, if the analyses methods are still valid for the assessment of the cask body. So it could be possible, that a actualized numerical analysis replace the analytical analysis for the cask body made in the past. In this case the influence of a numerical analysis of the package design safety has to be assessed. If during the licensing procedure a static numerical model was used and the state-of-the-art analysis shows the application of a dynamic numerical model is necessary, this influence on the safety report has to be checked. In figure 3 a static and dynamic Finite-Element-Model are presented. In Komann et al. [13] the approach of comparing static and dynamic Finite-Element-Analysis described in more detail.

Usually by using numerical models the local stresses has to be assessed (figure 4). If an analytical approach within the primary licensing procedure was applied, the influence of these local stresses compared with global loading scenarios has to be investigated. For this case additionally investigations regarding the material behavior could be needed. Furthermore the validity of the applied boundary conditions of the loading scenarios according to the regulations must be checked (e.g. temperature, pressure).

Figure 4: Local Stress Distribution in a Cask Body (1-m-Puncture Bar Drop)

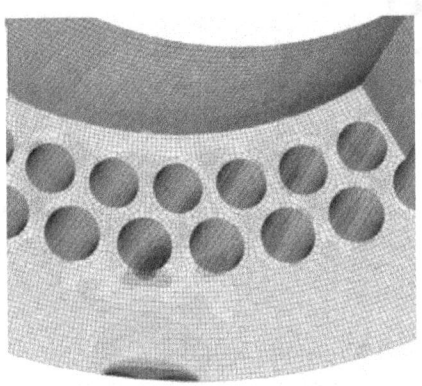
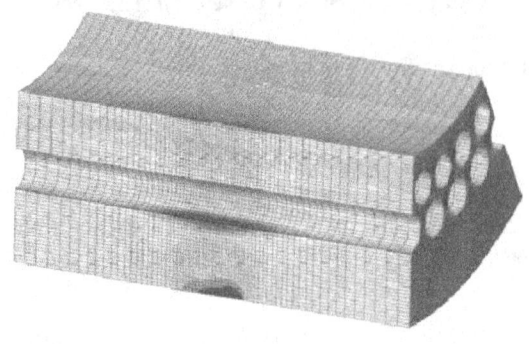

Lid System Structural Analysis

In Germany a DPC is normally equipped with a double barrier closure system consisting of a primary lid and a secondary lid (figure 5). This requirement results from the boundary conditions of the storage facility to ensure the monitoring of the DPC during the storage time. The lid system consists of the lid, normally made of stainless steel, the lid screws and the metallic or elastomerical seals.

Figure 5: Lid System of a Cask (Photo: GNS)

During the drop tests according to the regulations several loading conditions arise in the components of the lid system. For the determination of the activity release, respectively the leakage rate of the lid system, it is necessary to know the loading conditions in the components exactly. It means here for instance the screw pretension or the pressure in the cask flange, where the seals are applied. To assess the activity release after interim storage period the loading conditions must be checked regarding these mentioned aspects.

Impact Limiter Design Assessment

Impact limiters are necessary to decrease the impact loads resulting from drop test scenarios onto the packaging and the radioactive inventory. The construction of these components is therefore essential for the assessment of the package. The impact limiters are only part of the transport design approval of the package in Germany. Inside the interim storage facilities the impact limiters are not part of the cask storage design. To ensure the transportability during the interim storage period the impact limiters must be assessed periodically as well. The main materials of such impact limiters in Germany are wood and steel. The development in the knowledge in the field of the characteristic material behavior, e.g. stress-strain curves of timber regarding several temperatures, and the material condition is needed for the periodically assessment of the impact limiters. If the periodic structural assessment of the packaging results in phenomena like e.g. higher local stresses that could not have been identified by former calculations, compensatory measures could be a new impact limiter design leading to lower loads on the structure.

Load Attachment System

The purpose of the trunnions (figure 6) is the handling of the package inside a nuclear facility. Sometimes the trunnions will be used to store the package in a transport frame during transport over public traffic routes. Thus it is needed to assess these components for loading conditions resulting from regular handling procedures and transport conditions as well. The trunnions, often made of stainless steel, are robust against several loading conditions. But the conditions inside the trunnion screws are important for the safety assessment. The screw pretension depends amongst others on the decay heat of the package and the environmental temperature. So a change of the screw pretension can occur. This effect has an influence of the safety assessment of the trunnion system and the ability for handling and transportation.

Figure 6: Trunnion (Left: Real, Right: Numerical Model)

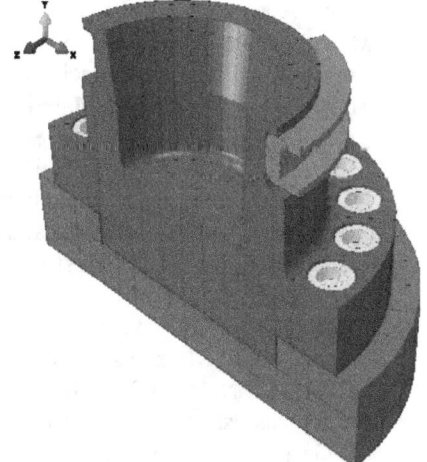

6. CONCLUSION

Safety assessment of spent nuclear fuel transport casks, as well as casks for spent nuclear fuel storage as dual-purpose casks is based on well established methods. For future applications a better harmonization of both licensing (transport and storage) areas is recommended.

The basic document, defining the transport safety conditions, is the package design safety report (PDSR) for the transport version of the DPC. To ensure a safe transport after interim storage period periodical reviews of the PDSR, in connection with periodic renewals of package design approval certificates, have to be carried out. The main reviewing tool is a gap analysis. A gap analysis for a PDSR is an assessment of the state of technical knowledge, standards and regulations regarding safety functions of structures, systems and components.

Furthermore the periodic review and update of the safety case and the package design approval is an important element of knowledge management.

References

[1] International Atomic Energy Agency (IAEA). Regulations for the Safe Transport of Radioactive Material, 2012 Edition, Specific Safety Requirements No. SSR-6, Vienna, 2012

[2] Guideline for the design approval procedure of packages for the transport of radioactive material, of special form radioactive material and low dispersible radioactive material (R003), Verkehrsblatt No. 23, 2004 December 15, p. 594.

[3] European PDSR Guide, ISSUE 2 (September 2012)
http://www.bfs.de/de/transport/zulassung_behaelter/grundlagen_zulassungsverfahren/European_PDSR_Guide_Issue_2_Sept12.pdf

[4] IAEA-TECDOC-DRAFT "Guidance for preparation of a safety case for a dual purpose cask containing spent fuel". Draft TM-44985, IAEA. Vienna, Austria, 2013 (to be published)

[5] D. Wolff, H. Völzke, F. Wille, and B. Droste „Extended Storage after Long-Term Storage", 17th International Symposium on the Packaging and Transport of Radioactive Materials (PATRAM 2013), San Francisco, CA, USA, (August 18-23, 2013).

[6] A. Erhard, H. Völzke, U. Probst, and D. Wolff, "Ageing Management for Long Term Interim Storage Casks", Paper#61, 16th International Symposium on the Packaging and Transport of Radioactive Materials (PATRAM 2010), London, UK (Oct. 03-08, 2010).

[7] H. Völzke, D. Wolff, U. Probst, S. Nagelschmidt, S. Schulz, „Long-term Performance of Metal Seals for Transport and Storage Casks ", 17th International Symposium on the Packaging and Transport of Radioactive Materials (PATRAM 2013), San Francisco, CA, USA, (August 18-23, 2013)

[8] M. Jaunich, W. Stark, D. Wolff, and H. Völzke, "Investigation of Elastomer Seal Behavior for Transport and Storage Packages", 17th International Symposium on the Packaging and Transport of Radioactive Materials (PATRAM 2013), San Francisco, CA, USA, (August 18-23, 2013)

[9] F. Nitsche, F.-M. Börst, I. Reiche: The German Regulatory Concept of Transport Package Design Approval for Dual Purpose Casks during Interim Storage. 17th International Symposium on the Packaging and Transport of Radioactive Materials (PATRAM 2013), San Francisco, CA, USA, (August 18-23, 2013)

[10] F. Wille et al.: German Approach and Experience Feedback of Transport Ability of SNF Packages after Interim Storage. 17th International Symposium on the Packaging and Transport of Radioactive Materials (PATRAM 2013), San Francisco, CA, USA, (August 18-23, 2013)

[11] RECOMMENDATION of the Nuclear Waste Management Commission (ESK) - Guidelines for dry cask storage of spent fuel and heat-generating waste (revised version of 10.06.2013), (in German), http://www.entsorgungskommission.de/englisch/statements---recommendations--letters/index.htm

[12] B. Droste, S. Komann, F. Wille, A. Rolle, U. Probst, S. Schubert: Considerations of Aging Mechanisms Influence on Transport Safety and Reliability of Dual Purpose Casks for Spent

Fuel Nuclear Fuel or HLW, PSAM12, Probabilistic Safety Assessment and Management, Honolulu, Hawaii, USA (June 22-27, 2014)

[13] S. Komann, M. Neumann, V. Ballheimer, F. Wille, M. Weber, L. Qiao, B. Droste: Mechanical assessment within type B packages approval: applications of static and dynamic calculations approaches, In: Packaging, Transport, Storage & Security of Radioactive Material, 2011, Vol. 22, No. 4, p. 179-183

The evolution of safety related parameters and their influence on long-term dry cask storage

Klemens Hummelsheim[a], Jörn Stewering[b], Sven Keßen[a] and Florian Rowold[a,*]

[a] Gesellschaft für Anlagen und Reaktorsicherheit (GRS) mbH, Garching, Germany
[b] Gesellschaft für Anlagen und Reaktorsicherheit (GRS) mbH, Cologne, Germany

Abstract: For spent nuclear fuel management in Germany, the concept of dry interim storage in dual purpose casks prior to direct disposal is applied. Due to current delay in site selection and exploration, the necessity of an extension of the storage period beyond the granted license time for 40 years seems inevitable. Compliance with safety requirements under consideration of aging effects, in particular safe confinement, radiation shielding, subcriticality and decay heat removal will be crucial for the extension of these operation licenses. Thermal loads, mechanical stresses, gamma and neutron radiation are considered to be the main contributors to aging and degradation effects of the fuel, its cladding and the cask over the period of long term storage including subsequent transport. In order to assess the long term safety of such a system, knowledge about the evolution of the influencing variables is required.

The paper at hand describes numerical investigations in the field of spent fuel long-term behavior, e.g. fuel clad temperature and hoop stress over a time period of 100 years. Analytical storage temperature and stress calculations for a generic cask with different burnups and loading patterns of UO_2 and MOX fuel will be presented. The gained results will be related to actual questions regarding long-term degradation effects. Furthermore, shielding analyses with regard to varying densities of the integrated neutron moderator of the cask will be discussed.

Keywords: Dry Cask Storage, PWR fuel, Temperature, Hoop Stress, Shielding.

1. INTRODUCTION

In Germany, the on-site and central facilities for dry storage of spent nuclear fuel have granted license time of 40 years. In the middle of 2013, the new Repository Site Selection Act [1] became effective, which regulates the course of actions in order to find a final disposal for heat generating waste in the Federal Republic of Germany. An underground disposal site is scheduled to be found by the end of 2031. Based on experience, the following application, licensing and legal actions by the public may take up several years until construction work may begin. Regarding a realistic construction time and the fact that the licenses of the first central storage facility Gorleben and the first on-site storage facility in Lingen expire in 2034 and 2042, an extension of the licensed storage period will be needed.

It is important to know that the temporary licenses of 40 years are based on administrative reasons and not on limiting physical or technical parameters. Related to this subject, the "Guidelines for dry cask storage of spent fuel and heat-generating waste" [2] stipulates that if the licensed storage period seems likely insufficient, further appropriate safety assessments concerning e.g. long-term behavior of fuel assemblies and cask components have to be provided by the licensee. Consequently, it has to be shown that the safety functions, in particular safe enclosure of the radioactive inventory, subcriticality, radiation shielding and decay heat removal will be fulfilled during the envisaged timeframe beyond 40 years. Further, important aspects for the strategy of long-term storage prior to final disposal are transportability of the casks and manageability of the fuel assemblies. Accordingly, additional knowledge and data about material and component performance in conjunction with predominant conditions are necessary for sufficient safety assessments generated by the applicants and safety evaluations conducted by the competent authority and its technical experts.

---

* Corresponding author email: *Florian.Rowold@grs.de*

Together with the Federal Institute for Materials Research and Testing (BAM) and Oeko-Institut e.V., GRS is currently working on a state of the art report on the safety aspects of long-term dry storage on behalf of the Federal Ministry for the Environment, Nature Conservation, Building and Nuclear Safety. The work of the GRS is focused on long-term safety aspects originated from the fuel and its enclosure. Investigations comprise on the one hand a variety of PWR fuels, which have been used in Germany from the past up to now, ranging from low to high burnup UO_2 and MOX. Important parameters are represented by time-dependent decay heat, released fission gases and subsequent temperature and hoop stress of the cladding. On the other hand, the state of the art in the field of possible fuel degradation mechanisms is a matter of particular interest for the GRS. Safe enclosure and manageability of fuel elements require that systematic failure of the fuel rods should not occur during the storage period and that the fuel assemblies have to keep their geometric arrangement. Evidence of conformity has been verified for the initial 40 years of storage by the licensees but by extending the storage periods, one has to account for new boundary conditions, such as the temperatures and stresses mentioned, and investigate its influence on the mechanical properties of the cladding. Further interest is devoted to the radiological behavior of the casks and the impact of the long-term application itself as well as postulated neutron moderator degradation due to radiation.

The paper at hand gives an overview of the work GRS performed in the field of safety of long-term storage until early 2014, consisting of burnup calculations, the development of generic models for heat transfer simulation, shielding analyses and cladding hoop stress prediction. First results of the numerical investigations for different cask loadings and fuel properties will be presented and discussed in the context of actual findings in the field of degradation mechanisms of dry storage systems.

2. MODELS AND METHODS

2.1. General Approach

The pursued dry storage of spent nuclear fuel in Germany in thick-walled dual purpose casks filled with helium and equipped with integrated neutron shielding components represents the underlying system of the described work and investigations. The generic model to simulate this system incorporates a typical cask designed to hold 19 PWR fuel assemblies of the 18x18-24 type (see **Fig. 1**). The maximum heat load of the cask is 39 kW. Further details will be given in the respective chapters.

Figure 1: Quarter section of the generic cask model

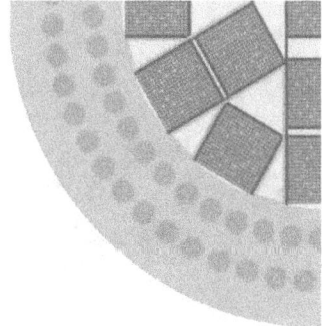

2.2. Burnup calculations

For the burnup and decay analyses, four types of fuel were chosen to cover the range of fuel used in German pressurized water reactors. An overview is provided in **Tab. 1**. Calculations were performed with the GRS in-house code OREST-V08. OREST-V08 is built upon the zero-dimensional burnup code ORIGEN, which is coupled with the one-dimensional HAMMER code for neutron flux spectrum and effective cross section determination. Beside burnup calculations, the code allows the determination of decay heat, activity and nuclide inventories for user-defined decay time steps. The fuel assembly is of the 18x18-24 type with an outside cladding diameter of 9,5 mm, pellet diameter of 8,05 mm and a pitch of 12,72 mm.

Table 1: Fuel data

Fuel	Enrichment [%]	Burnup [GWd/tHM]
UO_2	3,6	40,0
	4,4	55,0
HBU-UO_2	4,8	70,0
MOX	4,75 ($Pu_{fiss.}$)	55,0

2.3. Cask model and heat transport

Thermal analyses were performed with the GRS Code COCOSYS V.2.4.0, whose intentional purpose is the simulation of severe accident propagation in containment systems of nuclear power plants. Its internal heat transfer module was used to create two-dimensional models of a generic cask and fuel assemblies. The basic principle is the radial subdivision of the structures into heat slabs and the definition of the heat transfer mechanisms between them. The input value is the average pin power derived from the burnup calculations. The code then determines the corresponding temperatures at the boundaries of the heat slabs. By repeated calculations for different time-dependent pin powers, the generation of functions of storage time vs. fuel clad temperature is being enabled.

To reduce complexity and resulting computing time, several assumptions and simplifications were necessary. Between the fuel rods, only heat radiation was considered whereas heat conduction through the helium was only applied between the outer fuel rods and the basket wall. Natural convection of helium inside the fuel element was neglected because of the two-dimensional approach and preliminary investigations showing low impact on heat transmission from fuel to cask environment.

Analyses were performed for three types of cask loading patterns. In the first type, the cask is homogeneously loaded, meaning that all 19 fuel element are of the same type with a maximum burnup of 55 GWd/tHM. In the second and third type, the cask is loaded heterogeneously with 15 UO_2 fuel elements (max. burnup 55 GWd/tHM) and 4 HBU-UO_2 or MOX fuel elements. The positions of the HBU or MOX fuel elements are represented by the lighter green in **Fig. 2**.

Figure 2: Scheme of heterogeneous cask loading

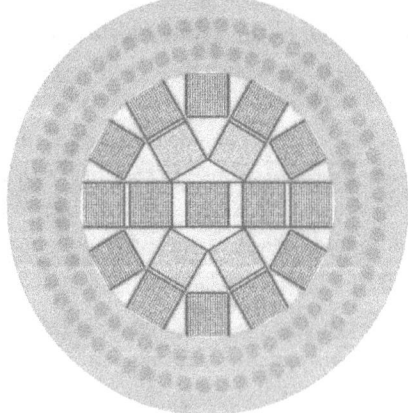

2.4 Fuel rod internal pressure and hoop stress

For the determination of the time-dependent fuel rod internal pressure, GRS developed a tool, which solves the Van der Waals equitation for the user-defined storage time steps:

$$p_k = \frac{n \cdot R \cdot T}{V - n \cdot b} - \left(\frac{n}{V}\right)^2 \cdot a \qquad (1)$$

The fuel rod internal pressure is determined by the amount of free gas, the free volume in the rod and the predominant temperature. As described, the temperatures derive from the thermal analyses, whereas the free gases are composed of the released fission gases Xenon and Krypton during operation and the fill gas Helium used for pressurization of the fuel rod in the manufacturing process. Fission gas amounts are given by the previously described burnup calculations in combination with a user-defined release rate. The amount of fill gas is calculated from the initial pressure level at room temperature in an iterative way. Free rod volume consists of plenum, gap and dishing volumes reduced by burnup dependent fuel swelling. Respective input data and quantities were based on publicly available data. The summation of the gas partial pressures leads to the internal gas pressure from which the cladding hoop stress σ_t can be calculated with Barlow's formula,

$$\sigma_t = \frac{p \cdot d}{2 \cdot s} \qquad (2)$$

where p stands for rod pressure, d for rod diameter and s for wall thickness.

2.5 Shielding analyses

A further matter of interest in our investigations is the evolution of the dose rate on top of the secondary lid during an extended storage period. Typically a storage cask is equipped with a polyethylene plate between the primary and secondary lid (see **Fig. 3**). The purpose of the plate is the moderation of neutrons. Thereby the absorption of the neutrons in secondary lid will be improved. Consequently the neutron shielding capability of the system improves as well.

Figure 3: Scheme of the double lid system

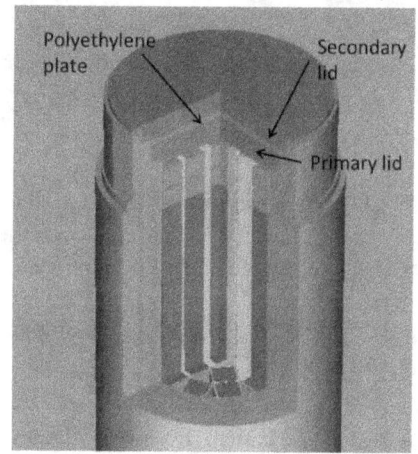

The equivalent dose rate outside the cask is made up by a source term that consists of neutrons and photons from fission and fission products and secondary gammas from capture processes. In addition, the activated parts of the fuel assemblies are considered. The source term resulting from the spent fuel was provided by OREST-V08 calculations. Activation was calculated with GRSAKTIV, an ORIGEN based code of GRS. The source is prepared in a 174 neutron and 39 gamma grouped spectrum. The shielding calculations were performed as follows: the fuel assemblies were segmented into homogenized fuel, plenum and head sections. The cask itself was modeled in detail. The particle transport calculations were done by MCNP5 with continuous energy cross section libraries based on ENDFB-VII data evaluation. To convert the calculated particle flux into dose equivalent rates, ICRP-74 conversion factors were used. The previously described homogeneous and heterogeneous cask loading patterns were investigated. The dose rate at the top of the secondary lid was calculated with a surface tally that gives an average value over the lid surface.

3. RESULTS

3.1. Evolution of temperatures, rod pressures and hoop stresses

Fig. 4 shows the decay power per fuel rod vs. time for the different fuel types, beginning five years after discharge. UO_2 fuel decay power depends on the final burnup. Between the HBU-UO_2 with 70 GWd/tHM and the moderate burnup UO_2 fuel with 40 GWd/tHM, decay power is about twice as high. For MOX fuel, the decay power level exceeds those of HBU-UO_2 and decreases to a lesser extent, e.g. the power level of the MOX fuel after 100 years is reached by the HBU fuel after 30 years.

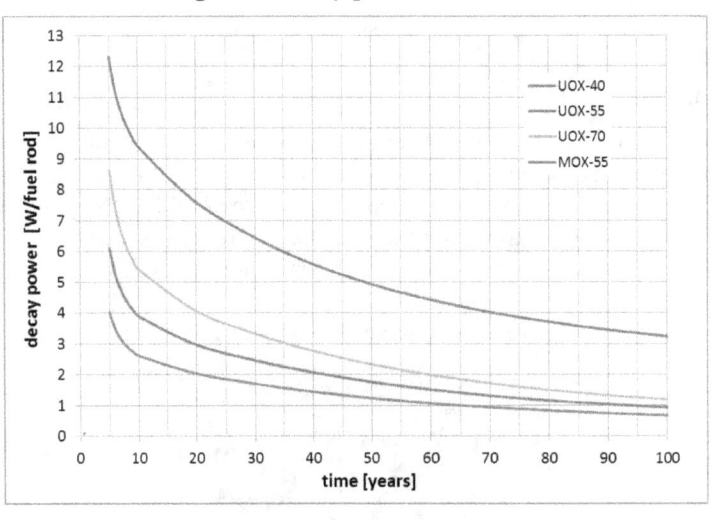

Figure 4: decay power vs. time

By gaining the specific depletion characteristics of the different fuel types, we were able to use the results for different time steps as inputs for the thermal model of the cask and fuel assemblies. It is important to note that due to time and resource constraints, our investigations focused on the hottest fuel pin which are
   (1) the central fuel pin of the central fuel assembly in the homogenously loaded cask and
   (2) the central fuel pin of the high burnup UO_2 and MOX fuel assembly in the heterogeneously loaded cask.

Nevertheless, this approach ensures conservatism by covering the most unfavorable conditions in terms of subsequent internal rod pressure and cladding stress determination.

Thermal analyses for the different cask configurations produced the results shown in **Fig. 5** with correspondent indexing:
   a) 19 UO_2 fuel assemblies (FA) with 40 GWd/tHM burnup,
   b) 19 UO_2 FA with 55 GWd/tHM burnup,
   c) 15 UO_2 FA with 55 GWd/tHM and 4 HBU-FA with 70 GWd/tHM burnup,
   d) 15 UO_2 FA with 55 GWd/tHM and 4 MOX-FA with 55 GWd/tHM burnup.

Figure 5: Cladding temperatures as a function of pin power

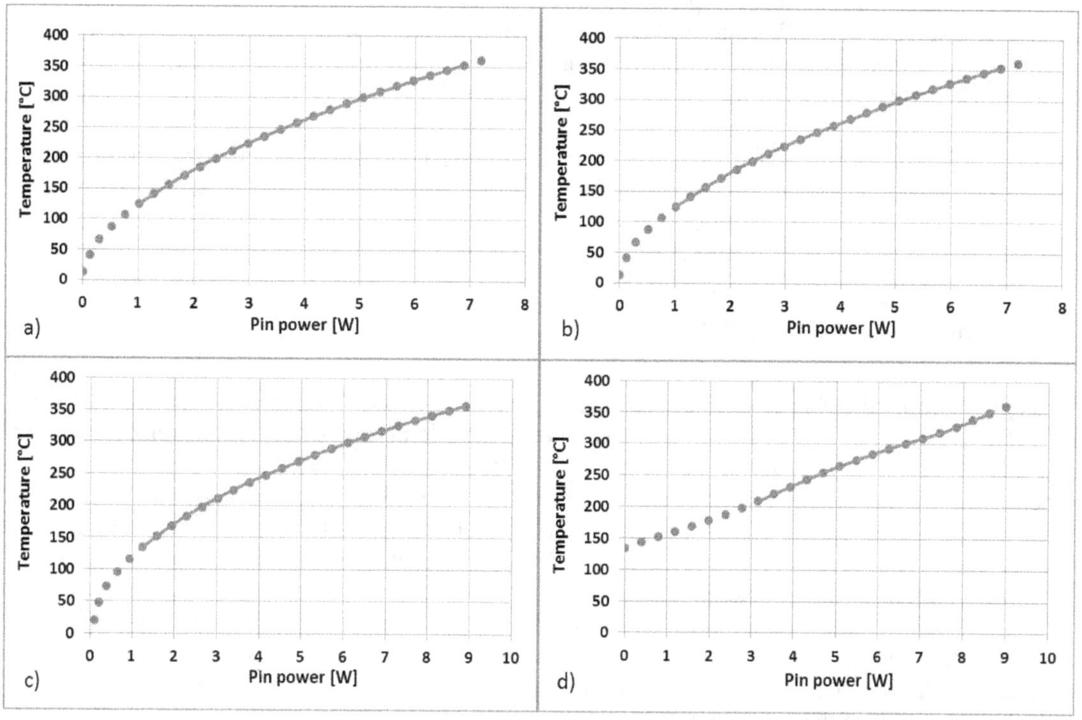

Fitting the curves in the pin power range for 100 years of storage (blue lines and **Fig. 3**) allows the setup of time dependent temperature functions. The difference in power ranges between the upper and lower diagrams is a result of the different loading patterns. With a maximum cask heat load of 39 kW, HBU and MOX FA are permitted to have higher heat loads at the expense of the remaining UO_2 FA in the cask, which reduces their required wet storage period. It is visible that the maximum temperature of the hottest pin is about the same for both configurations although the pin power is 30 % higher for HBU and MOX fuel. This effect is based on the positioning of the FA in the middle section of the basket (see **Fig. 2**), where additional heat transmission to central FA with lower power is provided.

As the temperature decreases very slowly, a quasi-static approach with a sufficient number of data points, e.g. every 10 years, was considered satisfactory for the following investigations. In order to calculate the time-dependent fuel rod internal pressure, knowledge about the initial free gas inventory in the rod is required. **Tab. 2** shows the total amount of gases generated till the end of reactor operation, resulting from the burnup calculations.

Table 2: Gas inventory at discharge

Fuel	Krypton	Xenon	Helium
	EOL amount [mol/tHM]		
UO_2 - 40 GWd/tHM	5,11	48,31	1,49
UO_2 - 55 GWd/tHM	6,87	66,85	2,11
UO_2 - 70 GWd/tHM	8,39	85,82	2,82
MOX-55 GWd/tHM	4,10	63,13	5,45

Fission gas release from the fuel increases with burnup but is mainly dependent on temperature and power history of the fuel. As a result, a broad range of fission gas release data exists. Taking into consideration the data range and recommendations given in [3], a pessimistic and at the same time conservative value of 15 % was chosen for the internal pressure calculations. Further input values comprise a fuel swelling rate of 1 % per 10 GWd/tHM burnup, a plenum volume of 16 cm³, a dishing volume fraction of 1,33 % and an initial pressurization level of 1,72 MPa. Most of the data originated from conversations with in-house experts of GRS. It is important to note that the data only represent one generic example of a PWR fuel rod. Fuel rods and fuel assemblies have been subject to steady research, development and optimizations. They are produced by a multitude of vendors which results in the existence of varying rod data.

In **Fig. 6** the results of the fuel rod pressure and cladding hoop stress calculations are presented. Additionally, the respective temperatures are provided for the
        a) UO_2 fuel rod with 40 GWd/tHM,
        b) UO_2 fuel rod with 55 GWd/tHM,
        c) HBU-UO_2 fuel rod with 70 GWd/tHM,
        d) MOX fuel rod with 55 GWd/tHM.

The diagrams show that wet storage periods had to be adjusted from 3 years for the low burnup UO_2 fuel to 10 years for the MOX fuel to reach maximum tolerable decay power levels. During that time, the cladding temperature was artificially held at 50 °C. At the beginning of dry storage the temperatures will increase rapidly and reach between 350 °C and 370 °C. Afterwards the temperatures decrease more slowly with higher burnup for UO_2 fuels. After 100 years, the UO_2 fuels have temperatures of 100 °C (a), 118 °C (b) and 131 °C (c). Due to the decay characteristic of MOX, the temperature decreases much more slowly and is at 211 °C (d) after 100 years. Governed by temperature, fuel rod pressure and hoop stress develop in the same manner but show a large variation across the different fuels. This is caused by the fission gas inventory and its release as main contributor to the rod pressure. Especially Xenon plays an important role with its fraction of 88 % of the fission gases. Since the generation of fission gases strongly depends on the burnup (see **Tab.2**), it is logical that HBU-UO_2 fuel rods will be affected the most. At the beginning of dry storage, the HBU-UO_2 (c) rod shows an internal pressure of 18,1 MPa and a clad hoop stress of 116 MPa. During a time span of 100 years, these values decrease to 10,8 MPa and 70 MPa. In comparison the low burnup UO_2 rod (a) starts at 9,3 MPa and 60 MPa and ends at 5,3 and 34 MPa. For the MOX (d) and UO_2 (b) rods with the same burnup of 55 GWD/tHM, the cladding hoop stresses at the beginning of dry storage are 81 MPa for MOX and 85 MPa for UO_2. Over the course of time, the hoop stress of the UO_2 rod decreases to 50 MPa whereas the hoop stress of the MOX rod decreases to only 60 MPa because of the higher temperature.

Figure 6: Evolution of temperatures, rod pressures and hoop stresses

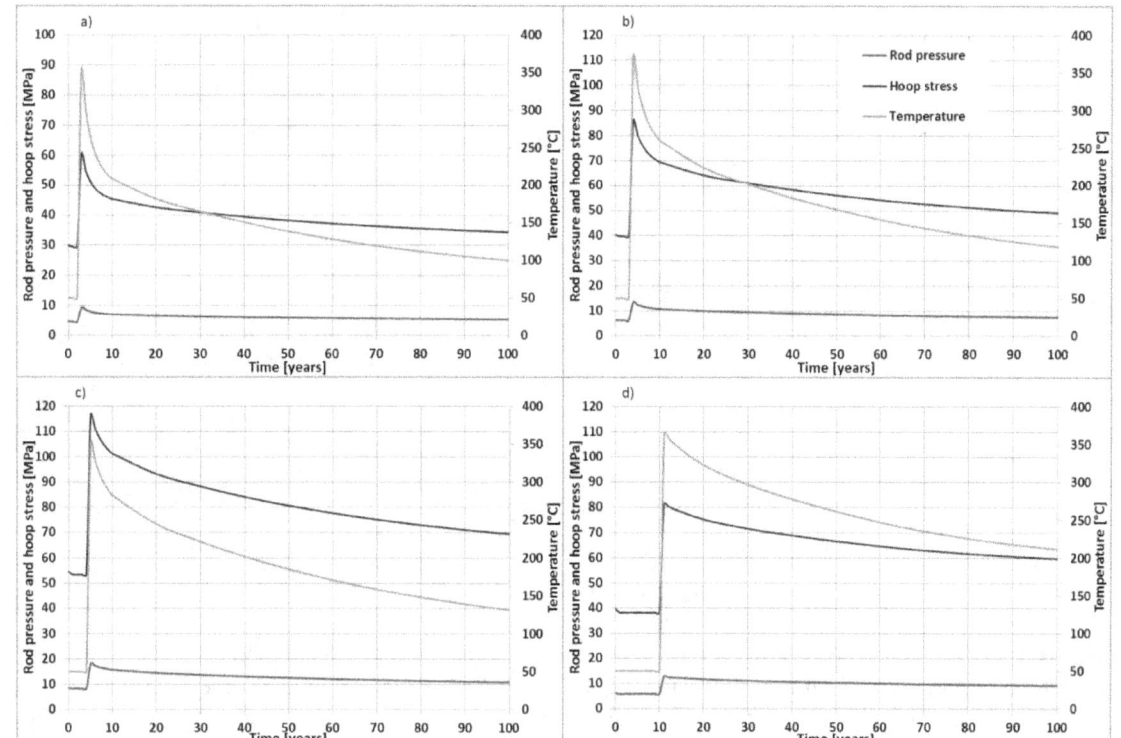

3.2. Evolution of dose rates

The shielding analyses were performed only for the homogenously loaded cask with spent fuel of 55 GWd/tHM burnup and for the heterogenic loading pattern with MOX fuel. It is important to note that for the spent UO_2 fuel a wet storage time of 5 years was used, the 4 MOX assemblies had a 10 years cooling time. The time axis in **Fig. 7** starts in the moment of cask loading. The presented results for the dose rates are divided in two parts: dose rates from neutrons and gammas. The neutron source data is dominated by spontaneous fission, the (α,n) part is nearly negligible. The gamma source originates from decay processes and from neutron induced reactions. Activation is a very small part and negligible after 50 years. The radiation transport calculation through about 35 cm of stainless steel in a very small angled direction cannot be done without variance reduction. Consequently, weight-windows were used to get statistically satisfying results. **Fig. 7** demonstrates that the highest dose rate on the lid will be expected for the heterogeneously loaded cask. Due to the higher neutron emission rate from MOX fuel (factor 8 to 9), the dose rate is about a factor of 3 higher than in the homogeneous case.

The gamma dose rate starts at a higher value for the uniform case because of a higher gamma source term. Then the dose rate decreases and both curves intersect. After 40 years of storage, the curves show similar characteristics and differ with constant ratio. In general, the gamma dose rate is more than one order of magnitude lower than the neutron dose rate.

The dose rate for neutrons will drop from the emplacement of the cask until a storage period of 80 years by a factor of 15. The gamma dose is a composition of fission product emission, activation, bremsstrahlung and secondary gammas. However, most contributors have no relevance to the final result. Fission products and secondary gammas solely contribute to the gamma dose rate. The gamma dose rates will decline with a factor of 80 for the homogeneously loaded cask and for the heterogeneously loaded cask with a factor of 18 respectively.

Additional calculations were done to show the impact of a reduced density of the moderator material in the previously described double lid system (see **Fig. 3**). The density was exemplarily reduced by 50 %. As a consequence, the neutron dose rate will rise by a factor of 3 to 4.

Figure 7: Dose rate evolution on the secondary lid of the cask

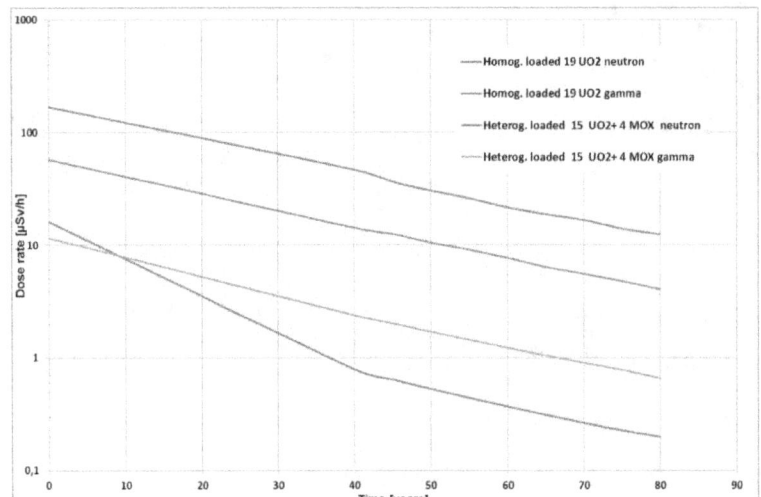

4. DISCUSSION

In this chapter, the results with respect to long-term safety of dry spent fuel storage will be discussed. Finding and erecting a final repository for spent nuclear fuel is a complex and time-consuming task. Many countries are adversely affected by significant delays, valid options, political issues or public resistance and consequently have to consider the option of long-term storage. Envisaged time frames extend up to 300 years in some countries. In this framework and under the aspect of safety, issues formerly considered less important such as aging management and long-term component behavior arose and are currently gaining more and more attention.

Regarding the spent fuel itself, the deterioration of cladding mechanical properties is an important aspect. One effect, which is often discussed in this regard, is hydride reorientation and embrittlement. During the reactor operation, the cladding absorbs hydrogen and after exceeding the solubility limit it precipitates in the form of circumferential hydrides. When the spent fuel is transferred into the dry storage cask and the cask is dried to remove residual water, the fuel heats up to around 400 °C where 200 ppm of hydrogen will dissolve again. During the storage period, the temperature will decrease and the hydrides will precipitate again. Under these conditions, a fraction of the hydrides could possibly precipitate as radial hydrides which are responsible for a reduction of ductility especially at low temperatures. This process is called hydride reorientation. The process is very complex because there is a multitude of parameters having an impact on the degree of reorientation e.g. the cooling rate, the thermal cycling, the temperature, stress levels and the cladding material itself. It has been reported that hydride reorientation was observed at stress levels above 70 MPa. For the ductile-to-brittle transition temperature it was observed that it strongly depends on the cladding material. Reported temperatures vary between 85 °C for M5, 150 °C for ZIRLO and 220 °C as the upper limit [4-8]. The results of the calculations show that the respective stress levels could prevail in fuel rods with a burnup higher than 55 GWd/tHM. By extending dry storage beyond 40 years, the possibility increases that the cladding temperatures fall below the ductile-to-brittle transition temperatures. With the concurrent decrease of the hoop stress and the mechanical load, adverse impacts should not emerge as long as the fuel is not exposed to additional loads. In the case of transport or accident scenarios, where additional loads are quite likely, this aspect should be considered.

Concerning radiation protection, an extension of the storage period will lead to further reduced dose rates on the surfaces of the cask. As a result, the working staff will be less exposed if work in the environment of the cask is necessary. Additionally, the postulated degradation of the neutron moderator after 40 years will not result in dose rates higher than those at the beginning of dry storage.

5. CONCLUSIONS

In order to assess the safety of long-term storage scenarios, the knowledge about the predominant conditions is required. These conditions originate from a multitude of variables which will vary for different dry storage systems. Our investigations aimed at the German concept of dry storage in thick-walled dual purpose casks. By using different computer codes and generic models, the GRS was able to cover most of the influencing variables. The analyses comprised different fuels, burnups, cladding temperatures and internal rod pressures. A set of results was produced with a conservative approach and allows the association with cladding degradation mechanisms. Since the long-term behavior of dry storage systems gains more and more attention on an international level, the presented work constitutes a contribution to the efforts.

Acknowledgements

This work was funded by the Federal Ministry for the Environment, Nature Conservation, Building and Nuclear Safety of Germany.

References

[1] *"Gesetz zur Suche und Auswahl eines Standortes für ein Endlager für Wärme entwickelnde Abfälle und zur Änderung anderer Gesetze (Standortauswahlgesetz - StandAG)"*, Bundesgesetzblatt 2013, Part 1 Nr. 41, (2013).
[2] Nuclear Waste Management Commission (ESK). *"Guidelines for dry cask storage of spent fuel and heat generating waste"*, Revised version of 10.06.2013.
[3] M.A. McKinnon and M.E. Cunningham. *"Dry Storage Demonstration for High-Bunrup Spent Nuclear Fuel - Feasibility Study"*, Report No. PNNL-14390. August 2003.
[4] M.C. Billone et al. *"Used Fuel Disposition Campaign, Phase I Ring Compression Testing of High-Burnup Cladding"* Prepared for U.S. Department of Energy Used Fuel Disposition Campaign, Argonne National Laboratory, December 31, 2011
[5] M. Aomi et al. *"Evaluation of Hydride Reorientation Behavior and Mechanical Properties for High-Burnup Fuel-Cladding Tubes in Interim Dry Storage"*, Journal of ASTM International, Vol. 5, No. 9 (2008).
[6] R.S. Daum et al. *„Radial-hydride Embrittlement of High-burnup Zircaloy-4 Fuel Cladding"*, Journal of Nuclear Science and Technology, Vol. 43, No. 9 (2006), 1054-1067.
[7] K.S. Chan. *"A micromechanical model for predicting hydride embrittlement in nuclear fuel cladding material"*, Journal of Nuclear Materials 227 (1996) 220-236.
[8] P. Bouffioux, A. Ambard et al. *"Hydride Reorientation in M5® Cladding and its Impact on Mechanical Properties"*, Top Fuel 2013, Charlotte, North Carolina, September 15-19, (2013).

Aging Management of Dual-Purpose Casks on the Example of CASTOR® KNK

Iris Graffunder[a,(*)], **Ralf Schneider-Eickhoff**[b] **and Rainer Nöring**[b]
[a] EWN Energiewerke Nord GmbH, Lubmin, Germany
[b] GNS Gesellschaft für Nuklear-Service mbH, Essen, Germany

Abstract: In 2010 the spent fuel of the German prototype fast breeder reactor KNK was returned from France to Germany. For the return and the interim storage 4 transport and storage casks of the type CASTOR® KNK were designed and fabricated by GNS Gesellschaft für Nuklear-Service mbH. The casks were transported to Germany in December 2010 and stored in the interim storage facility ZLN operated by Energiewerke Nord GmbH (EWN). Due to there dual-purpose all CASTOR® casks have to fulfill the requirements of both fields of operation – transport and storage. After a minimum storage period of 40 years, a last transport to the final repository has to be carried out with the same requirements as for new casks. To be sure that the cask can be transported after the storage period the authorities require the renewal of the package design approval, normally each 5 or 10 years. In case of CASTOR® KNK the approval expires at October 2014. EWN and GNS are planning an extension of the validity period of the package design approval to 10 years. For this purpose an aging management report is necessary considering all stress factors, which are crucial for the rate of aging: Radiation, thermal and mechanical loads and corrosion.

Keywords: CASTOR, Dual-purpose cask, KNK, aging management, Germany

1. INTRODUCTION

In Germany there is so far no repository for high active waste, such as spent fuel assemblies or vitrified waste from reprocessing plants. That is why such waste is loaded into casks which may be used both for transport and for storage over a period of several decades. For this purpose, the German Company GNS Gesellschaft für Nuklearservice GmbH has developed the special CASTOR® type cask since the 1970es. The development of this cask type continued during the following years, being adapted to the specific cases. In the mean time, there are more than 20 different types of casks, which differ both in geometry (height, diameter, wall thickness) and in basket design, according to the specification of different plants and types of fuel.

Before loading the CASTOR® casks, a multitude of verifications are required to prove that the casks fulfill both the regulations of international requirements for legal transport regulations, for Type B(U) packagings, in agreement with the IAEA regulations, and also with national storage regulations according to the Atomic Law, requiring at least 40 years of interim storage. The license as transport cask is independent from concrete utilizations; license as storage cask is always granted for a specific storage purpose. For this, a storage license application is filed according to the Atomic Law, for storing the concerned CASTOR® casks in a specific interim storage facility. Fulfillment of the requirements for a storage period of 40 years must be proven within the scope of this application, that is, all aging factors which may occur during the interim storage period must be taken into account. Once it has been granted, the license for storage is valid during the whole interim storage period.

Due to the fact that the validity of the package design approval according to transport regulations is limited in time as opposed to the former, a renewal of the approval requires that it must be verified every 5 years that cask complies with the actualized transport regulations in every case. This also will

---

(*) contact: iris.graffunder(at)ewn-gmbh.de

be true when the concerned casks are not being transported during the considered period. This is based on the fact that the storage authorization according to the Atomic Law requires that it must be possible to transport casks for removal at any time.

Interim storage of the casks is decentralized in Germany. The used fuel assemblies are stored in interim storage facilities at the nuclear power plants, in so called on-site interim storage facilities. The only type of storage which was carried out centrally so far was that of waste returning from reprocessing in France. As soon as a repository is available, all casks must be transported there.

Till mid 2013 it was planned to construct a repository in a salt stock near Gorleben, in the North of Germany. This boundary condition was basically changed through the Repository Site Selection Act, which came into force in mid 2013. The repository searching process was initiated anew according to a transparent procedure, on a "blank" map of Germany, involving the public. The repository site shall be found till 2031. The approval and construction of the repository will follow. Final construction of the repository is not expected before 2050.

Due to the increased period of interim storage which will follow as a result of this, the question of aging management is being considered with increasing intensity by the authorities responsible for interim storage. In Germany, a periodic safety inspection which must take place every 10 years, as is required for nuclear power plants, is now being required for interim storage facilities. This must include a control of aging management.

Independently from the relevance of the increased interim storage period for the storage facilities themselves, the question is arising as to what the behavior of casks will be under normal conditions of transport and due to hypothetical incidents during transport, after 50 to 70 years of interim storage. Presently, the competent German Authorities are requiring detailed information concerning aging management, in order to take this question into account and to establish a base for significantly increased periods of approval.

2. BACKROUND KNK

The Compact Sodium-Cooled Nuclear Reactor Facility KNK II, located at the Karlsruhe Institute for Technology (KIT), the former Research Center Karlsruhe, has been operated from 1977 to 1991 as a prototype Fast Breeder Reactor facility. The fuel of the KNK II consisted of fuel assemblies (FA) with highly enriched Uranium-/Plutonium-MOX fuel (up to 93 % ^{235}U enrichment and up to 35 % plutonium in the heavy metal).

The fuel rods were transported to C.E.A. (France) in 1993 for reprocessing. However, due to the low solubility of the MOX fuel 2413 fuel rods from 27 FA could not be reprocessed. They were encapsulated and stored in a pool of the French research centre Cadarache operated by the Commisariat à l'énergie atomique (C.E.A.).

In a German-French fuel return project these fuel rods was returned to Germany to be stored for long time storage in the interim storage facility ZLN operated by state-owned company Energiewerke Nord GmbH (EWN).

The project started in September 2001 and ended with the transport in December 2010 [1].

For the return and the interim storage of the fuel, 4 transport and storage casks of the type CASTOR® KNK were designed and fabricated by GNS Gesellschaft für Nuklear-Service mbH especially for this project.
The license for storage was granted in 2010 by Bundesamt für Strahlenschutz (BfS), valid for 40 years. The package design approval for transport was granted in 2009 by BfS, but only valid for 5 years.

3. CASK DESIGN

The transport and storage cask CASTOR® KNK consists of a thick-walled cylindrical cask body, which contains a basket for holding nine cans with the inventory, and is closed with a primary and a secondary lid with the associated bolts and seals. Figure 1 shows a 3D-model of the cask.

The cask body is made of ductile cast iron, the two lids are made of stainless steel. Above the secondary lid a protection plate of unalloyed structural steel is arranged to protect the lid system from external influences. For the crane handling, two trunnions each are arranged diametrically on the lid side and on the bottom side of the cask wall. As a transport package the cask equipped with lid side and bottom side shock absorbers. On the means of transport, the cask is covered by a transport hood that constitutes the readily accessible surface in the sense of the transport regulations.

Figure 1.
3D-Model of CASTOR® KNK

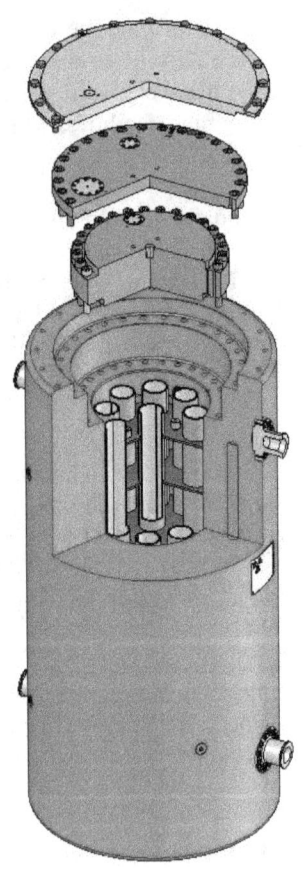

The main dimensions of the cask are approx.:
- outer diameter (without shock absorbers): 1380 mm
- width (with shock absorbers): 2090 mm
- height (without shock absorbers): 2784 mm
- height (with shock absorbers): 3906 mm
- inner diameter (cask body): 640 mm
- height of cask cavity: 2014 mm

The maximum mass (loaded, with shock absorbers) is approx. 32,500 kg, the mass without shock absorbers (loaded) is approx. 26,300 kg.

The inventory data of one cask are
- decay heat: 570 W
- total activity: 4.7×10^{15} Bq
- inventory mass: 500 kg
- heavy metal mass: 210 kg
- total mass U-235: 67 kg
- total mass Pu: 32 kg

4. REQUIREMENTS FOR RENEWAL OF PACKAGE DESIGN APPROVAL

The package design approval according to transport regulations for the CASTOR® KNK cask type expired in October 2014. The competent Authority is willing to grant a 10 years prolongation of the approval, provided the following conditions are fulfilled:
- the cask type is no longer being manufactured,
- all casks are loaded and stored in the interim storage facility,
- no transports are planned in the near future.

These boundary conditions are fulfilled for the CASTOR® KNK, so that it is endeavored to obtain a 10 years extension of the approval according to transport regulations.

For renewal of the package design approval it has to be demonstrated, that the Safety Analysis Report of the dual-purpose cask meets the state of the art. The state of the art permanently progresses. But new analysis methods require specific material parameters which often cannot be determined afterwards on the basis of available data. Due to this fact, the progression of the state of the art is one of the most challenging aspect in the long-term storage.

Active measures can be taken to increase the existing safety margins. Direct measures could be the re-design of the shock absorber or the introduction of an additional over-pack.

Indirect measures could be the mitigation of accident scenarios by new handling equipment/procedures or structural strengthening of the facility against outside impacts.

As mentioned in the beginning, the verification of the Safety Report has to be completed with an aging management report. The report has to reflect operation factors like handling and storage conditions and the experience during life time.

5. STORAGE

5.1 Actual storage situation

The interim storage facility ZLN is located near Greifswald at the Baltic Sea and is owned and operated by EWN GmbH, which is a 100 % daughter of the German Federal Ministry of Finance. As the KNK fuel is also owned by the German government and the Karlsruhe site has no storage facility for fuel any more, ZLN was chosen as interim storage. Figure 2 und 3 shows pictures of the premises, the ZLN facility and the storage hall.

Figure 2. Premises of EWN GmbH and Interim storage facility ZLN

Premises of Energiewerke Nord GmbH at the Baltic Sea   ZLN facility (red circle = storage hall 8)

Figure 3. 4 CASTOR® KNK casks between their "big brothers" in hall 8

The ZLN is a dry storage facility consisting of 8 storage halls. Halls 1-7 are used for non heat producing waste. CASTOR® casks may only be stored in hall 8. While air is being dried and heated by means of a ventilating system in halls 1-7, there is no supplementary ventilation system in hall 8. The heat from the casks is removed through natural convection. Special venting orifices in the walls and in the roof are provided for this purpose.

5.2 Possible Maintenance Actions to Control Aging

As a consequence of the storage requirements dual-purpose casks must have a double lid system, which is permanently monitored. Due to the double-lid design, the outer barrier (consisting of the secondary lid system and associated bolts and gaskets) can completely be removed without opening the inner containment of the radioactive material.

If the outer barrier would show any sign of intolerable aging effects, there is the opportunity to change or rework the affected components. Moreover, also the bolting of the primary lid system is completely accessible to test or to apply the specified torque of the bolting. The loss of pre-stress due to relaxation can be easily compensated.

For removable parts of the transport package aging effect can be eliminated by maintenance measures in advance. Only for the cask body, the basket and the primary lid system the influence of aging on specified characteristics must be ruled-out.

Some of the main components (cask body, basket, primary lid with bolts and metallic gasket and of course the fuel) cannot be substituted after the storage period and thus at least for those components aging management is already necessary in the casks design phase to avoid later interventions. Here the term aging means changes of characteristics of the cask, its components or the inventory with time or use. It is the task of the designer and user to take engineering, operation and maintenance actions to control the aging effects of the cask and its components within acceptable limits in order to ensure the specified characteristics and functionality over the complete lifetime cycle.

6. EVALUATION OF AGING INFLUENCES

6.1 Experiences by GNS

By now, more than 1,000 CASTOR® casks are in operation, about half of them in storage facilities operated by GNS. The overall storage time of CASTOR® casks worldwide sums up to more than 10,000 years (see Figure 4), while the storage period of single casks reaches up to 30 years. With this huge operational experience the CASTOR® design has got an in-service proof of its long-term operational reliability. [2]

On the basis of the operational experiences it can be concluded, that design and operation of CASTOR® casks is appropriate to keep the effects of aging within acceptable limits and to ensure the required functionality over the complete life cycle.

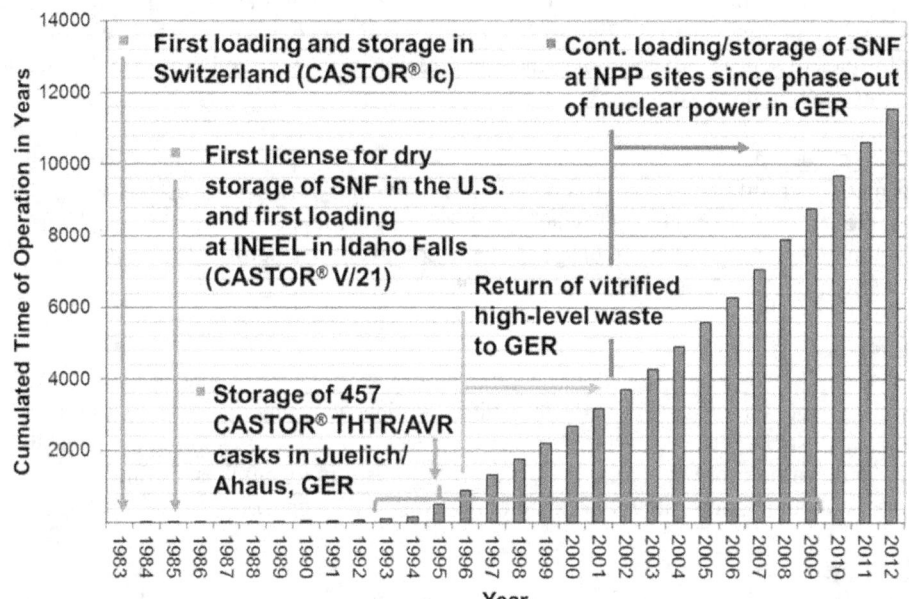

Figure 4. Cumulated time of operation of CASTOR® casks worldwide

6.2 Operational influence factors

The following influences are investigated within the scope of the evaluation of aging effects:

Environmental and handling influences
The following environmental and handling influences are being evaluated in relation to the interim storage facility:

- environmental temperature and change of temperature
- humidity
- aggressive media
- UV radiation
- mechanical stresses (dynamic and static stresses)

Inventory influences
The radioactive material in the casks has gone through a multitude of nuclear reactions and activation processes during reactor operation. Accordingly, during the following radioactive decay or during the cooling time, all types of radioactive radiation are being emitted by the inventory. As alpha and beta radiation merely have a short range, only neutron and gamma radiation are taken into account when considering the interactions with cask components. Due their long range, both types of radiation reach all cask components. Furthermore, the kinetic energy released during the radioactive decay of the inventory (especially beta minus decays and gamma radiation), the so-called decay heat must be removed passively over the cask components and transferred to the environment. The component and inventory temperatures during the corresponding operating phases are significant when evaluating material behavior.

The following parameters are taken into account:

- decay thermal power
- radioactive radiation
- residual humidity
- nuclides due to nuclear reactions

Influences due to design
Those influences resulting from the design and the dispatching of the casks are considered as influences due to design. These include on the one hand mechanical stresses resulting from the tightening torques of screwed connections, the compression of gaskets and pressurizations, and on the other, materials mating due to design.

Consequential influences
The above mentioned influences due to operation and inventory will cause effects which themselves may cause the release of materials, which in turn may influence the materials. Thus, different mechanisms may be the cause of hydrogen release, of hydrogen peroxide and of other aggressive media generation.

6.3 Aging effects

During storage, the casks are subject to effects which may modify their structural material characteristics. Possible effects and their consequences on the cask components shall now be explained. It will furthermore be explained which effects resulting from these influences will cause no impairments, so that they must not be evaluated in the following chapter.

Structural changes
Temperature effects and/or neutron irradiation may cause changes of the crystalline structure and of the microstructure of the materials. No structural changes of the metallic and inorganic materials caused as a consequence of thermal effects must be expected at the temperatures considered here. However, in principle, structural changes due to irradiation are possible.

Creeping/relaxation
When materials are submitted to mechanical stress, this will cause a diffusion of vacancies, to displacements of dislocations and/or to gliding processes along gliding planes. Due to this, plastic deformations may occur already below the apparent yielding point. Different effects may result from these creeping effects.
Screws may be subject to extension of length due to these creeping processes. This means a loss of the adjusted preliminary strain. This effect is called relaxation, and must only be considered for screws.
As only metallic gaskets and screws are subject to mechanical stress, the evaluation of creeping processes will remain limited to these two components.

Chemical reactions
Chemical reactions will cause changes of the electron shell of the participating atoms, resulting in changes of the linking conditions within the molecules. This may cause the breaking of bonds and the formation of new ones. The different reactions which may result are:

- solid state reactions
- gaseous phase reactions
- reactions in solutions

Anodic and cathodic partial reactions may occur separately on a surface or on different surfaces. This will lead to different mechanisms and corrosion effects, such as

- surface corrosion
  both the GGG material used as cask body material, as also the unalloyed and low-alloyed steels and zinc will present a tendency to surface corrosion under the boundary conditions considered here.
  As most of the used structural materials consist of these, the cask components must be evaluated for uniform corrosion.

- Galvanic corrosion (contact corrosion)
  The casks which must be evaluated are made of a multitude of different materials which are in contact with one another. Thus, galvanic corrosion must be evaluated.

- Pitting
  Pitting occurs in metals which may form a passive surface layer. Components made of stainless steel, of aluminum and aluminium alloys, are used in the casks. These materials generate native passive layers. However, as no aggressive media will be used, or be generated during interim storage, no pitting is expected.

7. EVALUATING OPERATING EXPERIENCE WITH CASTOR® KNK

The following operating phases are taken into account for evaluation:

- loading,
- transport to the interim storage facility,
- long term interim storage,
- transport to a nuclear facility after interim storage.

Before loading, it is assured by means of examinations that the cask is in the conditions required by specifications. The unloaded casks will be submitted neither to radiation nor to decay heat before being loaded, so that aging effects dating from the time before loading may be excluded.

7.1 Loading

Place and date of loading
The casks were loaded mid 2010 at Cadarache, in France. Nothing noteworthiness was observed during loading. Figure 5 shows the interior of a loaded CASTOR® KNK, and the exterior lid (protection plate) after lead sealing.

Figure 5. Pictures of CASTOR® KNK, basket with 9 cans and lid with seal

Inventory

- Specifying inventory.
  Due to the high enrichment of MOX fuel, criticality design of the cask was carried out using actual fuel data, that is, each of the four casks has its own inventory data sheet. During loading, it was controlled and recorded that the fuel rods cans mentioned in the inventory data sheet are putted into the correct casks.

- Thermal power at the time of loading, and comparison with design data.
  Layout values according to the maximum admissible decay power amounted to 567 W for CASTOR® KNK; the real decay powers of the 4 casks were below the layout value (max. 422 W) at the time of loading.

- Dose rate at the time of loading, and comparison with calculations / limit values.
  At the time of loading, the measured real values were lower than the layout values and the transport limit values.

- Temperature at the time of loading, and comparison with calculations / limit values.
  At the time of loading, the measured real temperatures were lower than the layout values or respectively the maximum admissible temperatures.

7.2 Public transports

<u>Date, duration, beginning, destination, means of conveyance</u>
Transport was carried out as a road transport from C.E.A. Cadarache to the transshipment station of Les Milles, followed by rail transport from Les Milles to Lubmin ZLN (Figure 6)
Departure Cadarache: Dec. 13, 2010
Transfer at Les Milles: Dec. 14, 2010
Arrival at ZLN: Dec. 16, 2010, 23.35 hours

Figure 6. Transport on rail from Les Milles, Southern France, to Lubmin, Northern Germany

<u>Transport configuration</u>
Transport configuration for road and rail, consisting of vehicle, adapter plates, holding down clamp, transport frame, cask, transport hood (Figure 7). The trunnions do not interfere with the transport frame.

Figure 7. Cask on the transport frame

7.3 Storage

<u>Bringing into storage</u>

The casks are raised to vertical position after arrival (Figure 8), and transported to the place of intervention, where necessary dose rate, neutron radiation and contamination measurements, as well as tightness tests of the protection plate (Figure 9) were carried out. The casks were then sealed and brought to their storage positions.

Figure 8. Cask handling at ZLN with horizontal and vertical lifting beam

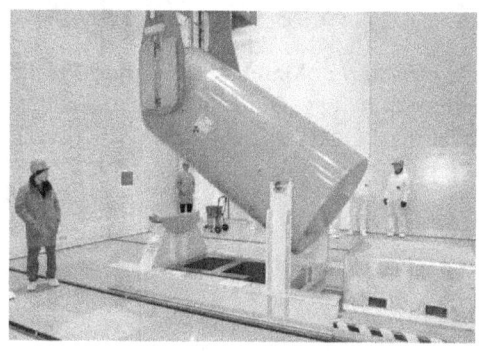

Figure 9. Neutron Measurement and leak test

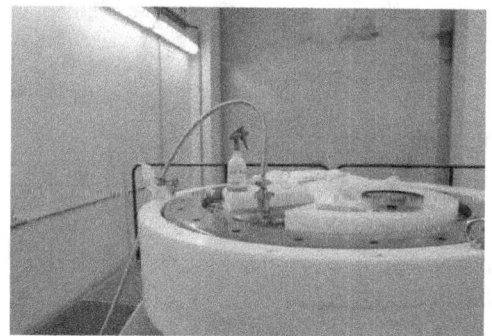

Position of the cask and of the surrounding casks
The casks were brought into storage according to the authorized storage position configuration. In 2011, the latter was changed, due to increased safety requirements. Neighboring casks are documented.

Temperature and humidity in the storage facility
The parameters temperature, relative ambient humidity, ambient pressure and dew point temperature are measured and recorded once an hour inside and for comparison outside the storage facility.

Exchange of air with environment outside the storage facility
Ventilation in hall 8 is designed as natural ventilation. Venetian shutters are installed in the north wall, to assure regulation of the volume stream. These are automatically opened or closed by means of servo motors, depending on temperature within the hall. The roof of hall 8 has 160 ventilator cowls arrayed in 32 rows, to assure ventilation. The maximum volume stream of this passive ventilating system amounts to 260,000 m³/h. Heat transfer is assured by convection. Temperature exchange over walls, roof and floor is insignificant.

Particular events during storage

- Failing pressure switch
  The pressure switch was replaced after the failure was identified. The failing pressure switch was sent to GNS for examination.

- Penetration of water into hall 8
  Due to unfavorable weather conditions, snow penetrated into hall 8 in March of 2013, through the ventilation cowls on the roof. Melting water covered the floor of hall 8, and the bottoms of some of the transport and storage casks. Water also was found under the 4 CASTOR® KNK casks. The cask bottoms however were not wet, because the casks stand on pads.

7.4 Findings resulting from storage to this day

Aging effects / corrosion
So far, CASTOR® KNK casks have been stored for 3 years. No aging effects or corrosion were found during this period.

Required conservation measures
Storage experience revealed that no supplementary conservation measures are necessary.

7.5 Repairs / Maintenance

Repair measures
No repairs were required during the storage period.

Maintenance
- Replacement of the pressure switch
- Changing the storage position of 2 casks due to operating conditions. The bottoms of the casks were examined visually during this procedure.

7.6 Periodic inspections carried out

So far, no periodic inspection was required for CASTOR® KNK.

8. CONCLUSION

Experience obtained during operation and storage showed that the conditions of transport and storage cask CASTOR® KNK fulfills the requirements of transport regulations, and that it may continue to be manufactured as planned, thanks to well defined and authorized measures.

Sufficient conservation measures were carried out for all casks, to assure protection against corrosive influences.

So far, no safety relevant aging effects were found for casks according to CASTOR® KNK design. Protection objectives were continuously assured during the total storage period.

Summarizing, it may be stated that the regulations and verification measures on which the safety technological layout of design package CASTOR® KNK as based on the safety report continue to be valid.

Evaluation of the operating experience shows that the design of package design CASTOR® KNK is sufficiently robust as far as storage and transport regulation requirements are concerned.

References

[1] D. Brauer, I. Graffunder, R. Vallentin, O. Pätzold. *"Return of the fuel from the German Compact Sodium Cooled Nuclear Reactor Facility KNK II with the Castor® KNK"*, Proceedings of the 16th International Symposium on the Packaging and Transportation of Radioactive Materials, PATRAM 2010

[2] R. Schneider-Eickhoff, R. Hüggenberg, L. Bettermann and H.P. Winkler. *„Aging Management in the Design and Operation of Dual-Purpose Casks"*, Proceedings of the 17th International Symposium on the Packaging and Transportation of Radioactive Materials, PATRAM 2013

Overview of New Tools to Perform Safety Analysis: BWR Station Black Out Test Case

D. Mandelli[a*], C. Smith[a], T. Riley[c], J. Nielsen[a], J. Schroeder[a], C. Rabiti[a], A. Alfonsi[a], J. Cogliati[a], R. Kinoshita[a], V. Pascucci[b], B. Wang[b], D. Maljovec[b]

[a] *Idaho National Laboratory, Idaho Falls (ID), USA*
[b] *University of Utah, Salt Lake City (UT), USA*
[c] *Oregon State University, Corvallis (OR), USA*

Abstract: The existing fleet of nuclear power plants is in the process of extending its lifetime and increasing the power generated from these plants via power uprates. In order to evaluate the impacts of these two factors on the safety of the plant, the Risk Informed Safety Margin Characterization project aims to provide insights to decision makers through a series of simulations of the plant dynamics for different initial conditions (e.g., probabilistic analysis and uncertainty quantification). This paper focuses on the impacts of power uprate on the safety margin of a boiling water reactor for a station black-out event. Analysis is performed by using a combination of thermal-hydraulic codes and a stochastic analysis tool currently under development at the Idaho National Laboratory, i.e. RAVEN. We employed both classical statistical tools, i.e. Monte-Carlo, and more advanced machine learning based algorithms to perform uncertainty quantification in order to quantify changes in system performance and limitations as a consequence of power uprate. We also employed advanced data analysis and visualization tools that helped us to correlate simulation outcomes such as maximum core temperature with a set of input uncertain parameters. Results obtained give a detailed investigation of the issues associated with a plant power uprate including the effects of station black-out accident scenarios. We were able to quantify how the timing of specific events was impacted by a higher nominal reactor core power. Such safety insights can provide useful information to the decision makers to perform risk-informed margins management.

Keyword: Dynamic PRA, SBO, BWR, data analysis, adaptive sampling, topological analysis

1. INTRODUCTION

In the RISMC [1] approach, what we want to understand is not just the frequency of an event like core damage, but how close we are (or not) to key safety-related events and how might we increase our safety margin through proper application of Risk Informed Margin Management. In general terms, a "margin" is usually characterized in one of two ways:
- A deterministic margin, typically defined by the ratio (or, alternatively, the difference) of a capacity (i.e., strength) over the load.
- A probabilistic margin, defined by the probability that the load exceeds the capacity.

A probabilistic safety margin is a numerical value quantifying the probability that a safety metric (e.g., for an important process observable such as clad temperature) is exceeded under accident conditions.

The RISMC Pathway uses the probabilistic margin approach to quantify impacts to reliability and safety. As part of the quantification, we use both probabilistic (via risk simulation) and mechanistic (via physics models) approaches. Probabilistic analysis is represented by the stochastic risk analysis while mechanistic analysis is represented by the plant physics calculations. Safety margin and uncertainty quantification rely on plant physics (e.g., thermal-hydraulics and reactor kinetics) coupled with probabilistic risk simulation. The coupling, which we call Computational PRA (CPRA), takes place through the interchange of physical parameters (e.g., pressures and temperatures) and operational or accident scenarios (e.g., the series of successes and/or failures representing a sequence of events).

This paper presents a case study in order to show the capabilities of the RISMC methodology [4] to assess limitations and performances of a Boiling Water Reactor (BWR) system during a Station Black

---
* Corresponding author: Diego Mandelli, P.O. Box 1625, MS 3850, Idaho Falls 83415 (ID); diego.mandelli@inl.gov

Out (SBO) accident scenario using a simulation-based environment also known as dynamic PRA [2]. Such assessment cannot be naturally performed in a classical ET/FT based environment.

We employ a system simulator code, one of the RELAP series of codes [3], coupled with a CPRA [2] code, RAVEN [5,6], that monitors and controls the simulation. The latter code, in particular, introduces both deterministic (e.g., system control logic, operating procedures) and stochastic (e.g., component failures, variable uncertainties, human actions) elements into the simulation.

This paper is structured as follows:
- Sections 2 and 3 describe the BWR system and the SBO scenario
- Section 4 presents the evaluation of the safety margins
- Section 5 describes adaptive sampling algorithms in order to reduce computational time
- Section 6 shows the stochastic analysis performed with RAVEN coupled with RELAP-5
- Section 7 presents the analysis of the data generated in Section 6.

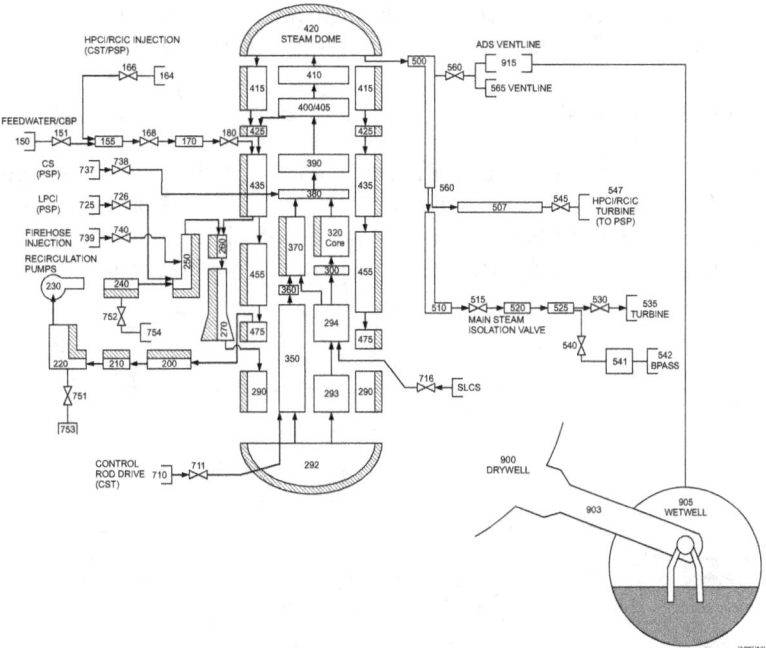

Figure 1: RELAP-5 nodalization scheme for the BWR system

2. BWR SYSTEM

The system considered in this test case is a generic BWR power plant with a Mark I containment as shown in Figure 1. The three main structures are the following:
1) Reactor Pressure Vessel (RPV), it is the pressurized vessel that contains the reactor core.
2) Primary containment includes:
   a) Drywell (DW): it contains the RPV and circulation pumps
   b) Pressure Suppression Pool (PSP) also known as wetwell: a large torus shaped container that contains a large amount of water; it is used as ultimate heat sink.
   c) Reactor circulation pumps

While the original BWR Mark I includes a large number of systems, we consider a subset of it:
- RPV level control systems: provide manual/automatic control of the RPV water level:
  1. Reactor Core Isolation Cooling System (RCIC): Provide high-pressure injection of water from the CST to the RPV. Water flow is provided by a turbine driven pump that takes steam from the main steam line and discharges it to the suppression pool. Alternatively, the water source can be shifted from the CST to the PSP.
  2. High Pressure Coolant Injection (HPCI): similar to RCIC, it allows greater water flow rates
- Safety Relief Valves (SRVs): DC powered valves that control and limit the RPV pressure.

- Automatic Depressurization System (ADS): separate set of relief valves that are employed in order to depressurize the RPV.
- Cooling water inventory:
  1. Condensate Storage Tank (CST) that contains fresh water that can be used to cool the reactor core.
  2. PSP water: PSP contains a large amount of fresh water that is used to provide ultimate heat sink when AC power is lost.
  3. Firewater system: water contained in the firewater system can be injected into the RPV when other water injection systems are disabled and when RPV is depressurized.
- Power systems:
  1. Two independent power grids that are connected to the plant station thorough two independent switchyards. Loss of power from both switchyards disables the operability of all system except: ADS, SRV, RCIC and HPCI (which require only DC battery).
  2. Diesel generators (DGs) which provide emergency AC power
  3. Battery systems: instrumentation and control systems need DC power.

2.1 BWR Containment Management

In an accident scenario, the set of emergency operating procedures requires the reactors operators to monitor not just the RPV but also the containment (both DW and PSP) thermo-hydraulic parameters (level, pressure and temperature). In particular, a set of limit curves is provided so that when they are crossed, the operators are required to activate the ADS system. These limit curves, also known as Heat Capacity Temperature Limits (HCTL), are shown in Figure 2 for both PSP and DW.

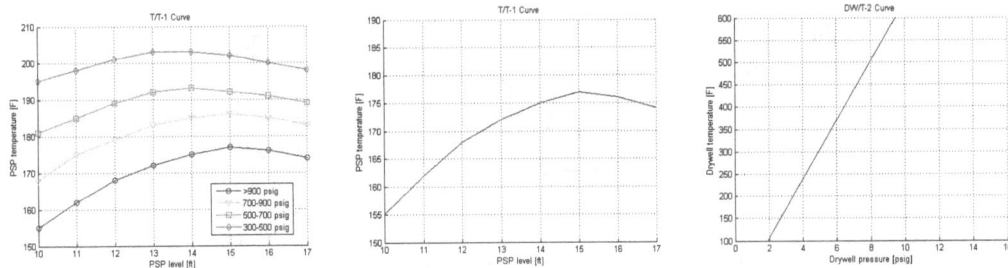

Figure 2: HTCLs for PSP (left and centre) and DW (right)

3. BWR SBO SCENARIO

The accident scenario under consideration is a loss of off-site power followed by loss of the DGs, i.e. SBO initiating event. In more details:
- At time $t = 0$: the following events occur:
  - LOOP condition occurs due to external events (i.e., power grid related)
  - LOOP alarm triggers the following actions:
    - Operators successfully scram the reactor and put it in sub-critical conditions by fully inserting the control rods in the core
    - Emergency DGs successfully start, i.e., AC power is available
    - Core decay heat is removed from the RPV through the RHR system
    - DC systems (i.e., batteries) are functional
- SBO condition occurs: due to internal failure, the set of DGs fails, thus removal of decay heat is impeded. Reactor operators start the SBO emergency operating procedures and perform:
  - RPV level control using RCIC or HPCI
  - RPV pressure control using SRVs
  - Containment monitoring (both drywell and PSP)
- Plant operators start recovery operations to bring back on-line the DGs while the recovery of the power grid is underway by the grid owner emergency staff
- Due to the limited life of the battery system and depending on the use of DC power, battery power can deplete. When this happens, all remaining control systems are offline causing the reactor core to heat until clad failure temperature is reached, i.e., core damage (CD)

- If DC power is still available and one of these conditions are reached:
  - Failure of both RCIC and HPCI
  - HCTL limits reached
  - Low RPV water level

  then the reactor operators activate the ADS system in order to depressurize the RPV
- Firewater injection: as an emergency action, when RPV pressure is below 100 psi plant staff can connect the firewater system to the RPV in order to cool the core and maintain an adequate water level. Such task is, however, hard to complete since physical connection between the firewater system and the RPV inlet has to made manually.
- When AC power is recovered, through successful re-start/repair of DGs or off-site power, RHR can be now employed to keep the reactor core cool

3. STOCHASTIC MODELING

The choice of the set of stochastic parameters to consider in the analysis was based on the preliminary PRA model results obtained for a typical BWR SBO case. For all basic events (e.g., DG fail to run) we have considered the following indexes:
- Fussell-Vesely and Birnbaum importance
- Event-tree structure for a LOOP-SBO

The most relevant basic events obtained from the PRA model are listed in Table 1. Ultimately, we also included uncertainties associated with two additional parameters:
- Clad damage temperature
- Reactor initial power (ranging from 100% to 120%)

Table 1: Basic Events obtained from the PRA model

1	Failure time of DGs	5	RCIC fails to run time
2	Offsite AC power recovery time	6	SRVs stuck open time
3	Recovery time of DGs	7	Battery life
4	HPCI fails to run time		

3.1 Human interventions

The probabilistic modeling of the five human interventions was done by looking at the SPAR-H [13] model from a generic BWR PRA. In this respect, we have identified 3 actions:
1. Manual ADS activation: operator manually depressurizes the RPV by activation the ADS system after HCTL limits are reached
2. Extended ECCS operation: operators may extend RCIC/HPCI and SRVs control even after the batteries have been depleted. This action actually summarizes two events:
   a. Manually control RCIC/HPCI by acting on the steam inlet valve of the turbine
   b. Alternate DC power availability through spare batteries
3. Firewater injection availability time (measured after ADS has been activated)

In general, SPAR-H characterizes each operator action though eight parameters – for this study we focused on just three factors: stress level, task complexity and time available to perform such task.

These three parameters are used in the SPAR model to compute the probability that such action will happen or not. However, from a simulation point of view we are not seeking *if* an action is performed but rather *when* such action is performed. Thus, we used the three factor mentioned above to determine the characteristic parameters (i.e., mean and standard deviation) of probability distribution function (assumed to be lognormal) that such action will occur as function of time.

4. SAFETY MARGIN ANALYSIS

This section shows some of the preliminary results regarding the effect of power uprates on SBO accident scenario. A higher value of thermal power generated in the core causes the following:
1. Faster heating of the PSP and, thus, a reduction of the time interval between ADS activation time and loss of DG time, i.e., $T_{ADS}-T_{SBO}$

2. A faster core temperature increase rate after ADS activation; thus leading to less time available to the plant staff to align the firewater

In summary, we expect that a power uprate reduces the time available to the plant staff to recover AC power and the time available to the plant staff to align FW (see Figure 3)

Scope of this section is to measure such reductions.

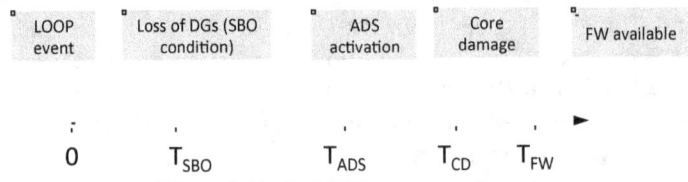

Figure 3: Typical SBO sequence of events

We performed an initial evaluation of the impact of power uprate by observing the PSP temperature increase rate as function of the thermal power generated by the core (see left image of Figure 4). In particular, we looked at the time to reach the PSP temperature limits for different values of core power (ranging from 100% to 120%). These results are shown on the right image of Figure 4. For this set of simulations we fixed T_{SBO} = 1h and we, thus, measured $T_{ADS}-T_{SBO}$.

As expected, by increasing the core power, the time to reach the PSP heat capacity limits decrease. In the left graph in Figure 4, the PSP temperature can be seen increasing in small steps as the SRVs open and close, and remaining relatively flat for a longer period of time whenever HPCI/RCIC activates and it is unnecessary to open the SRVs for a longer period of time. The sudden large increase in PSP temperature in each simulation is when the PSP heat capacity limit is reached and the ADS activates, dumping a huge amount of steam from the RPV into the PSP. Note that (Figure 4 right), if reactor power is increased to 110% and 120%, the time to reach core HCTL limits decrease from 4.5 h (16300 s) to 3.9 h (14100 s) and 3.5 h (12400 s) respectively.

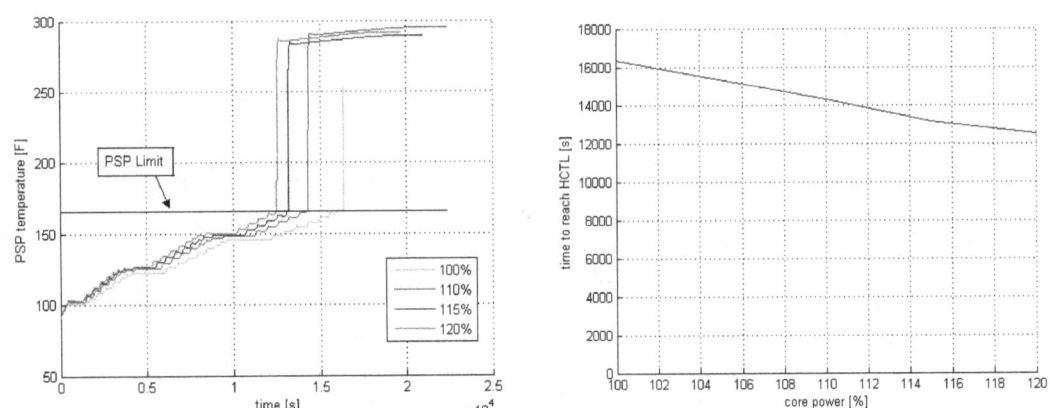

Figure 4: Impact of reactor power uprate on time to reach PSP heat capacity limits HCTL

We then considered the impact of power uprate for the following cases:
- Time to activate ADS vs. DG failure time (see Figure 5 left)
- Time to reach core damage vs. DG failure time (see Figure 5 right)

From Figure 5 note the following:
- We selected, for each power level (100%, 110% and 120%), a set of values for T_{SBO}. We then run a set of simulation runs and identified that time at which the reactor operators needs to activate the ADS. Compared to what is presented in Figure 4, this analysis considered not just PSP temperature as indication to trigger ADS activation but all the curves shown in Figure 2. In addition, AC power is not recovered and FW is never available.
- Figure 5 (left) shows T_{SBO} (x axis) vs. $T_{ADS}-T_{SBO}$ (y axis). By increasing T_{SBO}, we expect that the reactor operators are required to activate ADS much later. Again, a reactor power increase negatively affects ADS activation time.
- Figure 5 (right) shows T_{SBO} (x axis) vs. $T_{CD}-T_{SBO}$ (y axis). If AC power is available for a long time, the PSP HCTL limits are reached further in time. This allows reaching CD much later.

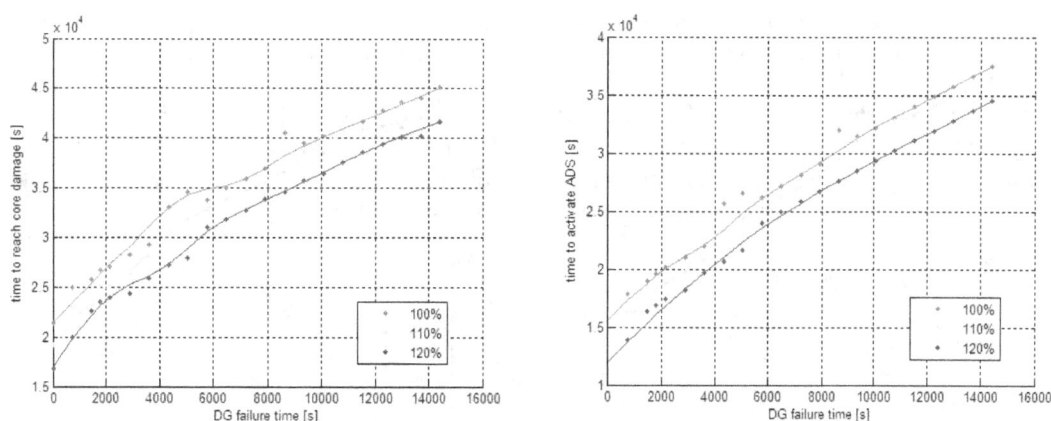

Figure 5: Time to activate ADS vs. DG failure time (left) and time to reach core damage vs. DG failure time (right) curves for 100% 110% and 120% power

Our second set of experiments focused on the determination on the "limit surfaces" [7,8], i.e. boundaries in the state space that separates failure from success. As a first step we focused on considering a 2-dimensional state space: FW availability time (measured after ADS activation, i.e. $T_{FW}-T_{ADS}$) and battery life. By randomly changing these two parameters we observe the outcome of each simulation (failure or success) and, by using a Support Vector Machines (SVM) based classifier [7] (see Appendix A), we determine the limit surface.

Results are shown in Figure 6 for two different values of power: 100%, and 120%. As expected, a longer battery life and a shorter firewater injection alignment time lead to success, while a short battery life and long firewater alignment time failed. The slope at the left end of the success space represents situations where battery power cuts out, the SRVs de-energize and close, and the RPV re-pressurizes just before the firewater can be aligned. In order to guarantee success, a minimum battery life was needed, and a higher core power allowed for the core to remain protected with a shorter battery life where a lower power core would have failed. This is due to the fact that the simulation did not account for the possibility of a manual ADS activation, and required that the heat curves for the plant be exceeded before ADS activation. In a higher power simulation, the heat curves are exceeded more quickly, ADS is activated sooner and less battery life is needed. The trade-off to this is that the firewater must be aligned more quickly in the higher power simulations than the lower power simulations, which is not a worthwhile trade-off in a real situation, as the ADS can be activated early in a real situation if the firewater injection is ready before the heat curves are exceeded.

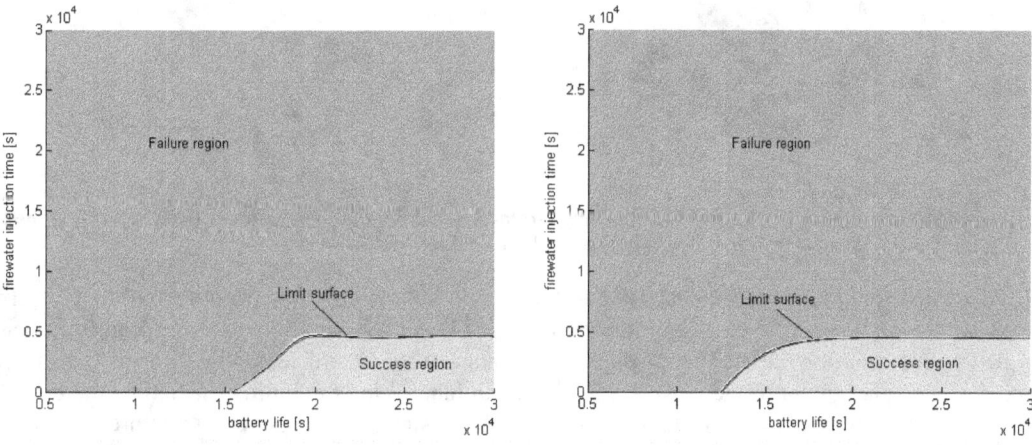

Figure 6: FW availability time vs. Battery life: limit surface for 100% (left) and 120% (right) power

Similarly, we determine the limit surface for a different 2-dimensional state space: DG failure time vs. AC power recovery time (either DG recovery or off-site power recovery). Thus, we sample these two parameters uniformly over the space and we observed the final outcome of the simulation (success or failure). Using the same SVM classifier we thus determined the limit surface.

For this case, we expect that core damage occurs for early DG failure time (i.e., early T_{SBO}) and late AC recovery time. In other words we expect failure for long time interval between AC power lost and AC power recovery events.

Limit surfaces are shown in Figure 7 for three different values of power levels: 100% (left) and 120% (right). As expected, failures occur when AC power is lost for a long time and for early failure of DG.

Note that if reactor power increases, time to reach PSP HCTL limits and time to reach core damage decreases. Thus, the time that the plant operators have to recover AC power shrinks.

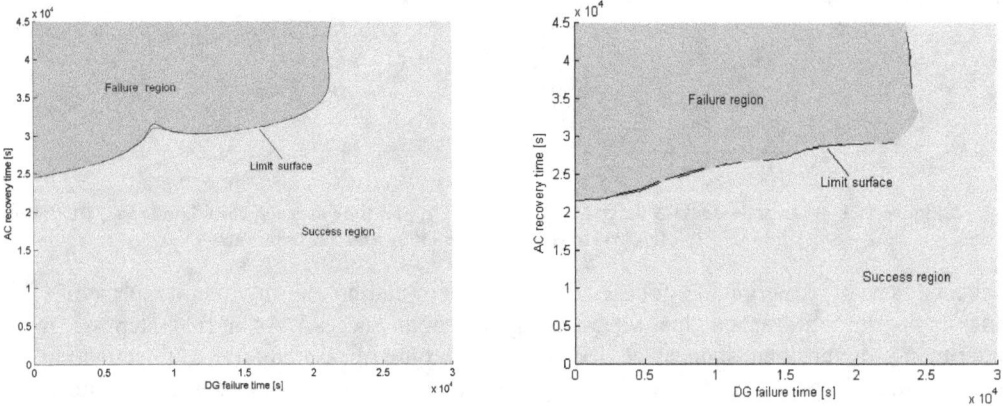

Figure 7: AC recovery time vs. DG failure time: limit surface for 100% (left) and 120% (right) power

5. ADAPTIVE SAMPLING

Nuclear simulations can be computationally expensive, time-consuming, and high dimensional with respect to the number of input parameters. Thus, exploring the space of all possible simulation outcomes is infeasible using finite computing resources. This limitation is a typical context for performing adaptive sampling where a few observations are obtained from the simulation, a surrogate model is built in order to predict behaviour of the system (e.g., maximum core temperature), and new samples are selected based on the model constructed (see Figure 8).

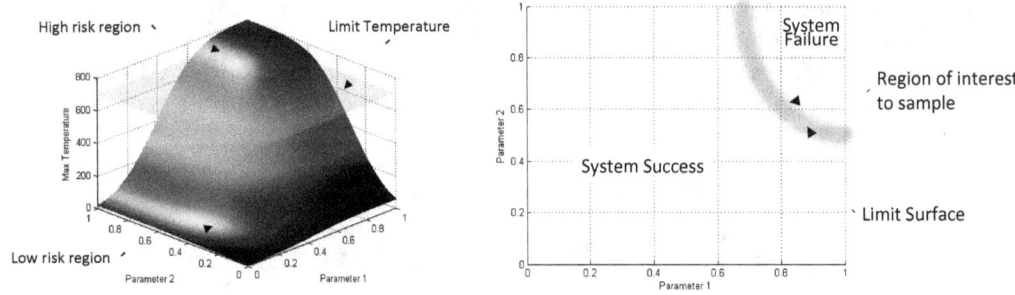

Figure 8: Max core temperature as function of 2 parameters and limit/fail temperature (left) and plot of their intersection: limit surface (right)

The surrogate model is then updated based on the simulation results of the sampled points [7,8]. In this way, we attempt to gain the most information possible with a small number of carefully selected sampled points, limiting the number of expensive trials needed to understand features of the simulation space. From a safety point of view, we are interested in identifying the limit surface, i.e., the boundaries in the simulation space between system failure and system success. The generic structure of an adaptive sampling algorithm is shown in Figure 9 (left).

For this paper, we have implemented a graph-based adaptive sampling scheme [8] (see Figure 9 right). This algorithm begins by directly building a neighbourhood structure as the surrogate model (e.g. a relaxed Gabriel graph) on the initial training data. It then creates a candidate set by first obtaining linearly interpolated points along all spanning edges of the graph, and introducing a random perturbation along all dimensions to these points.

Note that this algorithm does not employ any mathematical model (e.g., Gaussian Process Model as shown in [8]) to infer the location of the limit surface but only relies on the data point location. The graph obtained during each round changes only slightly, such that without a random perturbation, the candidate points are generally located linearly along edges of the graph, which is less desirable.

We performed a set of preliminary tests to evaluate the performance of adaptive sampling schemes. In particular, in this report, we focused on the evaluation of the limit surfaces presented in Figures 6 and 7. The results shown in Table 1 indicate a great reduction in terms of simulation runs needed in order to identify such limit surfaces.

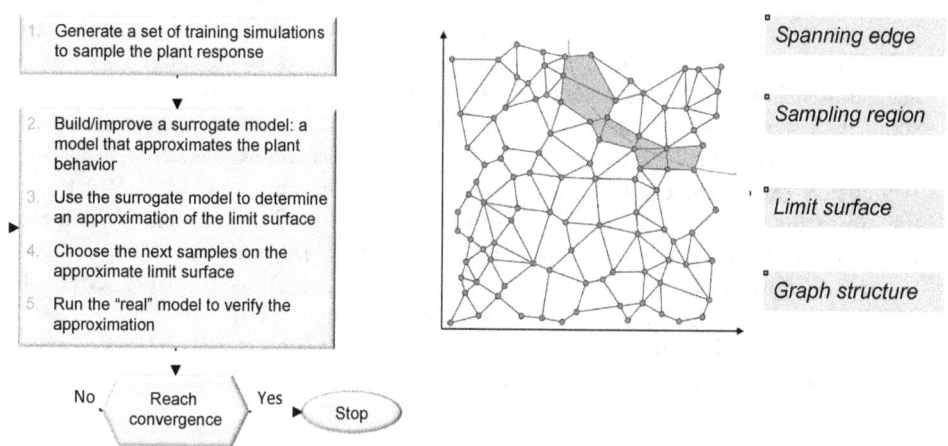

Figure 9: Generic scheme for adaptive sampling algorithms (left) and scheme of Graph base adaptive sampling algorithm (right)

Table 1: Preliminary adaptive sampling results

Test case	Monte-Carlo sampling	Adaptive sampling	Time reduction
Figure 6	700	~ 60	91.5 %
Figure 7	800	~ 60	92.5 %

6. STOCHASTIC ANALYSIS

The stochastic analysis for the BWR SBO test case has been performed using the code RAVEN [5,6] that is currently under development at INL. Originally, RAVEN was designed to control the code RELAP-7 but its capabilities have been extended to include also stochastic analysis methodologies (also known as dynamic PRA) such as Monte-Carlo and Dynamic Event Tree algorithms.

In addition, recently RAVEN has been coupled to RELAP-5 and RELAP-7. Such coupling allows performing multiple RELAP runs (through Monte-Carlo sampling).

The stochastic analysis (Monte-Carlo) when using RELAP-5 is performed through the following steps:
1. Probability distributions for the considered stochastic parameters are obtained from the PRA
2. A link between the considered stochastic parameters and the parameters coded in the RELAP-5 input file is established by RAVEN
3. A set of N RELAP-5 input files are generated and values for the considered stochastic parameters are randomly sampled from their own distributions and plugged in the input files
4. Through the use of high performance computing capabilities of INL, all RELAP-5 runs are distributed on all available nodes and cores
5. When all simulation runs are completed, RAVEN generates an output file (in .csv format) for each simulation for the original RELAP-5 output file
6. All .csv files generated can now be analyzed using state-of-the-art data analysis algorithms which include such as multi-dimensional data visualization tools

In order to evaluate the impact of the uncertain parameters listed in Section 3 on the simulation outcome we performed an extensive Monte-Carlo analysis that consisted of generating 20,000 Monte-Carlo runs.

7. UNCERTANTIES ANALYSIS AND VISUALIZATION

We apply clustering algorithms [9,10] based on the Morse-Smale complex [11,12] on the dataset obtained from RAVEN for the BWR SBO analysis (see Appendix B). In essence, we aim to reconstruct the response surface (i.e., max clad temperature) topological structure in a d-dimensional space where d is the number of uncertain parameters.

We further obtain a topological summary for each cluster and try to infer the correlations between simulation parameters and system observations. The objective is to find the combination of conditions (in the form of input simulation parameters) that can cause core damage.

Before analysing the data we performed a series of pre-processing procedures:

- Data standardization: The above data is pre-processed with a standardization process. Since different parameters may be measured on different scales and the range of values differ from each dimension, some parameters may dominate the results of the analysis. We employ a z-score data standardization process so that all dimensions are on the same scale. For values of each dimension, we subtract the mean and divide by the standard deviation.
- Dimension reduction: Upon further observations of the nature of the simulation, we further transform the data by reducing the number of dimensions. In particular, we introduce 3 new dimensions by combining 3 pairs of dimensions from the raw dataset:
  - ACPowerRecoveryTime: min {RecoveryTimeDG; OFFsitePowerRecoveryTime}.
  - SRVstuckopen: min {SRV1stuckopen; SRV2stuckopen}.
  - CoolingFailtoRunTime: max {HPCIFailToRunTime; RCICFailToRunTime}.

The 9D case includes then the following input variables:

1. FailureTimeDG
2. ACPowerRecoveryTime
3. SRVstuckOpenTime
4. cladFailureTemperature
5. CoolingFailtoRunTime
6. Reactor power
7. ADSactivation-TimeDelay
8. firewaterTime
9. TotalBatteryLife

The output variable is the maxCladTemp (MT).

Using HDViz [11,12,14] (see Appendix C), from 9D-MT-all-3C, we were able to obtain 3 clusters as shown in Figure 10. The topological structure of the clad max temperature as a 9-dimensional surface was characterized by a single local minimum and 3 local maxima as indicated in Table 3.

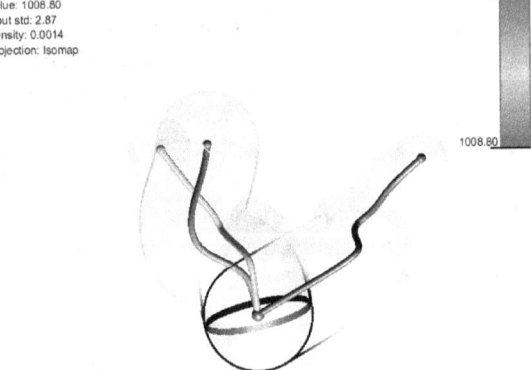

Figure 10: Topological summary

Figure 11 shows the projection of the three crystals for each dimension including their regression curves: x-axis corresponds to output variable (maximum clad temperature) while y-axis corresponds to input variable. From Figure 11, by looking at the regression curve obtained, we can see that a high value of clad temperature is reached, for all 3 crystals, for a late AC recovery time. As expected a late AC recovery time is a necessary condition to reach core damage. The same conclusions can be drawn for FW injection time, a late FW injection time guarantees core damage as well.

Failure time of DGs differentiates the three crystals, i.e., a late DG failure time is not sufficient to guarantee system success. In fact, by looking at the green crystal regression curve, a late failure time of DGs coupled with an early SRV stuck-open event and an early failure of the high-pressure injection system (both RCIC and HPCI) leads to core damage. By looking at the regression curve of the purple crystal, core damage condition was reached for an early DGs failure time and an early failure of the high-pressure injection system (both RCIC and HPCI).

Table 2: Minima and maxima of the crystals of Figure 10.

Crystal colour (see Figure 10)	Min	Max
Red	1008.80	2600.09
Green	1008.80	2597.20
Blue	1008.80	2534.16

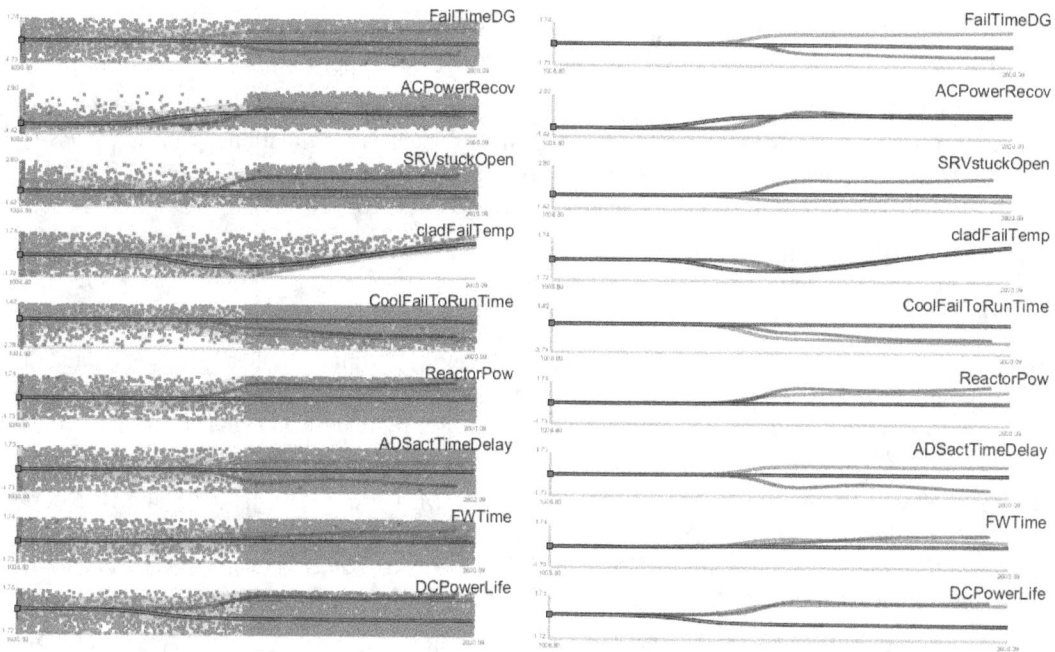

Figure 11: Inverse coordinate plots with (left) and without (right) points projection

8. CONCLUSIONS

In this paper we have summarized a series of methodologies/algorithms that are being implemented within the RISMC project. Its main scope is to provide stakeholders with risk-informed information when power uprates and life extension of an existing plant is being considered. While some of this risk-informed information has been generated in a classical fashion, others have been produced using more advanced methodologies based upon CPRA. Lastly, we introduced the value of the limit surface associated to adaptive sampling and data analysis/visualization algorithms.

We have presented their application for a detailed BWR SBO test case in order to evaluate the impact of power uprates (to 110% and 120%) on system dynamics.

REFERENCES

[1] Smith, C., Rabiti, C., and Martineau, R., "Risk Informed Safety Margins Characterization (RISMC) Pathway Technical Program Plan", INL/EXT-11-22977, Idaho National Laboratory (2012).

[2] N. Siu, "Risk assessment for dynamic systems: an overview," *Reliability Engineering and System Safety*, **43**, no. 1, pp. 43-73 (1994).

[3] RELAP5 Code Development Team. (2012). RELAP5-3D Code Manual. INL, (2012).

[4] D. Mandelli, C. Smith, C. Rabiti, A. Alfonsi, J. Cogliati, and R. Kinoshita, "New methods and tools to perform safety analysis within RISMC," in *Proceeding of American Nuclear Society (ANS)*, Washington (2013).

[5] Alfonsi, C. Rabiti, D. Mandelli, J. Cogliati, and R. Kinoshita, "Raven as a tool for dynamic probabilistic risk assessment: Software overview," in *Proceeding of M&C2013 International Topical Meeting on Mathematics and Computation*, CD-ROM, American Nuclear Society, LaGrange Park, IL (2013).

[6] A. Alfonsi, C. Rabiti, D. Mandelli, J. Cogliati, and R. Kinoshita, "Raven as a tool for dynamic probabilistic risk assessment," in *Proceeding of American Nuclear Society (ANS)*, Atlanta (GA), vol. 108, pp. 555-558, (2013).

[7] D. Mandelli and C. Smith, "Adaptive sampling using support vector machines," in *Proceeding of American Nuclear Society (ANS)*, San Diego (CA), vol. 107, pp. 736-738 (2012).

[8] D. Maljovec, B. Wang, V. Pascucci, P.-T. Bremer, and D. Mandelli, "Adaptive sampling algorithms for probabilistic risk assessment of nuclear simulations," in *ANS PSA 2013 International Topical Meeting on Probabilistic Safety Assessment and Analysis*, Columbia, SC, on CD-ROM, American Nuclear Society, LaGrange Park (IL) (2013).

[9] D. Mandelli, A. Yilmaz, T. Aldemir, K. Metzroth, and R. Denning, "Scenario clustering and dynamic probabilistic risk assessment," *Reliability Engineering & System Safety*, **115**, pp. 146-160 (2013).

[10] D. Mandelli, C. Smith, A. Yilmaz, and T. Aldemir, "Mining nuclear transient data through symbolic conversion," in *ANS PSA 2013 International Topical Meeting on Probabilistic Safety Assessment and Analysis*, Columbia, SC, on CD-ROM, American Nuclear Society, LaGrange Park, IL (2013).

[11] D. Maljovec, B. Wang, V. Pascucci, P.-T. Bremer, M. Pernice, D. Mandelli, and R. Nourgaliev, "Exploration of high-dimensional scalar function for nuclear reactor safety analysis and visualization," in *Proceeding of M&C2013 International Topical Meeting on Mathematics and Computation*, CD-ROM, American Nuclear Society, LaGrange Park, IL (2013).

[12] D. Maljovec, B. Wang, D. Mandelli, P.-T. Bremer and V. Pascucci, "Analyzing Dynamic Probabilistic Risk Assessment Data through Topology-Based Clustering," in *ANS PSA 2013 International Topical Meeting on Probabilistic Safety Assessment and Analysis*, Columbia, SC, on CD-ROM, American Nuclear Society, LaGrange Park (IL) (2013).

[13] D. Gertman, H. Blackman, J. Marble, J. Byers, and C. Smith, "The SPAR-H Human Reliability Analysis Method", NRC (2005).

[14] S. Gerber, P.-T. Bremer, V. Pascucci, and R. T. Whitaker, "Visual exploration of high dimensional scalar functions", IEEE Transactions on Visualization and Computer Graphics, vol. 16, pp. 1271–1280 (2010).

APPENDIX A: LIMIT SURFACE DETERMINATION

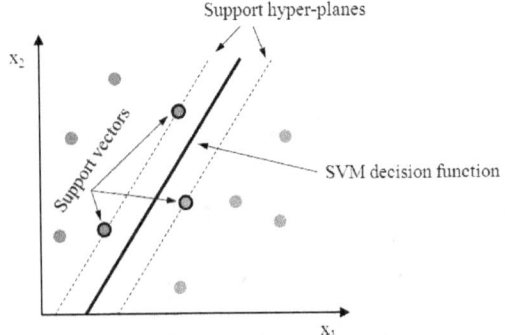

Figure 12: Limit surface evaluation using SVMs

Given a set of N multi-dimensional samples x_i and their associated results $y_i = \pm 1$ (e.g., $y_i = +1$ for system success and $y_i = -1$ for system failure), the SVM finds the boundary (i.e., the decision function) that separates the set of points having different y_i. The decision function lies between the support hyper-planes that are required to: pass through at least one sample of each class (called support vectors) and not contain samples within them

For the linear case, see Figure 12, the decision function is chosen such that distance between the support hyper-planes is maximized.

Without going into the mathematical details, the determination of the hyper-planes is performed recursively and updated every time a new sample has been generated. Figure 12 shows the SVM decision function and the hyper-planes for a set of points in a

2-dimensional space having two different outcomes: $y_i = +1$ (green) and $y_i = -1$ (red). The transition from a linear to a generic non-linear hyper-plane is performed using the kernel trick. This process involves the projection of the original samples into a higher dimensional space known as feature space generated by kernel functions $K(x_i, x_j)$.

APPENDIX B: MULTI-DIMENSIONAL DATA VISUALIZATION

The need for software tools able to both analyze and visualize large amount of data generated by Dynamic PRA methodologies has been emerging only in recent years. In the past 2 years, INL and the University of Utah have developed a software tool able to analyze multi-dimensional data: HDViz.
HDViz models the relations between output variables (e.g., maximum clad temperature) and stochastic/uncertain parameters as high-dimensional functions. In this respect, HDViz segments the domain of these high-dimension functions into regions of uniform gradient flow by decomposing the data based on its approximate Morse-Smale complex (see Figure 13). Points (i.e., simulation runs) belonging to a particular segmentation have similar geometric and topological properties, and from these it is possible to create compact statistical summaries of each segmentation. Such summaries are then presented to the user in an intuitive manner that highlight features of the dataset that are otherwise hidden.

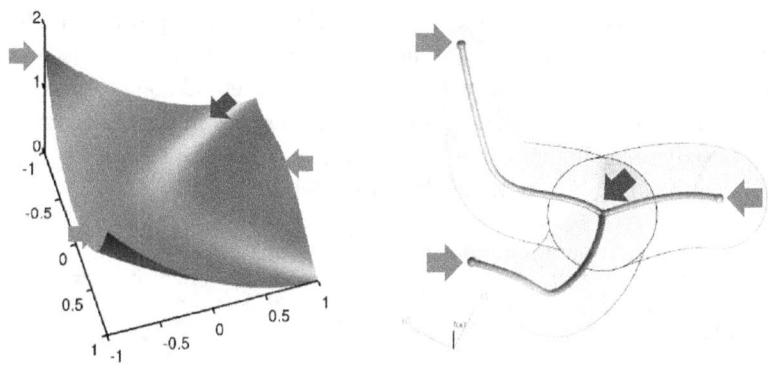

Figure 13: Representation example of a 2-dimensional function in terms of crystals that connect local minima to local maxima. In this case, a single minima (blue arrow) and 3 maxima (red arrows) have been identified. Three crystal have also been determined; each one showing the path that connect a local minima to a local maxima

Simulation Methods to Assess Long-Term Hurricane Impacts to U.S. Power Systems

Andrea Staid[a][*], Seth D. Guikema[a], Roshanak Nateghi[a,b], Steven M. Quiring[c], and Michael Z. Gao[a]

[a]Johns Hopkins University, Baltimore, MD USA
[b]Resources for the Future, Washington, DC USA
[c]Texas A&M University, College Station, TX USA

Abstract: Hurricanes have been the cause of extensive damage to infrastructure, massive financial losses, and displaced communities in many regions of the United States throughout history. The electric power distribution system is particularly vulnerable; power outages and related damages have to be repaired quickly, forcing utility companies to spend significant amounts of time and resources during and after each storm. Being able to anticipate outcomes with some degree of certainty allows for those affected to plan ahead, thus minimizing potential losses. This is true for both very short and very long time scales. In the context of hurricanes, utility companies try to correctly anticipate power outages and bring in the repair crews necessary to quickly and efficiently restore power to their customers. A similar type of planning can be applied to a long time horizon when making decisions on investments to improve grid reliability, resilience, and robustness. We present a methodology for assessing long-term risks to the power system while also incorporating possible changes in storm behavior as a result of climate change. We describe our simulation methodology and demonstrate the assessments for regions lying along the Gulf and Atlantic Coasts of the United States.

Keywords: Hurricane impact, Simulation, Power outages, Climate change.

1. INTRODUCTION

Tropical cyclones, and hurricanes in particular, have historically been some of the most destructive natural hazards for coastal areas [1, 2, 3]. Power outages caused by tropical cyclones are usually one of the biggest concerns as a storm approaches a region; widespread or long-lasting power outages can result in huge financial losses and health impacts. Utility companies are responsible for maintaining the electricity distribution system, and they are greatly concerned with the reliability of power for their customers and service area. Power outages have to be dealt with quickly, and this often comes at great expense to the utilities in the aftermath of a storm. Anticipating power outages in advance can help minimize these costs by ensuring that crews and equipment are available and sufficient. Any information that can assist in planning for oncoming storms is highly valued by utilities.

There is recent research into developing power-outage prediction models, and they have proven to be remarkably useful for local utility companies [4, 5, 6, 7, 8, 9]. This research addresses the problem of planning for tropical cyclones at a very short time scale: on the order of days as a storm is approaching. Planning for storms on a longer time scale is also critically important, but this problem presents very different challenges. A stronger, more robust power grid would better withstand strong winds from hurricanes and would suffer fewer power outages during a storm. Long-term planning raises questions regarding new investments in stronger utility poles or moving power lines underground, for example. These investments are potentially very costly and decision-makers want to ensure that they have the best available information in order to make appropriate decisions. The measured historical hurricane

---

[*] Corresponding author, staid@jhu.edu

record is relatively short by long-term planning standards. New investments in infrastructure should be designed to withstand strong storms. The question of "how strong?" is coupled with uncertainty about what future storms may look like. Analysis based on the historical tropical cyclone record is useful if the future climate remains stationary, but this technique can fail when the historical record is not long enough, since it doesn't necessarily represent the full range of possible outcomes. Such is the case when planning for hurricanes.

To mitigate this issue, we present a simulation methodology that can be used to assess the expected impact of hurricanes on power systems in the United States. We simulate a large number of replications of hurricane seasons in order to assess the overall expectations of impacts in various regions of the coastal states. The simulation methodology can also be used to assess the potential for a future affected by climate change. The average behavior of tropical cyclones is expected to change as the climate warms, and these changes can be modeled to generate a range of possible impacts from hurricanes in a changing climate. We apply this methodology to the U.S. power system along the Gulf and Atlantic Coasts and present preliminary results of the capabilities of the model to assess local and regional changes in hurricane impacts.

By simulating a large number of virtual storms, we can assess the resulting impacts from tropical cyclones under a variety of climate assumptions. The baseline case assumes a static climate and uses static population data. This represents the current state of things, assuming that the observed storm behavior represents the natural variability in the climate and tropical cyclone behavior. We also simulate scenarios with stronger or weaker storms, scenarios with a higher or lower average annual frequency of storms, and scenarios in which the average distribution of where storms make landfall is shifted.

2. BACKGROUND

Planning for infrastructure investments involves an assessment of the conditions in which the system will operate. Many infrastructure systems have operating lives on the order of multiple decades, and accurately assessing conditions that far in the future is challenging. There is a high degree of uncertainty, and many forecast techniques make assumptions based on projections of the current conditions. Planning for tropical cyclones is especially challenging for two reasons: (1) our knowledge of the current conditions is based on a very limited history and (2) the uncertainty is compounded by the possible effects of climate change, which could change storm behavior significantly. Accurate recordings of hurricane activity are fairly recent. Prior to the 1960's, there was not a lot of reliable scientific data on tropical storm behavior. Even with the older observations, the record is relatively short for tasks such as estimating the 100-year storm.

In addition to the limited data, planning for the future brings the challenge of unknown conditions. As the climate changes and warms, it is speculated that tropical cyclone behavior will also change. The relationship between tropical cyclones and climate has been studied extensively as researchers attempt to make projections about what future tropical cyclone seasons may look like around the world. However, there is still a great deal of uncertainty behind the results [2, 10, 11, 12, 13, 14, 15, 16]. Different models produce different projections, and different climate projections offer a variety of possible outcomes. For example, the results of the physics-based models developed by Knutson et al. suggest that the frequency of Atlantic hurricanes and tropical storms will likely be reduced in the future [13]. Results obtained by downscaling IPCC AR4 simulations also suggest a reduction in the global frequency of hurricanes in a warmer future climate scenario, with a potentially large increase in intensity in some locations [10]. Some statistical models developed suggest that the intensity and frequency of tropical cyclones will likely increase with a warmer future climate[17, 18].

The inherent uncertainty in future climate projections is coupled with uncertainty in the relationship between climate and tropical cyclones, especially since the relationship may vary in different regions of the world. Many traditional risk and decision analysis methods break down under such deep uncertainty [19]. Instead, for problems with such wide-reaching uncertainty, more robust planning tools are needed to deal with uncertainties that are changing with time. Without a strong understanding of the nature of the uncertainty, planning for the most likely scenario should be replaced with planning for short-term actions that perform well under a range of possible scenarios and that can be modified over time as conditions change and new information comes to light [20]. Long-term planning for major infrastructure projects, such as updates to the electric grid, should use scenarios to assess the robustness of possible actions and not as parameters to determine an optimal solution.

3. METHODOLOGY

Our simulation uses historical hurricane data dating back to 1900 gathered from the National Hurricane Center as the main input [21]. In addition, we use a wind-field simulation model and a power outage prediction model to assess the long-term impact of hurricane damage to the power grid. To explain the methods used, we first describe how virtual storms are created in the baseline case, in which future storms are assumed to resemble past storms. Next, we describe how scenarios influenced by potential climate change are incorporated into the simulations.

3.1. Baseline Simulation

For the baseline case, the simulation uses historical hurricane and tropical storm data as initial inputs. For each independent replication, we first randomly sample from a Poisson distribution to determine the number of storms that make landfall in the U.S. in that replicated year. The mean of the Poisson distribution represents the average number of tropical cyclones that make landfall each year, and this value is set equal to the historical mean as calculated from the measured storm record. Within each replication, each storm is treated independently. For each storm, we randomly sample a landfall location from a smoothed probability distribution that assigns a probability to each 50 km stretch of coastline from Texas through Maine. The historical record was smoothed so that each segment of coast had a non-zero probability of a storm making landfall. The smoothing was done to maintain the general shape of the distribution. From there we generate the path that the storm travels along, or the track. Again, we use historical hurricane tracks as a starting point, but simply sampling from past storm movement would severely limit the options of storm movement and would bias the simulation results. Instead, we use a simple statistical model to generate realistic track movement. Based on which section of the coastline the storm hits, we subset the historical tracks, keeping only those that made landfall in the same region. These remaining tracks are then used to train a random forest model, which predicts the x- and y-direction movement based on the previous direction of travel and previous location for each six-hour time step.

Concurrently with sampling a landfall location for each storm, we randomly sample a maximum wind speed. This represents the wind speed when the storm makes landfall, and the values are drawn from the maximum landfall wind speeds from the historical record. For each time step associated with the track movement, the wind speed decays according to the hurricane decay models of Kaplan and Demaria until the wind speeds fall below the tropical cyclone classification level [22].

Once the storm track and intensity for each point along the track are determined, these parameters are fed into a wind field model that generates the wind parameters for the entire storm as it moves along its track. For all areas within range of the storm, we estimate the maximum 3-second gust wind speed and the duration of wind speeds above 20 m/s for the centroid of each census tract. This wind field model is based on that developed by Huang *et al.* and used in Han *et al.* [1, 8, 9]. This wind data is then passed to a statistical outage prediction model, which uses a random forest model to predict the number of customers without power as a result of wind-induced power outages. This model has been trained and validated on past hurricanes in different areas of the U.S. It estimates outages from strong winds, but we do not account for outages caused by storm surge or flooding, so the estimates would be conservative for areas that are particularly prone to flooding, for example.

The outage predictions are compiled for each storm in each replication. We run the simulation for a large number of replications until we reach convergence. We run 1600 replications in order to achieve a 99% confidence on at least the 96th percentile according to the method proposed by Morgan and Henrion [23]. The aggregated results from 1600 simulated years worth of tropical cyclones allow us to calculate expected return periods for the output values. We calculate the 100-year, 50-year, 25-year, and 4-year, and 2-year return periods for maximum wind speed, duration of winds above 20 m/s, and the fraction (and number) of customers without power for each census tract. We also calculate the probability of each census tract having at least 5% of customers losing power in a given year.

3.2. Alternate Storm Scenarios

The simulation can also be used to assess the impacts and changes in impacts if the tropical cyclone behavior changes. Planning for a future with climate change, for example, will necessitate assessments of what the impacts could be if the warmer climate affects tropical cyclones. As mentioned in section refsec:background, climate change could affect the behavior of tropical cyclones in the North Atlantic Basin. Our simulation methodology can be modified to study these potential effects in terms of storm intensity, frequency, and location. We demonstrate this by selecting plausible scenarios in which we vary intensity, frequency, and location and running the simulation for each scenario. An overview of the methodology is shown in Figure 1.

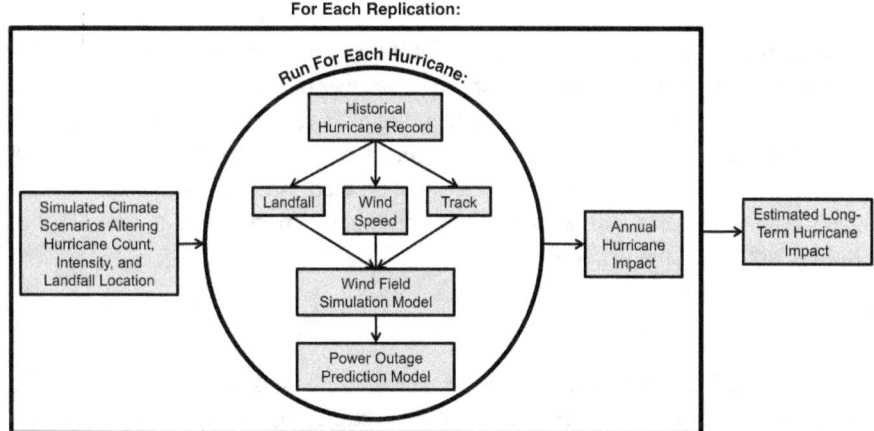

Figure 1: Schematic of the simulation methodology.

Running the simulation under these various scenarios requires very simple modifications. Each factor is varied individually within the simulation, and the overall structure is the same as that described for the baseline case above. We vary intensity by taking the randomly sampled maximum wind speed for each storm and multiplying it by a factor. We simulate scenarios for intensity factors of 0.8, 1.2, and 1.4, meaning a decrease in strength of 20%, an increase of 20%, and an increase of 40%. The results of these simulation runs show possible range of impacts for the intensity changes modeled. For scenarios of varying frequency, we adjust the mean of the Poisson distribution that is used to sample the number of storms in each replicated year. The baseline case has a mean of 2, and we simulate scenarios for means of 0.5, 1, 3, and 4. If there is a reduction or increase in the number of storms, the simulation can produce estimates for the expected impacts in each case.

The location scenarios are more subjective; starting from the smoothed baseline distribution, we adjust the probabilities for each 50-km segment of coastline. Each new distribution still retains the general shape of the baseline distribution, since it is based on actual geographical characteristics, but the individual probabilities are shifted. For example, some land areas are simply more prone to landfalling hurricanes because they jut out into the path of oncoming storms. We created four modified distributions to assess the changing impacts as storm genesis location may change. The first scenario shifts storms further up along the mid-Atlantic coast, the second shifts them further down into the Gulf of Mexico, the third spreads the distribution out more evenly to reduce the natural peak around Florida, and the fourth concentrates the peak around Florida, thereby reducing the probabilities in the Gulf and in the Northeast. The distributions are plotted along the coastline in figure 2.

(a) Baseline   (b)   (c)   (d)   (e)

Figure 2: Demonstrating the probability distributions used for shifting the landfall location of tropical cyclones. The distribution is shifted towards the mid-Atlantic in 2(b), towards the Gulf of Mexico in 2(c), spread more evenly along the coastline in 2(d), and concentrated on the Florida peninsula in 2(e).

For each scenario, we again ran the simulation for 1600 replications in order to reach convergence. From these results, we calculate the same wind speed and power outage parameters discussed previously. The scenario runs offer insight into the expected range of changes that could be brought on by climate change and its influences on tropical cyclones. For each scenario, the results from 1600 replicated years of hurricane seasons allows us to calculate the expected impacts in terms of wind speeds and power outages for all of the affected regions. These parameters portray the potential climate change impacts on both a large and small spatial scale.

4. RESULTS

We applied this simulation methodology to hurricane-prone states in the U.S. along the Gulf and Atlantic coasts. We use all coastal states and some states stretching inland if they lay within potential hurricane impact areas. The baseline case models the impact for the current state of both meteorological and geographical conditions. Although we replicate 1600 virtual years of hurricane activity, these years do not represent any sort of future time period. The replications all use current conditions, and so the aggregation of this large number of replications represents the average impact for the current state of hurricane activity in the United States, subject to the assumptions and simplifications incorporated into the model. We use 2010 population numbers for all census tracts and assume that the electric grid is the same as its current state. For these reasons, our results offer useful insight into plausible impact scenarios, but they do not necessarily represent future impacts because we do not take into account possible infrastructure upgrades or population changes. This is a major assumption, but it allows us to simply demonstrate the simulation methodology and the capabilities of modeling changes in storm impacts. For example, we can look at the changes in wind speeds if the intensity of tropical storms changes in the North Atlantic Basin. Figure 3 plots the expected 50-year wind speed for the baseline case as well as scenarios with the average storm intensity is adjusted by a factor of 0.8, 1.2, or 1.4.

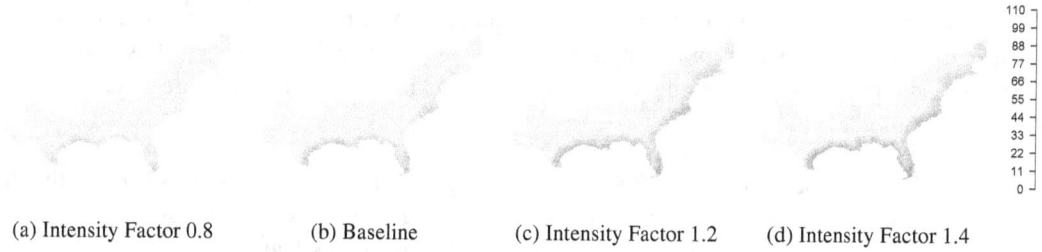

(a) Intensity Factor 0.8   (b) Baseline   (c) Intensity Factor 1.2   (d) Intensity Factor 1.4

Figure 3: 50-year wind speed [m/s] for scenarios of varying tropical cyclone intensity.

Similarly, we can look at the expected impact for a different measure under different scenarios. Figure 4 plots the expected probability of each census tract having at least 5% of its customers losing power in a given year. This value is plotted for scenarios of varying storm frequency. We vary this by adjusting λ, the mean of the Poisson distribution that is used to sample the number of storms in each replication. The baseline value from the historical record is $\lambda = 2$, and we model scenarios for $\lambda = 0.5, 1, 3$, and 4 to assess the impacts from changes in tropical cyclone frequency.

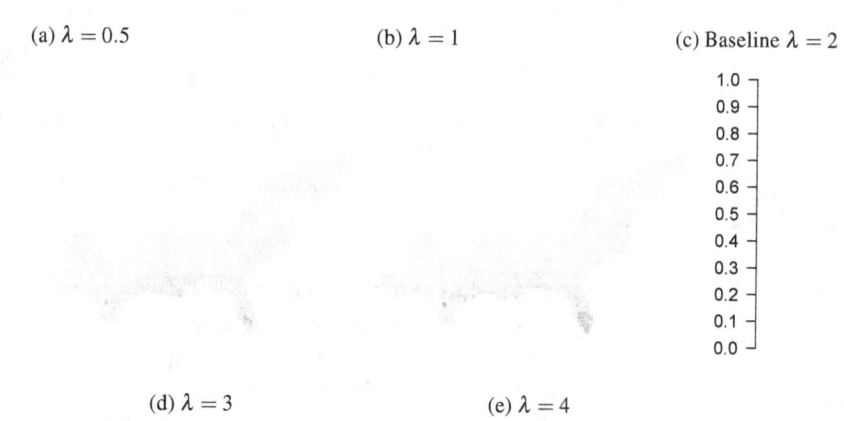

(a) $\lambda = 0.5$ (b) $\lambda = 1$ (c) Baseline $\lambda = 2$

(d) $\lambda = 3$ (e) $\lambda = 4$

Figure 4: Annual probability of at least 5% of customers without power for scenarios of varying tropical cyclone frequency.

4.1. Local Impacts

One of the biggest advantages of this simulation methodology is the ability to look at local-area effects. We use census tracts as our area of analysis, and this allows us to study changes on a small scale. This is especially useful for agencies that are responsible for decisions regarding their own, often small, local area of influence. Utility companies, for example, are responsible for the poles and lines that make up the distribution system in the region that they serve. Any decisions that they make about strengthening the grid should be based on projections of conditions in their service area and not for the country as a whole.

To demonstrate this, we take a closer look at several metropolitan areas. Based on the scenarios that we selected, we can assess the impacts in terms of wind speeds and customers without power for scenarios of changing intensity, frequency, or location. For example, in figures 5 and 6 we plot the 100-year fraction of customers without power for the New York, NY and New Orleans, LA metropolitan areas as the average storm intensity varies.

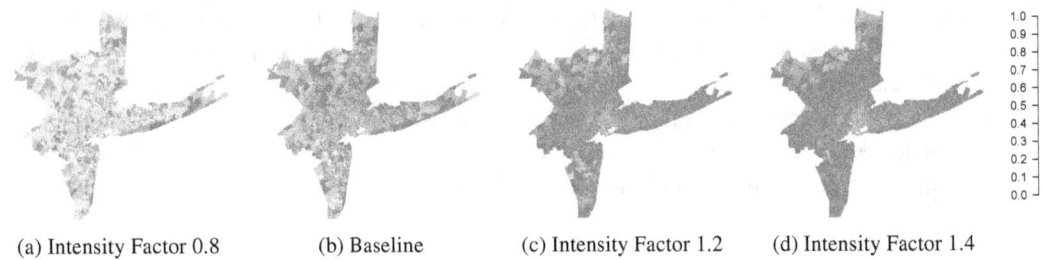

(a) Intensity Factor 0.8   (b) Baseline   (c) Intensity Factor 1.2   (d) Intensity Factor 1.4

Figure 5: 100-Year fraction of customers without power for the New York, NY metropolitan area for scenarios of varying storm intensity.

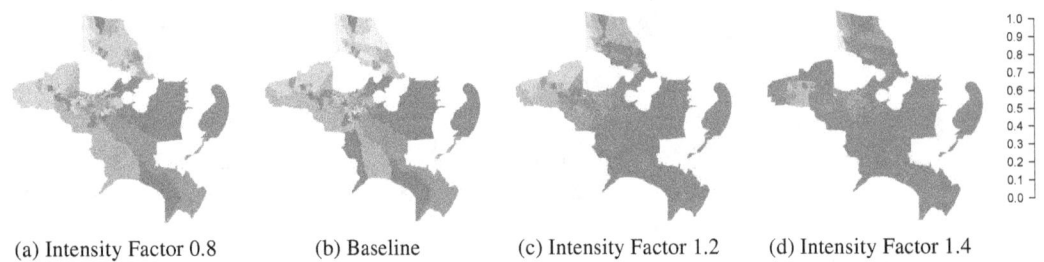

(a) Intensity Factor 0.8   (b) Baseline   (c) Intensity Factor 1.2   (d) Intensity Factor 1.4

Figure 6: 100-Year fraction of customers without power for the New Orleans, LA metropolitan area for scenarios of varying storm intensity.

Some regions of the country are more affected by certain scenarios that were evaluated here. Overall, a large increase in storm intensity will cause the greatest changes in the number of people losing power. However, the strongest effects will be felt primarily in coastal areas where the wind speeds will be highest. Some inland areas may be more concerned with an increase in storm frequency, which will result in a greater chance of suffering power outages (although the overall number of customers out may be lower.) A shift in landfall location could change the expected impacts significantly in some regions, i.e. any region whose local probability of landfall shifts from low to high. While the impacts can be analyzed locally, it is also important to understand that different areas of the country may be more or less severely impacted depending on the climate-influenced scenario being assessed.

5. CONCLUSION

The simulation methodology presented here can serve as a tool for understanding the consequences of possible scenarios. This type of analysis is useful for quantifying the effects of the scenarios that may be used as part of a long-term planning process. To start, one can gain a deeper understanding of the potential impact of the current state of tropical cyclones and their impact on the United States. Looking at the big picture, the recorded hurricane record is very limited. Planning for future hurricanes based on the historical record alone would leave big gaps in terms of possible impacts and impacted regions. Decision-makers, such as utility companies, may be planning for upgrades to their local distribution grid to better withstand hurricanes. They need to incorporate information on the expected impact over the lifetime of the system. The wind-speed data produced as outputs from the simulation model will be critical in such planning decisions. Beyond this, our simulation allows for input conditions to be modified in order to model any possible changes to parameters including the meteorology as a result of climate change or population growth. While we demonstrated some plausible scenarios in which tropical cyclones are affected by climate change, this methodology is more general. It can be used for analyzing a large range of changing input conditions in terms of storm behavior. It can also be coupled with projections of population growth in the United States to study potential impacts as populations grow or change within each census tract. This tool can provide valuable assistance to decision-makers faced with long-term decisions about power system investments and upgrades.

Acknowledgements

This work is funded in part by NSF CMMI Grant 1149460, NSF CBET SEES Grant 1215872, and NSF CMMI Grant 0968711. Thanks also to the National Oceanic and Atmospheric Administration from which much of the original data was gathered.

References

[1] Z. Huang, D. Rosowsky, and P. Sparks, "Hurricane simulation techniques for the evaluation of wind-speeds and expected insurance losses," *Journal of Wind Engineering and Industrial Aerodynamics*, vol. 89, no. 7, pp. 605–617, 2001.

[2] R. A. P. Jr, C. Landsea, M. Mayfield, J. Laver, and R. Pasch, "Hurricanes and global warming," *Bulletin of the American Meteorological Society*, vol. 86, no. 11, pp. 1571-1575, 2005.

[3] P. J. Vickery, F. J. Masters, M. D. Powell, and D. Wadhera, "Hurricane hazard modeling: The past, present, and future," *Journal of Wind Engineering and Industrial Aerodynamics*, vol. 97, no. 7, pp. 392–405, 2009.

[4] H. Liu, R. A. Davidson, D. V. Rosowsky, and J. R. Stedinger, "Negative binomial regression of electric power outages in hurricanes," *Journal of Infrastructure Systems*, vol. 11, no. 4, pp. 258–267, 2005.

[5] H. Liu, R. A. Davidson, and T. Apanasovich, "Statistical forecasting of electric power restoration times in hurricanes and ice storms," *Power Systems, IEEE Transactions on*, vol. 22, no. 4, pp. 2270–2279, 2007.

[6] R. Nateghi, S. D. Guikema, and S. M. Quiring, "Comparison and validation of statistical methods for predicting power outage durations in the event of hurricanes," *Risk Analysis*, vol. 31, no. 12, pp. 1897–1906, 2011.

[7] J. Winkler, L. Duenas-Osorio, R. Stein, and D. Subramanian, "Performance assessment of topologically diverse power systems subjected to hurricane events," *Reliability Engineering & System Safety*, vol. 95, no. 4, pp. 323–336, 2010.

[8] S.-R. Han, S. D. Guikema, and S. M. Quiring, "Improving the predictive accuracy of hurricane power outage forecasts using generalized additive models," *Risk Analysis*, vol. 29, no. 10, pp. 1443–1453, 2009.

[9] S.-R. Han, S. D. Guikema, S. M. Quiring, K.-H. Lee, D. Rosowsky, and R. A. Davidson, "Estimating the spatial distribution of power outages during hurricanes in the gulf coast region," *Reliability Engineering & System Safety*, vol. 94, no. 2, pp. 199–210, 2009.

[10] K. Emanuel, R. Sundararajan, and J. Williams, "Hurricanes and global warming: Results from downscaling ipcc ar4 simulations," *Bulletin of the American Meteorological Society*, vol. 89, no. 3, pp. 347–367, 2008.

[11] A. Henderson-Sellers, H. Zhang, G. Berz, K. Emanuel, W. Gray, C. Landsea, G. Holland, J. Lighthill, S.-L. Shieh, and P. Webster, "Tropical cyclones and global climate change: A post-ipcc assessment," *Bulletin of the American Meteorological Society*, vol. 79, no. 1, pp. 19–38, 1998.

[12] T. R. Knutson, J. L. McBride, J. Chan, K. Emanuel, G. Holland, C. Landsea, I. Held, J. P. Kossin, A. Srivastava, and M. Sugi, "Tropical cyclones and climate change," *Nature Geoscience*, vol. 3, no. 3, pp. 157–163, 2010.

[13] T. R. Knutson, J. J. Sirutis, S. T. Garner, G. A. Vecchi, and I. M. Held, "Simulated reduction in atlantic hurricane frequency under twenty-first-century warming conditions," *Nature Geoscience*, vol. 1, no. 6, pp. 359–364, 2008.

[14] M. E. Mann and K. A. Emanuel, "Atlantic hurricane trends linked to climate change," *EOS, Transactions American Geophysical Union*, vol. 87, no. 24, p. 233, 2006.

[15] R. Mendelsohn, K. Emanuel, S. Chonabayashi, and L. Bakkensen, "The impact of climate change on global tropical cyclone damage," *Nature Climate Change*, vol. 2, no. 3, pp. 205–209, 2012.

[16] T. Yonetani and H. B. Gordon, "Simulated changes in the frequency of extremes and regional features of seasonal/annual temperature and precipitation when atmospheric co2 is doubled," *Journal of Climate*, vol. 14, no. 8, pp. 1765–1779, 2001.

[17] K. Emanuel, "Increasing destructiveness of tropical cyclones over the past 30 years," *Nature*, vol. 436, no. 7051, pp. 686–688, 2005.

[18] M. A. Saunders and A. S. Lea, "Large contribution of sea surface warming to recent increase in atlantic hurricane activity," *Nature*, vol. 451, no. 7178, pp. 557–560, 2008.

[19] N. Ranger, T. Reeder, and J. Lowe, "Addressing 'deep' uncertainty over long-term climate in major infrastructure projects: four innovations of the thames estuary 2100 project," *EURO Journal on Decision Processes*, vol. 1, no. 3-4, pp. 233–262, 2013.

[20] W. E. Walker, M. Haasnoot, and J. H. Kwakkel, "Adapt or perish: a review of planning approaches for adaptation under deep uncertainty," *Sustainability*, vol. 5, no. 3, pp. 955–979, 2013.

[21] N. H. Center, "Nhc data archive," 2014.

[22] J. Kaplan and M. DeMaria, "A simple empirical model for predicting the decay of tropical cyclone winds after landfall," *Journal of Applied Meteorology*, vol. 34, no. 11, pp. 2499–2512, 1995.

[23] M. G. Morgan and M. Henrion, *Uncertainty: a Guide to dealing with uncertainty in quantitative risk and policy analysis*. Cambridge University Press, 1990.

Towards Reliability Evaluation of AFDX Avionic Communication Systems With Rare-Event Simulation

Armin Zimmermann[a]*, Sven Jäger[a], and Fabien Geyer[b]

[a]Software and Systems Engineering, Ilmenau University of Technology; Ilmenau, Germany
[b]Airbus Group Innovations, Dept. TX4CP; Munich, Germany

Reliability is a major concern for avionic systems. The risks in their design can be minimized by using model-based systems engineering methods including simulation and mathematical analysis. However, there are non-functional properties that are computationally expensive to evaluate, for instance when rare events are important. Rare-event simulation methods such as RESTART can be used, leading to speedups of several orders of magnitude. We consider AFDX (avionic full-duplex switched Ethernet) networks as an application example here, where the end-to-end delay and buffer utilizations are important for a safe and efficient system design. The paper proposes generic model patterns for AFDX networks, and shows how very low probabilities can be computed in acceptable time with the presented method and software tool.

Keywords: Rare-Event Simulation, AFDX, Avionic Networks, Stochastic Petri Nets

1. INTRODUCTION

Reliability and safety are important non-functional requirements of many man-made systems, especially when failures may lead to catastrophic events. Common examples include automotive systems, train control, and avionics. The resulting effect of local design decisions on overall system properties are not obvious, because there are numerous specialists working on design details. Mathematical models can help to describe such systems and to compute their system properties with the help of appropriate software tools ("model-based design" and "model-based systems engineering" [1]).

Unavoidable faults may be masked or tolerated by static or dynamic redundancy measures. The main task is to design a system such that its reliability and safety requirements are achieved with the least amount of resources. Classic models and tools for static analysis such as Fault Trees and Reliability Block Diagrams [2] are wide-spread in these domains, but are not able to cover systems in which the complex behavior influences failures, or if dynamic reconfigurations are applied. In avionic system design, fly-by-wire systems, flight control and management, maintenance processes, as well as future communication architectures are examples in which dynamic reliability models are necessary.

Avionic networks are an increasingly important element of distributed sensing, processing, and control architectures on board modern aircrafts [3]. A modern avionic communication system is AFDX ([4], more details in Section 2). Its general reliability aspects based on hardware failures is important for system design and certification, and can be analysed with models [5]. It uses a dual-layer hardware redundancy setup (static redundancy) to survive single failures of hardware elements.

In this paper we are interested in the more complex question of network guarantees for the served real-time applications, which require a model-based analysis of end-to-end delays [6, 7, 8]. There are several methods for worst-case end-to-end delay analysis (simulation, network calculus, and model checking [6]), all with their individual advantages and drawbacks. Besides the concentration on guaranteed bounds on maximum end-to-end delays, an upcoming question for network and buffer

---

* Corresponding author, armin.zimmermann@tu-ilmenau.de

sizing are probabilistic end-to-end measures such as quantiles of the distribution [9]. For instance, the pessimism of bounds may lead to system designs in which the guaranteed maximum delay may be 20ms, while the actually observed delay rarely exceeds 1ms.

The end-to-end delay evaluation can be reduced to an analysis of buffer levels and their probabilities (c.f. Section 2), which this paper concentrates on. Buffer overflow with packets is guaranteed to not happen in switches designed based on guarantees, but it is interesting to check buffer utilization and what the actual probabilities of utilized buffers are. Switch design can benefit from knowing how many buffer elements are needed. Moreover, if very rare packet losses or delays exceeding the guarantee are acceptable by the served applications, how much buffer space can be saved? There is obviously a trade-off between resource utilization (and cost) vs. end-to-end delays (c.f. Sections 2 and 4).

The necessary dynamic models need to consider discrete events, states, probabilistic choice and stochastic delays. Depending on the complexity of the system behavior and the corresponding size of the state space, simulation programs, Markov chains, and stochastic Petri nets (SPNs) are applied to reliability problems in the literature [2], among others. The latter two are attractive as long as the underlying assumption of a Markov behavior is realistic, because then a direct numerical solution is possible [10]. Petri nets have been suggested for reliability engineering of complex systems in an international standard recently [11].

However, non-Markovian delay distributions are necessary, for instance, in the case of periodic events typical of AFDX networks and embedded systems in general. The numerical analysis of models incorporating them is very restricted, only allowing the application to special cases [10]. An alternative evaluation technique is simulation, but the problem here is that the computational effort to generate enough failure states to achieve statistical confidence in the estimated results is usually intractable — it simply takes too long until significant events are generated.

This problem is well-known as rare-event simulation, and there are two main approaches used: *importance sampling* and *splitting*. They have the common goal to increase the frequency of the rare event in order to gain more significant samples out of the same number of generated events. Among methods that can be automated and implemented in a software tool for industrial applications, the splitting technique has the advantage of requiring less insight into the model details. A variant is the RESTART algorithm [12], which has been shown to work robustly and efficiently for many applications. Considerable speedups of several orders of magnitude can be achieved even for non-trivial system models. Rare-event simulation of general communication networks with importance sampling is, for instance, presented in [13].

A brief description of AFDX networks and related work on its model-based design and analysis is given in the subsequent section. After a short coverage of stochastic Petri nets in Section 3, generic patterns for AFDX network modeling with SPNs are proposed in Section 3.1. The topology of an AFDX application example network is presented in Section 3.2 together with its SPN model, which has been constructed modularly with the patterns. Section 4 explains how the used rare-event simulation technique RESTART works, points out the used software tool TimeNET [14], and presents numerical results of simulation experiments carried out for the example with it.

The contribution of the paper are realistic SPN model patterns for AFDX, and to show that existing methods in rare-event simulation can help to compute reliability measures that are otherwise computationally intractable. To the best of the authors' knowledge, this technique has not been applied in avionics reliability evaluation before. Results are presented for an example with non-trivial size.

AFDX network modeling and performance evaluation using stochastic Petri nets has been tried before [15, 16]. However, no network structure has been taken into consideration; the transmission delay is assumed to be a sequence of exponential transitions only, independent of the actual number of links. The load model is assumed as a mix of periodic and sporadic message generations that alternate. The mean end-to-end delay is analyzed based on this overly simplified model in [15]; however, more detailed information about its distribution such as quantiles or maximum values are of much higher interest.

2. MODEL-BASED DESIGN OF AFDX NETWORKS

AFDX (Avionics Full-DupleX Ethernet) is a data network based on Ethernet, developed by Airbus and created during the development of the A380. It was standardized in Part 7 of the ARINC 664 specifications [4], and has since then been used in other Airbus projects.

This network technology is based on switched Ethernet twisted pair 100 Mbps full duplex technology. It attempts at addressing the issue of non-deterministic network, best-effort and lack of bandwidth guarantees of traditional Ethernet. It aims at providing a redundant deterministic network, adapted to safety-critical applications used in aircrafts. The main differences compared to Ethernet, are the redundancy property where frames are duplicated and sent on two separate networks, a frame identifier at layer 2 to avoid packet duplication, and a verification of flow properties (packet size and frequency) by the switches. The nondeterministic effect of message collisions on standard Ethernet is avoided by connecting only two nodes with each physical link and using a dedicated link for each direction (full duplex). Thus there will be no collisions on the physical level, and the only nondeterminism can arise from the possible waiting times in output queues of switches because of temporary link contention.

An AFDX network is composed of *end-systems* and *switches* as nodes. End-systems serve as source and destination nodes in the network, over which applications may send data according to bandwidth restrictions to avoid overloading. One fundamental building block of AFDX is the notion of *virtual link* (VL), which can be seen as rate-constrained network tunnels. The parameters describing a VL are: the emitter end-system of this VL, the list of receiving end-systems, static routes between emitter and receivers, the Bandwidth Allocation Gap (BAG), as well as minimum and maximum frame length (s_{min} and s_{max}). The BAG is defined as the minimum time interval between the first bit of two consecutive frames from the same VL and has a value of $2^k ms$ with $k \in \{1..7\}$.

The packet structure follows Ethernet and contains 67 Bytes overhead (including the inter-frame gap) in addition to the possible $17\ldots1471$ Bytes payload (between which s_{min} and s_{max} can be chosen). Assuming a transmission bandwidth of 100 Mbps, each packet will thus require a per-link transmission time between $6,72\mu s$ and $123.04\mu s$.

The elements of the AFDX network are deterministic, the only source of randomness is in the times that end-systems have to send packets (or the offsets between them). Even if every application would be sending periodic messages only with the maximum frequency given by its BAG value, there is no globally synchronized clock and thus any offset between end-systems may occur already because of clock drift. Sporadic message generations can happen at arbitrary times, as they can be sent immediately after generation if the last message has been sent more than the BAG value before.

Important properties of an avionic network are safety against packet loss (by avoiding buffer overruns and redundant hardware) as well as a maximum end-to-end delay (specified dependent on the network architecture [4]). The guaranteed worst-case behavior of AFDX comes from the encapsulation of every network flow in a VL, and the fact that the VL properties are enforced by the switches in the network.

If an end-system does not send packets according to the VL specifications (BAG and frame size), the packets are dropped, which avoids overloading the network and guarantees the end-to-end latencies of the other flows.

Elements of the end-to-end delay that a packet experiences are discussed in [17]. There are unavoidable deterministic parts: 1) the transmission delay over the statically predefined set of links for a VL and 2) processing delays in switches between their input and output ports (hardware- and implementation-dependent, but guaranteed not to exceed $16\mu s$). The sum of these delays constitutes a minimum transmission delay in the case of no queuing.

However, temporarily the network may be populated because of resource sharing: if a packet is put into a switches' output buffer and finds the subsequent transmission link busy, or even other packets in front of it in the queue, there will be a delay before the packet may be transmitted. These delays lead to jitter in the overall end-to-end delay, and are thus the subject of several analysis approaches in the literature. The most important property for a certification of an AFDX network for flight-critical applications is a guaranteed maximum end-to-end delay.

The most prominent method are standard and stochastic network calculus, which allow to compute safe upper bounds on the maximum end-to-end delay for industrial-size network topologies [7, 18, 6]. The algorithm can be improved with the trajectory approach [8]. For small-size systems the state space may be manageable, allowing to compute an actual maximum delay with model checking [19]. Simulation is another choice [20, 17], but there is no guarantee that the visited parts of the stochastic process will include the worst-case delay. It will, however, give a lower bound on possible maximum delays.

The quality of computable bounds is discussed in the literature [7]: simplified, the derived bounds are less tight (the pessimism increases) when the network topology becomes larger, and with higher network loads [17]. They are best if only one switch is used, but may be off by a factor of up to 20 otherwise. However, industrial-size networks contain paths with up to 4 switches [17]. Unfortunately, the worst case cannot be derived by simply assuming worst-case input values; the end-to-end delay increases for some cases, when the BAG occupation of another VL is decreased [7].

The downside cost of a provably safe network setup with some remaining pessimism that is never needed in reality leads to a bad utilization of network resources. The maximum utilization of real-life AFDX networks is usually around or below 20%. Another issue is that even if a computed bound is tight, the probability that a packet will actually experience it may be marginally small and acceptable for the applications waiting for it. There is thus an interest in not only computing bounds on the worst case, but also the actual end-to-end delay distribution or its quantiles, as well as the connected probabilities of a certain buffer utilization [21]. It is, however, still an open problem how resources can be better utilized depending on how rare the computed maximum delays are [9]. A possible solution approach is presented in this paper in Section 4.

3. AFDX NETWORK MODELING WITH STOCHASTIC PETRI NETS

Stochastic Petri nets (see [22, 23], e.g., for an overview) represent a graphical and mathematical method for the specification of processes with concurrent, synchronized and conflicting or nondeterministic activities. The graphical representation of Petri nets comprises only a few basic elements. They are therefore useful for documentation and a figurative aid for communication between system designers. Complex systems can be described in a modular way, where only local states and state changes need to be considered. The mathematical foundation of Petri nets allows their qualitative analysis based on state equations or reachability graph, and their quantitative evaluation based on the reachability graph or by simulation.

Petri nets contain *places* (depicted by circles), *transitions* (depicted by boxes or bars) and directed *arcs* connecting them. Places may hold *tokens*, and a certain assignment of tokens to the places of a model corresponds to its model state (called *marking* in Petri net terms). Transitions model activities (state changes, events). Just like in other discrete event system descriptions, events may be possible in a state — the transition is said to be *enabled* in the marking. If so, they may happen atomically (the transition *fires*) and change the system state.

In stochastic Petri nets, activities may take some time, thus allowing the description and evaluation of performance-related issues. Basic quantitative measures like the throughput, loss probabilities, utilization and others can be computed. A *firing delay* is associated to each transition, which may be stochastic (a random variable) and thus described by a probability distribution. It is interpreted as the time that needs to pass between the enabling and subsequent firing of a transition. In the net class *extended deterministic and stochastic Petri nets* (eDSPN [10]) that is used here, transition delays may be zero (immediate), exponentially distributed, deterministic, or a general distribution can be specified.

The dynamics of a Petri net are defined as follows. A transition is said to be *enabled* in a marking, if there are enough tokens available in each of its input places. Whenever a transition becomes newly enabled, a *remaining firing time* (RFT) is randomly drawn from its associated firing time distribution. The RFTs of all enabled transitions decrease with identical speed until one of them reaches zero (race enabling semantics). The fastest transition (in case of multiple ones, a probabilistic choice decides) will *fire*, and change the current marking to a new one by removing the necessary number of tokens from the input places and adding tokens to output places.

3.1. Petri Net Patterns for AFDX Network modeling

One of the advantages of Petri net models over, e.g., automata, is their modular way of describing complex systems. The following text proposes generic Petri net patterns for AFDX elements along their functional details as described in Section 2.

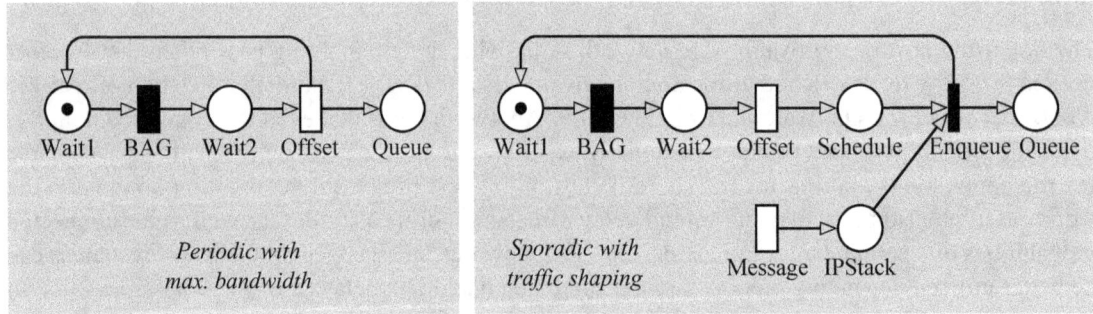

Figure 1: Petri net model variants of AFDX source end-system with packet regulation

Figure 1 introduces two variants of source end-systems. The left part shows how applications and virtual links generating periodic messages should be modeled. Every time both transitions `BAG` and `Offset` have fired sequentially, one packet will be generated and the according token is added to place `Queue`. FIFO behavior of queuing is not significant here, as there is no way of (and no need to) differ between tokens resp. messages under certain simplifying assumptions.

It may seem strange at first that a periodic behavior is modeled by a deterministic plus an exponential transition. This is however necessary to create a stochastic process covering all possible end-system offsets in a steady-state simulation run. The exponential transition `Offset` models drift and other possible inaccuracies between end-system clocks in the overall distributed system, because there is no global time synchronization [24].

The right part of Figure 1 shows a model variant in which an application sends sporadic messages over a virtual link (transition `Message` adds a token to `IPStack`), and thus the end-system needs to apply traffic shaping to ensure a minimum BAG time between messages. This is done similarly to the left-hand side model. When a message may be added to the output queue of the end-system, a token is in `Schedule`, and transition `Enqueue` will fire immediately when a message arrives in `IPStack`. After that, transitions `BAG` and `Offset` have to fire subsequently before a new message may pass.

In the case of a VL carrying periodic messages, but with a longer time between subsequent message generations than the BAG, the left-hand model in Figure 1 can still be used, but the firing delay of transition `BAG` would have to be set to this fixed time between messages. In any case, as long as the period does not exceed the allowed BAG value, traffic shaping will not change the behavior and is thus not explicitly modeled.

In general, transition delays in the models have to be set according to the chosen model time as well as the delays of the described actions. We assume a base time of 1 model time unit equal to $1\mu s$ in the following. Possible BAG values in the range $1ms\ldots 128ms$ will then result in transition delays $1,000\ldots 128,000$. The actual value of the random offset is not that important, but should be small compared to the BAG[1]. The average time between sporadic message generations in this case is associated with transition `Message`.

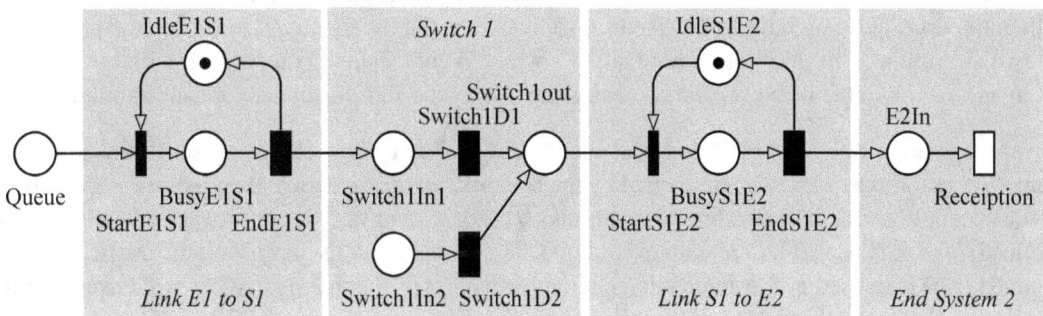

Figure 2: Link, switch, and destination end system model

Petri net patterns for links between nodes, switches, and destination end-systems are proposed in Figure 2. Larger models can be constructed by merging corresponding places such as `Queue` from both model figures. A link is simply a mutually exclusive resource for queued waiting messages. It is either in state `Idlexy` or `Busyxy`. A transmission begins immediately (`Startxy` fires), when a message is waiting and the link is idle. The transmission time for each message is modeled by the delay of the deterministic transition `Endxy`, which is selected based on message length and bandwidth. For the minimum and maximum message lengths (84 and 1538 Bytes), the delay has thus to be chosen as 6.72 and 123.04 in our μs-based timing. We assume identical message lengths here.

The switch has input and output ports (places `SwitchInx` and `SwitchOutx` in Figure 2, for instance). Any number of input ports for incoming links can be added similar to the two shown in

---

[1] For an even better approximation, the mean delay could be chosen depending on the variance in the end-system clocks, and its half subtracted from the BAG to compute the firing delay of transition `BAG`.

the model. The $16\mu s$ delay inside the switch to process a message and enter it into an output port is modeled by transitions `SwitchDx`. A second link connects the switch in the model with a destination end-system.

3.2. An Application Network Setup

Figure 3 sketches a sample topology of an AFDX network that has been chosen as an application example. Its design resembles network architecture examples in the literature [8] with a little longer switch sequence but without VL paths that split away from the considered flows at switches[2].

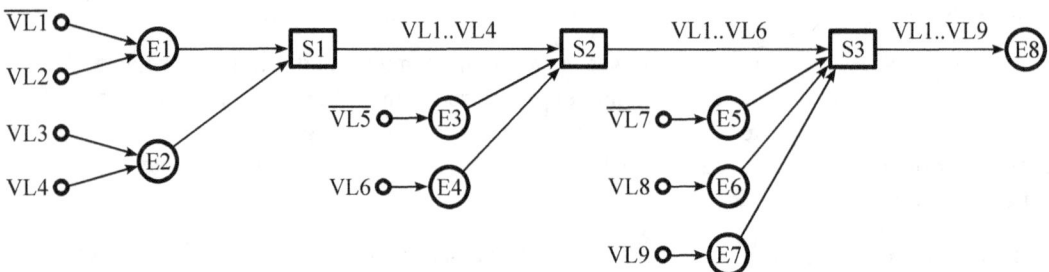

Figure 3: Topology of a sample AFDX network

A real-life topology would not necessarily contain longer paths (a real-life size avionic network with several thousand VLs analyzed in [17] had only four switches on the longest paths), but many more VLs and paths. However, following the findings of [17], only VLs on paths that *directly influence* the VL or path under consideration have to be taken into account, as the others do not influence the end-to-end delay distribution. This allows to ignore all VLs which do not share an output port with the analyzed VL at any switch in the network, and decreases the size of the significant network substantially.

The example topology contains 8 end-systems `E1...E8`, 3 switches `S1...S3`, and 10 links between them. 9 virtual links `V1...V9` are carried by the network, their association to end-systems and links is obvious from Figure 3. The overlined virtual links $\overline{VL1}$, $\overline{VL5}$ and $\overline{VL7}$ symbolize periodically sending applications; i.e., a packet is sent with every BAG. The remaining VLs carry sporadic traffic, which is assumed to be requested by the application randomly (at a lower rate than 1/BAG to avoid overloading), but needs to be controlled and possibly delayed before entering the network.

Figure 4 depicts the full eDSPN model constructed with the tool TimeNET [14]. It is designed in a modular way using the building blocks proposed in Section 3. BAG values are set to the minimal value of $1ms$ equal to a transition delay of 1000 (with an additional jitter of $5\mu s$ at transitions `Offset`). Messages on sporadic virtual links are sent on average every $1.2ms$. Packet lengths are chosen as 8000bits, corresponding to a transmission delay of $80\mu s$. The delay spent by each message in a switch between arrival and queuing or transmission is 16 (transitions `Switchx`).

4. RARE-EVENT SIMULATION FOR AFDX BUFFER SIZING

We are interested in probabilities that the number of packets in a network buffer exceeds certain values in stationary operation, and these states of interest only happen rarely in comparison to the rapid state changes inside the network. Thus a regular simulation would have to execute huge numbers

---

[2] The reason for this choice is that for such a system setup, packets arriving from one input of a switch but belonging to different VLs would have to be queued at different output queues. This would require additional information which is not available with the standard Petri nets chosen here.

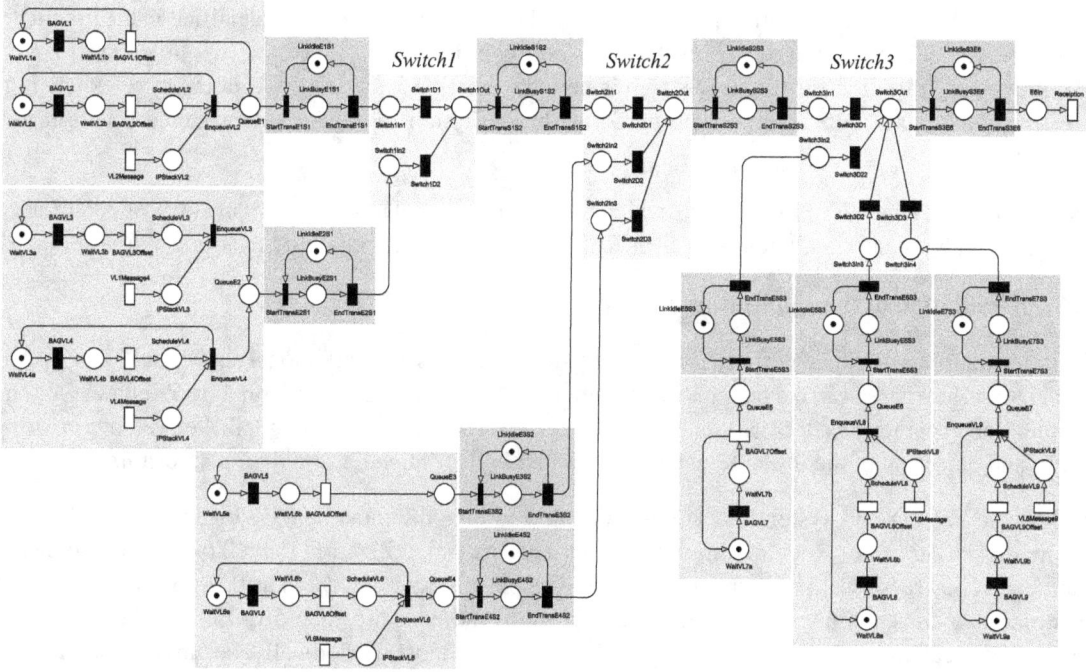

Figure 4: A Petri net model of the example avionic communication system

of events until a sufficiently large number of significant events has been collected and the estimated performability values have converged with a predefined error margin.

The section briefly touches the RESTART method used in this paper to solve this issue, and shows how it can be applied to our application model.

4.1. The RESTART Algorithm

The RESTART algorithm [12] cuts the reachability space of a model into enclosing regions of increasing probability to hit a rare event or state. Upon entering a state in a region of higher probability ("closer" to the region of interest), the state is stored and may be restored later when the simulation state is about to leave the region. Applying this simple rule on every border between regions leads to a simulation that hits the rare events of interest much more often. Some changes have to be applied to the accumulation of performance measures during the simulation to reverse the bias introduced by this way of controlling the simulation trajectory. A *weight* is maintained by each simulation path, which starts with 1.0 and is divided by the *splitting factor* whenever the path crosses a region border upwards. The factor equals the number of times that the path is retried from a stored state, which leads to a setup similar to the splitting of particles. The final surviving path after a split will be continued, and regains the previous weight.

A user-specified *importance function* returns a number for each state, which should be related to the distance of the current state from the region of interest. In our application model, simply the number of tokens in the buffer `Switch3Out` is used, which is actually not a very good measure for our RESTART application, but it turns out that the speedup is still very high - underlining the robustness of the approach. RESTART for stochastic Petri nets has been proposed in [25] and improved later [23, 26].

The issue of rareness in simulation is comparable to the problem that has been identified in the literature for improvements of network calculus that considering trajectories with expected long delays [6, 27]. An interesting question for future research is how the existing knowledge about trajectories leading to near-worst-case behavior could be used for the definition of a better importance function for RESTART.

We use the tool TimeNET [14, 23] here, which implements several model classes of stochastic Petri nets and their analysis and simulation, as well as RESTART splitting for reliability measures [28, 26].

4.2. Evaluation of the Application Example

This section presents some performability results for the AFDX network model shown in Figure 4. All simulations have been carried out on an Intel core i5 2.4GHz laptop computer running Windows 7 64bit. Simulation accuracy has been chosen as confidence level 95%; the maximum allowed relative error is 10%, and detection of the initial transient is activated in the TimeNET simulation algorithms.

All traffic ends at end-system E8 in our model, and thus the most heavily loaded link is the one from switch S3 to it. Regular simulation computes the link utilization to be 0.6507, and the mean number of messages waiting in the queue (tokens in place Switch3Out) as 0.2963 (i.e., waiting in addition to a currently transmitted message). To validate the results, the utilization can be compared to a theoretical approximation: there are 3 VLs sending (almost) periodically, occupying the link for $80\mu s$ within a time period of $1005\mu s$. In addition, there are 6 sporadically sending VLs, transmitting an $80\mu s$-packet every $1200\mu s$ on average. This corresponds to a theoretical joint utilization of 0.6376, resulting in an error of about 2% compared to the simulation result.

For our buffer sizing problem and an analysis of the variable waiting time of messages because of queuing, we are interested in the probabilities of having at least n messages waiting in place Switch3Out. The results of this analysis are shown in Table 1.

Number of	Standard Simulation		RESTART	
Messages	Probability	CPU time	Probability	CPU time
1	2.6221E-01	0:00:01	2.30463E-01	0:00:03
2	4.3991E-02	0:00:05	3.00958E-02	0:00:15
3	2.1414E-03	0:04:56	2.32636E-03	0:04:39
4	—	>24h	8.54219E-21	1:04:37

Table 1: Probabilities of buffered messages exceeding bounds (time in h:min:sec)

The table shows the net required CPU time computed by normal and RESTART simulation. As the tool uses a master/slave process architecture with 3 slaves to decrease variance with multiple independent replications, the multiple cores of the computer can be utilized, and thus the actual run time of each simulation experiment is only about a third of the shown value.

The results show that up to three tokens / messages, the probabilities are getting smaller but stay in a range that is not rare. Therefore, both normal and RESTART simulation are able to compute the values with comparable and acceptable CPU times. The reason for this is that this value is related to the probability of having an arriving packet on a subset of the four input links to switch 3 during overlapping time intervals. However, for token numbers above three, the probability drops considerably, and cannot be computed with normal simulation in an acceptable time[3]. The probability that the number of waiting messages at switch 3 is bigger than 3 is less than 10^{-20}, and this very rare value can be derived by the RESTART simulation method after about one hour with the same convergence requirements as for the other experiments. This shows that even for complex system models which are unfavorable for the RESTART algorithm, considerable speedups can be achieved.

The results show that while for certification purposes of safety-critical applications a mathematical proof of worst-case assumptions may be legally necessary, it is possible to evaluate the actual probabilities of exceeding certain buffer utilization values and to decide how many buffers are actually needed. In our example, if the output buffer of switch 3 would be restricted to 3 slots, the resulting loss probability of incoming packets would be estimated in the order of magnitude of 10^{-18}, corresponding to one lost message within about every 4 million years operation time. This will most probably be acceptable because of the other layers of redundancy in the used avionic hard- and software. Moreover, the value is negligible small compared to the packet error rate of Ethernet, which is around $10^{-9} \ldots 10^{-12}$.

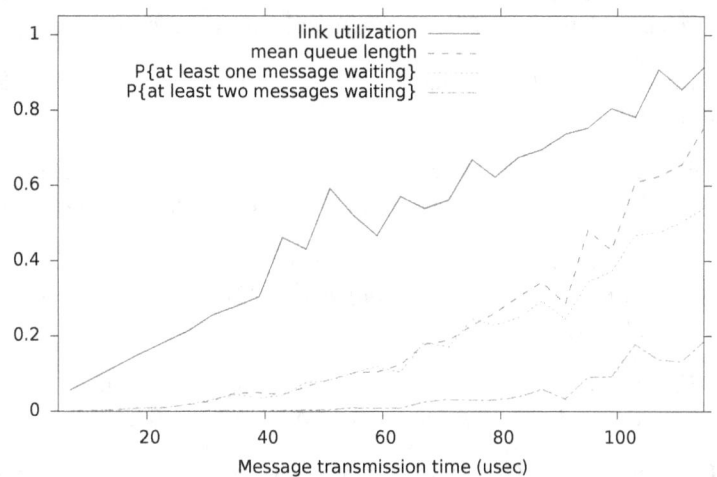

Figure 5: Various performance measures of link `S3E6` vs. message length

A second experiment was conducted to find out the dependency of basic network performance measures on the overall utilization and validate model as well as simulation. Link utilization can be changed without adding more VL structural models by simply varying the message length of the existing model within the allowed range. The results are shown in Figure 5. The expected result of a linear dependency of link utilization is visible (within the bounds of simulation inaccuracy). Queue lengths and probabilities of having at least one or two messages waiting at the output port of switch 3 are increasing nonlinearly, as it can be expected from queuing theory.

5. CONCLUSION

The paper shows how AFDX networks can be modeled with stochastic Petri nets and proposes a set of realistic patterns. It demonstrates use and advantages of the splitting rare-event simulation method RESTART by applying it to a non-trivial AFDX network example and deriving probabilities in the range of 10^{-20} in acceptable simulation time.

---

[3] The corresponding experiment has been stopped after more than 8 hours without hitting even one significant event, which is equal to more than 24 hours of CPU time.

Acknowledgements

The authors would like to acknowledge the work of Alexander Wichmann and Timur Ametov, who implemented the current RESTART algorithms in TimeNET. This work has been supported by a TU Ilmenau internal excellency grant in funding period 2013/14.

REFERENCES

[1] A. Ramos, J. Ferreira, and J. Barcelo, "Model-based systems engineering: An emerging approach for modern systems," *IEEE Trans. on Systems, Man, and Cybernetics, Part C: Applications and Reviews*, vol. 42, no. 1, pp. 101–111, January 2012.

[2] K. S. Trivedi, *Probability and Statistics with Reliability, Queuing and Computer Science Applications*, 2nd ed. Wiley, 2002.

[3] T. Schuster and D. Verma, "Networking concepts comparison for avionics architecture," in *IEEE/AIAA 27th Digital Avionics Systems Conference (DASC 2008)*, 2008, pp. 1–11.

[4] "Arinc 664, aircraft data network, part 7: Avionics full duplex switched Ethernet (AFDX) network," Jun. 2005.

[5] K. Wang, S. Wang, and J. Shi, "Integrated reliability theory and evaluation methodology of AFDX," in *10th IEEE Int. Conf. on Industrial Informatics (INDIN)*, 2012, pp. 657–662.

[6] J.-L. Scharbarg and C. Fraboul, "Methods and tools for the temporal analysis of avionic networks," in *New Trends in Technologies: Control, Management, Computational Intelligence and Network Systems*, M. J. Er, Ed. Sciyo, 2010.

[7] H. Charara, J.-L. Scharbarg, J. Ermont, and C. Fraboul, "Methods for bounding end-to-end delays on an AFDX network," in *Proc. 18th Euromicro Conf. on Real-Time Systems (ECRTS06)*, 2006.

[8] H. Bauer, J.-L. Scharbarg, and C. Fraboul, "Improving the worst-case delay analysis of an AFDX network using an optimized trajectory approach," *IEEE Trans. Industrial Informatics*, vol. 6, no. 4, pp. 521–533, Nov. 2010.

[9] C. Fraboul and J.-L. Scharbarg, "Trends in avionics switched Ethernet networks," in *Proc. 1st Workshop on Real-Time Ethernet (RATE) at the IEEE Real-Time Systems Symposium*, Vancouver, Canada, 2013.

[10] R. German, *Performance Analysis of Communication Systems, Modeling with Non-Markovian Stochastic Petri Nets*. John Wiley and Sons, 2000.

[11] "Analysis techniques for dependability — Petri net techniques," IEC 62551:2012, Sep. 2013.

[12] M. Villén-Altamirano and J. Villén-Altamirano, "Analysis of RESTART simulation: Theoretical basis and sensitivity study," *European Transactions on Telecommunications*, vol. 13, no. 4, pp. 373–385, 2002.

[13] J. K. Townsend, Z. Haraszti, J. A. Freebersyser, and M. Devetsikiotis, "Simulation of rare events in communications networks," *IEEE Communications Magazine*, vol. 36, no. 8, pp. 36–41, 1998.

[14] A. Zimmermann, "Modeling and evaluation of stochastic Petri nets with TimeNET 4.1," in *Proc. 6th Int. Conf on Performance Evaluation Methodologies and Tools (VALUETOOLS)*. IEEE, 2012, pp. 54–63.

[15] Z. Jiandong, L. Dujuan, and W. Yong, "Modelling and performance analysis of AFDX based on Petri net," in *2nd Int. Conf. on Future Computer and Communication (ICFCC)*, vol. 2, May 2010, pp. 566–570.

[16] D. Li, J. Zhang, and B. Liu, "Periodic message-based modeling and performance analysis of AFDX," in *IEEE Int. Conf. on Wireless Communications, Networking and Information Security (WCNIS)*, 2010, pp. 162–166.

[17] J.-L. Scharbarg, F. Ridouard, and C. Fraboul, "A probabilistic analysis of end-to-end delays on an AFDX avionic network," *IEEE Trans. Industrial Informatics*, vol. 5, no. 1, pp. 38–49, February 2009.

[18] T. Lv, N. Hu, Z. Wu, and N. Huang, "The analysis of end-to-end delays based on AFDX configuration," in *9th Int. Conf. on Reliability, Maintainability and Safety (ICRMS)*, 2011, pp. 1296–1300.

[19] M. Adnan, J.-L. Scharbarg, J. Ermont, and C. Fraboul, "An improved timed automata approach for computing exact worst-case delays of AFDX sporadic flows," in *Proc. 16th IEEE Int. Conf. Emerging Technologies and Factory Automation (ETFA)*, Sep. 2011, pp. 1–4.

[20] J.-L. Scharbarg and C. Fraboul, "Simulation for end-to-end delays distribution on a switched Ethernet," in *Proc. IEEE Int. Conf. on Emerging Technologies and Factory Automation (ETFA 2007)*, 2007, pp. 1092–1099.

[21] H. Bauer, J. Scharbarg, and C. Fraboul, "Worst-case backlog evaluation of avionics switched Ethernet networks with the trajectory approach," in *24th Euromicro Conference on Real-Time Systems (ECRTS)*, July 2012, pp. 78–87.

[22] M. Ajmone Marsan, G. Balbo, G. Conte, S. Donatelli, and G. Franceschinis, *Modelling with Generalized Stochastic Petri Nets*, ser. Series in parallel computing. John Wiley and Sons, 1995.

[23] A. Zimmermann, *Stochastic Discrete Event Systems*. Springer, Berlin Heidelberg New York, 2007.

[24] R. Alena, J. Ossenfort, K. Laws, A. Goforth, and F. Figueroa, "Communications for integrated modular avionics," in *IEEE Aerospace Conference*, March 2007, pp. 1–18.

[25] C. Kelling, "Rare event simulation with RESTART in a Petri net modeling environment," in *Proc. of the European Simulation Symposium*, Erlangen, 1995, pp. 370–374.

[26] A. Zimmermann and P. Maciel, "Importance function derivation for RESTART simulations of Petri nets," in *9th Int. Workshop on Rare Event Simulation (RESIM 2012)*, Trondheim, Norway, Jun. 2012, pp. 8–15.

[27] E. Heidinger, "Rare events in network simulation using MIP," in *Proc. 23rd Int. Teletraffic Congress (ITC 2011)*, 2011, pp. 314–315.

[28] A. Zimmermann, "Dependability evaluation of complex systems with TimeNET," in *Proc. Int. Workshop on Dynamic Aspects in Dependability Models for Fault-Tolerant Systems (DYADEM-FTS 2010)*, Valencia, Spain, Apr. 2010.

Extension of DMCI to heterogeneous infrastructures: model and pilot application

Paolo Trucco[a], Massimiliano De Ambroggi[a], Pablo Fernandez Campos[a], Ivano Azzini[b], and Georgios Giannopoulos[b]

[a] Department of Management, Economics and Industrial Engineering, Politecnico di Milano, Milan, Italy
[b] European Commission - DG Joint Research Centre (JRC), Ispra, Italy

Abstract: Since the adequate functioning of critical infrastructures is crucially sustaining societal and economic development, the understanding and assessment of their vulnerability and interdependency become more and more important for improving resilience at system level. The paper proposes an extension of DMCI (Dynamic Functional Modelling of vulnerability and interoperability of CIs) to modelling the vulnerability and interdependencies of heterogeneous infrastructures, i.e. the interactions between electric power infrastructure and the transport infrastructure system have been modelled. The simulation tool has been implemented with the Matlab platform Simulink in order to overcome some computational limitations, that affect the first DMCI version implemented in Matlab, in quantifying the propagation of inoperability and logical interdependencies related to demand shift, and to obtain a modular and user friendly solution, even for users who are not expert at simulation.
The new DMCI model has been tested with a pilot application that comprised more than 200 vulnerable nodes and covered both power transmission grid and transportation systems of the province of Milan (Italy). The most vital and vulnerable nodes have been identified under different blackout scenarios, for which specific data on vulnerable nodes has been collected directly from the operators.

Keywords: CIP, interdependencies, simulation, electricity, transportation.

1. INTRODUCTION

Critical Infrastructures are those assets, systems or parts thereof, which is essential in the provision of services that are deemed to be vital for the functioning of society, "including the supply chain, health, safety, security and economic or social well-being of the people" [1]. Since the adequate functioning of those infrastructures is crucially sustaining societal and economic development, the understanding and assessment of their vulnerability and interdependency become more and more important for improving resilience at system level.
CIs are highly interconnected and mutually interdependent [2] [3]. On the one hand, interdependencies have allowed a greater availability of resources but, on the other hand, they have increased their vulnerability; indeed, even a relatively small malfunctioning of an infrastructure can have considerably large and long-lasting effects on the whole infrastructural network due to largely unpredictable domino effects [4]. Recent worldwide events such as the 2001 World Trade Center Attack, the 2003 Italian and North America blackouts, the 2005 hurricane Katrina, the 2008 UK floods, and the 2011 Japan earthquake have sensibly influenced both the public opinion and the governments all around the world, enhancing the interest in how those destruction or disruption events propagate within the network of Critical Infrastructure.
Existing simulation approaches to vulnerability and interdependency analysis of complex CIs networks refer to different scientific fields, e.g. physical network modelling, network economics, etc. [5]. In the last decade, many approaches to CI protection have been developed. In this regard, Ouyang [6] provides a comprehensive review on modeling and simulation of interdependent critical infrastructure systems and groups the modeling approaches into six types: empirical approaches, agent based approaches, system dynamics based approaches, economic theory based approaches, network based approaches, and other approaches.

The present paper proposes an extension of the Dynamic functional Modelling of vulnerability and interoperability of CI (DMCI) formalism [7] to the problem of heterogeneous infrastructure modeling, e.g. the interdependencies between electric power infrastructure and the transport infrastructure systems. The paper is organized as follows: Section 2 summarizes the approach and key features of the first DMCI model; Section 3 describes the new modularized and enhanced implementation of DMCI formalism, with a special focus on how it can be applied to heterogeneous infrastructure systems; Section 4 presents the pilot application in the area of the province of Milan (Italy); Section 5 draws general conclusions on achieved results, major limitations and suggestions for further research.

2. FUNCTIONAL DYNAMIC MODELLING OF CRITICAL INFRASTRUCTURE

Trucco et al. in 2012 [7] developed a new integrated formalism for the Dynamic functional Modelling of vulnerability and interoperability of CI (DMCI). The proposed modelling formalism is characterised by some distinctive features:

- specification of vulnerable nodes defined as *"a large functional part of a CI that assures the satisfaction of a considerable part of service demand at regional or local level (e.g. part of a pipeline network, a railway station, a portion of a highway, an underground line) and that does not need further disaggregation for the sake of the analysis."* A vulnerable node has to be homogeneous (i.e. uniform in structure and function with respect to service demand), service self-providing (i.e. a system able to supply a value-added service through own means), and vulnerable (i.e. susceptible to threats that could decrease its functional integrity). Vulnerable nodes are mutually connected to create intra- and inter-infrastructure interdependencies;
- specification of threat nodes, characterised by time-variant intensity and specific potential impact on different vulnerable nodes;
- quantification of both functional and logic interdependencies thanks to the use of both service demand and service capacity for each node of the considered CI;
- time dependent specification of the main parameters of the model: node functional integrity, interoperability, service demand and loss, etc.;
- propagation of both inoperability and demand variations throughout the nodes of the same infrastructure and between interdependent CI.

Figure 1: Causes of service disruption

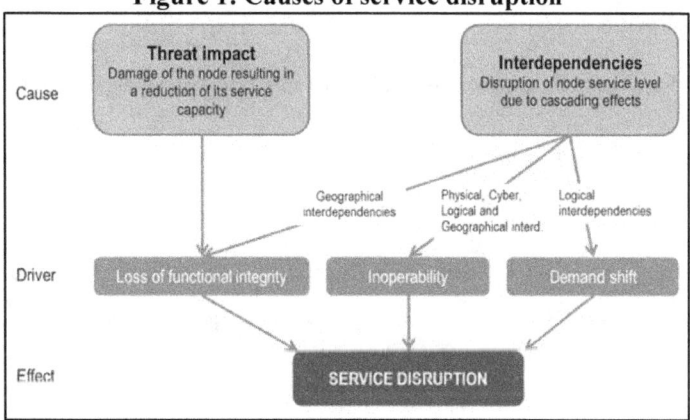

The model, firstly implemented in a software code by Matlab®, is able to assess the propagation of impacts due to a wide set of threats. Therefore, the disservice can be propagated within the same infrastructure or to other CI exploiting the model capability to represent functional, cybernetic, geographical, physical as well as logical interdependencies.

Service level can be reduced either by a threat impact on node or through interdependencies (Figure 1). Functional integrity quantifies the direct impact of threats on the node service capability, i.e. the reduction of its maximum service capacity over time (the direct effect). Inoperability quantifies how disturbances coming from the CIs network through interdependencies (physical, cyber, geographical, logical – [8]) reduce the maximum service level of a node starting from its actual service capacity. As a consequence, service disruption, globally, is due to combined effects of loss of functional integrity on some nodes and propagation of inoperability between nodes.

In order to test the capability of the model to represent all the types of interdependencies and to give an overview of the possible outcome of the model a pilot study has been carried out in the metropolitan area of the province of Milan (Italy) in which the considered CI referred to the transportation system (road, rail, underground, and airport systems) [7]. In particular, for the road system, it has been considered highways, beltways and the national roads.

Afterwards, Cagno et al. [9] applied DMCI to analyze a real scenario – the ex-post analysis of the overall impact on the transportation system of the severe snowfall that took place in the Northern part of Italy in December 2009.

3. THE EXTENDED DMCI TO HETEROGENEOUS CRITICAL INFRASTRUCTURE

To fully test and eventually prove the potentials of DMCI its application to the case of heterogeneous interdependent CI is crucial. Indeed, more complex, unpredictable and as such potentially impacting domino effects are those due to interdependencies (more specifically inoperability propagation) among infrastructure of different nature. On this regard, an important stream in literature is the study of cyber interdependencies between Electricity and ICT infrastructure through SCADA systems [10] [11] [12] [13] [14] [15] [16]. Similarly relevant, but relatively less studied is the coupling of Transportation and Electricity infrastructure. As a matter of example, logical interdependencies are particularly relevant in both infrastructure systems. In the former, logical interdependencies are mainly established by the shift of demand between two infrastructure that can provide the same or fully/partially replaceable mobility service (e.g. two different transportation means to connect the same towns); in the latter, they are established by the way different power generation sources or line sections are used to maintain the overall grid balance.

ortation infrastructure systems.

Figure 2 shows how the Electricity infrastructure, from power generation to distribution, can be functionally modeled into vulnerable nodes under the DMCI formalism. Whereas Figure 3 highlights key functional (a) and logical (b) interdependencies between vulnerable nodes of the Electricity and Transportation infrastructure systems.

Figure 2: Functional modeling of electricity infrastructure

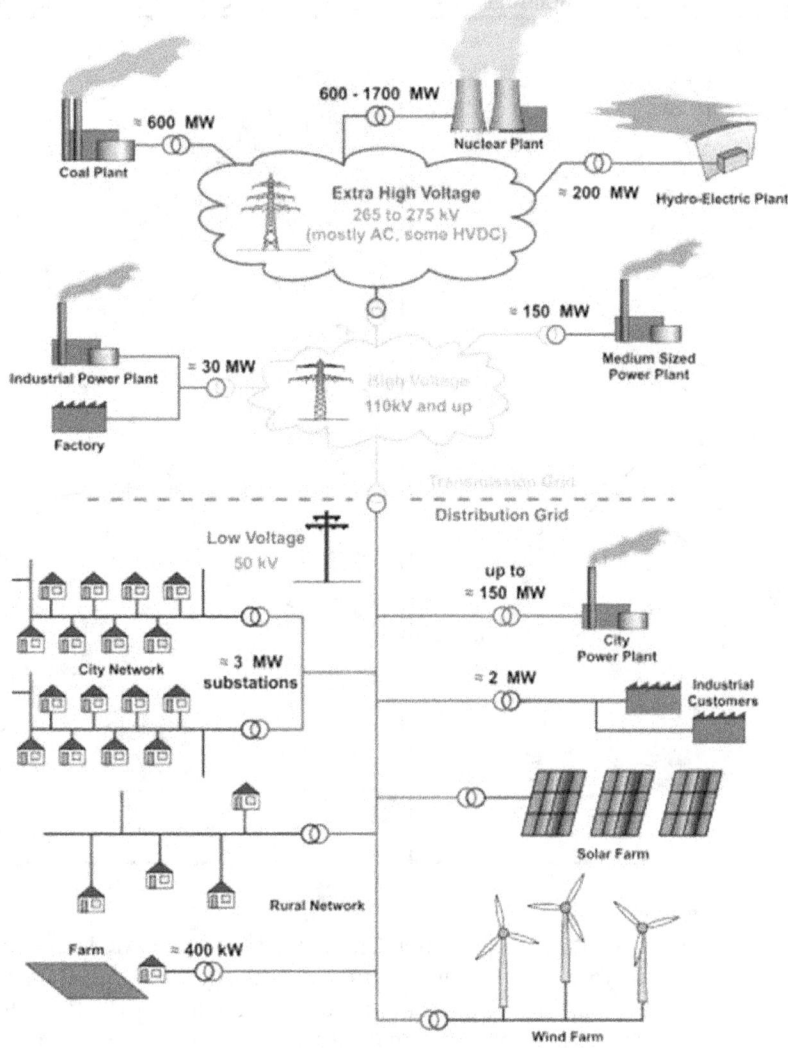

a. General layout of electricity grid

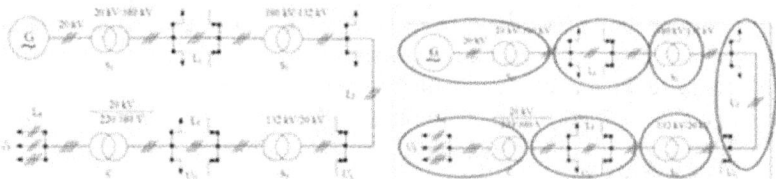

b. Diagram of electricity grid   c. Nodes of electricity grid (in DMCI)

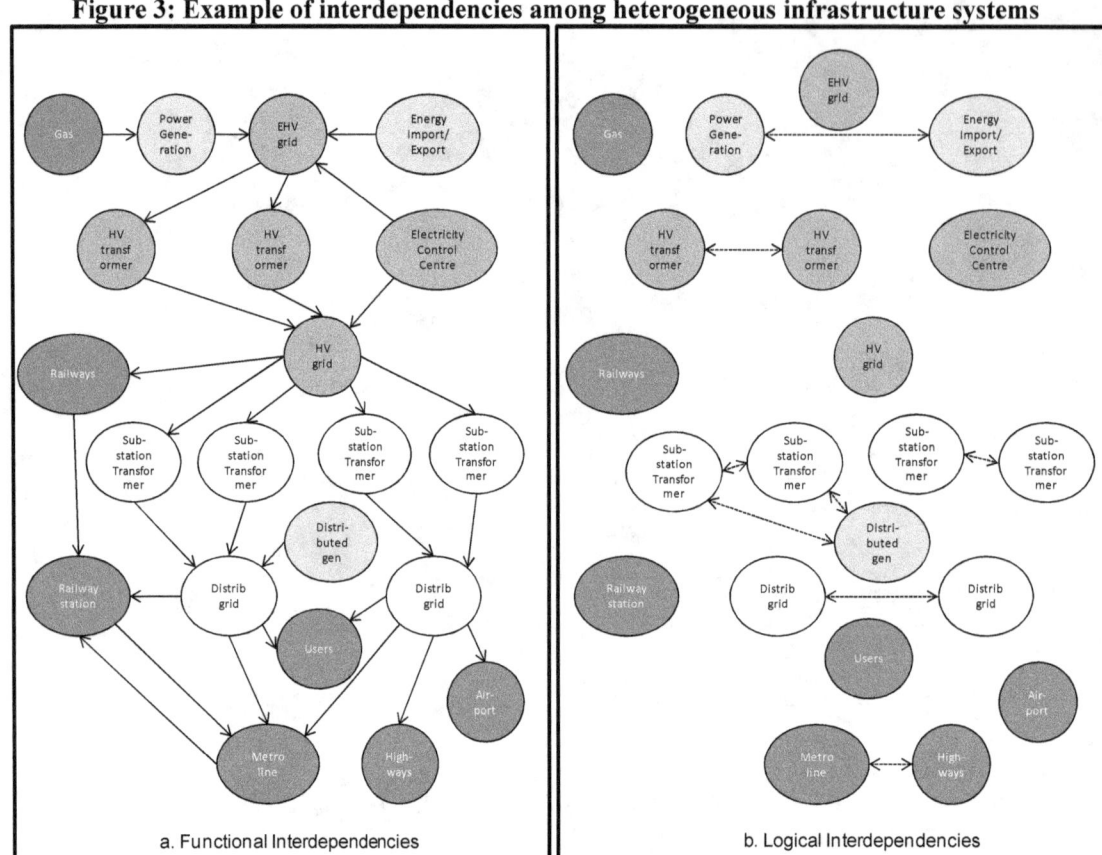

Figure 3: Example of interdependencies among heterogeneous infrastructure systems

a. Functional Interdependencies   b. Logical Interdependencies

3.1. DMCI Modularization

In order to overcome some computational limitations, that affect the first DMCI version implemented in Matlab®, mainly in quantifying the propagation of inoperability and logical interdependencies related to demand shift and to obtain a modular and user friendly solution, even for users who are not expert at simulation, such as policy makers or risk manager, the simulation tool has been implemented with the Matlab platform Simulink®.

Simulink® is a block diagram environment for multidomain simulation and Model-Based Design. It supports system-level design, simulation, automatic code generation, and continuous test and verification of embedded systems. Simulink provides a graphical editor, customizable block libraries, and solvers for modeling and simulating dynamic systems. It is integrated with MATLAB®, enabling you to incorporate MATLAB® algorithms into models and export simulation results to MATLAB® for further analysis. The key aspect of modularization is the definition of a standard vulnerable node, whether if it is a transportation node or an electricity node.

The standard vulnerable node developed in Simulink® (Figure 4) is built up of 4 principal modules: node vulnerable module, node inoperability module, demand interdependency module, node disservice delivery module.

Exploiting the capabilities of Simulink®, the simulator automatically builds up the network topology of vulnerable nodes and quantifies major model parameters starting from a simple user interface.

3.2. Node Vulnerability module

The Node Vulnerability module assesses the potential damage of a vulnerable node impacted by a threat. According to the first DMCI model, the threat impacting at time t on the k-th vulnerable node, causes a reduction of its functional integrity $F(k,t)$. The reduction depends on the functional integrity modulation function that defines the functional integrity steady reduction after the impact and the temporal modulation function that defines the dynamic through which the functional integrity reaches its new steady value [7]. The cause-effect relationship between threats and vulnerable node is modeled by means of three Simulink blocks: one is responsible for the temporal modulation of the threat, another for the intensity modulation and a third combining both modulations and calculates the functional integrity.

Figure 4: Vulnerable node developed in Simulink®

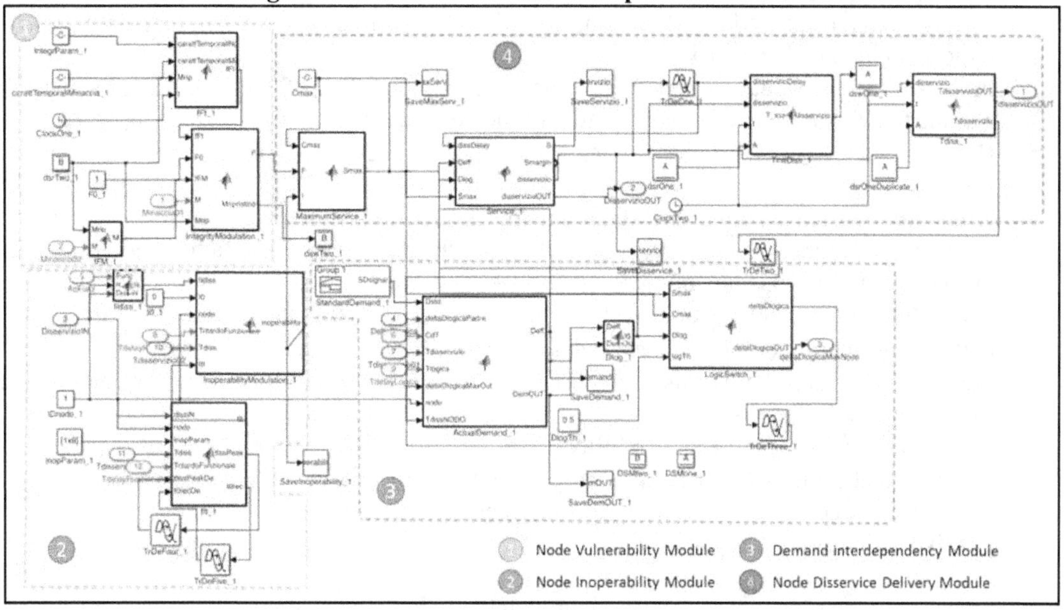

From a mathematical point of view, the threats are temporary signals with values between 0 and 1.
The main variables and parameters considered of the module are:
- Functional Integrity $F(k,t)$ – quantifies how the threats directly impacting on the k-th node reduce its maximum service level $S_{max}(k,t)$ at time t (the direct effect), and it is comprised between 0 (the k-th node is completely blocked) and 1(optimal state);
- Parameters of $F(k,t)$ – is a vector of four components that defines the dynamic response of the node to the threat. The four components are:
  - Buffer Time: time elapsed since the node is impacted until it begins to suffer any effect;
  - Propagation Time – is the duration of the first transient over which the $F(k,t)$ reached the functional integrity steady reduction after the impact of the threat;
  - Organizational Time – is a minimum time required for setting-up a countermeasure after an integrity loss due to the threat. Thus, it is the time elapsed since the node is impacted until it begins to recovery its functional integrity;
  - Recovery Time – is the duration of the transient over which the node recovers its full functional integrity;

The profile of the transients are exponential functions.

3.3. Node Inoperability Module
The node inoperability module assesses the inoperability of the node due to physical and cybernetic interdependencies. The physical and cybernetic interdependencies operate when the father node transfer a disservice to the child exploiting a supplier-customer relationship. The inoperability of the node depends on the intensity modulation function $fI_{diss}(j,i)$ that defines the inoperability steady reduction of the j-th child node due to the disservice of the i-th father node and the temporal modulation function fI_t that defines the profile curve of the transient regime through which the j-th child node reaches its inoperability steady value as well as recovers from it.

The intensity modulation function fI_{diss} depends on the disservice of the father node calculated in the previous iteration and on the parameter $F_{prox}(i,j)$, of the functional interdependency matrix $F_{prox}(NxN)$, that indicates the sensibility of the j-th child node to the disservice of the i-th father node in terms of maximum reduction of its service capacity.

The temporal modulation function is composed by five phases:
- Functional buffer period: is the first phase runs from the father node begins to indicate disservice until the child node begins to suffer the effects in terms of inoperability; it is indicated into the T_{func} matrix;
- Inoperability propagation: period in which the inoperability of the node is growing. We assume that the profile is a typical feature of the node response;
- Higher steady state: in this phase the time modulation function is 1 until the recovery transient starts;
- Recovery transient: period in which the node recovers its service capacity lost due to functional relationships. The recovery starts when the disservice of the father node is less than the percentage of its maximum value defined by the parameter *recovery threshold*. The parameter $diss_{peak}$ is used to store for each iteration the maximum value of the disservice of nodes. The duration of the recovery is described by the parameter *recovery time* t_{rec}.
- Lower steady state: during this last phase the value of the temporal modulation function is 0. The duration of this phase is not predefined, but can be followed by a new functional buffer period that results in a new inoperability cycle.
- The transient periods (inoperability propagation and recovery) allows four types of profile: linear, step function of constant value between 0 and 1, negative exponential function, or positive exponential function.

The inoperability of the j-th node is calculated through the following equation:

$$I(j) = \begin{cases} 0, & \sum_j fI_t(i,j) * fI_{diss}(i,j) < 0 \\ \sum_i fI_t(i,j) * fI_{diss}(i,j), & 0 < \sum_j fI_t(i,j) * fI_{diss}(i,j) \leq 1 \\ 1, & \sum_j fI_t(i,j) * fI_{diss}(i,j) > 1 \end{cases} \quad (1)$$

3.4. Demand Interdependency Module
The demand interdependency module assesses the logical interdependencies and computes the actual demand of each node. The module is composed by three computational steps:
- actual demand $D_{act}(i,t)$ – depends on the standard demand $D_{std}(i,t)$ of the node, the total incoming demand $D_{IN}(i,t)$ and the total demand out $D_{OUT}(i,t)$. The $D_{IN}(i,t)$ is the demand incoming from other nodes j that provide the same service or fully/partially replaceable service but they are not able to completely satisfy at time t their actual demand and therefore send a percentage $CdT(j,i)$ of their unsatisfied demand in the previous iteration $\Delta D_{log}(j, t-1)$. In the same way, $D_{OUT}(i,t)$ is the $CdT(i,j)$ part of the demand the i-th node is not able to satisfied at time t that is sent to the j-th nodes;
- logical demand $D_{log}(i,t)$ – is the maximum amount of demand that the i-th node can send to others at time t;
- $\Delta D_{log}(i)$ – is defined as the difference between the logical demand $D_{log}(i,t)$ and the maximum service $S_{max}(i)$.

All the demand shift mechanisms can be triggered after a logical buffer time $T_{log}(i,j)$ that is the time after which the demand of the i-th father node begins to switch to the j-th child node (e.g. the time after which the railway users decide to switch to the road transportation system).

3.5. Node Disservice Delivery Module
The node disservice delivery module aims at computing all the variables related to the service level of the node. The module is setup of three blocks: the maximum service block, the actual service block, and disservice time block.

The entire computational process followed by the simulator at each iteration to determine the overall state of a node is:
- calculation of functional integrity F through the node vulnerability module;
- calculation of inoperability I through the node inoperability module;
- execution of the maximum service block to determine S_{max};
- calculation of the variables actual demand D_{act}, total incoming demand D_{IN}, and total demand out D_{OUT} through the demand interdependency module;
- calculation of the disservice $diss$, the marginal service S_{margin} that is the percentage of the S_{max} which is not currently used, and the disservice time $T_{diss}(i,t)$ through the execution of the actual service block, and disservice time block of the node disservice delivery module.

4. PILOT APPLICATION

4.1. Case description and simulation settings
In order to demonstrate the potential of the new DMCI, this section shows its application to the heterogeneous CI system that takes place in Milan, Italy. For this pilot application, the same system as described in [9] has been used. The system consists of over 200 nodes which describe several critical infrastructures of both the transportation and energy distribution networks. The first include road, railway and suburban transportation networks. The airports of Malpensa and Linate are also considered. The latter mainly covers the electricity distribution grid and some parts of the gas distribution network. Figure 5 shows the approximate layout of the electric grid nodes in the system and highlights the most important nodes involved in the scenario simulated in this pilot application.

Figure 5: Electrical grid nodes considered in the pilot application.

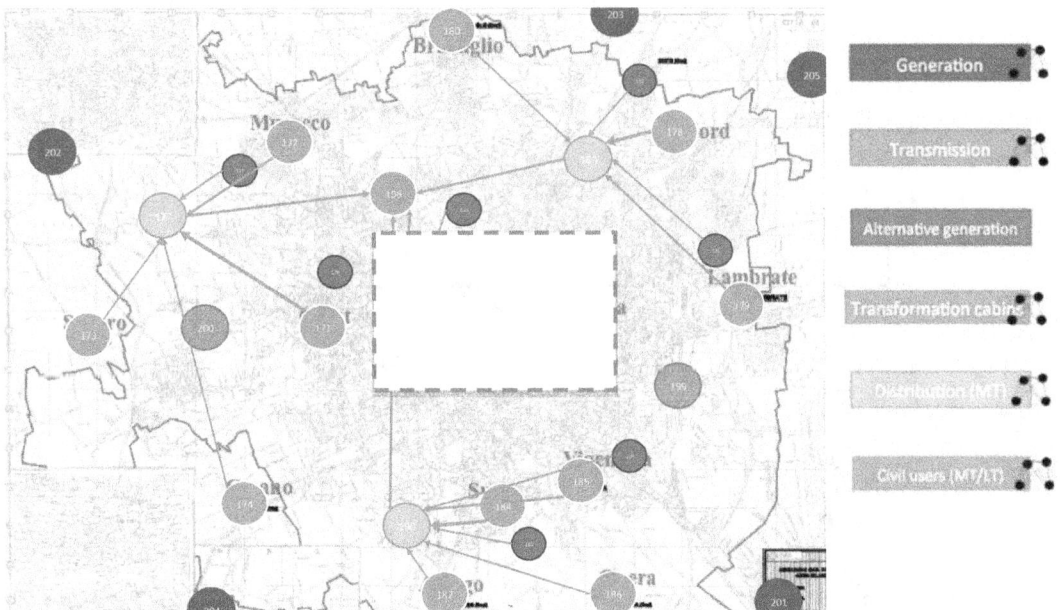

Figure 5 also shows the functional interrelations within the electricity distribution network. In addition to these, as shown in figure 3, there are also other important functional interdependencies with other CI such as railway stations or airports. Functional interdependencies are also used in the DMCI to model information flows or cyber interdependencies within CI. Hence, they are used to model the interdependencies between the electric grid and its control rooms.

Logical interdependencies have also been defined as presented in figure 3, therefore allowing for each subset of transformation cabins to take over each other's demand when one has been damaged. Distribution stations can satisfy demand from other distribution stations in the same exact way. These relations will affect greatly the resilience of the system, as will be seen in the results.

The scenario considered in the pilot application implements a cascading failure that affects the three transformation cabins (nodes 191-193) that supply the distribution station situated in the center of the city (node 197). The threat node then consists of three step signals that hamper each cabin's service capacity for six hours, the first starting at 12:00, second at 14:00 and third and last at 16:00. The response of the electric grid and the effects the event has on other CI can then be quantified and measured, allowing a comprehensive resilience study.

4.2. Analysis and discussion of results

Figure 6 shows the results obtained for the third transformation cabin. The results accurately show how, at 12:00, the cabin first satisfies the demand of the fallen cabin, and how it reaches its full capacity when trying to satisfy the demand from the second fallen cabin. At this point, the three cabins together are no longer able to fully supply its distribution station. The failure then propagates through the electric grid and other CI.

Figure 6: Node's 193 dynamic response.

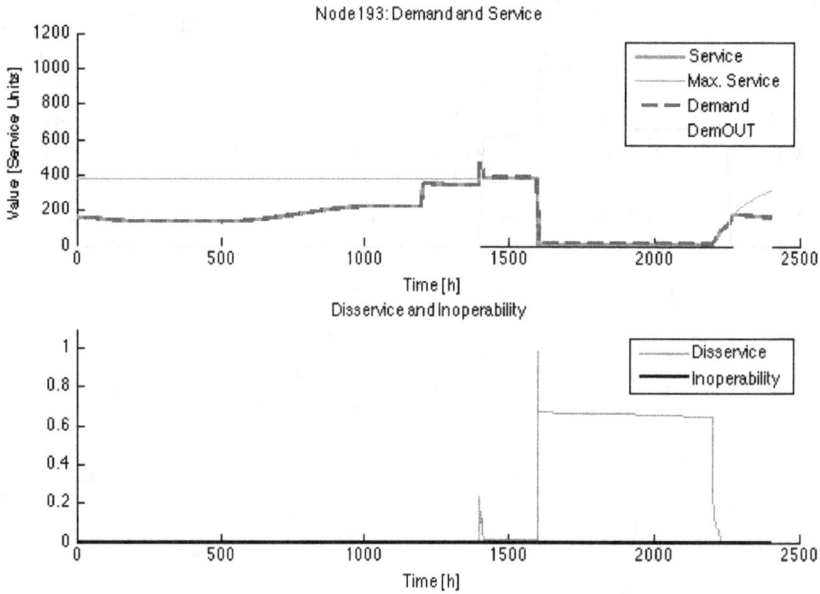

Figure 7 shows the failure of the second and third cabins affect the distribution station's service capacity. However, due to the existing logical interdependencies between distribution stations, the demand shifts allow its disservice to be minimum. This overall behavior accurately resembles that of the modeled system, in which operators rebalance the network to minimize disservice of the grid.

The complete results of the simulation also show several other CI affected in various ways, such as the train station, some national roads and some subway lines, and allow for critical interdependencies between infrastructures to be identified. Furthermore, second order indirect interactions between transportation networks can also be observed.

In conclusion, the results give relevant information regarding the resilience of the electrical network, its new balanced configuration and the detailed dynamic response of parts of the transportation and energy distribution grids. This information offered by the DMCI can then be processed in various ways to show the overall system performance.

Figure 7: Node's 197 dynamic response.

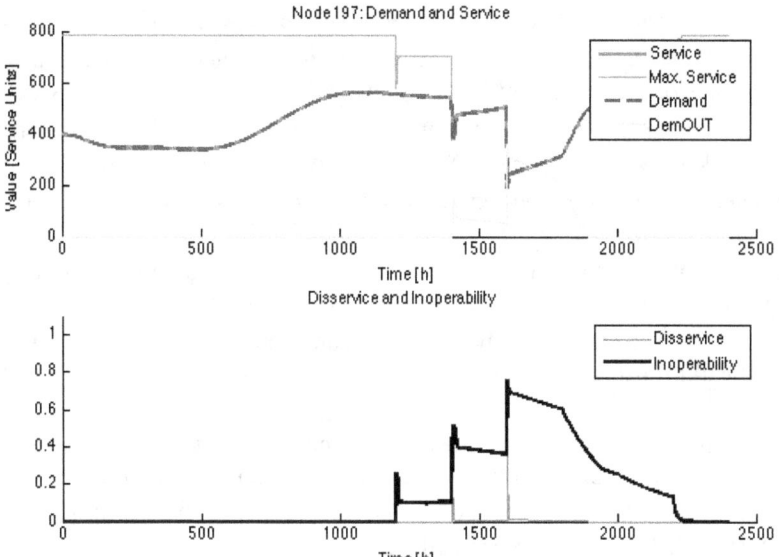

5. CONCLUSIONS

The study provides useful results and original contributions at both methodological and practical levels.

According to Ouyang's classification and review [6], the DMCI model [7] belongs to the category of flow-based network simulation models, considered to be the most capable to capture system dynamics and cover all the types of interdependencies, along with agent-based methods. In the present study we tested the capability of DMCI to be applied also to heterogeneous systems, specifically to analyze interdependencies between electricity and transportation infrastructure systems. Consequently, the major result of the study is a demonstration of the applicability of DMCI to a broad spectrum of CI systems and CIP/R problems. This generalization lies on a proper use of logical interdependencies also between nodes of the same infrastructure (e.g. electricity grid) as a way to model the control logic of the system. In this regard, the simulation results achieved through a pilot application in the metropolitan area of Milan shows the ability of DMCI to simulate and analyze complex cascading effects connected with the dynamic balancing of the transmission and distribution grids.

However, the proposed approach still suffer for some limitations. Despite the pilot application is based on characterization of nodes' performance (disruption, response time and recovery) shaped by real technical and organizational capabilities of operators and public agencies, the contribution (either positive or negative) of real-time decisions and possible changes in organization and strategy are not covered by the simulation model. Finally, further developments are needed to allow a real-time implementation of DMCI as a decision support tool during a large emergency and provide the analyst with a larger set of resilience indexes and enhanced reporting.

Acknowledgements

The work presented in this paper has been developed within MATRICS (www.matrics.it), a research project jointly co-funded by the Italian Ministry of Research and Education (MIUR) and the Lombardy Region Government (Italy). The financial support is gratefully acknowledged.

References

[1] "European Council Directive 2008/114/EC of 8 December 2008 on the identification and designation of European critical infrastructures and the assessment of the need to improve their protection. L 345/75," *Official Journal of the European,* 23 12 2008.

[2] H. A. M. Luiijf, H. Burger and M. Klaver, "Critical Infrastructure Protection in Netherlands: A Quickscan," in *EICAR Conference Best Paper Proceedings*, Copenhagen, 2003.

[3] B. Robert, M. H. Senay, M. E. P. Plamondon and J. P. Sabourin, "Characterization and ranking of links connecting life support networks," in *Public safety and emergency preparedness*, 2003.

[4] S. M. Rinaldi, J. P. Peerenboom and T. Kelly, "Identifying, understansing and analyzing critical infrastructure interdependencies," *IEEE Control System Magazine,* pp. 11-25, 2001.

[5] S. Bouchon, "The vulnerability of interdependent critical infrastructures systems: epistemological and conceptual state-of-art," IPSC Joint Research Centre, Ispra, 2006.

[6] M. Ouyang, "Review on modeling and simulation of interdependent critical infrastructure systems," *Reliability Engineering and System Safety,* vol. 121, pp. 43-60, 2014.

[7] P. Trucco, E. Cagno and M. De Ambroggi, "Dynamic functional modelling of vulnerability and interoperability of Critical Infrastructures," *Reliability Engineering and System Safety,* vol. 105, pp. 51-63, 2012.

[8] S. M. Rinaldi, J. P. Peerenboom and T. Kelly, "Identifying, understanding and analyzing critical infrastructure interdependencies," *IEEE Control System Magazine,* pp. 11-25, 2001.

[9] E. Cagno, P. Trucco and M. De Ambroggi, "Interdependency analysis of CIs in real scenarios," in *Proceedings of ESREL 2011*, Troyes (France), 2011.

[10] M. Beccuti, S. Chiaradonna, F. Di Giandomenico, S. Donatelli, G. Dondossola and G. Franceschinis, "Quantification of dependencies between electrical and information infrastructures," *International Journal of Critical Infrastructure Protection,* vol. 5, no. 1, pp. 14-27, 2012.

[11] I. Eusgeld, C. Nan and S. Dietz, ""System-of-systems" approach for interdependent critical infrastructures," *Reliability Engineering & System Safety,* vol. 96, no. June, pp. 679-686, 2011.

[12] C. Nan, I. Eusgeld and W. Kröger, "Analyzing vulnerabilities between SCADA system and SUC due to interdependencies," *Reliability Engineering & System Safety,* vol. 113, no. May, pp. 76-93, 2013.

[13] I. Onyeji, M. Bazilian and C. Bronk, "Cyber Security and Critical Energy Infrastructure," *The Electricity Journal,* Available online 5 March 2014.

[14] I. L. Pearson, "Smart grid cyber security for Europe," *Energy Policy,* vol. 39, no. 9, pp. 5211-5218, 2011.

[15] S. Gold, "The SCADA challenge: securing critical infrastructure," *Network Security,* no. 8, pp. 18-20, 2009.

[16] A. V. Kumar, K. K. Pandey and D. K. Punia, "Cyber security threats in the power sector: Need for a domain specific regulatory framework in India," *Energy Policy,* vol. 65, no. February, pp. 126-133, 2014.

A Longitudinal Analysis of the Drivers of Power Outages During Hurricanes: A Case Study with Hurricane Isaac

Gina Tonn[a*], Seth Guikema[a], Celso Ferreira[b], and Steven Quiring[c]
[a] Johns Hopkins University, Baltimore, MD, US
[b] George Mason University, Fairfax, VA, US
[c] Texas A&M University, College Station, TX

Abstract: In August 2012, Hurricane Isaac, a Category 1 hurricane at landfall, caused extensive power outages in Louisiana. The storm brought high winds, storm surge and flooding to Louisiana, and power outages were widespread and prolonged. Hourly power outage data for the state of Louisiana was collected during the storm and analyzed. This analysis included correlation of hourly power outage figures by zip code with wind, rainfall, and storm surge using a non-parametric ensemble data mining approach. Results were analyzed to understand how drivers for power outages differed geographically within the state. This analysis provided insight on how rainfall and storm surge, along with wind, contribute to power outages in hurricanes. By conducting a longitudinal study of outages at the zip code level, we were able to gain insight into the causal drivers of power outages during hurricanes. The results of this analysis can be used to better understand hurricane power outage risk and better prepare for future storms. It will also be used to improve the accuracy and robustness of a power outage forecasting model developed at Johns Hopkins University.

Keywords: Power Outages, Hurricanes, Random Forest

1. INTRODUCTION

Hurricane Isaac hit Louisiana in August 2012 and caused substantial power outages. It was a Category 1 hurricane at landfall and 47% of the state's electric customers lost power. The storm was large, slow-moving, and had significant storm surge associated with it. In comparison with other hurricanes, Isaac ranks fourth in customer power outages, behind Hurricanes Katrina, Gustav, and Rita, for the Entergy service area in Louisiana, Mississippi, Texas and Arkansas [1].

Power outages result in direct repair and restoration costs for utility companies, and can also result in loss of services from other types of critical infrastructure that are reliant on power service such as water, transportation, and communications systems. This can delay recovery times for a community that is impacted by a hurricane [2]. Accurate predictions of power outages prior to a storm can benefit both utility companies and government agencies by making planning and recovery more efficient [3].

Power outage prediction is often accomplished through the development of models based on wind field estimates, along with other covariates such as power system data, soil moisture levels, land use and topographical indicators [3]. A number of such statistical models have been developed [2,3,4]. While these models can be very accurate for some storms, they are less accurate for others.

In addition to accuracy of models varying from storm to storm, the causes of the outages can vary geographically across a region, and the existing models do not include some potential causes of power outages, particularly high rainfall. The main goal of this paper is to obtain a better understanding of the causes and geographic variance of power outages, both to improve basic understanding and to provide a stronger basis for improving outage forecasting models. Are the causes the same for a coastal area as an inland area? How important are rainfall and surge relative to wind? Damage to power systems is recorded by utilities, but good data on causes of outages is not generally available, making a longitudinal approach necessary.

---

[*] Corresponding author: Gina Tonn (e-mail: gtonn2@jhu.edu)

Can statistical analysis of power outage data and covariate data provide a better understanding of the causes of outages? The purpose of this study was to look at power outages longitudinally across the state of Louisiana for Hurricane Isaac to identify how the importance of covariates changes geographically. The results of this analysis may inform power outage prediction models and help to build more resilient infrastructure through improved understanding of the causes of outages.

In Section 2, the data used for the analysis as well as the statistical analysis methods are presented. Results and Discussion are included in Section 3, and Conclusions in Section 4.

2. METHODS AND DATA

2.1 Overview of Methods and Data

We focused on variables related to three key physical hazards associated with hurricanes: wind, storm surge, and rainfall in order to gain a better understanding of relative contribution of these three drivers. We analyzed all variables on an hourly basis, and so included variables that change over time as the storm progresses. We obtained data for the covariates of interest from publically available sources or modeled them based on publically available data. A summary of the covariates, abbreviations used for covariate names, data sources, and a description of each covariate is provided in Table 1. While data was available in varying time increments for each covariate, we performed interpolation to obtain hourly estimates. We chose the hourly change in outages as the response variable, and hours that did not have a positive increase in outages were removed from the analysis to focus the analysis on only the outage occurrence portion of the storm, not the outage restoration part of the storm. A more detailed description of each category of data and the data interpolation are provided in Sections 2.2 through 2.6.

Table 1: Summary of Covariates

Covariate	Abbreviation	Source	Description
Cumulative Precipitation	cumprecip	NCDC	Total precipitation amount during storm duration to hour of analysis in inches
Hourly Precipitation Total	precip	NCDC	Precipitation amount in hour of analysis in inches
Wind Speed	windspeed	Texas A&M model	Wind speed in meters/second for zip code in hour of analysis
Wind Gust Duration	windduration	Texas A&M model	Duration of wind gust >20 meters/second for zip code in hour of analysis
Previous Outages	prevout	Entergy website	Number of outages in previous hour of analysis for zip code
Population	population	US Census Bureau	Population estimate for zip code
Average Surge	surgeAVG	George Mason University model	Average storm surge depth for zip code in hour of analysis
Minimum Surge	surgeMIN	George Mason University model	Minimum storm surge depth for zip code in hour of analysis
Maximum Surge	surgeMAX	George Mason University model	Maximum storm surge depth for zip code in hour of analysis
Surge Variance	surgeVAR	George Mason University model	Variance of storm surge depth for zip code in hour of analysis

After completing the data collection and interpolation, we ran a Random Forest model for the entire data set. The most important variables were identified through Random Forest based importance measures for use in additional analysis as described further below. Using this reduced set of covariates, we ran the Random Forest analysis for each zip code separately. We plotted the results in map format for analysis of spatial trends. Then we used a Quantile Regression Forest for selected zip codes to gain insight into model accuracy. The modeling and analysis methods are described in more detail in Sections 2.7 through 2.9.

2.2 Outage Data

Power outage data was collected during the duration of the storm from August 27 to September 5, 2012. The data was harvested from the Entergy Louisiana website by a team of researchers at Johns Hopkins University [5]. The data was collected on a half-hourly basis during periods of peak outages, and was collected less frequently during non-peak outage periods. Data collected included the number of current customer outages by zip code. In order to standardize the data for use in analysis, we performed linear interpolation to estimate the number of outages for each zip code at the top of each hour for the duration of the storm.

We chose the change in outages (termed delta outages in this paper) for each hour of analysis for each zip code as the response variable for this analysis. Total power outages for the previous hour of analysis for each zip code was included as a covariate to account for the fact that the number of customers already without power impacts the number of power outages occurring in a given hour.

2.3 Precipitation Data

Precipitation data were obtained from the National Climatic Data Center (NCDC) website. Data was available for 36 stations in Louisiana. The time intervals at which the precipitation data were recorded varied by station, but were typically hourly or half-hourly. The data obtained was the hourly total rainfall [6]. In order to standardize the data for use in analysis, we interpolated the data set to estimate the hourly precipitation (precipitation that occurred in the previous 60 minute period) at the top of the hour for each station. Because our analysis was performed on a zip code basis, we needed rainfall estimates for each zip code. Based on the geographic coordinates of the zip code centroids and on the locations of the stations, we generated hourly rainfall estimates for each zip code using inverse distance weighted interpolation based on the spatially sparser set of rainfall stations that we had available.

2.4 Storm Surge Model

We used the coupled version of the 2-Dimensional Depth Integrated version of the Advanced Circulation (ADCIRC) model and the wave model SWAN [7] to simulate hurricane storm surge. The ADCIRC model [8] is a finite element, shallow water model that solves for water levels and currents at a range of scales and is widely used for storm surge modeling (e.g., Ferreira *et al.* 2013) [9]. This version of the program solves the Generalized Wave Continuity Equation (GWCE) and the vertically integrated momentum equations. SWAN is a third generation spectral wave model [10] that computes the time and spatial variation of directional wave spectra. We used the pre-validated numerical mesh *SL15* presented in Bunya *et al.* (2010) [11] and validated by Dietrich *et al.* (2010) [12] with resolution up to 30 meters in some areas. The hurricane surge model was forced by wind and pressure fields developed by a parametric asymmetric wind model [13] that computes wind stress, average wind speed and direction inside the Planetary Boundary Layer (PBL) based on the National Hurricane Center (NHC) best track data [14] meteorological conditions (e.g., central pressure, forward speed and radius to maximum wind). The simulations for Hurricane Isaac included tides (Tidal potential components M2, S2, N2, K2, K1, O1 and Q1) and neglected rivers inflows. Simulation results were recorded at 15 minute intervals for every model node in the study region. The water levels for each model node within each zip code were extracted from the entire model domain and inundation levels were converted to the NAVD88 vertical datum. Covariates based on the storm surge model include average storm surge, maximum storm surge, minimum storm surge, and storm surge variance.

2.5 Wind Model

The parametric wind field model of Willoughby et al. (2006) [15] is used to generate wind estimates for the duration of the hurricane at the zip code level for Hurricane Isaac. Parametric hurricane models are formulated from a physical understanding of hurricane wind fields. That is, winds are calm in the eye of the hurricane and they are typically at a maximum in the eyewall. Outside the eyewall the wind decreases with radius, although not always monotonically, and become near zero at some distance from the center of circulation. This wind field model was previously used in Han et al. [2, 16]. Two of the covariates are based on output from this model. The first is maximum wind speed in meters per second in the previous hour. The second is wind gust duration greater than 20 meters per second, with duration being taken cumulatively over the life of the storm for each zip code. Both of these covariates are simulated for the centroid of each zip code polygon based on running the wind field model every 60 minutes over the duration of the storm.

2.6 Other Data

Population estimates for each zip code were obtained from the US Census Bureau American Community Survey. These estimates were based on the US Census Bureau data for the year 2011. Because the US Census bureau does not track population on a zip code basis, the population data is an estimate based on census tract data (US Census) [17].

2.7 Random Forest and Quantile Regression Forest Methods Overview

A Random Forest is a non-parametric ensemble data mining method [18]. In the method, a large number of regression trees are developed, with each tree based on a bootstrapped sample of the data set. Random Forest models are good for data sets with non-linear data, outliers, and noise. Two types of output from the Random Forest model fit very nicely with the objectives of this analysis. The first is variable importance, which is a measure of the contribution of a given covariate to the model prediction accuracy. The second is the partial dependence plot. These plots show the marginal effect of a covariate on the response variable. The randomforest package in R was used for this analysis [19].

Quantile Regression Forests provide a non-parametric way of estimating conditional quantiles based on an underlying Random Forest model [20]. Quantiles give more information about the distribution of the response variable as a function of the covariates than just using the conditional mean as a standard Random Forest model does. In this method, regression trees are grown as in the Random Forest method. Then the weighted distribution of the observed response variables is used to estimate a conditional distribution. The difference between Random Forest models and Quantile Regression Forest models is that Random Forest models keep only the mean predictions and disregard other information. Quantile regression forests estimate the quantiles of the predictions based on the trained forest (Meinshausen, 2006). The quantregForest package in R was used for this analysis [21].

2.8 Statistical Analysis

A statewide Random Forest model was run using the data for all covariates and zip codes. Variable importance was reviewed to identify the variables that are most significant for predictive accuracy. Based on the variable importance, one covariate from each category of covariates (precipitation, wind, storm surge, and outages) was retained for individual zip code analysis in order to better understand the influences of the different variables. Partial dependence plots were generated for each of these covariates, and were reviewed to understand the marginal effects of these covariates on the response variable.

In order to understand the relative importance of the four covariates, and how that importance varies geographically, plots of importance for each of the covariates were generated. Because the magnitude of variable importance is not the same for each Random Forest run, comparing the variable importance between zip codes would not be useful. Instead, we calculated a percent variable importance for each zip code. The variable importance for the four covariates (wind speed, cumulative precipitation, maximum storm surge, and previous outages) was summed to calculate the total importance value for each zip code. Then the percent of total importance accounted for by each covariate was calculated. For each of the four covariates, we plotted the percent variable importance by zip code. We visually reviewed these plots to identify how the percent importance for each covariate differed geographically. The plots were also evaluated in light of the plots of the covariate values (Figure 4), so that the magnitude of the covariates was accounted for in evaluating the percent importance trends.

In order to better understand the predictive accuracy of the random forest model, a Quantile Regression Forest model was run on three selected zip codes. The zip codes were chosen so that different geographic areas in the state were represented. These zip codes include 71220 (north), 70129 (southeast), and 70525 (southwest).

3. RESULTS AND DISCUSSION

3.1 Variable Importance

The variable importance results for the Random Forest model with all covariates included are shown in Figure 1. Cumulative precipitation, wind speed, and previous outages are the most important variables, followed by population and hourly precipitation. All of the surge variables, along with wind gust duration had considerably lower variable importance. This differs from some previous work where wind gust duration was shown to be an important variable (e.g., Han et al. 2009) [16] and may be specific to this hurricane for which wind speeds were lower than for the hurricanes included in the Han et al. work [2].

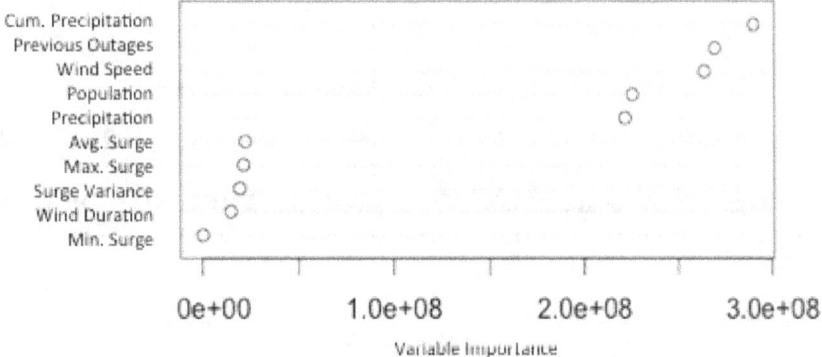

Figure 1: Variable Importance, all covariates included

Based on these results, four variables were selected as part of a reduced covariate set to be used for the remainder of the analysis. These variables were: cumulative precipitation, wind speed, previous outages, and maximum surge. Maximum surge depth was selected over average surge depth in each zip code because it had a clearer physical interpretation than the average surge depth yet had nearly the same importance score. Population was not included because the remainder of the analysis was done on an individual zip code basis wherein population is constant. The Random Forest model for the entire state was rerun with this reduced set of covariates. The resulting variable importance plot is included as Figure 2. In this model, the previous outages covariate has the highest variable importance, followed closely by cumulative precipitation and wind speed. Maximum surge has a lower importance, as should be expected since only a small portion of the state was impacted by storm surge.

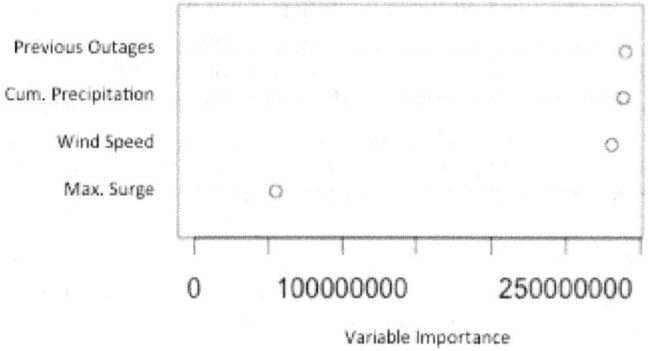

Figure 2: Variable Importance, reduced covariate set

3.2 Partial Dependence

Partial dependence plots were generated for the four covariates in the reduced set, and are provided as Figure 3. Partial dependence provides insight into the marginal impact of the covariate on the response variable, increase in outages.

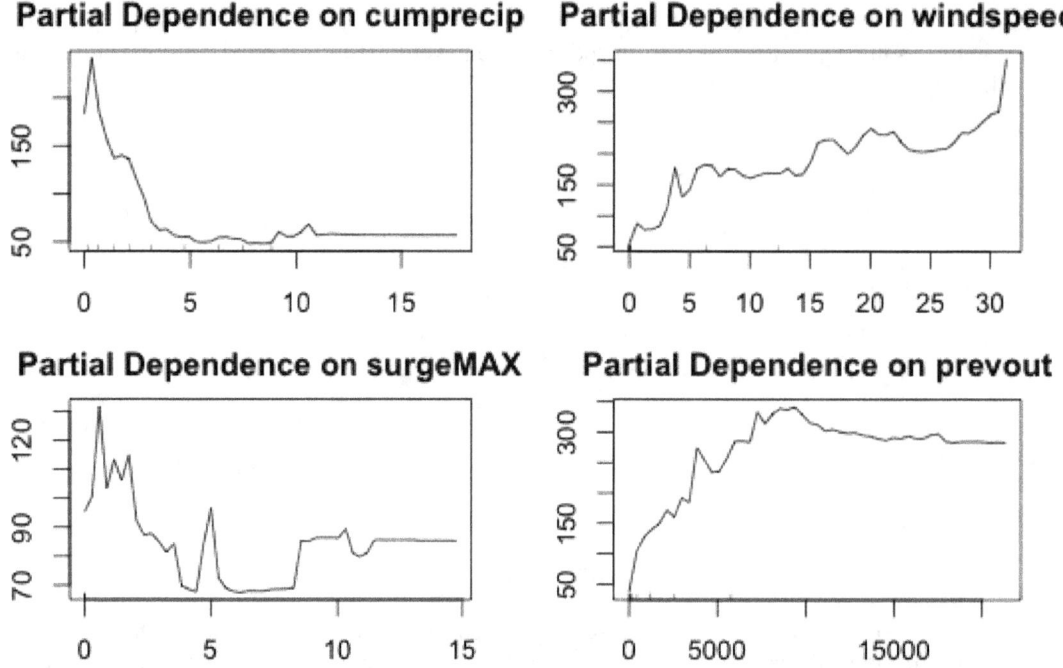

Figure 3: Partial Dependence Plots: a) partial dependence on cumulative precipitation, b) partial dependence on wind speed, c) partial dependence on maximum surge, and d) partial dependence on number of previous outages. The x-axis represents the value of the covariate and the y-axis represents the marginal influence of the covariate on delta outages.

The marginal influence of the cumulative precipitation covariate is highest for about 0 to 4 inches of precipitation. The marginal influence of wind speed generally increases with increasing wind speed. The influence of maximum surge is more variable, which may be due to the fairly low number of zip codes that experience storm surge. The marginal influence of the previous outages covariate increases up to around 10,000 outages, and then slightly decreases, since once a high number of outages occurs in a zip code, additional outages may be small in magnitude.

3.3 Geospatial Analysis

In order to analyze spatial trends across the state, we generated plots to get a sense of the magnitude of precipitation, wind speed, storm surge, and outages, and how the magnitude varied across the state. These plots are presented in Figure 4. Total precipitation (cumulative precipitation) was highest in the southeast part of the state, with more than 12 inches of precipitation recorded in some locations. Maximum wind speed was also highest in the southeast part of the state, where the hurricane made landfall. Maximum storm surge was highest in several zip codes bordering the Gulf of Mexico, as well as in several zip codes bordering the Mississippi River. The maximum numbers of power outages were observed in zip codes in the southeast, around New Orleans, where the population is greatest and the storm impacts were more significant.

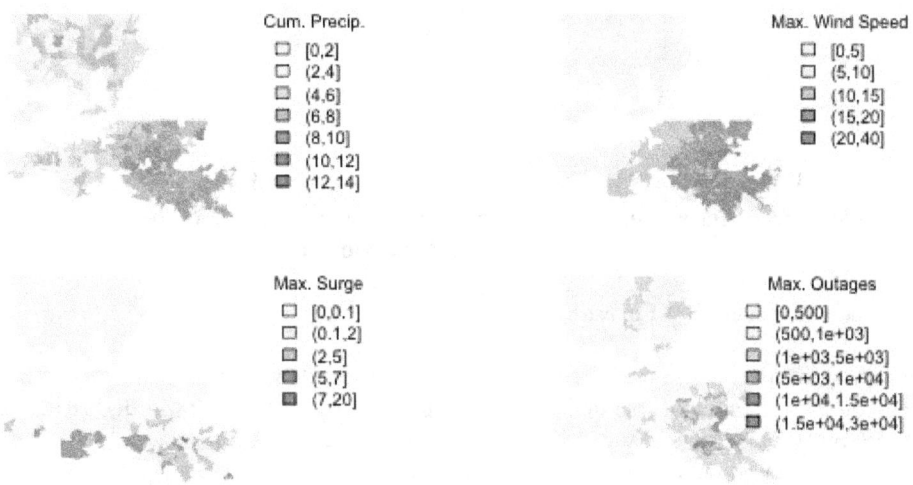

Figure 4: Covariate values for a) cumulative precipitation (inches), b) maximum wind speed (meters/second), c) maximum surge (meters), and d) maximum number of outages. Zip codes not colored are not part of the utility's service area.

Figure 5 illustrates the percent importance for cumulative precipitation, wind speed, maximum storm surge, and previous outages for all zip codes analyzed in Louisiana. In the northern part of the state, both cumulative precipitation and previous outages had the highest percent importance. Wind speed had the highest percent importance in the southeast part of the state. In the southwest portion of the state, percent importance varied from zip code to zip code, with cumulative precipitation, wind speed, and previous outages having varying importance. With the exception of a few zip codes, the percent importance for maximum storm surge was less than 30%, even in coastal areas.

These results indicate that the importance of covariates varies geographically. This is due to the storm's track and characteristics, but also potentially due to the interaction of other factors pertaining to the topography and power system. Both wind speed and cumulative precipitation were highest in the southeast, due to the storm's track; however, wind speed generally had greater importance in that area than precipitation. In the northern part of the state, where precipitation was moderately high, but wind speeds were low, precipitation was of greater importance. The previous outages covariate was generally more important in areas that had a lower maximum outages value.

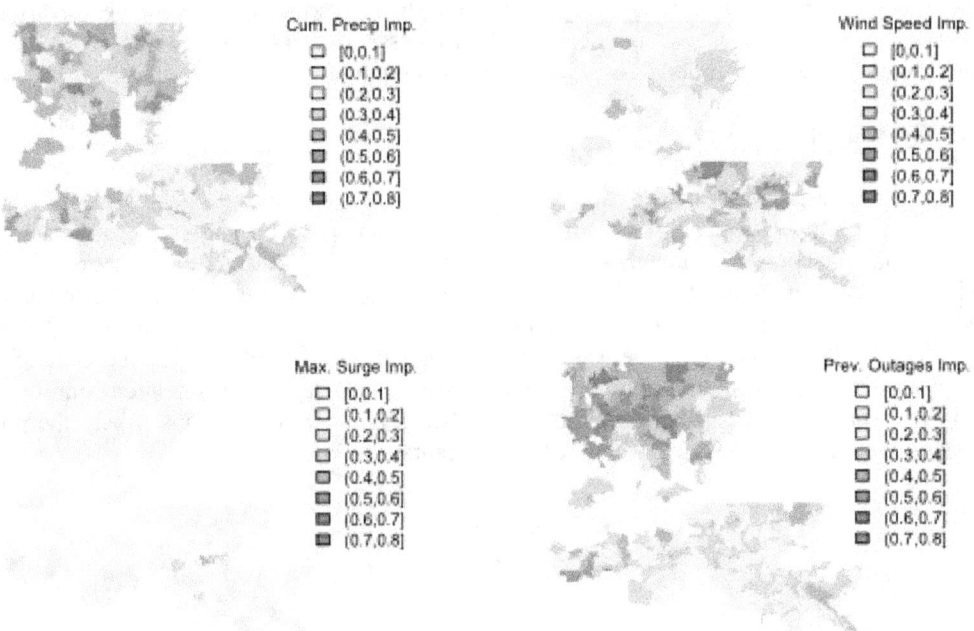

Figure 5: Percent Importance Plots for a) cumulative precipitation, b) wind speed, c) maximum surge, and d) previous outages

3.4 Quantile Regression Forest

We ran a Quantile Regression Forest model on three selected zip codes in order to better understand the predictive accuracy of the model. These zip codes include 71220 (north), 70129 (southeast), and 70525 (southwest). Plots of the Quantile Regression Forest results for these three zip codes are shown in Figure 6. These plots show the 90% prediction confidence intervals and whether predictions using out-of-bag data fall inside or outside of the prediction intervals. As shown on the plots, the majority of the predictions fall within the prediction intervals.

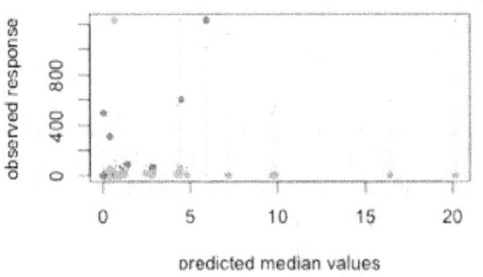
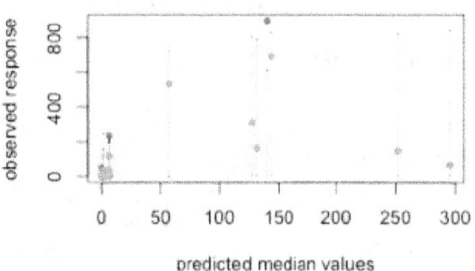
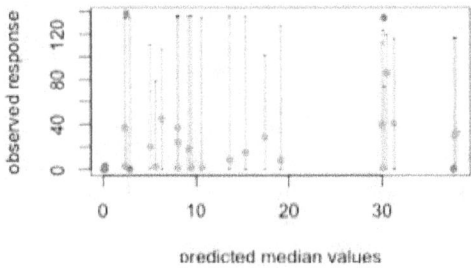
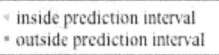

Figure 6: Quantile Regression Forest plots for a) zip code 70129, b) zip code 71220, and c) zip code 70546

Table 2 shows the range of predictions that fall between the 10% and 90% quantiles. For low values of delta outages (0 to 2), the coverage of the 90% interval is very low; the model has little reliability at the lowest level of delta outages. For middle of the range values of delta outages (2 to 75), the model confidence interval coverage is fairly high, ranging from 76% to 100% for a 90% interval for the three zip codes analysed. At the high end of the delta outages range (75+), the coverage accuracy varies significantly. This makes sense given the nature of power outages and the covariates used in the model. Very low increases in power outages are not likely well correlated to wind, precipitation, or previous outages, and are more likely caused by random events occurring at individual houses. Very high increases in power outages can sometimes be correlated with high precipitation or wind, but could also occur due to sudden problems in the power grid.

Table 2: Percent of Predictions within 80% Confidence Interval

Delta Outage Range	Zip Code 70129	Zip Code 71220	Zip Code 70546
0-2	0%	0%	3%
2-75	80%	100%	76%
75+	58%	N/A	100%

4. CONCLUSION

The purpose of this analysis was to provide insight on how rainfall and storm surge, along with wind, contribute to risk of power outages in hurricanes. By conducting a longitudinal study of outages at the zip code level, we were able to gain insight into the causal drivers of power outages during hurricanes. Our analysis showed that the drivers of power outages and the importance of the drivers can vary geographically. In Louisiana, during Hurricane Isaac, rainfall and previous outages were the most important covariates in the north, while wind was more important in the southeast. Rainfall, wind, and previous outages were all relatively important in the southwest. With the exception of a few zip codes, storm surge was generally not an important variable in predicting power outages. The reason the drivers vary geographically is likely due to characteristics of the location and of the storm. In areas where the highest wind speeds are experienced, wind is likely to be the most important covariate. Elsewhere, the importance of covariates differs geographically.

Acknowledgements

The authors would like to acknowledge Michael Gao, a Johns Hopkins University undergraduate researcher for his GIS assistance and members of Seth Guikema's research group at JHU for their assistance with collecting power outage data during Hurricane Isaac. This work was supported by a fellowship to Gina Tonn by the Environment, Energy, Sustainability and Health Institute (E2SHI) and the NSF IGERT Water, Climate, and Health program at Johns Hopkins University as well as by NSF grant number 1149460 from CMMI.

References

[1] "UPDATE 1-Entergy estimates Hurricane Isaac damage at $500 mln." *Chicago Tribune*. 18 Sept. 2012.
[2] S.R Han, et al. "Estimating the spatial distribution of power outages during hurricanes in the Gulf coast region." *Reliability Engineering & System Safety,* vol. 94.2, p. 199-210, 2009.
[3] R. Nateghi, S. Guikema, and S. Quiring. "Power outage estimation for tropical cyclones: Improved accuracy with simpler models." *Risk analysis,* 2013.
[4] S.D. Guikema, R. Nateghi, and S. Quiring. "Predicting Infrastructure Loss of Service from Natural Hazards with Statistical Models: Experiences and Advances with Hurricane Power Outage Prediction," in Proceedings, ESREL 2013, Amsterdam, October 2013.
[5] *Entergy View Outages*. Entergy. http://www.entergy.com/storm_center/outages.aspx, Accessed 27 Aug. – 5 Sept. 2012.
[6] National Climatic Data Center (NCDC). http://www.ncdc.noaa.gov, Accessed 16 Nov. 2012.
[7] J.C. Dietrich, M. Zijlema, J.J. Westerink, L.H. Holthuijsen, C. Dawson, R.A. Luettich, R. Jensen, J.M. Smith, G.S. Stelling, and G.W. Stone. "Modeling Hurricane Waves and Storm Surge using Integrally-Coupled, Scalable Computations," *Coastal Engineering*, vol. 58, p. 45-65, 2011.
[8] R. Luettich, and J. Westerink. "Formulation and numerical implementation of a 2D/3D ADCIRC Finite Element Model Version 4.46." http://adcirc.org/adcirc_theory_2004_12_08.pdf. Accessed 13 Nov. 2010.
[9] C.M. Ferreira, J. Irish, F. Olivera. "ArcStormSurge: Integrating GIS and Hurricane Storm Surge." *Journal of the American Water Resources Association*, 2013.
[10] N. Booij, R.C. Ris, and L. H. Holthuijsen. "A third generation wave model for coastal regions. Model Description and Validation." *Journal of Geophysical Research,* vol. 104, p. 7649-7666, 1999.
[11] S. Bunya, J. Dietrich, J. Westerink, B. Ebersole, J. Smith, J. Atkinson, R. Jensen, D. Resio, R. Luettich, C. Dawson, V. Cardone, A. Cox, M. Powell, H. Westerink, and H. Roberts. "A High-Resolution Coupled Riverine Flow, Tide, Wind, Wind Wave, and Storm Surge Model for Southern Louisiana and Mississippi. Part I: Model Development and Validation." *Monthly Weather Review*, p. 345-377, 2010.
[12] J. Dietrich, S. Bunya, J. Westerink, B. Ebersole, J. Smith, J. Atkinson, R. Jensen, D. Resio, R. Luettich, C. Dawson, V. Cardone, A. Cox, M. Powell, H. Westerink, and H. Roberts. "A High-Resolution Coupled Riverine Flow, Tide, Wind, Wind Wave, and Storm Surge Model for Southern Louisiana and Mississippi. Part II: Synoptic Description and Analysis of Hurricanes Katrina and Rita." *Monthly Weather Review*, p. 378-404, 2010.
[13] C. Mattocks, and C. Forbes. "A real-time, event-triggered storm surge forecasting system for the state of North Carolina", *Ocean Modelling*, vol. 25, p. 95-119, 2008.
[14] National Oceanic and Atmospheric Administration (2013) Atlantic basin hurricane database (HURDAT). http://www.aoml.noaa.gov/hrd/hurdat/, Accessed 08 Jul. 2013.
[15] H.E. Willoughby, R.W.R. Darling, and M.E. Rahn. "Parametric representation of the primary hurricane vortex. Part II: A new family of sectionally continuous profiles." *Monthly Weather Review,* vol. 134.4, 2006.

[16] S.R. Han, S.D. Guikema, and S. Quiring. "Improving the predictive accuracy of hurricane power outage forecasts using generalized additive models." *Risk analysis,* vol. 29.10, p. 1443-1453, 2009.

[17] *American Community Survey*. US Census Bureau. http://www.census.gov/acs/www, Accessed 7 Jul. 2013.

[18] T. Hasti, R. Tibshirani, and J. Friedman. The Elements of Statistical Learning; Data Mining, Inference and Prediction, 1st ed. New York: Springer, 2001.

[19] A. Liaw, and M. Wiener. "Classification and Regression by randomForest." R News, vol. 2(3), p. 18-22. 2002.

[20] N. Meinshausen. "Quantile regression forests." *The Journal of Machine Learning Research* 7, p. 983-999, 2006.

[21] N. Meinshausen. "quantregForest: Quantile Regression Forests." R package version 0.2-3, 2012.